he Elements

						III		IV		V		VI		VII		O
															2	2 He 4.0026
					2 3	5 B 10.811	2 4	6 C 12.01115	2 5	7 N 14.0067	2 6	8 O 15.9994	2 7	9 F 18.9984	2 8	10 Ne 20.183
					2 8 3	13 Al 26.9815	2 8 4	14 Si 28.086	2 8 5	15 P 30.9738	2 8 6	16 S 32.064	2 8 7	17 Cl 35.453	2 8 8	18 Ar 39.948
28 Ni 58.71	2 8 18 1	29 Cu 63.546	2 8 18 2	30 Zn 65.37	2 8 18 3	31 Ga 69.72	2 8 18 4	32 Ge 72.59	2 8 18 5	33 As 74.9216	2 8 18 6	34 Se 78.96	2 8 18 7	35 Br 79.904	2 8 18 8	36 Kr 83.80
46 Pd 106.4	2 8 18 18 1	47 Ag 107.868	2 8 18 18 2	48 Cd 112.40	2 8 18 18 3	49 In 114.82	2 8 18 18 4	50 Sn 118.69	2 8 18 18 5	51 Sb 121.75	2 8 18 18 6	52 Te 127.60	2 8 18 18 7	53 I 126.9044	2 8 18 18 8	54 Xe 131.30
78 Pt 195.09	2 8 18 32 18 1	79 Au 196.967	2 8 18 32 18 2	80 Hg 200.59	2 8 18 32 18 3	81 Tl 204.37	2 8 18 32 18 4	82 Pb 207.19	2 8 18 32 18 5	83 Bi 208.980	2 8 18 32 18 6	84 Po (210)	2 8 18 32 18 7	85 At (210)	2 8 18 32 18 8	86 Rn (222)
63 Eu 151.96	2 8 18 25 9 2	64 Gd 157.25	2 8 18 27 8 2	65 Tb 158.924	2 8 18 28 8 2	66 Dy 162.50	2 8 18 29 8 2	67 Ho 164.930	2 8 18 30 8 2	68 Er 167.26	2 8 18 31 8 2	69 Tm 168.934	2 8 18 32 8 2	70 Yb 173.04	2 8 18 32 9 2	71 Lu 174.97
95 Am (243)	2 8 18 32 25 9 2	96 Cm (247)	2 8 18 32 26 9 2	97 Bk (247)	2 8 18 32 27 9 2	98 Cf (249)	2 8 18 32 28 9 2	99 Es (254)	2 8 18 32 29 9 2	100 Fm (253)	2 8 18 32 30 9 2	101 Md (256)	2 8 18 32 31 9 2	102 No (254?)	2 8 18 32 32 9 2	103 Lw† (257)

Atomic weights are based on carbon-12;
values in parentheses are for the most stable or the most familiar isotope.
† Symbol is unofficial.

Organic Chemistry: A Brief Course

Organic Chemistry: A Brief Course

JOHN WILEY & SONS, INC.
NEW YORK LONDON SYDNEY TORONTO

This book was set in Baskerville by York Graphic Services, Inc. It was printed and bound by Halliday Lithograph Corp. The drawings were designed and executed by John Balbalis with the assistance of the Wiley Illustration Department. Deborah Herbert was the copy editor. Marion Palen was the production manager.

Cover art—Pierre Cordier: Chimigramme 19/2/71 II, D4

Library of Congress Cataloging in Publication Data

Holum, John R
Organic chemistry.

Includes index.
1. Chemistry, Organic. I. Title.
QD251.2.H64 547 74-20773
ISBN 0-471-40849-2

Printed in the United States of America

10 9 8 7 6 5 4 3 2 1

To Mary
and to our children:
Elizabeth, Ann,
and Kathryn

Preface

Most students in one-term course in organic chemistry plan careers in such sciences as medical technology, nursing, nutrition and dietetics, agricultural sciences, biology, and other allied health science fields. Many students study biochemistry later. Some, of course, become chemical technicians and analysts. All live in a world that is becoming more aware of the many ways in which organic chemicals are important in both industry and medicine. These students actually will promote that awareness. I kept these thoughts in mind as I selected the topics, arranged them for study, and set the level of this textbook.

I tried to lay out a path for students who intend to enter allied health science careers. The path, however, is not set only between high walls; there are other sights to see and study. On one side are important organic fuels, petroleum refining, petrochemicals, polymers and plastics. On the other side are several interesting biochemicals—lipids, carbohydrates, proteins, vitamins, enzymes, nucleic acids, and drugs (both licit and illicit). Around the bend are modern techniques for purifying, analyzing, and identifying organic compounds—GLC, TLC, IR, mass, NMR spectroscopy. Down the middle are most of the reactions of major importance in organic synthesis. Side by side are repeated references to ways in which structures relate to physical properties. And we have frequent pauses to examine specific chemicals of one kind of importance or another—solvents, pesticides, pharmaceuticals, derivatizing agents, fluorocarbons, and organometallics. It is a survey of organic chemistry that does full justice to this great field as a subject in itself, and it is presented with concrete goals of students in mind.

The book was also written with the needs of beginning students of organic chemistry in mind. The important question about a beginning textbook is not whether it is the most intellectually satisfying to the experienced chemist (who has usually forgotten the first struggles), but whether it is teachable to the beginner. The experienced chemist, looking down on familiar terrain, will tend to reorganize his field to settle it on a foundation of quantum mechanics, statistical thermodynamics, and kinetics. With very few exceptions, we cannot teach beginners that way. Hence, I organized the content of this text in a traditional way, around families of compounds rather than families of reactions. For beginners to organic chemistry, this organization makes the best pedagogical sense—or so it seems to me.

Beginners also need self-help devices that are strategically located. There are many "in-chapter" questions, all with answers at the back of the book. There are summary statements and chapter reviews. The first chapter itself is a review of basic concepts of chemical bonds. Not wanting to overload students with this topic all at once, and wanting to get them on as quickly as possible to organic substances,

I resisted the intellectually satisfying tactic of putting everything about bonds in organic compounds in one place. I have found that students enjoy the mix of topics, chapter by chapter—a mix of something about bonds, about physical properties, names, ways to synthesize, chemical reactions, how reactions happen, important individual compounds, and now and then low-level surveys of interesting fields (e.g., spectroscopy, drugs, or nucleic acids, to name just a few).

One of the novel features of the text is the unit in Chapter 5 on world supplies of fossil fuels. The important organic chemicals of industry come from them. Their supplies and costs directly affect supplies and costs of all petrochemicals, polymers, synthetics, pesticides, pharmaceuticals, dyes, and others. Not enough people realize this. A short course in organic chemistry can be one vehicle for disseminating information about the world's fossil fuel situation and how it bears on issues besides the cost of gasoline. Very little time is needed to do this. No other chemistry text of this nature does it.

A number of people helped me directly and indirectly in preparing this text. I particularly thank Dr. Earl Alton and the whole Augsburg administration for an atmosphere that was supportive of my efforts. I thank Robert Rogers and Gary Carlson, the former and the present chemistry editors at Wiley, and John Balbalis of the Wiley Illustration department. I also thank Mrs. Arlys Bjornlie for her usual superb skills in preparing final copies of manuscripts.

My most important partner, however, is my wife, Mary. I doubt whether this book would have been completed without her and our three girls, Liz, Ann, and Kathryn.

On pedagogical matters I thank these experienced teachers:

NEIL R. COLEY, Chabot College
MANFRED NAUMANN, City College of San Francisco
ARNE LANGSJOEN, Gustavus Aldolphus College
DUANE FISH, West Valley College
JERRY FLOOK, Gavilan College.
RICHARD MCDONALD, Fullerton Junior College
JOHN SULLIVAN, Eastern Michigan University
TERENCE C. MORRILL, Rochester Institute of Technology

Minneapolis, Minnesota

John R. Holum
Augsburg College

What To Do for the Accompanying Laboratory

Professor Arne Langsjoen of Gustavus Adolphus College has prepared a laboratory manual, to accompany this text. All of the experiments have been tested by students. Many are of an "unknown" variety. For an examination copy write to Director of Marketing, John Wiley & Sons, Inc., 605 Third Ave., New York, N. Y. 10016.

Writing on department letterhead, instructors may ask also for the Solutions Guide for the end-of-chapter-questions for this text.

Contents

Organic Chemistry: A Brief Course

chapter one

Introduction to Organic Chemistry

1.1 Concerning Things Organic

Organic chemistry is the study of the chemical compounds of carbon. With the exception of carbonates, bicarbonates, cyanides, and a few other simple carbon compounds, all the compounds of carbon are classified as organic compounds. This family comprises the majority of the known chemical compounds in the world. The food you had for breakfast, the clothes you are wearing, the paper in this book, the drugs you take when you are sick, the vitamins that help keep you well, the plastics in things too numerous to name, and the gasoline in the car or bus that may have carried you to school—all these are organic substances.

Within your eyes scanning this page, in your brain processing what they see, in your heart and lungs silently busy with their vital tasks, in and among all these parts, down at the subcellular level, there are countless organic molecules.

All of life has a molecular basis. Its molecular level is a realm of beauty and wonder but also a realm of labyrinths and traps to any who would enter without a map. That map is a basic knowledge and understanding of organic chemistry.

Some of the air pollutants and many of the water pollutants are organic substances. They vary from hydrocarbons that are emitted when we use gasoline, to pesticides that drift and drain off farm land, to chemicals that are released during the manufacture or use of industrial organic substances.

Out of the huge group of industrial organic chemicals—the "industrial" in this phrase would have been unthinkable 150 years or so ago—we today receive virtually all of our dyes, drugs, and detergents, all of our "synthetics"—nylon, Dacron, Orlon, polyesters, epoxies—and most of our gasoline and rubber.

1.2 Vital Force Theory

Prior to the middle of the nineteenth century scientists widely believed that one of the laws of nature was that we could never duplicate nature's feat of making those chemicals isolated from plants and animals that contain carbon. As with all scientific laws, this one was also based on experience, even if it was negative experience. Try as they did scientists failed in all their efforts to use mineral sources of carbon to make any then-known organic compounds—compounds at that time obtained exclusively from organisms. Living things, they decided, possessed a vital force, a special source of energy or ability that could not be transferred from living thing to living thing (except by procreation) or from man to test tube. The chasm between the mineral and the organic realms seemed unbridgeable. Only an accidental event involving a trained scientist could have made the connection, and it occurred in 1828 to Friedrich Wöhler[1], a professor at the University of Göttingen in Germany.

1.3 Wöhler's Urea Synthesis

Cyanogen, C_2N_2, and ammonia were considered in Wöhler's time to be inorganic materials. Cyanogen reacts with water:

$$C_2N_2 + H_2O \longrightarrow \underset{\text{CYANIC ACID}}{HCNO} + \underset{\text{HYDROGEN CYANIDE}}{HCN}$$

If ammonia is added, ammonium ions and cyanate ions form:

$$\underset{\text{AMMONIA}}{NH_3} + \underset{\text{CYANIC ACID}}{HCNO} \longrightarrow \underset{\text{AMMONIUM ION}}{NH_4^+} + \underset{\text{CYANATE ION}}{CNO^-}$$

Wöhler's plan was to heat a solution made by these steps to expel both the water and the hydrogen cyanide. He expected the residue to be a crystalline solid—the inorganic salt, ammonium cyanate, NH_4CNO.

Wöhler obtained a white crystalline solid, all right, but it wasn't ammonium cyanate, which decomposes when heated to 60°. Instead, it was urea, which melts at 133°.

$$\underset{\text{UREA}}{NH_2-\overset{\overset{\displaystyle O}{\|}}{C}-NH_2}$$

Urea had been isolated half a century earlier. We now know that urea is an end product of the body's metabolism of protein. It is made in the liver, removed from circulation by the kidneys, and it is excreted in the urine. Even though its molecules are small and simple, urea was, to scientists of the day, clearly of animal origin and therefore an organic compound. Wöhler had made it from nonorganic materials. Much more work, as it happened, was needed to clinch the defeat of the vital force theory, but by the middle of the nineteenth century chemistry had made

[1]Friedrich Wöhler (1800–1882), while remembered by most chemists as having been instrumental in overthrowing the vital force theory, was at least as important a figure in inorganic chemistry. He was the first to isolate in reasonably pure form the elements aluminum, beryllium, boron, silicon, and titanium.

a number of organic compounds without the internal aid of living organisms. While scientists dropped the idea of a vital force—at least in this area—they kept the name "organic" for classifying the compounds of carbon in general (with a few exceptions noted earlier).

In addition to opening the great doors to the realm of "synthetics," Wöhler's work helped establish once and for all the fact that organic compounds obey all the laws of chemistry. This meant, for example, that the architecture of organic molecules—their molecular structures—became part of the general problem of the structures of all molecules and the chemical bonds that are within them. To understand the properties of organic substances we must become familiar with their molecular structures, and that is why we turn next to a review of chemical bonds. There are three types of importance to us, the covalent, the ionic and the hydrogen bond. We shall review the first two next and study the hydrogen bond in Chapter 6. Both the ionic and the covalent bonds are explained at the simplest level by the octet theory.

1.4 The Octet Theory

The electronic configurations of several elements without regard to subshells and orbitals are given in Table 1.1. Included are the elements in the family of "noble

Table 1.1
Configurations of Ions and Noble Gas Structures

Group	Common Element	Nuclear Charge (atomic number)	ATOMS Electronic Configurations 1	2	3	4	5	6
I Alkali metals	Li	3^+	2	1				
	Na	11^+	2	8	1			
	K	19^+	2	8	8	1		
II Alkaline earth metals	Mg	12^+	2	8	2			
	Ca	20^+	2	8	8	2		
	Ba	56^+	2	8	18	18	8	2
VI	O	8^+	2	6				
	S	16^+	2	8	6			
VII Halogens	F	9^+	2	7				
	Cl	17^+	2	8	7			
	Br	35^+	2	8	18	7		
	I	53^+	2	8	18	18	7	
O Noble gases	He	2^+	2					
	Ne	10^+	2	8				
	Ar	18^+	2	8	8			
	Kr	36^+	2	8	18	8		
	Xe	54^+	2	8	18	18	8	

Ion	IONS Electronic Configurations 1	2	3	4	5	6	Comparable Noble Gas
Li^+	2						Helium
Na^+	2	8					Neon
K^+	2	8	8				Argon
Mg^{2+}	2	8					Neon
Ca^{2+}	2	8	8				Argon
Ba^{2+}	2	8	18	18	8		Xenon
O^{2-}	2	8					Neon
S^{2-}	2	8	8				Argon
F^-	2	8					Neon
Cl^-	2	8	8				Argon
Br^-	2	8	18	8			Krypton
I^-	2	8	18	18	8		Xenon

The noble gases do not form stable ions.

gases." With the exception of the first member of this family, helium, the atoms of the noble gases have eight electrons in their highest occupied main energy levels. We call this condition an *outer octet.*

There have been few successful attempts to make compounds from the noble gas elements. An atom's having an outer octet somehow means the particle is resistant to chemical change. We don't know why this is so, but it apparently is. This is the central fact behind the octet theory.

According to the octet theory, the chemical behavior of simple elements (at least those from 1–20, and others in Groups I, II, and VI and VII) can be successfully correlated by this rule:

> "The atoms of simple elements will usually tend to change in whatever direction leads most directly to an outer octet."

If the outside shell happens to be the first, which can hold only two electrons anyway, then the rule states that it is an outer, filled first-shell of two electrons that is the condition of stability, not an outer octet.

1.5 Ionic Bonds and Ionic Compounds

The atoms of the alkali metals (Table 1.1)—Group I—all have one electron in their outside shells. If this lone electron were lost, the remaining particle would have a charge of +1; it would be a positive ion (a cation). Most importantly, it would have an outer octet. (For lithium, the ion would have an outer first-shell of two electrons.) And that is exactly the tendency for the alkali metals; they form ions with +1 charge.

Atoms of the alkaline earths (Table 1.1)—Group II—all have two outside shell electrons. For them we need a loss of two electrons to leave behind an outer octet. The resulting cations have charges of +2 each (e.g., Ca^{2+}, Mg^{2+}). For Group III elements atoms must lose three electrons to strip down to having outer octets. This actually does not happen often; losing three is costly in energy. Yet it can, and the aluminum ion, Al^{3+}, is common.

Over on the other side of the periodic table are the halogens, Group VII, whose atoms have seven electrons each in outside shells. Their atoms are but one electron away from outer octets. When they form ions, therefore, they are ions with charges of −1 (e.g., Cl^-, Br^-). Energetically it is vastly easier to pick up one electron than to lose seven electrons. With these principles let us now review one way ions can be made and thereby study the idea of an ionic bond.

If sodium atoms, to take a classic example, are to lose one electron each, two conditions must be met: there must be an acceptor for the electron, and the newly formed cations, Na^+, must become surrounded, engulfed, and smothered by oppositely charged particles. Even though ions can exist, they do not readily do so in isolation. Ions must normally be surrounded by something that is reasonably close to being oppositely charged.

Let us provide sodium atoms an environment rich in electron acceptors—let us imagine we have atoms of chlorine.[2] The sodium atoms, donate one electron

[2]You no doubt know that elemental chlorine exists as a collection of diatomic molecules, Cl_2. Our purpose here, however, is not to explain how metallic sodium and gaseous chlorine interact. We are simply after an understanding of the ionic bond.

each; chlorine atoms accept the electrons. We may picture the electron transfer as follows:

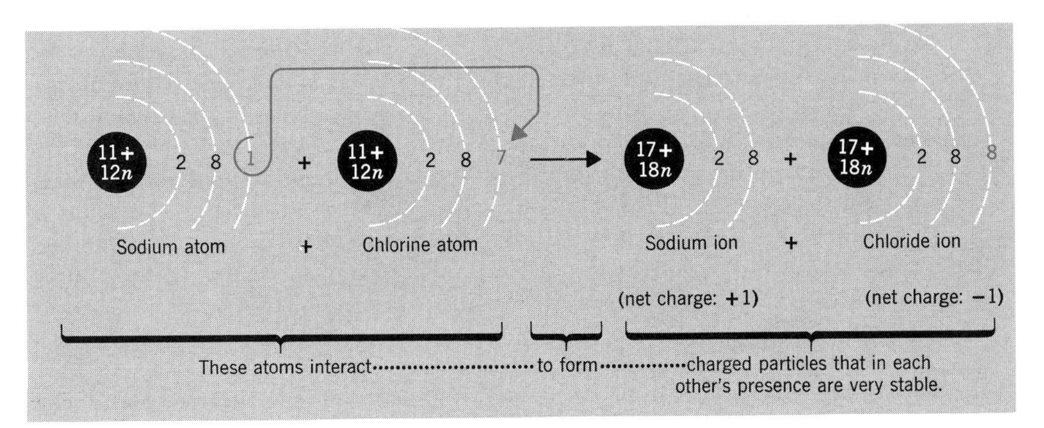

Of course, any interaction between real samples of sodium and chlorine produces thousands of billions of sodium ions and chloride ions. It is quite a storm of ions, at first, because ions of like charge repel each other. Sodium ions repel each other; chloride ions repel each other. What eventually (and quickly) happens is most reasonable in terms of an elementary rule of physics: like charges repel; unlike charges attract. Sodium ions gather, as nearest neighbors, as many chloride ions as possible. Chloride ions get, as nearest neighbors, as many sodium ions as possible. The result is a crystal built up from an orderly stacking of ions, illustrated in Figure 1.1.

Within the crystal there are strong forces of attraction between sodium ions and chloride ions. *It is this force of attraction between oppositely charged ions that we call the ionic bond.*

Ionic bonds are extremely strong; we may infer this from the fact that all ionic compounds are crystalline solids at room temperature. Only at a high temperature—generally above 300°—is there sufficient thermal energy available to get the ions vibrating so hard that the crystal collapses and changes to the liquid state.[3]

[3]Unless otherwise stated, all temperatures in this book are in degrees Celsius (formerly called centigrade).

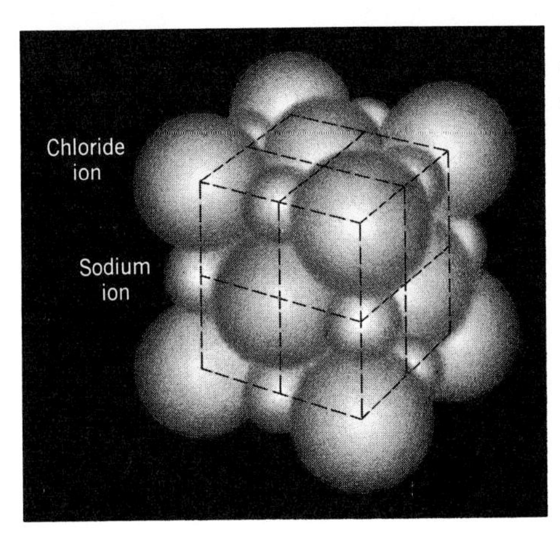

Figure 1.1
Orderly aggregation of ions in crystalline sodium chloride.

Sodium chloride, for example, melts at 801°; it boils at 1413°. Evidently, a sodium ion in a salt crystal does not easily give up its electron-rich surroundings, its neighboring chloride ions. There is one circumstance where this will readily happen at room temperature! That is when we dissolve sodium chloride in water. Water can do at room temperature what over 1400° Celsius is otherwise required—get sodium ions pried loose from chloride ions. It happens because water molecules can serve as substitutes for chloride ions. As we shall see, a water molecule is electron rich around its oxygen atom and electron poor around its hydrogens. Water molecules surround sodium ions in the solution; they also surround the chloride ions. To understand this property of water we need to know more about its structure. Water is a molecular substance; it consists of molecules, not oppositely charged ions. Therefore we next review the covalent bond, the principal bond in organic molecules.

Unlike ions, molecules are electrically neutral particles that contain at least two atomic nuclei, usually many more. To understand how electrons and nuclei hold together in these particles, we need a more detailed picture of energy levels. We have to review the concept of an atomic orbital.

The Covalent Bond

1.6 Atomic Orbitals

An atomic orbital is a tiny volume of space near an atom's nucleus. It's a region where one or two electrons can be. At the first level there is just one of these orbitals, called the 1*s* orbital. See Figure 1.2. The "1" specifies the first main energy level. The "*s*" is a label that specifies the kind of shape or symmetry of the orbital. The *s* orbital has the symmetry of a sphere (with the nucleus in the center where the

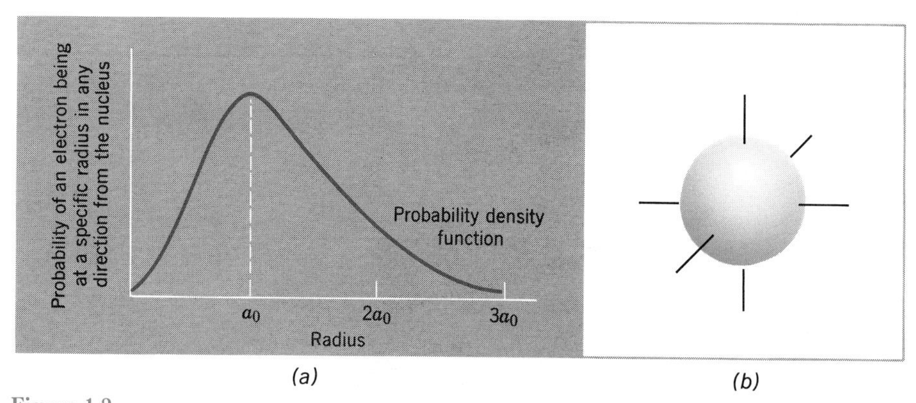

Figure 1.2
The 1*s* orbital. (*a*) Imagine the space surrounding a proton to be made up of layer on layer of infinitesimally thin concentric, spherical shells. What is the probability that an electron can be in each of these? This curve is the answer. At radius equal to a_0 (0.529Å), the probability is a maximum. (One Angstrom (Å) is defined as 1Å = 10^{-8} cm.) (*b*) One of the microspheres in (*a*) will enclose a space within which the total probability of finding an electron will be 90%. The contour of such a "probability envelope" is shown in (*b*), and this depicts the shape and symmetry of the 1*s* orbital. If the electron of a hydrogen atom is going to be in its lowest energy state it will be in a space such as this 90% of the time.

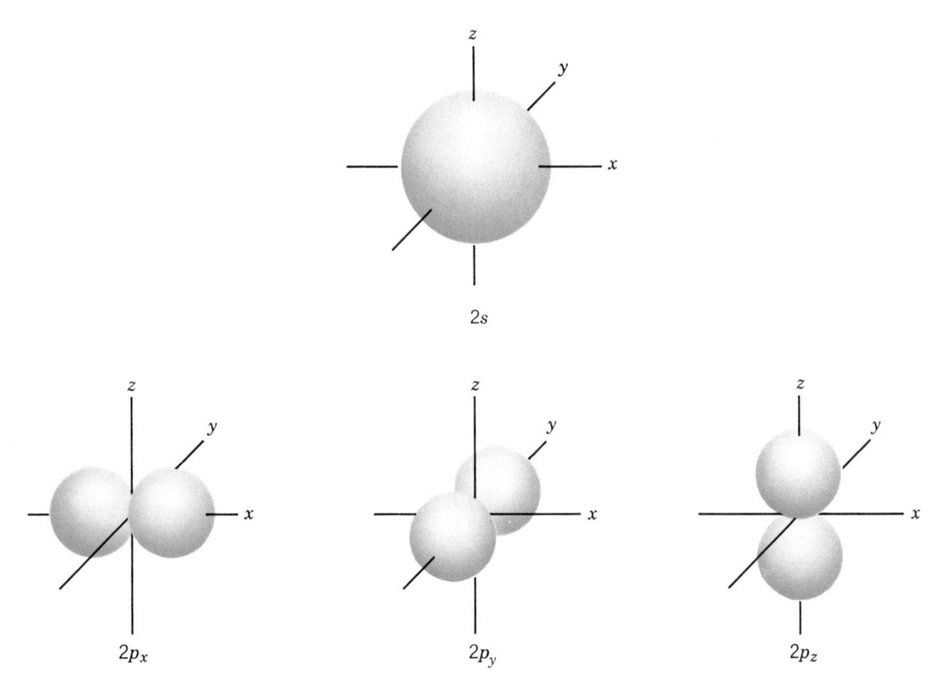

Figure 1.3
Orbitals at principal energy level number 2. For the hydrogen atom, the electron may be at the second level in any one of four ways: in the 2*s* orbital or in one of the mutually perpendicular 2*p* orbitals.

x-y-z axes intersect.) We say the 1*s* orbital has spherical symmetry. Every main level has one of these kinds of orbitals. They are, to make a crude analogy, concentric with each other in the way layers of an onion are concentric. There are 1*s*, 2*s*, 3*s*, etc. orbitals. There are no other orbitals at the first main level, just the 1*s*. Therefore, the maximum number of electrons an atom can have in the first level is 2.

At the second energy level, there are two sublevels. One sublevel is called the 2*s*, and the other sublevel is called the 2*p*. The *p* sublevel has three equivalent orbitals, p_x, p_y and p_z. See Figure 1.3. Their equivalence is in energy; they differ in orientation in space. The subscripts, *x*, *y*, and *z*, sort the three *p* orbitals into their three different projections into space relative to each other. Each *p* orbital has two lobes, and neither includes the intersection of the axes.

From the second level and out, there is a set of three *p* orbitals at each main level: 2*p*, 3*p*, 4*p*, etc.

At the third main energy level still another type of sublevel appears, the set of so-called *d* orbitals. There are five *d* orbitals per set, and from level three and out there is a set at each. Because they will not be important in our study, we do no more than mention them. Similarly, we mention the last type of sublevel, the sets of seven *f* orbitals that make their first appearance at principal level 4.

1.7 Pauli's Principle and Hund's Rule

The electronic configurations of elements 1 to 20 are shown in Table 1.2. Two rules of quantum chemistry are illustrated by these configurations, rules we can

Table 1.2
Electronic Configurations of Elements 1 to 20 Showing Distributions Among the Orbitals[a]

Element	Atomic Number	1s	2s	$2p_x$	$2p_y$	$2p_z$	3s	$3p_x$	$3p_y$	$3p_z$	4s
H	1	↑									
He	2	↑↓									
Li	3	↑↓	↑								
Be	4	↑↓	↑↓								
B	5	↑↓	↑↓	↑							
C	6	↑↓	↑↓	↑	↑						
N	7	↑↓	↑↓	↑	↑	↑					
O	8	↑↓	↑↓	↑↓	↑	↑					
F	9	↑↓	↑↓	↑↓	↑↓	↑					
Ne	10	↑↓	↑↓	↑↓	↑↓	↑↓					
Na	11	↑↓	↑↓	↑↓	↑↓	↑↓	↑				
Mg	12	↑↓	↑↓	↑↓	↑↓	↑↓	↑↓				
Al	13	↑↓	↑↓	↑↓	↑↓	↑↓	↑↓	↑			
Si	14	↑↓	↑↓	↑↓	↑↓	↑↓	↑↓	↑	↑		
P	15	↑↓	↑↓	↑↓	↑↓	↑↓	↑↓	↑	↑	↑	
S	16	↑↓	↑↓	↑↓	↑↓	↑↓	↑↓	↑↓	↑	↑	
Cl	17	↑↓	↑↓	↑↓	↑↓	↑↓	↑↓	↑↓	↑↓	↑	
Ar	18	↑↓	↑↓	↑↓	↑↓	↑↓	↑↓	↑↓	↑↓	↑↓	
K	19	↑↓	↑↓	↑↓	↑↓	↑↓	↑↓	↑↓	↑↓	↑↓	↑
Ca	20	↑↓	↑↓	↑↓	↑↓	↑↓	↑↓	↑↓	↑↓	↑↓	↑↓

[a] From 21 to 30, electrons now go into the available 3*d* orbitals until they are filled and then 4*p* orbitals start to fill. Elements 21 to 30 make up the first group of transition elements.

only state without detailed development. First, the Pauli exclusion principle: if there are to be two electrons in the same orbital, they must be of opposite spin. (The arrows pointing oppositely in Table 1.2 indicate electrons of opposite spin.) Second, Hund's rule: two electrons will not crowd into the same orbital if there is another empty orbital at the same sublevel. We see this in operation among the 2*p* orbitals of carbon, nitrogen, and oxygen.

1.8 Molecules

If you could get a supersubmicroscopic view of a sample of hydrogen gas, you would see a tumbling collection of particles. Each is electrically neutral; they are not ions. Each consists of *two* atomic nuclei and two electrons. Therefore, these particles are not atoms, either. They are called *molecules* (from the Greek, "little mass").

A molecule is an electrically neutral particle having two or more atomic nuclei together with just exactly enough electrons to "cancel" the total positive charge from all the nuclei. In hydrogen molecules there are two electrons and two protons. (The proton is all there is to the nucleus of a hydrogen atom.) In fluorine molecules, F_2, there are two nuclei, each with a positive charge of 9 (giving a total charge of +18 in the molecule), and 18 electrons. Now, atomic nuclei naturally (that is, spontaneously) repel each other; they are like charged. Electrons also repel each other. How then does a fluorine molecule hang together? Is it because this particle is actually a pair of ions, F^+ and F^-? The ion F^+, however, is actually $:\ddot{F}^+$, and it does not have an outer octet. It does not exist in compounds of fluorine. How can we then have particles of F_2 or of H_2? How, for that matter, can we have particles of CH_4, methane, if we cannot have the +4 carbon ion, $\cdot C \cdot^{4+}$? If we're going to understand organic compounds, we obviously have to find out the answers to these questions.

1.9 Molecular Orbitals

Figure 1.4 illustrates a simple model of how forces of attraction in the hydrogen molecule can arise.[4] Two atomic orbitals from the original hydrogen atoms overlap or interpenetrate. This overlap of the spaces of two atomic orbitals produces a new space called a *molecular orbital*. A molecular orbital is a region in space surrounding two nuclei (sometimes more) within which electrons can be found. We say that the electrons in a molecular orbital are shared. By this we mean that they move about two (or more) nuclei. A molecular orbital can hold a maximum of two electrons. Each shared electron is attracted to two nuclei instead of only one. That is what "sharing" means. The nuclei now can't repel each other. Instead, they are attracted toward the region between them where the shared electrons most frequently are.

For the expression, "most frequently are" we could better say, where the *electron density* is greatest. This will develop into our most useful way of looking at electrons in orbitals. In this view or model, an orbital with electrons in it is like a "cloud", an *electron cloud* or a negative-charge cloud.

Because sharing of electrons produces net forces of attraction, we say that sharing produces a bond, called a *covalent bond*. Among the simpler elements the covalent bond neatly fits the octet rule, provided we count shared electrons for both atoms when we check octets.

Any kind of atomic orbital—*s*, *p*, *d*, or *f*— can participate in overlapping to form a covalent bond. Figure 1.5 pictures the formation of the covalent bond in the fluorine molecule, F_2, where *p* orbitals overlap.

[4]Teachers familiar with molecular orbital theory and valence bond theory will recognize that the discussion eventually borrows from both. This seems to work best for introductory organic chemistry.

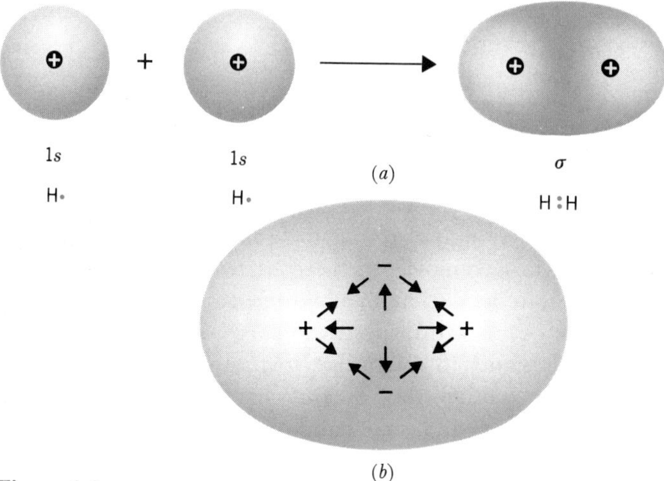

Figure 1.4
The hydrogen molecule and the covalent bond. (*a*) The overlap or interpenetration of two 1*s* atomic orbitals produces the new molecular orbital shown here. Two electrons are in this new orbital. Their extremely rapid movements would seem like a blur; we say that an "electron cloud" exists. It is most dense where the shading is darkest—between the nuclei. (*b*) The strength of the covalent bond results from the forces of attraction (shown with arrows pointing directly toward each other) being greater than forces of repulsion (arrows pointing away). This cluster of two nuclei (protons) and two electrons clings strongly together. In chemical terms we say that a bond—a covalent bond—exists.

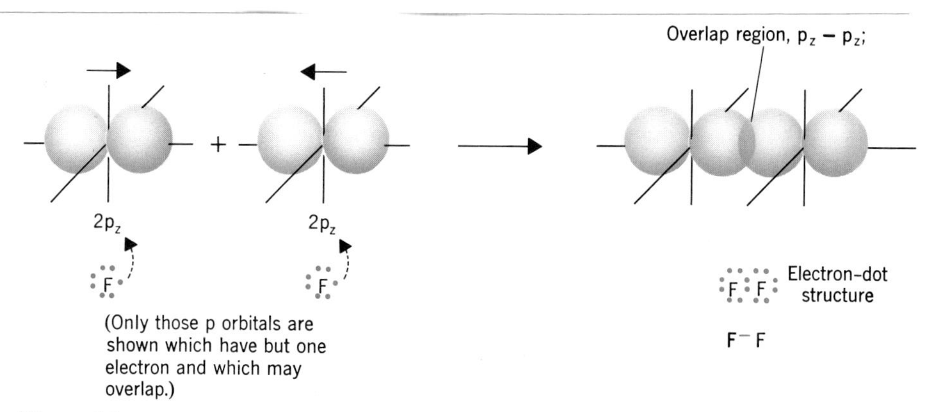

Figure 1.5
The overlap of the two p_z orbitals of two fluorine atoms gives a covalent bond between the two fluorines. The shape of the resultant molecular orbital is oversimplified, above. A better model would be more like the following:

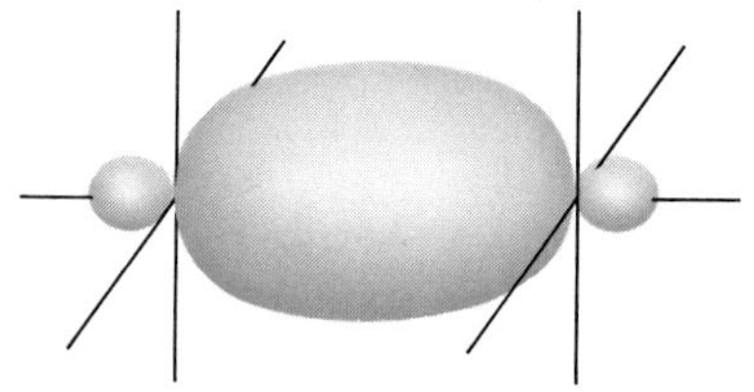

In other words, there is more of a concentration of electron density between the nuclei than indicated in the more simplified picture.

For the routine writing of structures, we simply use a straight line to represent a shared pair of electrons. Thus, for H_2, we write H—H. For F_2, F—F. The outer electrons are not shown unless we have a special reason to do that. Fluorine, for example, may be written as $:\ddot{\underset{..}{F}}—\ddot{\underset{..}{F}}:$ or (in older usage $:\ddot{\underset{..}{F}}:\ddot{\underset{..}{F}}:$. With this kind of formula it is easy to check outer octets.

1.10 Covalence Numbers

Nonmetals can form covalent bonds with each other. We now ask, how many bonds can a particular nonmetal have? Consider this series of compounds of hydrogen with first-row nonmetals:

H—$\ddot{\underset{..}{Cl}}$:	H—$\ddot{\underset{..}{O}}$—H	H—$\ddot{N}$—H (with H below N)	H—C—H (with H above and below C)
HYDROGEN CHLORIDE	WATER	AMMONIA	METHANE

In hydrogen chloride, the chlorine has only one shared pair, one bond. In water, oxygen has two bonds. Nitrogen has three in ammonia, and carbon has four in methane. In Table 1.1 we see that a chlorine atom has seven outer-shell electrons. It needs just one or a share of one more electron for an outer octet. Hence its covalence is 1, which is the covalence of the other halogens as well. Oxygen atoms, with six outer-shell electrons, require a share of two more and have a covalence of 2. (So does sulfur, in the same family as oxygen.) Nitrogen, where atoms need three more electrons for an octet, has a covalence of 3. Carbon atoms need four more to have octets and that is why carbon has a covalence of 4. Hydrogen's covalence is always 1. These simple numbers, collected in one place below, make it easy to figure out (by trial and error) what might be possible structures for a given molecular formula. A "reasonable" structure is one in which each atomic symbol has as many lines going from it as its covalence number.

Atom	A Common Covalence
F, Cl, Br, I; H	1
O, S	2
N, P	3
C, Si	4

Exercise 1.1 Within the rules of covalence, what would be a reasonable structure for each of the following?

(a) C_2H_6 (b) H_2O_2 (c) CNH_5 (d) N_2H_4

1.11 Multiple Bonds

Atoms are not restricted to forming only single bonds in attaining octets. Double and triple bonds are common as these examples illustrate:

$$\ddot{O}=C=\ddot{O} \qquad H_2C=CH_2 \qquad H-C(=O)-\ddot{O}-H \qquad H-C\equiv C-H \qquad :N\equiv N:$$

CARBON DIOXIDE — ETHYLENE — FORMIC ACID — ACETYLENE — NITROGEN

Each atom in each structure has the correct number of covalent bonds according to their common covalences. Just *how* these multiple bonds arise—what orbitals overlap to make them—will be studied in later chapters as we take up the study of each type. We are interested here only in the fact that such multiple bonds exist and that they fit the general pattern of covalence numbers based on the octet rule.

Exercise 1.2 Within the rules of covalence what would be reasonable structures for each of the following?

(a) HCN (b) CH_2O (c) HNO_2 (d) C_2F_4

1.12 Hybrid Orbitals

Methane, CH_4, is the simplest organic compound, and there are two important experimental facts about its structure. (We state them without proof here.) First, there are four equivalent bonds from carbon. Second, these bonds point toward the corners of a regular tetrahedron as pictured in Figure 1.6. We call the carbon in methane a **tetrahedral carbon.** In fact, in any molecule where there are four single bonds from a carbon, the bonds will have this tetrahedral arrangement around carbon, or nearly so.

The electronic configuration of carbon is $1s^2 2s^2 2p_x 2p_y$. In the outside shell of carbon there are only two unpaired electrons, those in the $2p_x$ and $2p_y$ orbitals. The $2p_z$ orbital is empty. We obviously could not develop four equivalent orbitals arranged tetrahedrally from this configuration. To have four equivalent *bonds* from carbon we need four equivalent *atomic orbitals.* What now follows is the way we understand how such orbitals arise from the atomic state of carbon.

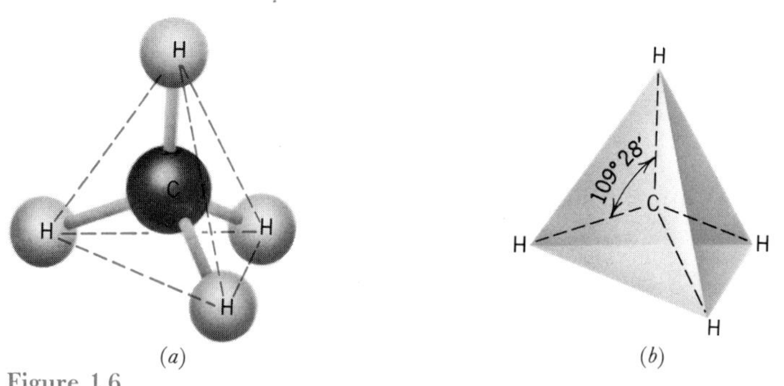

Figure 1.6
Tetrahedral carbon atom. Part *a* shows a common ball-and-stick model of methane with dotted lines added to bring out the tetrahedron. A different perspective is given in part *b*.

We begin with all the *empty* atomic orbitals of a carbon atom at the first and second level. We picture each orbital as a box on an energy diagram. Then we imagine that all four orbitals at level 2—the 2*s* and the three 2*p*—become reorganized as follows, where we reorganize *orbitals,* not electrons:

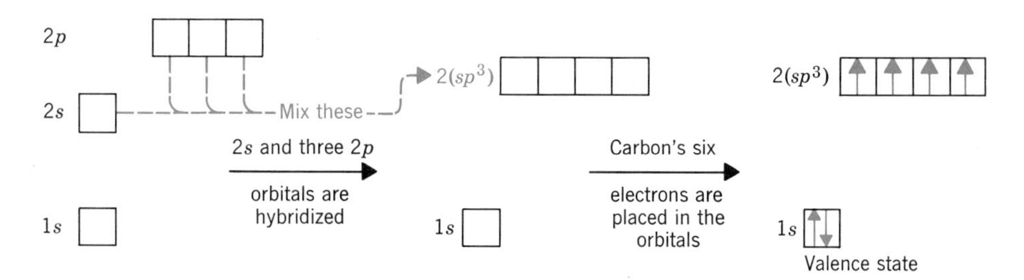

Now we distribute carbon's six electrons into the new orbitals, remembering from Hund's rule to distribute them as widely as possible among the available orbitals of a given sublevel. When we have done this we have carbon in a new state called its *valence state.*

This "mixing" of orbitals is called *hybridization.* Borrowing a term from genetics, we call the new orbitals *hybrid orbitals*; they are hybrids of the *s* and the three *p* atomic orbitals. Hybridization is actually done mathematically. Equations for atomic orbitals are "mixed" to generate new equations describing the new orbitals. This is a standard technique in theoretical chemistry, but also one well beyond the scope of our study. This ought not stop us, however, from using the results.

Since there are ultimately several kinds of hybrid orbitals, we have to give them names. Whenever hybrid orbitals are made from one *s* and three *p* orbitals, they are called sp^3 orbitals (pronounced "s-p-three"). (In Chapter 3 we shall study two more kinds of hybrid orbitals for carbon, orbitals used in making double and triple bonds. These will be called the sp^2 and the *sp* orbitals.)

The shape of an sp^3 hybrid orbital is shown, in cross section, in Figure 1.7. It has both "*s*-character" and "*p*-character" in these ways. There is *s*-character because the orbital encloses the intersection of the *x-y-z* axes. There is *p* character because there are two lobes. The carbon atom will have four of these hybrid orbitals arranged about it—Figure 1.8—each with negative charge in it. Since like charges repel, these four orbitals must get as far from each other as they can while not leaving their carbon. That is why they take up the tetrahedral arrangement, and that is why the carbon in methane is tetrahedral. See Figure 1.9. Each bond in methane comes from the overlap of an sp^3 orbital on carbon and a 1*s* orbital of a hydrogen. The new molecular orbital has symmetry with respect to the axis of

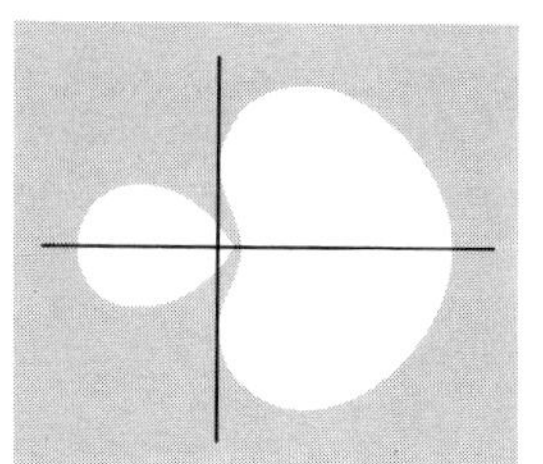

Figure 1.7
Cross section of an sp^3 hybrid orbital.

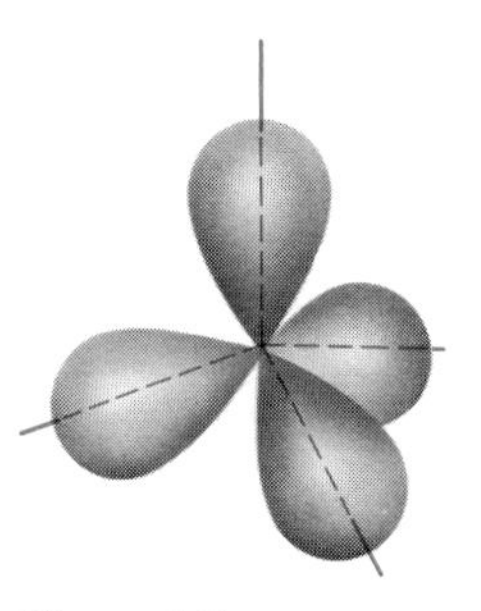

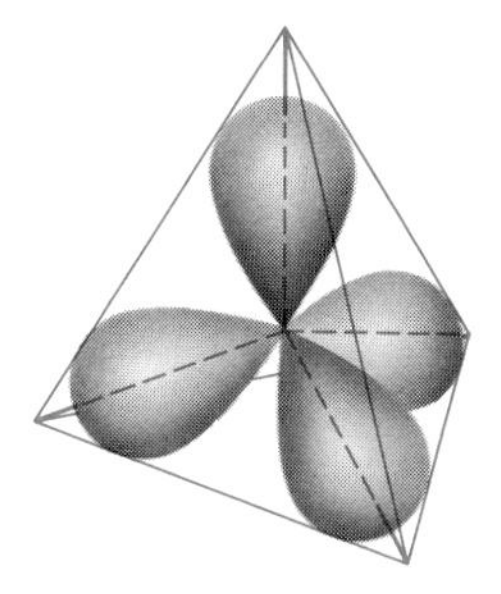

Figure 1.8
The four equivalent sp^3 hybrid orbitals of carbon are positioned so that their axes point to the corners of a regular tetrahedron. Only the large lobes of the sp^3 orbitals are indicated in this figure.

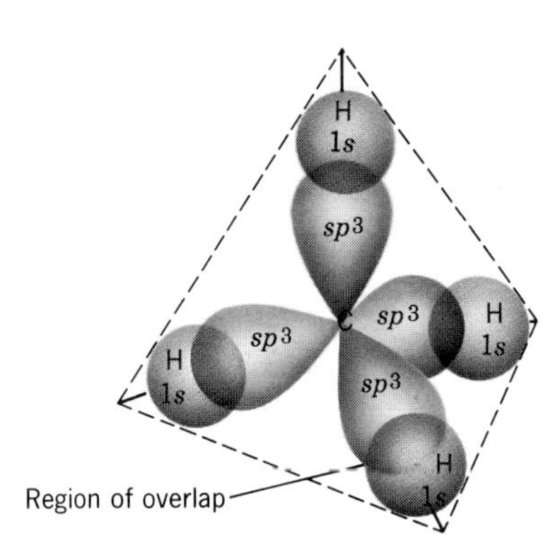

Figure 1.9
Bonds in methane. Each carbon-to-hydrogen covalent bond is thought of as forming from the overlap of an sp^3 hybrid orbital from carbon and a 1*s* orbital from hydrogen. Because there is symmetry about the bonding axis, each bond is a σ-bond. The principle of maximum overlap ensures that the nucleus of each hydrogen is at the corner of the regular tetrahedron outlined by the dotted lines.

the bond, the kind of symmetry we find in a regular cylinder. Any bond with this kind of symmetry is called a **sigma bond** (abbreviated σ-bond).

1.13 Hybrid Orbitals in Oxygen and Nitrogen

Since atoms of these elements are in the molecules of many families of organic compounds, we need to see how oxygen and nitrogen form single bonds. We do exactly as we did with carbon; we take empty orbitals at the second level and hybridize them to give four equivalent sp^3 hybrid orbitals. Then we put in the electrons belonging to each atom, and we have their valence states:

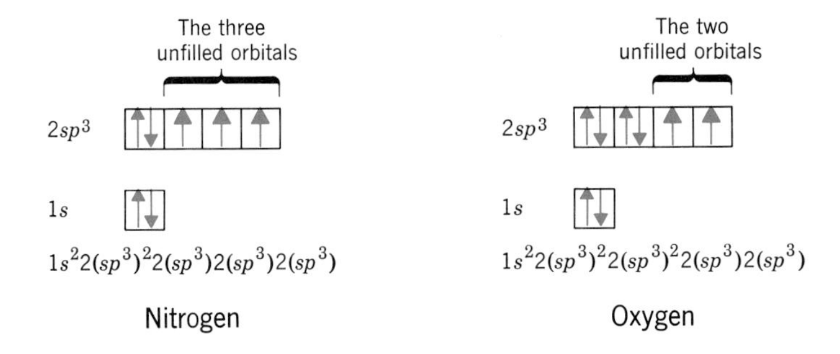

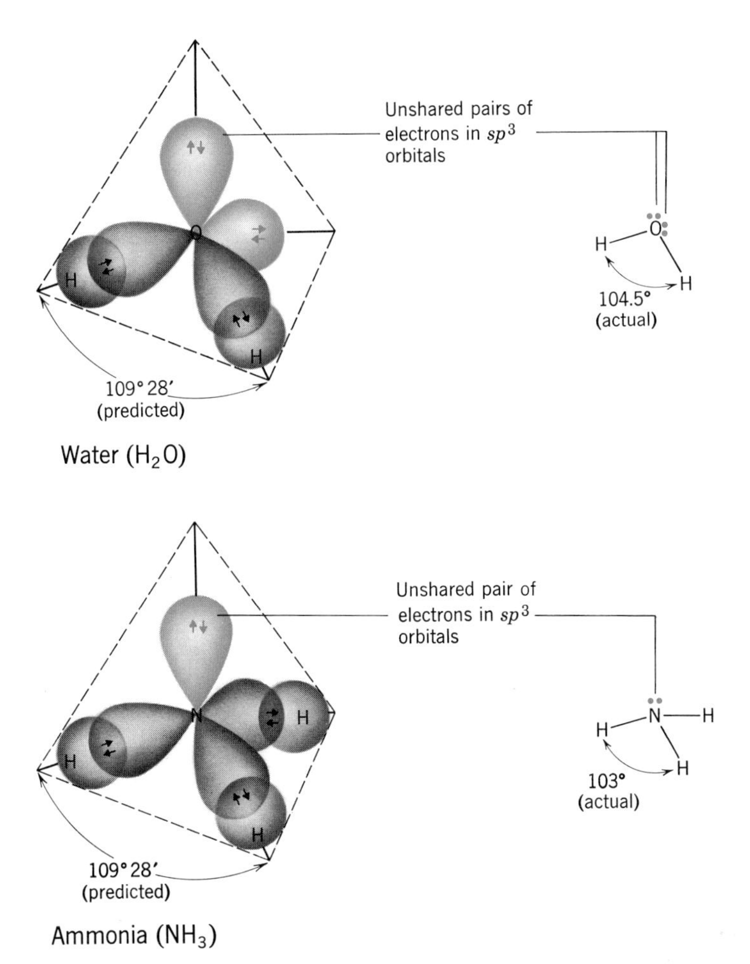

Figure 1.10
Molecular orbital models for molecules of water and ammonia.

The hybrid orbitals point toward the corners of a regular tetrahedron. Figure 1.10 shows how the unfilled orbitals of these atoms overlap with hydrogen 1*s* orbitals to make molecules of water and ammonia. The bond angles in each are very close to the tetrahedral value of 109° 28′ predicted by this model.

1.14 Coordinate Covalent Bonds

In the atoms of oxygen and nitrogen in water and ammonia there are filled hybrid orbitals. These could not overlap with a 1*s* orbital of a hydrogen *atom*, which contains an electron. We could not get three electrons into the new molecular orbital. They could overlap, however, with an empty 1*s* orbital of a hydrogen *ion*, H^+. The new molecular orbital would contain two electrons. The only difference is that both were provided by one atom of the pair that bond. Otherwise, the new bond is indistinguishable from a covalent bond. When we want to make a distinction, we call it a *coordinate covalent bond*. Its formation is sometimes called *coordination*. We say, for example, that a water molecule coordinates a proton in changing to a hydronium ion. The ammonia molecule coordinates a proton in changing to an ammonium ion.

$$H_2\ddot{O}: + H^+ \longrightarrow H_3O^+$$

HYDRONIUM ION

$$H_3\ddot{N} + H^+ \longrightarrow H_4N^+$$

AMMONIUM ION

The central atom in these two ions now seems to "violate" its covalence number. Such numbers, however, do not hold when coordinate covalent bonds form.

The positive charge is associated with the central atom, not with the hydrogens. In the sharing process the central atom suffers a *net* loss of just *one* electron, leaving a +1 charge. The hydrogen ion gets a net gain of one electron as the new bond forms. (Two electrons are shared more or less 50 : 50. Fifty percent of 2 is 1. Hence, as H^+ becomes bound to the central atom it gets a net of one electron, and it becomes neutral.)

The ability of unshared electron pairs on oxygen and nitrogen to coordinate a proton is a fact of major importance in all of organic chemistry and biological chemistry. It is this ability that figures in nearly all proton transfers.

1.15 Polar Molecules. Electronegativity

When electrons are shared between two *different* atoms, they are seldom shared exactly 50 : 50. The electron density in the covalent bond is not so evenly distributed that the positive charge on the nucleus at one end of the bond is *exactly* neutralized by the electron density immediately near it. The electron density may be a little too low, for example. This leaves some positive charge by that end of the bond. The other end has the slightly richer electron density. The atomic nucleus there has drawn toward itself more electron density from the bond than it really needs to neutralize exactly the effect of its positive charge. We say that such an atom is more electronegative than the atom at the other end of the bond. The more electronegative atom has a slight excess of negative charge about it. The less electronegative atom, having been partly drained of electron density, will be partially positively charged. We use the Greek lower-case delta, δ, as our symbol for "partial," and we illustrate the situation just described as follows, where Y is more electronegative than X.

$$\overset{\delta+}{X}—\overset{\delta-}{Y}$$

Because individual bonds can be polar, individual molecules can be polar. If molecules could not be polar, they could not "stick" to each other in substances. They could not attract each other. Their substances would therefore be gases or, at best, volatile liquids. Substances whose molecules are very polar are solids at room temperature. To understand physical properties of organic compounds, we have to be able to look at a molecular structure and decide if the molecule will likely be polar. We do that by first deciding if individual bonds are. For that we use a scale of relative electronegativities:

$$F > O > N > Cl > Br > C \gtrsim H$$

(This series should be learned.) Note that carbon and hydrogen have about the same electronegativities. Therefore C—H bonds will not be polar, a fact of major importance in organic chemistry. The bonds in the water molecule are quite polar. So, too, are the bonds in carbon tetrachloride. However, because of the symmetrical tetrahedral array, the individual bond polarities cancel out. Therefore, the carbon

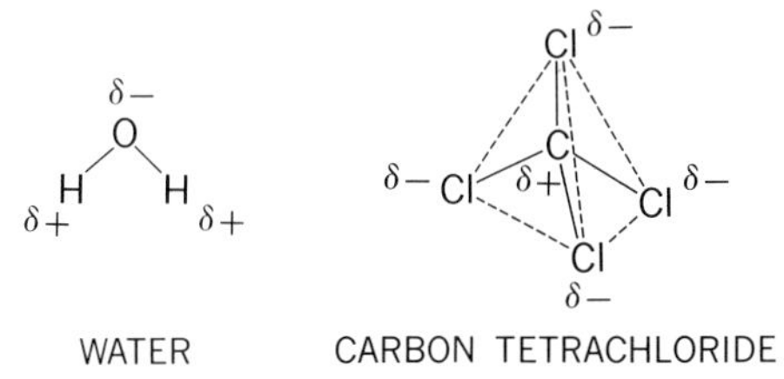

WATER   CARBON TETRACHLORIDE

tetrachloride molecule is not polar. It is important to realize that having polar *bonds* does not guarantee polar *molecules*. Molecular geometry must also be considered.

Exercise 1.3 The carbon dioxide molecule, O=C=O, is linear as written. Are the carbon-oxygen bonds polar? Is the molecule polar?

1.16 The Carbon-Carbon Covalent Bond

Molecules of ethane, C_2H_6, contain two carbons joined by a single bond, as pictured in Figure 1.11*a*. This bond forms by the overlap of two sp^3 hybrid atomic orbitals. Part *b* of Figure 1.11 pictures "ball-and-stick" models of ethane. All the bonds are sigma bonds; all have the cylindrical symmetry that defines such a bond. Regardless of how the two ends of the ethane molecule are twisted relative to each other, the overlap that created the bond between the carbons is not affected, at least not appreciably so. If the overlap is not affected, the strength of the bond is not affected. Therefore it costs no energy (or almost no energy) to twist or rotate the two groups joined by the carbon-carbon bond. We say there is *free rotation* about the bond. This is generally true of single bonds—free rotation. The exceptions, as they occur, will seem normal in their own way, and we shall take them as they come in our study.

The carbon-carbon covalent bond is one of the truly unusual bonds in nature, and carbon is a most unique element. Unlike any other element, atoms of carbon can bond to each other, successively, an unlimited number of times. While doing this, carbon atoms can use their other valences to form strong bonds to atoms of other elements. These may range through the entire group of nonmetallic ele-

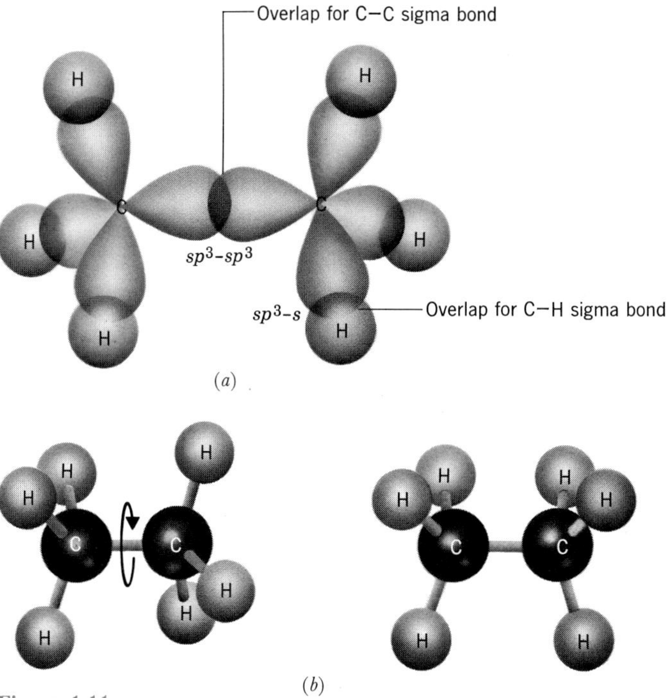

Figure 1.11
Ethane. (a) The molecular orbitals in ethane, CH_3CH_3. (b) Free rotation is possible about single bonds.

ments—such as hydrogen, oxygen, the halogens, and nitrogen—and even some metallic elements—such as lead, mercury, copper, and lithium. No other element has this versatility. It partly accounts for another phenomenon about carbon compounds, isomerism.

1.17 Isomerism

Ammonium cyanate and urea have the same molecular formula, CH_4N_2O. When Wöhler tried to make ammonium cyanate but obtained urea instead he added to the small but growing list of substances that were obviously different yet had identical molecular formulas. In 1832 Jöns Jakob Berzelius proposed that such substances be called *isomeric,* from the Greek *isos,* equal, and *meros,* part. He further suggested that the difference between compounds having identical molecular formulas must lie in a different arrangement of their parts, a prediction that was later confirmed.

Consider the molecular formula, C_2H_6O. In how many ways within the rules of valence can the nuclei of two carbons, six hydrogens, and one oxygen be arranged? Trial and error will show just two, given as ball-and-stick models in Figure 1.12. There are substances consisting of molecules of each kind—ethyl alcohol and dimethyl ether. Table 1.3 has a summary of some of their properties.

Compounds that have identical molecular formulas but different structures are

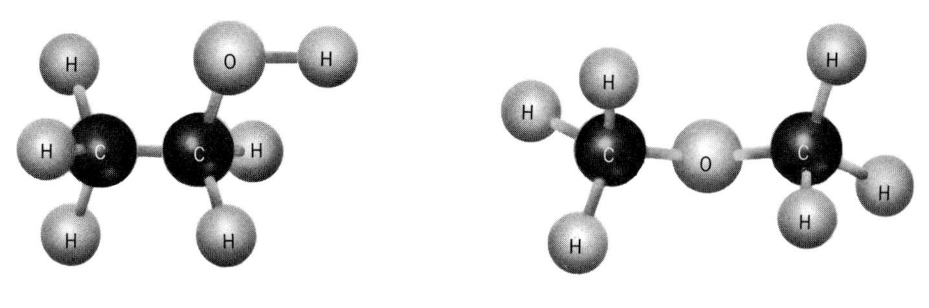

Ethyl alcohol

Dimethyl ether

Figure 1.12
Isomers of C_2H_6O, ethyl alcohol and dimethyl ether, in ball-and-stick models. Each isomer has the same molecular formula, but they differ in the sequences in which these atoms are joined together.

isomers of each other. The phenomenon of the existence of isomers is called *isomerism.* The structural differences between isomers must be such that simple rotation about single bonds will not convert one into the other.

1.18 Condensed Structural Formulas

On the bottom line of Table 1.3 the structure of ethyl alcohol is given as CH_3CH_2OH, a condensed version of the full structural formula. This structure, the "condensed" structure, CH_3CH_2OH, is obviously easier to write and it saves space and time. Equally important, we sacrifice no information. To convert "full"

Table 1.3
Properties of Two Isomers: Ethyl Alcohol and Dimethyl Ether

Property	Ethyl Alcohol	Dimethyl Ether
Molecular formula	C_2H_6O	C_2H_6O
Boiling point	78.5°C	−24°C
Melting point	−117°C	−138.5°C
Density	0.789 g/cc	2 g/liter
Solubility in water	Completely soluble in all proportions	Slightly soluble
Action of metallic sodium	Vigorous reaction, hydrogen evolved	No reaction
Structural formula	H H H—C—C—O—H H H	H H H—C—O—C—H H H
Condensed structural formula	CH_3CH_2OH	CH_3OCH_3

structures into condensed structures we use just a few simple guides, and we introduce three of them here.

1. We write CH_3— whenever there are three hydrogens on one carbon.
2. We write —CH_2— when two hydrogens are on one carbon.
3. We write —CH— (with a third bond upward) when only one hydrogen is on a carbon.

These rules are illustrated by the several ways of writing the following structure:

```
    H
    |
H H—C—H H     H           CH3                        CH3
|   |   |     |            |                          |
H—C———C———C———C—H  or  CH3—CH—CH2—CH3   or      CH3CHCH2CH3
|   |   |     |
H   H   H     H
```

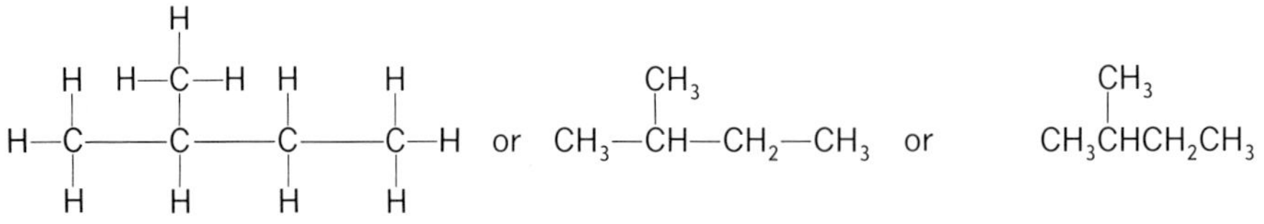

ALL BONDS ARE SHOWN

ALL C—H BONDS ARE "UNDERSTOOD"

ALL C—H AND ALL C—C BONDS THAT OTHERWISE WOULD APPEAR ON A HORIZONTAL LINE ARE "UNDERSTOOD."

We never leave as "understood" any bonds that would appear at any angle from the horizontal. We never leave double or triple bonds "understood." We do, however, "understand" that the bond angle at a saturated carbon is roughly 109°.

1.19 Organizing Organic Substances for Study

Although there are millions of organic compounds, there are only a few important classes. The situation resembles zoology and the animal kingdom. There are millions of animals, but just a few families based on structural (anatomical) similarities.

Some of the most important organic families are described in Table 1.4. Our study will be organized around these and similar families.

Problems and Exercises

1. Check each of the following structures to see whether—within the rules of covalence—they represent possible compounds.

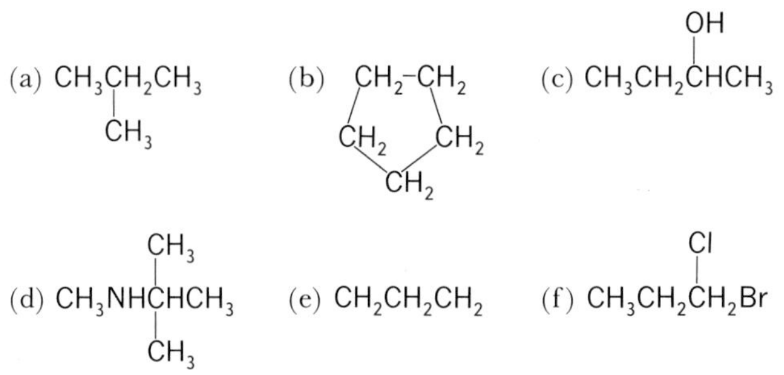

```
(a) CH3CH2CH3        (b)  CH2-CH2           (c)           OH
      |                  /      \                          |
      CH3              CH2      CH2              CH3CH2CHCH3
                          \    /
                           CH2

        CH3                                                Cl
        |                                                  |
(d) CH3NHCHCH3       (e) CH2CH2CH2          (f) CH3CH2CH2Br
        |
        CH3
```

Table 1.4
Some Important Classes of Organic Compounds

Class	Characteristic Structural Features of Molecules	Example
Hydrocarbons	Contain only carbon and hydrogen; may have carbon chains or carbon rings. Subclasses according to presence of multiple bonds	
	Alkanes: all single bonds	CH_3CH_3, ethane
	Alkenes: at least one double bond	$CH_3CH{=}CH_2$, propylene
	Alkynes: at least one triple bond	$HC{\equiv}CH$, acetylene
	Aromatic: at least one benzene-like ring system	, benzene
Alcohols	At least one —OH joined to a tetrahedral carbon	CH_3CH_2OH, ethyl alcohol
Aldehydes	$-\overset{O}{\overset{\|}{C}}-H$ or aldehyde group	$CH_3\overset{O}{\overset{\|}{C}}H$, acetaldehyde
Ketones	$-\underset{\|}{\overset{\|}{C}}-\overset{O}{\overset{\|}{C}}-\underset{\|}{\overset{\|}{C}}-$ or keto-group	$CH_3\overset{O}{\overset{\|}{C}}CH_3$, acetone
Carboxylic acids	$-\overset{O}{\overset{\|}{C}}-OH$ or carboxyl group	$CH_3\overset{O}{\overset{\|}{C}}OH$, acetic acid
Amines	$-NH_2$, or $-NH-$ or $-\overset{\|}{N}-$	CH_3NH_2, methylamine CH_3NHCH_3, dimethylamine

2. Convert each of the following structural formulas into a condensed structure. Straighten out the longest continuous carbon chain so that it is written on one line. (Cyclic structures, of course, are exceptions.)

(a) 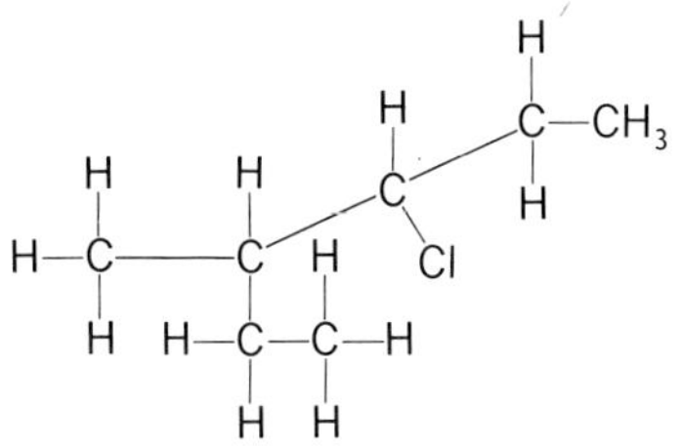

(b) 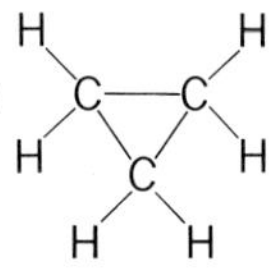

```
                          H
          H H             |
           \|          H—C—H
     H      C————————C—H
     |      |           \
(c) H—C—————C—H          C—H   H
     |      |           /   \ /
     H      H          H     C—C—H
                            /  |\
                           H   H H

           Cl
           |
(d) H—C—Cl
           |
           Cl
```

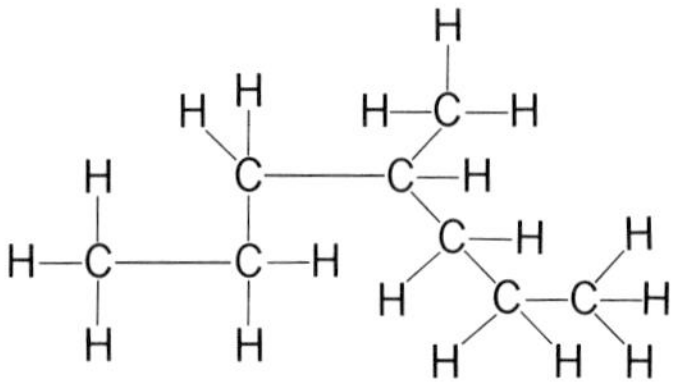

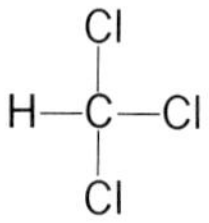

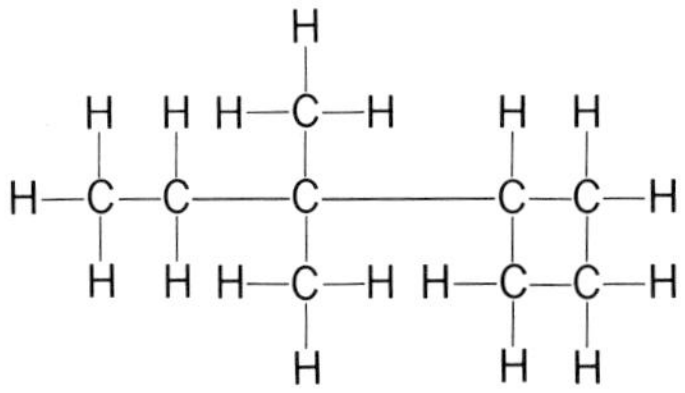

```
                 H
                 |
     H  H  H—C—H       H  H
     |  |    |         |  |
(e) H—C—C————C—————————C—C—H
     |  |    |         |  |
     H  H  H—C—H  H—C—C—H
                 |     |  |
                 H     H  H

        H  H  H  H  H
        |  |  |  |  |
(f) H—N—C—C—C—C—H
           |  |  |  |
           H  |  H  H
              |
            H—C—H
              |
            H—C—H
              |
              H
```

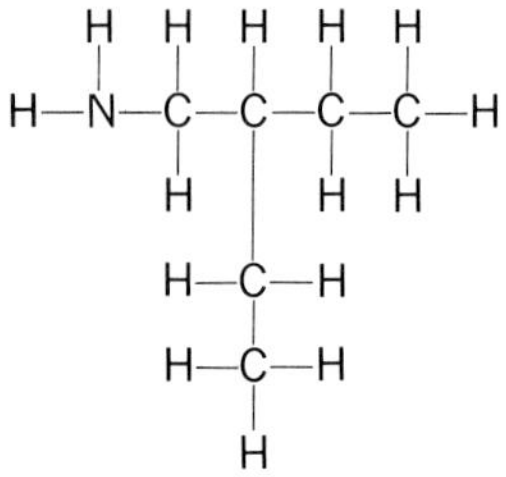

3. For each of the following pairs of structures, decide whether the two are *identical, isomers,* or *neither.* With some pairs the only difference is the relative orientation of the same structure in space (i.e., on paper). Others differ only in relative rotational position within the molecule. Assume free rotation about all single bonds except those that are part of a cyclic system.

```
         CH2—CH2              CH3—CH2
         |    |                    |
(a) CH3—CH2  CH3                   CH2—CH2—CH3

(b) CH3—O—CH2CH3              CH3CH2—O—CH3

                                    CH3
                                    |
(c) H—O—CH2CH2CH2CH3          CH3CHCH2OH

        CH3                         CH3
        |                           |
(d) CH3CHCH2CH2Br             CH3CHCH2CH2CH2Br

     CH3CHCH2CH2CH2               CH3              CH3
        |        |                |                |
(e) CH3CCH3     CHCH3         CH3CHCH2CH2CH2CHC—CH3
        |        |                           |  \
        CH3      CH3                         CH3  CH3

(f) CH3CH=CHCH3               CH2=CHCH2CH3
```

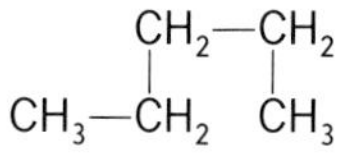

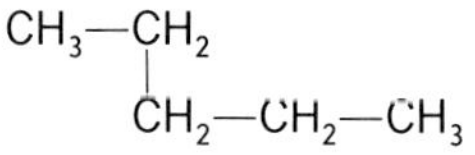

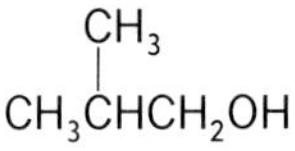

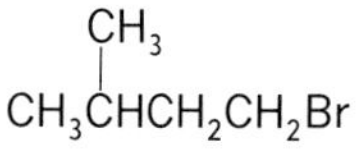

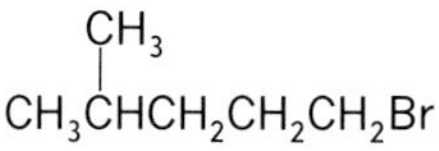

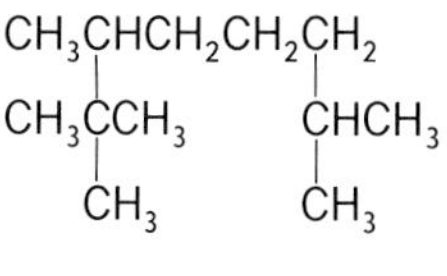

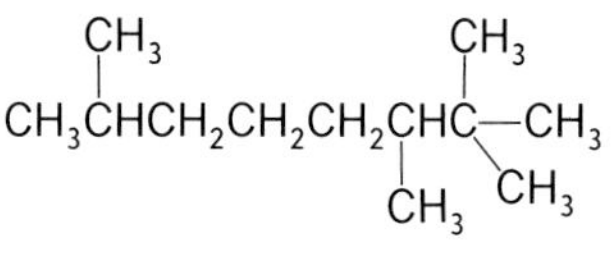

4. State very carefully and explicitly the key differences between the entities in each of the following pairs.
   (a) atom and molecule
   (b) atom and ion
   (c) ion and molecule
   (d) atom and element
   (e) ionic compound and molecular compound
   (f) ionic bond and covalent bond
   (g) covalent bond and coordinate covalent bond
   (h) atomic orbital and molecular orbital

chapter two
Saturated Hydrocarbons

2.1 Families of Hydrocarbons

Their family name discloses their composition; hydrocarbons contain only carbon and hydrogen[1]. Covalent bonds hold their molecules together, not ionic bonds. Depending on the presence or absence of one or more double or triple bonds in its molecules, a hydrocarbon may be a member of one of several families of hydrocarbons. These are defined in Figure 2.1.

2.2 Saturated and Unsaturated Organic Compounds

If the molecules of an organic substance contain only *single* bonds, the substance is said to be *saturated.* The alkanes are saturated hydrocarbons. If a substance's molecules have even one double or triple bond, the substance is unsaturated. Alkenes, alkynes, and aromatic hydrocarbons are all unsaturated hydrocarbons.

2.3 Sources of Hydrocarbons

Petroleum and coal, two of the fossil fuels, are our principal sources of hydrocarbons. We shall study these natural resources in detail in Chapter 5. Mention of the fossil fuels here, however, reminds us that organic chemistry, as with all the natural sciences, is a study of nature, at least an important part of it.

[1]This is, of course, only a manner of speaking. While chemists and advanced students of chemistry know that the phrase "contain only carbon and hydrogen" is not to be taken literally, beginners sometimes miss this subtle point and endure a certain amount of confusion. Hydrocarbon molecules *contain* carbon and hydrogen only in the sense that they are made from the *pieces* of the atoms of these elements and do not contain the intact atoms.

The Alkanes

2.4 Homologous Series

The structures and important physical properties of the 10 smallest "straight chain" members of the alkanes are given in Table 2.1. Straight chain means only that the carbon nuclei follow one another as links in a chain. The expression does not describe the geometry of the molecule. With free rotation about single bonds, these chains may flex, coil, and otherwise become kinky.

The series in Table 2.1 is said to be a homologous series because members differ from each other in a consistent, regular way. Here each member differs from the one just before or after it by one CH_2 unit. In chemical terminology butane, for example, is "the next higher *homolog* of propane."

2.5 Isomerism Among the Alkanes

Among the alkanes, from butane and on to higher homologs, isomerism is possible. Table 2.2 lists several examples. There are two isomers of formula C_4H_{10}, three of formula C_5H_{12}, and five of formula C_6H_{14}. As the homologous series is ascended, the number of possible isomers approaches astronomical figures. There are 75 possible isomers of decane, $C_{10}H_{22}$, and an estimated 6.25×10^{13} possible isomers of $C_{40}H_{82}$. Not all possible isomers have actually been prepared in the pure state and studied, for no useful purpose would be served and time would not permit. The occurrence of isomerism again emphasizes how limited is the information in a molecular formula. Only with a structural formula can the uniqueness of a molecule be understood and correlated with its properties.

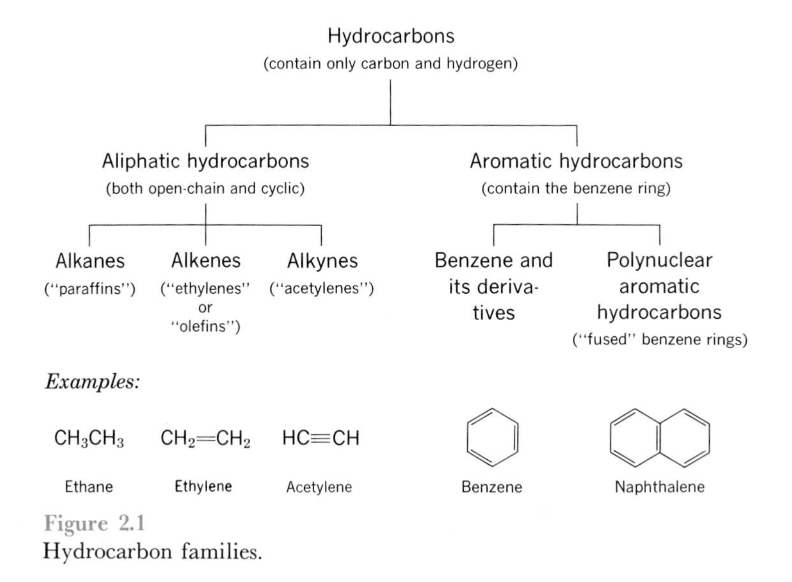

Figure 2.1
Hydrocarbon families.

2.6 Nomenclature

The earliest known organic compounds were named after their source, for example, formic acid (Latin *formica,* ants) can be made by grinding ants with water and distilling the result. Hundreds of compounds were named after their sources, but the system becomes impossibly difficult to extend to all compounds. These common names, however, are still used, and the beginning student is faced with the necessity of learning some of them. In addition, to be able to read and talk about more complicated structures without common names, the beginner must learn rules for formal or systematic nomenclature. (A third system, the derived system of nomenclature, applies in only a few situations, and we shall omit them from our study.)

2.7 Common Names of Alkanes

The straight-chain isomers are designated the *normal* isomers, and their common names include *n*- for normal. Examples are *n*-butane, *n*-octane, but not *n*-ethane because ethane has no isomer. In Table 2.1 butane and all subsequent names would be common if *n*- were placed before each. As they stand, they are formal names, to be described later. Except where noted, the names in Table 2.2 are also common names. The names of all the normal alkanes through *n*-decane as well as the names of the isomers through the five-carbon series must be learned. The total carbon content of the alkane should be associated with the prefix portion of its name. Thus "eth-" is a word part for a two-carbon unit; "but-" signifies four carbons whether in *n*-butane or isobutane. The "-ane" ending is characteristic of all alkanes.

Table 2.1
Straight-Chain Alkanes

Name	Number of Carbon Atoms	Molecular Formula[a]	Condensed Structural Formula	Bp (°C at atmospheric pressure)	Mp (°C)	Density (in g/cc at 20°C)
Methane	1	CH_4	CH_4	−161.5		
Ethane	2	C_2H_6	CH_3CH_3	−88.6		
Propane	3	C_3H_8	$CH_3CH_2CH_3$	−42.1		
Butane	4	C_4H_{10}	$CH_3CH_2CH_2CH_3$	−0.5	−138.4	
Pentane	5	C_5H_{12}	$CH_3CH_2CH_2CH_2CH_3$	36.1	−129.7	0.626
Hexane	6	C_6H_{14}	$CH_3CH_2CH_2CH_2CH_2CH_3$	68.7	−95.3	0.659
Heptane	7	C_7H_{16}	$CH_3CH_2CH_2CH_2CH_2CH_2CH_3$	98.4	−90.6	0.684
Octane	8	C_8H_{18}	$CH_3CH_2CH_2CH_2CH_2CH_2CH_2CH_3$	125.7	−56.8	0.703
Nonane	9	C_9H_{20}	$CH_3CH_2CH_2CH_2CH_2CH_2CH_2CH_2CH_3$	150.8	−53.5	0.718
Decane	10	$C_{10}H_{22}$	$CH_3CH_2CH_2CH_2CH_2CH_2CH_2CH_2CH_2CH_3$	174.1	−29.7	0.730

[a] The molecular formulas of the open-chain alkanes fit the general formula C_nH_{2n+2}, where n is the number of carbons in the molecule.

Table 2.2
Properties of Isomeric Alkanes

Family	Common Name (except where noted)	Structure	Bp (°C at atmospheric pressure)	Mp (°C)	Density (in g/cc)
Butane isomers	*n*-Butane	$CH_3CH_2CH_2CH_3$	−0.5	−138.4	0.622 (−20°C)
	Isobutane	CH_3CHCH_3 | CH_3	−11.7	−159.6	0.604 (−20°C)
Pentane isomers	*n*-Pentane	$CH_3CH_2CH_2CH_2CH_3$	36.1	−129.7	0.626 (20°C)
	Isopentane	$CH_3CHCH_2CH_3$ | CH_3	27.9	−159.9	0.620
	Neopentane	CH_3 | CH_3CCH_3 | CH_3	9.5	−16.6	0.591
Hexane isomers	*n*-Hexane	$CH_3CH_2CH_2CH_2CH_2CH_3$	68.7	−95.3	0.659
	3-Methylpentane (no common name)	CH_3 | $CH_3CH_2CHCH_2CH_3$	63.3		0.664
	Isohexane	CH_3 | $CH_3CHCH_2CH_2CH_3$	60.3	−153.7	0.653
	2,3-Dimethylbutane (no common name)	CH_3 | $CH_3CHCHCH_3$ | CH_3	58.0	−128.5	0.662
	Neohexane	CH_3 | $CH_3CCH_2CH_3$ | CH_3	49.7	−99.9	0.649

Octane isomers, C_8H_{18}—total of 18
Decane isomers, $C_{10}H_{22}$—total of 75
Eicosane isomers, $C_{20}H_{42}$—total of 366,319
Tetracontane isomers, $C_{40}H_{82}$—estimated total of 6.25×10^{13} isomers

2.8 Formal Names. Geneva System. IUPAC Rules for Alkanes

Representatives from chemical societies all over the world, meeting as the International Union of Pure and Applied Chemistry, with sessions held usually in Geneva, have recommended adoption of the following rules for naming alkanes.

1. The general name for saturated hydrocarbons is *alkane*.
2. The names of the straight-chain members of the alkanes are those listed in Table 2.1. (The designation *n*- is not included. Names going beyond the 10-carbon alkanes are, of course, available, but we shall not need them.)
3. For branched-chain alkanes, base the root of the name on the alkane that

corresponds to the longest continuous chain of carbons in the molecule. For example, in the compound

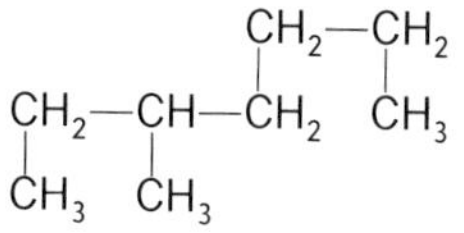

$$
\begin{array}{l}
\qquad\qquad\quad CH_2{-}CH_2 \\
\qquad\qquad\quad | \qquad\quad | \\
CH_2{-}CH{-}CH_2 \quad CH_3 \\
| \qquad\quad | \\
CH_3 \quad CH_3
\end{array}
$$

that when "straightened" is

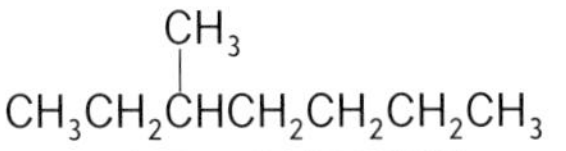

$$
\begin{array}{c}
\qquad CH_3 \\
\qquad | \\
CH_3CH_2CHCH_2CH_2CH_2CH_3 \\
\text{3-METHYLHEPTANE}
\end{array}
$$

the longest continuous chain totals seven carbons. The *last* part of the complete name for this compound will therefore be *heptane*. We next learn how to specify the location of the small CH_3-branch.

4. To locate branches, assign a number to each carbon of the longest continuous chain. Begin at whichever end of the chain will result in the smaller number or set of numbers for the carbon(s) holding branches. In our example,

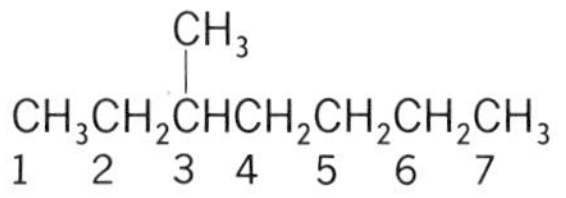

$$
\begin{array}{c}
\qquad CH_3 \\
\qquad | \\
CH_3CH_2CHCH_2CH_2CH_2CH_3 \\
1 \quad 2 \quad 3 \;\; 4 \quad 5 \quad 6 \quad 7
\end{array}
$$

if this chain had been numbered from right to left, the carbon holding the branch would have the number 5. Having located the branch(es), we must now be able to name it (them).

5. If a side chain or branch consists only of carbon and hydrogen linked with single bonds, it is called an *alkyl group;* "alkyl" comes from changing the "-ane" ending of "alkane" to "-yl." This change is the key to making up names for alkyl groups, and Table 2.3 lists the names and structures for the most common ones. Their structures must be learned. The table shows four butyl groups, two related to *n*-butane and two to isobutane. All are butyl groups because each has four carbons. To distinguish them from each other, additional word parts are tacked on. The words "secondary" and "tertiary" (abbreviated *sec-* and *tert-* or simply *t-*) denote the condition of the carbon in the group having the unused bond. In the *sec*-butyl group this carbon is directly attached to *two* other carbons and is therefore classified as a *secondary* carbon. In the *t*-butyl group this carbon has direct bonds to three other carbons and is classified as a *tertiary* carbon. When a carbon is attached directly to but one other carbon, it is classified as a *primary* carbon. These classifications of carbons should be learned.

   The prefix "iso-" has a special meaning. It can be used to name any group with the following feature:

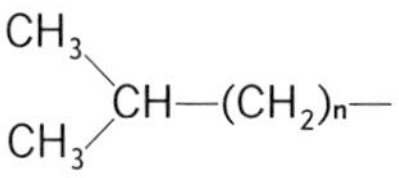

$$
\begin{array}{l}
CH_3 \diagdown \\
\qquad\quad CH{-}(CH_2)_n{-} \quad (n = 0, 1, 2, 3, \text{etc.}) \\
CH_3 \diagup
\end{array}
$$

$n = 0$	ISOPROPYL	$n = 2$	ISOPENTYL
$= 1$	ISOBUTYL	$= 3$	ISOHEXYL

Note that the total carbon content of each group is also disclosed in the name.

Table 2.3
Alkyl Groups—Names and Structures

Parent Alkane	Structure of Parent Alkane	Structure of Alkyl Group from Alkane	Name of Alkyl Group
Methane	CH_4	$CH_3—$	Methyl
Ethane	CH_3CH_3	$CH_3CH_2—$	Ethyl
Propane	$CH_3CH_2CH_3$	$CH_3CH_2CH_2—$	*n*-Propyl[a]
		$(CH_3)_2CH—$	Isopropyl
n-Butane	$CH_3CH_2CH_2CH_3$	$CH_3CH_2CH_2CH_2—$	*n*-Butyl[b]
		$CH_3CH_2CH(CH_3)—$	Secondary butyl (*sec*-butyl)
Isobutane	$CH_3CH(CH_3)CH_3$	$CH_3CH(CH_3)CH_2—$	Isobutyl
		$CH_3C(CH_3)_2—$	Tertiary butyl (*t*-butyl)
Any normal alkane:	If the "free valence" extends from the *end* of the unbranched chain, change the "-ane" ending of the alkane to "-yl"; e.g., $CH_3CH_2CH_2CH_2CH_2CH_2CH_2—$ is *n*-heptyl		
Any alkane in general:	R—H	R—	Alkyl

[a]The IUPAC name is simply "propyl."
[b]The IUPAC name is "butyl."

The specific groups given here are official names in the IUPAC system. (When **n** is more than 3, the IUPAC system uses a different strategy that we shall not study. However, for common names we may extend this use of "iso-" to higher values of **n**.)

Having learned rules for locating and naming side chains, we next consider situations in which identical groups are located on the same carbon of the main chain.

6. Whenever two identical groups are attached at the same place, numbers are supplied for each group. If there are two identical groups, we add the prefix "di-"; if three, "tri-"; etc. For example:

$$CH_3C(CH_3)_2CH_2CH_2CH_2CH_3$$

Correct name: 2,2-Dimethylhexane
Incorrect: 2-Dimethylhexane
Incorrect: 2,2-Methylhexane

7. Whenever two or more different groups are affixed to a chain, two ways are acceptable for organizing all the name parts into the final name. The last word

part is always the name of the alkane corresponding to the longest chain.

(a) The word parts can be ordered by increasing complexity of side chains (e.g., in order of increasing carbon content): methyl, ethyl, propyl, isopropyl, butyl, isobutyl, *sec*-butyl, *t*-butyl, etc.

(b) They can be listed in simple alphabetical order. (This corresponds to the indexing system of *Chemical Abstracts,* a publication of the American Chemical Society.)

In this text we shall not be particular about the matter of order. For the following compound both names given are acceptable.

```
 CH3
 |
CH3CHCH2CHCH2CH2CH3        2-Methyl-4-isopropylheptane
        |
        CHCH3           or 4-Isopropyl-2-methylheptane
        |
        CH3
```

Note carefully the use of hyphens and commas in organizing the parts of names. Hyphens always separate numbers from word parts, and commas always separate numbers. The intent is to make the final name one word.

8. The formal names for several of the more important nonalkyl substituents are as follows:

-F	fluoro	$-NO_2$	nitro
-Cl	chloro	$-NH_2$	amino[2]
-Br	bromo	-OH	hydroxy[2]
-I	iodo		

Several examples of compounds correctly named according to these rules are given below. Some common ways in which incorrect names are often devised are also shown. As an exercise, describe how each incorrect name violates one or more of the rules.

```
      CH3
      |
CH3—C—CH3
      |
      CH2
      |
      CH3
```

2,2-DIMETHYLBUTANE
Not 2-METHYL-2-ETHYLPROPANE

```
             CH3 CH3
               \ /
CH3—CH2     CH
       |       |
       CH2—CH—CH3
```

2,3-DIMETHYLHEXANE
Not 2-ISOPROPYLPENTANE

```
      CH3  CH3
      |    |
CH3—C——CH—CH2—CH3
      |
      CH3
```

2,2,3-TRIMETHYLPENTANE
Not 2,3-TRIMETHYLPENTANE
Not 2-*t*-BUTYLBUTANE

```
CH3—CH2—CH—Cl
         |
         Cl
```

1,1-DICHLOROPROPANE
Not 3,3-DICHLOROPROPANE
Not 3,3-CHLOROPROPANE
Not 1-DICHLOROPROPANE
Not 1,1-CHLOROPROPANE

[2]These two are used only in special circumstances. As we shall see in later chapters, amino compounds are usually named as amines and hydroxy compounds as alcohols, with special IUPAC rules.

```
          CH3
          |
   CH3—CH
          |
          CH3
```

2-METHYLPROPANE
Not 1,1-DIMETHYLETHANE
Not ISOBUTANE, WHICH IS ITS *COMMON* NAME

```
                                  CH3
                                  |
CH3CH2CH2CHCH2CHCH3
               |
       CH3—C—CH3
               |
               CH3
```

2-METHYL-4-*t*-BUTYLHEPTANE
Not 4-*t*-BUTYL-6-METHYLHEPTANE
But 4-*t*-BUTYL-2-METHYLHEPTANE IS ACCEPTABLE

Exercise 2.1 Write the structures (condensed) of each of the following.

(a) 1-bromo-2-nitropentane
(b) 2,2,3,3,4,4-hexamethyl-5-isopropyloctane
(c) 2,2-diiodo-3-methyl-4-isopropyl-5-*sec*-butyl-6-*t*-butylnonane
(d) 1-chloro-1-bromo-2-methylpropane
(e) 4,4-di-*sec*-butyldecane

Exercise 2.2 Write IUPAC names for each of the following.

```
(a) CH3—CH2
            \
             CH—CH3
            /
    CH2—CH2
    |
    CH3
```

```
                          CH3
                          |
(b)           CH3  CH3—C—CH3
                 \        |
                  CH—CH—CH2—CH2—CH3
                 /
         CH3—CH
              |
              CH3
```

```
                  CH3       CH3          CH3
                  |         |            |
(c) CH3—CH2—CH—CH—CH—CH2—CH—CH3
                      |
                      CH2—CH2—CH2—CH3
```

```
                       I
                       |
(d) Cl—CH2—CH—CH2—Br
```

```
                       CH3—CH—CH3
                            |
(e) CH3—CH2—CH2—CH—CH—CH2—CH2—CH3
                       |
                 CH3—C—CH3
                       |
                       CH3
```

```
                         CH3
                         |
(f) NO2—CH2—C—CH2—NO2
                         |
                         CH3
```

```
                        CH2—Cl
                        |
(g) CH3—CH2—CH—CH—CH3
                  |
                  CH3
```

```
          CH3                    CH2—CH3
          |                      |
(h) CH3—CH—CH2—CH2—CH—CH—CH3
                                      |
                                      CH3
```

```
(i) CH3
        \
         CH—CH2—CH2
        /        |
    CH3          CH3
```

```
(j) CH3—CH2—CH2          CH2—CH3
                 |         |
                 CH2—CH—CH—CH2CH2CH3
                      |
                      CH3
```

2.9 Physical Properties of Hydrocarbons—Saturated and Unsaturated

We expect the molecules of a substance to be polar whenever they contain polar bonds whose polarities do not cancel. We expect a covalent bond to be polar whenever it joins atoms of different electronegativities. (Cf. Chapter 1.) Hydrogen and carbon differ only slightly in relative electronegativities. Carbon-hydrogen bonds, therefore, must be almost nonpolar. Carbon-carbon bonds, of course, are nonpolar, also. The result is that molecules of any of the families of hydrocarbons—saturated or otherwise—are only weakly polar at best. That is how we understand certain trends in some of the physical properties of hydrocarbons, especially in boiling points and solubilities.

Molecules of methane (formula weight 16) and water (formula weight 18) have nearly the same masses. Yet methane, being nonpolar, boils at $-161.5\,^{\circ}C$. The higher the alkane is in the homologous series of normal alkanes (Table 2.1), the higher its boiling point, melting point, and density. All hydrocarbons are insoluble in water, a very polar solvent, and soluble in the relatively nonpolar solvents that include some common hydrocarbons or mixtures of hydrocarbons (benzene, gasoline, ligroin, and petroleum ether) and halogen derivatives (carbon tetrachloride, chloroform, and dichloromethane). The rule of thumb is that "likes dissolve likes," where "likes" refers to polarities. Because molecules of a polar solvent "stick" to polar molecules, such as each other, they cannot let in nonpolar molecules.

What we are seeking here are general correlations between structure and property. Just as we have now learned to associate hydrocarbons with low water solubility, we shall in the future be able to predict that *any molecule* of *any family* substantially *hydrocarbonlike* will also have this property.

2.10 Gas-Liquid Chromatography. GLC

One of the experimental situations in which it is helpful to understand the structural factors that affect boiling points and solubilities is in the problem of separating mixtures of compounds into their various components. Fractional distillation, fractional crystallization, and liquid-liquid extraction are three long-established techniques that you no doubt will study in the laboratory. Chromatography is another technique of great power that is available in several variations, as outlined in Table 2.4. Some of these are discussed in your laboratory manual too.

In gas-liquid chromatography the analyst injects a very small sample of a liquid mixture into a slowly moving stream of a heated, inert gas such as helium, the "carrier gas." See Figure 2.2. Because the injection chamber is hot, the injected sample vaporizes at the moment of injection. The resulting mixture of vapors and gases is carried on the helium stream into a tube—the column—that has been filled with a specially treated powdery material. Coating each tiny particle of the powder is a very thin film of a liquid with such a high boiling point that it will not vaporize under conditions of the experiment. This liquid is called the adsorbent. Silicone oil is an example. It is deposited onto powdered firebrick, to mention one inert "support" material. Properly prepared, the coated support or "packing" remains powdery or granular. Molecules of the various vapors dissolve in the thin film of adsorbent, some being more soluble than others. If the adsorbent is a polar

Table 2.4
Some Common Types of Chromatography

General Class of Chromatography	The principle behind a successful separation is that the different components of the mixture . . .	Examples
Partition chromatography	. . . dissolve to different degrees in a component of the column packing. (They become differently partitioned between the moving phase and the stationary phase.)	Gas-liquid chromatography (GLC) Paper chromatography (partly both types)
Adsorption chromatography	. . . become adsorbed onto the more or less polar surface of a stationary, powdery "adsorbant."	Column chromatography Gas-solid chromatography (GSC) Thin layer chromatography (TLC)

liquid, then polar molecules in the vapor tend to lag behind nonpolar molecules that are swept along in the helium at a faster rate. Otherwise, molecules of higher boiling components will lag behind lower boiling components. Eventually, the various components of the original mixture are strung out along the column. Each is in its own zone, and each zone moves at its own rate. One by one they come off the column into the detector.

There are a variety of devices for detecting when a compound mixed with helium emerges from the column. Whatever the device, when one component does emerge, the detector activates a pen that moves back and forth across a strip of unrolling chart paper. See Figure 2.3.

2.11 Chemical Properties

Alkane chemistry is quite simple. Very few chemicals react with alkanes. That's why they're called *paraffins* or paraffinic hydrocarbons, after the Latin *parum affinus,* meaning "little affinity." To illustrate, at room temperature alkanes are not attacked by water, by strong, concentrated acids (e.g., sulfuric acid, hydrochloric acid), strong, concentrated alkalies (e.g., sodium hydroxide), active metals (e.g., sodium), most strong oxidizing agents (e.g., potassium permanganate, sodium dichromate), or any of the reducing agents. These facts explain why alkanelike portions of the molecules of other families of compounds are called *nonfunctional groups*. When molecules that do have functional groups also have alkanelike portions, those parts of them generally ride through chemical events unchanged. The compound 1-pentanol, an alcohol, illustrates what we mean. See Figure 2.4.

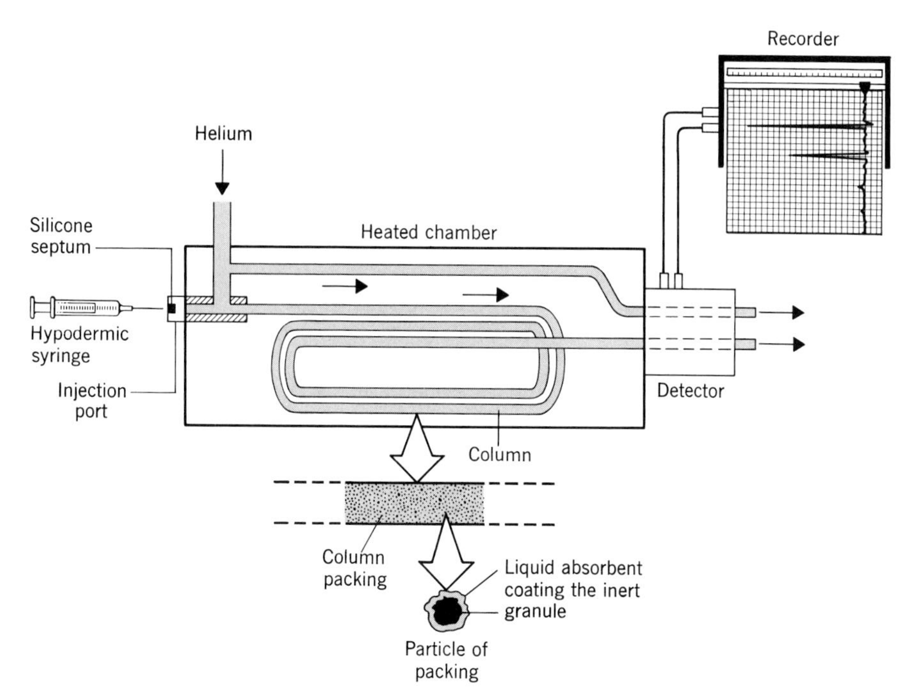

Figure 2.2
Gas liquid chromatography or GLC. Columns are made from stainless steel or aluminum tubing and are filled with the packing before being bent or coiled to fit the heating chamber. The detector compares the helium that bypasses the column with the helium (plus components) coming from the column.

The alkanes do have some very important chemical properties, of course. They burn, for example. (Virtually all organic compounds do.) They can be "cracked" to alkenes and hydrogen. Substances such as ethane, propane and the butanes are among our most important industrial chemicals, and we shall use part of Chapter 5 for this facet of alkane chemistry. Some alkanes can be partly cracked and

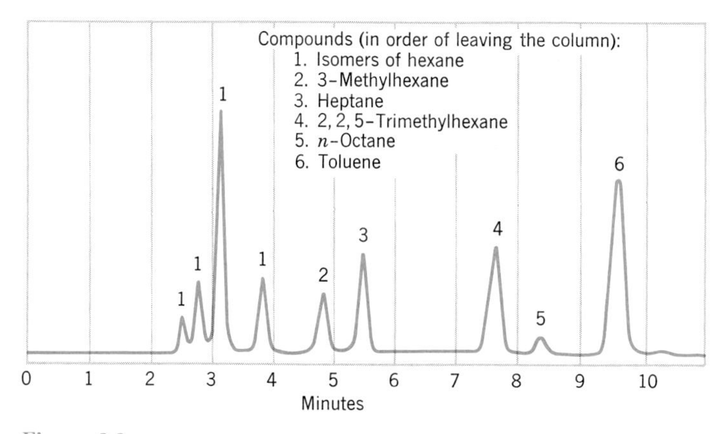

Figure 2.3
GLC chromatogram of a mixture of hydrocarbons using dinonyl phthalate (a high boiling ester) as the liquid adsorbent. (Done on a Carle Model 8000 instrument. Starting temperature, 40°; ending, 120°.)

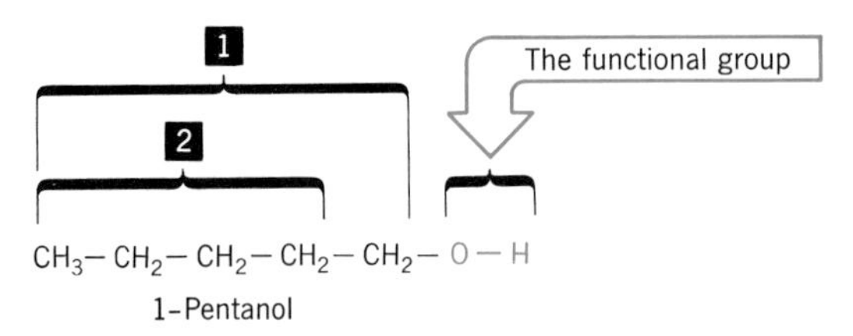

Figure 2.4
Functional and nonfunctional groups. Usually the entire alkane-like group 1 is called the nonfunctional group in a molecule such as this. Parts of this group near the functional group, however, sometimes enter into a reaction. Thus, only 2 is left alone when this alcohol is oxidized (Section 6.29).

rearranged into important aromatic compounds such as benzene, toluene, and the xylenes. Alkanes are also attacked by fluorine, chlorine, and bromine and by hot nitric acid. We shall look briefly here at combustion and at chlorination.

2.12 Combustion

We burn certain alkanes or their mixtures to obtain energy. Natural gas is mostly methane. Gasoline, diesel fuel, and heating oil are all mixtures of hydrocarbons.

If enough oxygen is available, what we call *complete combustion* occurs. Complete combustion of any hydrocarbon or any oxygen derivative of a hydrocarbon (e.g., ethers, alcohols, esters, etc.) produces solely carbon dioxide, water and energy. To illustrate, using propane:

$$CH_3CH_2CH_3 + 5O_2 \longrightarrow 3CO_2 + 4H_2O + 531 \text{ kilocalories energy per mole of propane}$$

2.13 Chlorination

Chlorine, Cl_2, will not attack an alkane unless energy is provided—heat or ultraviolet light. When the energy is provided, the alkane molecule will have one (or more) of its hydrogens pulled off and replaced by an atom of chlorine, as we see in the case of methane and chlorine:

$$CH_4 + Cl_2 \xrightarrow[\text{ultraviolet light (e.g., sunlight)}]{\text{Heat or}} \underset{\text{METHYL CHLORIDE (bp } -24°)}{CH_3Cl} + HCl$$

The energy initiates the reaction. It is absorbed by molecules of chlorine—only a very small percent need be affected—to break them into chlorine atoms:

$$\underline{\text{Initiation}} \quad :\ddot{\underset{..}{Cl}}:\ddot{\underset{..}{Cl}}: \xrightarrow[\text{Energy}]{} \underset{\text{TWO CHLORINE ATOMS (FREE RADICALS)}}{:\ddot{\underset{..}{Cl}}\cdot + \cdot\ddot{\underset{..}{Cl}}:}$$

Now we have particles—chlorine atoms without octets—very reactive toward alkanes. They can pull off hydrogen atoms; by generating atoms of chlorine we have initiated what we call a *free radical chain reaction.* Let us see what that means by following the fate of the chlorine atom. The next two steps are called *chain propagation.*

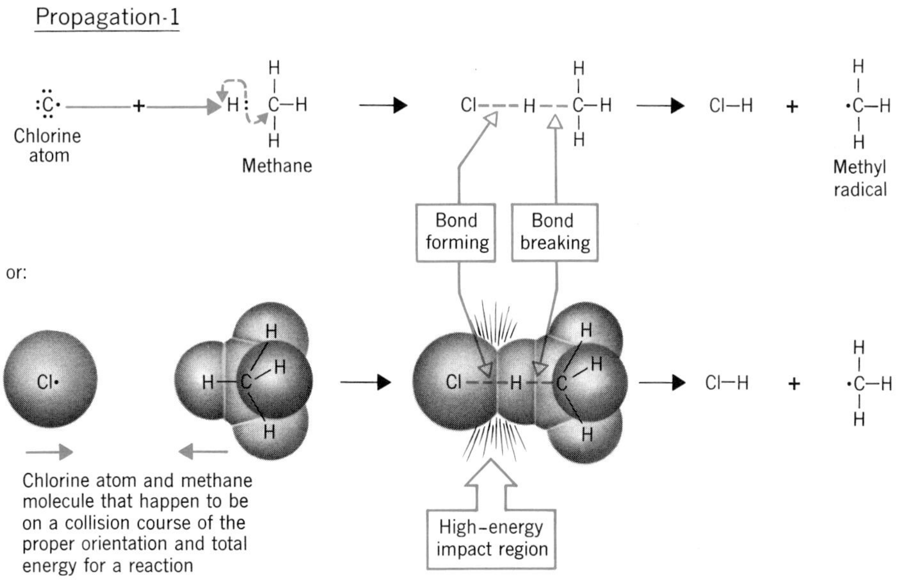

Propagation-1

Chlorine atom + Methane → Cl---H---CH3 → Cl—H + Methyl radical

Bond forming

Bond breaking

or:

Chlorine atom and methane molecule that happen to be on a collision course of the proper orientation and total energy for a reaction

High-energy impact region

We still have a particle without an octet, the methyl radical. The word "radical" in this context—in chemistry—means any atomic or molecular sized particle having an unpaired electron. It can pull molecules apart, too. It's a collision of a methyl radical with a chlorine molecule that leads toward the products:

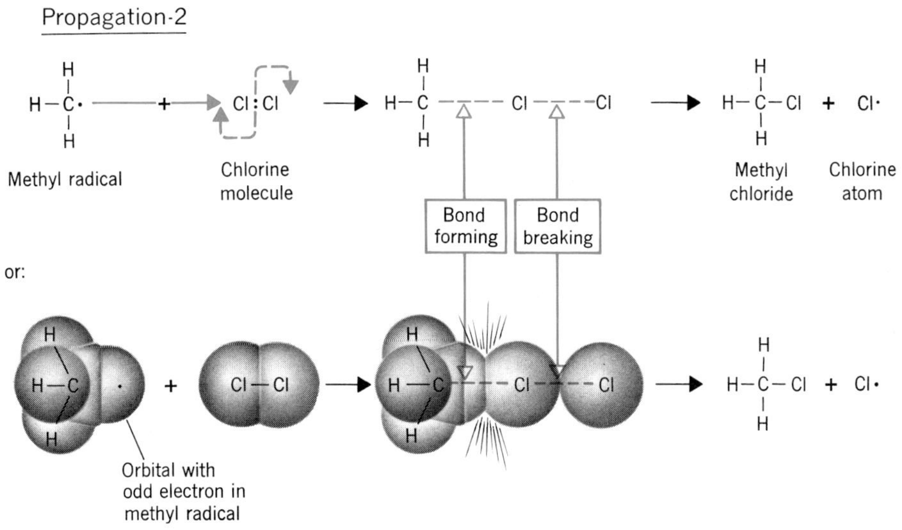

Propagation-2

Methyl radical + Chlorine molecule → H3C---Cl---Cl → Methyl chloride + Chlorine atom

Bond forming

Bond breaking

or:

Orbital with odd electron in methyl radical

The second propagation step gave a molecule of product, methyl chloride. Just as importantly, it gave a chlorine atom. This new atom will now cause another first propagation step. Then that leads into the second, which gets another first going again, and so on. What we have is a chain reaction. Each chlorine atom that is produced by action of heat or light (initiation step) starts a chain that keeps cranking out molecules of methyl chloride plus new chlorine atoms to keep the chain going. That's why only a small percent of starting chlorine molecules need be broken apart by the initiation step. We get the other chlorine atoms from the chain itself.

Once in awhile chains are stopped; in the case of methane and chlorine, about once in roughly 5000 repeats of the propagation steps. Any time a chlorine atom collides with another chlorine atom or a methyl radical they will combine. In so doing, *both* reactive particles are taken out of the chain. New chains have to be initiated.

Chain Termination Steps

These occur too seldom to be useful in making a product, but when one happens a whole chain is stopped.

$$Cl\cdot + \cdot CH_3 \longrightarrow Cl—CH_3$$

$$Cl\cdot + Cl\cdot \longrightarrow Cl—Cl$$

$$CH_3\cdot + \cdot CH_3 \longrightarrow CH_3—CH_3$$

What we have just studied is our first example of a **mechanism** of a reaction. To the organic chemist, a reaction mechanism is an answer to a "how?" question—"*how* does the reaction take place?" Not "what is the overall result?" (We have to know that, too, of course.) Certainly not "*why* does the reaction take place?" (At a nonphilosophical level, the reaction happened because someone mixed the chemicals and added energy to get it started. At a deeper level, we don't know why! Not anymore than we know *why* we have a universe.)

2.14 Side-Reactions, By-Products, and Unbalanced Equations

If we add energy to a 1 : 1 mole mixture of methane and chlorine, we shall certainly generate hydrogen chloride and methyl chloride, as we have just studied. We also will get methylene chloride, chloroform and carbon tetrachloride, too! We couldn't stop it if we tried—not if we have a 1 : 1 mole mixture at the start. Forget for a moment the equations we have studied. Think about the mixture of methane and chlorine. Think of the molecules, atoms, and radicals intermingling and colliding. A short while into the reaction a little methyl chloride has appeared. Its molecules have C—H bonds just as do those of methane. That is why the following reaction is possible and begins to take place:

$$CH_3—Cl + Cl—Cl \longrightarrow \underset{\text{METHYLENE CHLORIDE (bp 40°)}}{Cl—CH_2—Cl} + H—Cl$$

Methyl chloride, in other words, can compete against methane for chlorine. So can methylene chloride; it still has C—H bonds:

$$\text{Cl—CH}_2\text{—Cl} + \text{Cl—Cl} \longrightarrow \underset{\substack{\text{CHLOROFORM} \\ \text{(bp 61°)}}}{\text{CHCl}_3} + \text{H—Cl}$$

Chloroform still has a C—H bond:

$$\text{CHCl}_3 + \text{Cl—Cl} \longrightarrow \underset{\substack{\text{CARBON} \\ \text{TETRACHLORIDE} \\ \text{(bp 77°)}}}{\text{CCl}_4} + \text{H—Cl}$$

This ends it. Assuming it was methyl chloride we wanted, then these other reactions are called *side reactions* and their products are called *by-products.*

Because of side reactions it is seldom possible to write the kind of balanced equations you became accustomed to in earlier courses in chemistry. Sometimes we can, but often an equation will be "balanced" only in the sense that the proportions of starting materials used are correctly shown.

Exercise 2.3 How can we adjust the initial conditions (1:1 mole ratio of reactants) to maximize the formation of methyl chloride and keep production of higher chlorination products at a minimum? (*Hint.* How can a chemist adjust the initial ratio to increase considerably the probability that chlorine molecules will encounter *unchanged methane* molecules rather than newly formed molecules of methyl chloride?) It is not practical to remove the methyl chloride as it forms, and no matter what is done, higher chlorinated products will form. The problem is to minimize these.

Exercise 2.4 How can the initial conditions be adjusted to maximize the production of carbon tetrachloride?

2.15 Chlorination of Higher Homologs of Methane

The hydrogens in ethane form an equivalent set, which means that all members of the set (hydrogens in this case) are the same with respect to being replaced by some other group. No matter which of the six hydrogens on ethane is replaced by chlorine, only one monochloroethane is possible:

$$\text{CH}_3\text{CH}_3 + \text{Cl}_2 \longrightarrow \underset{\substack{\text{ETHYL CHLORIDE} \\ \text{(CHLOROETHANE, bp 13°C)}}}{\text{CH}_3\text{CH}_2\text{Cl}} + \text{HCl}$$

Although ethyl chloride has no isomer, if a second chloro group is put into ethyl chloride, a mixture of isomeric dichloroethanes forms:

$$\text{CH}_3\text{CH}_2\text{Cl} + \text{Cl}_2 \xrightarrow[\text{ultraviolet light}]{\text{Heat or}} \underset{\substack{\text{1,2-DICHLOROETHANE} \\ \text{(ETHYLENE CHLORIDE, bp 84°C)}}}{\text{ClCH}_2\text{CH}_2\text{Cl}} + \underset{\substack{\text{1,1-DICHLOROETHANE} \\ \text{(bp 57°C)}}}{\text{CH}_3\text{CHCl}_2} + \text{HCl}$$

This equation is not balanced and tells us only that the action of macroscopic samples of chlorine on ethyl chloride produces a mixture of the two isomers shown.

How much of each forms depends on the precise experimental conditions, and the ratio of isomers must be measured experimentally.

The chlorination of propane will produce two isomeric monochloropropanes (again, an unbalanced "reaction sequence" is shown),

$$CH_3CH_2CH_3 + Cl_2 \xrightarrow[\text{Energy}]{} \underset{\substack{n\text{-PROPYL CHLORIDE} \\ \text{(bp 47°C)}}}{CH_3CH_2CH_2Cl} + \underset{\substack{\text{ISOPROPYL CHLORIDE} \\ \text{(bp 35°C)}}}{CH_3\underset{\substack{|\\Cl}}{C}HCH_3} + HCl$$

because propane, unlike ethane, has two *different* sets of equivalent hydrogens, one set of six (from the two CH_3 groups) and one set of two (from the middle CH_2 group). Compounds whose molecules have two or more chloro substituents also form.

Exercise 2.5 (a) How many monochloro derivatives of *n*-butane are possible? Give both the common and the IUPAC names. *Hint.* How many different sets of equivalent hydrogens does a *n*-butane molecule have? (Be concerned only about *mono*-chloro compounds.) (b) How many monochloro derivatives of isobutane are possible? Name them according to the common and the IUPAC systems.

2.16 Preparation of Alkanes

In addition to natural sources such as crude oil, several laboratory methods for making pure alkanes have been developed. These syntheses require the conversion of some compound in another family to an alkane. We shall defer our discussion of these syntheses until they are encountered in the reactions of other families.

2.17 Alkanes from Lithium Dialkylcopper Compounds

Many metals will react directly with alkyl halides, R—X (X = Cl, Br, or I). Compounds called alkylmetals form, which we shall study further in Chapter 7. These compounds vary widely in stability and usefulness, but if the metal and the alkyl halides are carefully chosen we have good syntheses of alkanes. The overall result is:

$$R\text{—}X + X\text{—}R' \xrightarrow{\text{Metal(s)}} \underset{\substack{\text{AN ALKANE} \\ \text{(R MAY BE THE SAME AS R')}}}{R\text{—}R'} + \text{Metal halide(s)}$$

The best combination of metals is lithium and copper. First, one of the alkyl halides is allowed to react with metallic lithium:

$$R\text{—}X + 2Li \longrightarrow \underset{\text{An alkyllithium}}{R\text{—}Li} + LiX$$

Then copper(I) iodide and the second alkyl halide are added. Copper(I) iodide reacts with the alkyllithium (which, itself, is not reactive toward the second alkyl halide):

$$2R{-}Li + CuI \longrightarrow \underset{\text{A LITHIUM DIALKYLCOPPER}}{R_2CuLi} + LiI$$

The lithium dialkylcopper reacts as it forms with the second alkyl halide:

$$R_2CuLi + 2R'{-}X \longrightarrow 2R{-}R' + \text{mixed salts}$$

For example (written as a "flow scheme," not as consecutive balanced equations, a common practice in organic chemistry):

$$CH_3I \xrightarrow{Li} CH_3Li \xrightarrow{CuI} (CH_3)_2CuI$$

$$(CH_3)_2CuI + \underset{n\text{-HEXYL IODIDE}}{CH_3(CH_2)_4CH_2I} \longrightarrow \underset{n\text{-HEPTANE}}{CH_3(CH_2)_5CH_3} + \text{salts}$$

The second halide used, R′—X, should be a primary halide; the first one may have any type of alkyl group.

Exercise 2.6 Select two alkyl halides that could be used to make each of these alkanes by the lithium dialkylcopper method.
(a) ethane (b) propane (c) isobutane (d) isopentane

Alkylsodiums will also form by action of sodium on alkyl halides, but they are so reactive that they attack still unchanged alkyl halide as soon as they form. Therefore, the molar proportions of reagents are adjusted to use this property to make symmetrical alkanes, R—R. This is one of the old, time-honored "named reactions" (see box), the *Wurtz reaction.* In general:

$$2R{-}X + 2Na \xrightarrow[\text{Heat}]{} \underset{\text{A SYMMETRICAL ALKANE}}{R{-}R} + 2NaX$$

Specific example:

$$\underset{n\text{-BUTYL BROMIDE}}{2CH_3CH_2CH_2CH_2Br} + 2Na \xrightarrow[\text{Heat}]{} \underset{n\text{-OCTANE}}{CH_3CH_2CH_2CH_2CH_2CH_2CH_2CH_3} + 2NaBr$$

2.18 Alkanes from Indirect Reduction of Alkyl Halides

By this strategy we simply exchange X in R—X for H; the product is R—H. We can do this by action of zinc and acetic acid:

$$R{-}X \xrightarrow[H^+]{Zn} R{-}H + \text{Zinc salts}$$

We may also do this by action of magnesium *followed by* acid.

$$R{-}X + Mg \xrightarrow{\text{Ether}} R{-}Mg{-}X \xrightarrow{H^+} R{-}H + \text{Magnesium salts}$$

The intermediate R—Mg—X, is not isolated. We show here only one of the many reactions of this, one of the most versatile single reagents in all of synthetic organic chemistry. We shall meet it again in Section 6.16.

NAMED REACTIONS

Charles Adolphe Wurtz (1817–1884) discovered "his" reaction. Naming the more important reactions in organic chemistry after their discoverers was a rather common practice until recently. The designation *Wurtz reaction* serves as a highly abbreviated "symbol" for all the basic features of the general reaction. In oral or written communication all the pertinent facts, for example, types of reactants, other conditions, and types of products, are indicated in the one brief expression. As we encounter other named reactions in our study, the usefulness of this symbolism will become more and more apparent. Fewer reactions are now named after their discoverers only because learning so many became a burden and because rather large teams of scientists are frequently involved in new discoveries, making it difficult to establish priority.

2.19 Cycloalkanes

A carbon skeleton can take the form of a closed ring as well as of an open chain. The simplest of such cyclic compounds are called the cycloalkanes or cycloparaffins. The following examples are illustrative.

CYCLOPROPANE: CH_2—CH_2 ring closed by CH_2

CYCLOBUTANE: CH_2—CH_2, CH_2—CH_2 (four-membered ring)

CYCLOPENTANE: ring of five CH_2 groups

CYCLOHEXANE: ring of six CH_2 groups

Rings of higher carbon content (C_7 to over C_{30}) are also known. Since the molecules in these compounds have only single bonds, they generally exhibit the same types of physical and chemical properties as the open-chain alkanes. Cyclopropane and cyclobutane are exceptions. Their molecules are "strained."

2.20 Ring Strain

As written, the C—C—C bond angle for cyclopropane should be that of the internal angle of an equilateral triangle, that is, 60°. But the normal bond angle for tetravalent carbon is much larger than that, 109°28′. The internal angle in cyclobutane should be 90°, also smaller than the angle for open-chain systems. The internal angle of a regular pentagon (cf. cyclopentane) is 108°, very close to 109°28′.

The internal energy of a system involving tetravalent carbon is minimized when bond angles are 109°28′. Maximum overlap between bonding orbitals is achieved with this geometry. Any deviation from this angle must therefore imply departure from maximum overlap (Figure 2.5), which means weaker bonds. The bonds in cyclopropane are weaker, and so are those in cyclobutane, to a lesser extent. Cyclopentane, however, in which deviation from the normal bond angle is very slight, is normal. The relative reactivities of cyclopropane, cyclobutane, and cyclopentane reflect these considerations. To illustrate,

$$\text{cyclo-}(CH_2)_3 + H_2 \xrightarrow[80^\circ C]{Ni} H\text{—}CH_2CH_2CH_2\text{—}H$$

CYCLOPROPANE → PROPANE

The sp^3-sp^3 overlap to form a C—C bond is maximized at a bond angle of 109° 28′

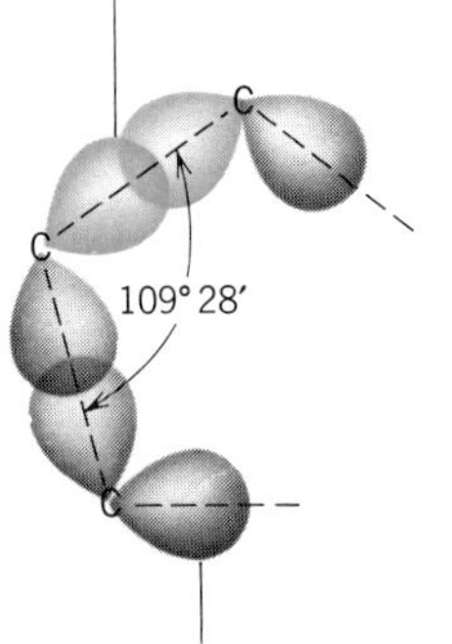

The sp^3 hybrid orbital (only two shown for each carbon)

The sp^3-sp^3 overlap is poorer here when the bond angle deviates much from 109° 28′

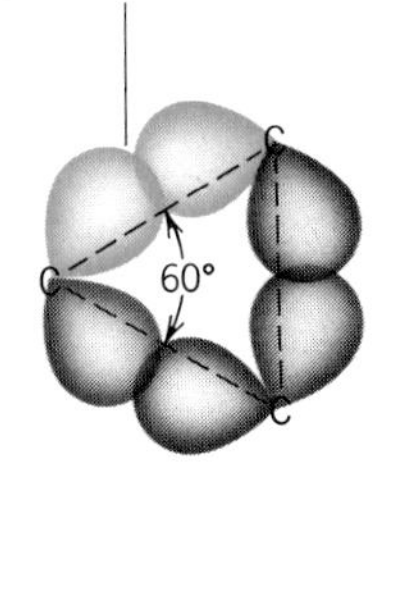

Ball and stick model of cyclopropane in which springs must be used to represent the C—C bonds in the ring

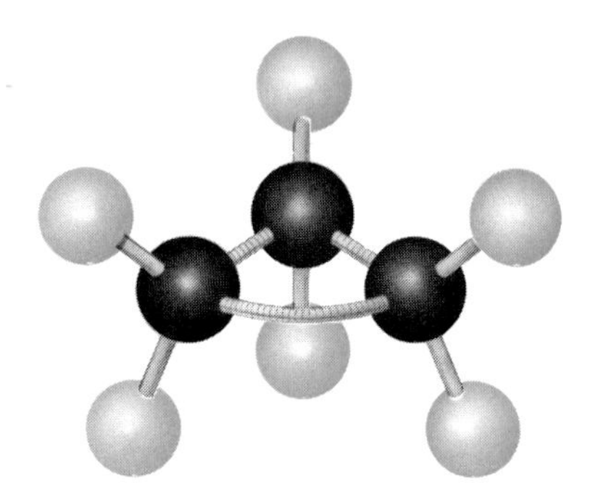

Figure 2.5
The ring strain in cyclopropane is attributed to relatively poor overlap between sp^3 hybrid orbitals of the carbons.

$$\underset{\text{CYCLOBUTANE}}{\begin{array}{c}CH_2\text{—}CH_2\\ |\qquad\ \ |\\ CH_2\text{—}CH_2\end{array}} + H_2 \xrightarrow[200^\circ C]{Ni} \underset{n\text{-BUTANE}}{\begin{array}{c}CH_2CH_2CH_2CH_2\\ |\qquad\qquad\quad |\\ H\qquad\qquad\quad H\end{array}}$$

$$\underset{\text{CYCLOPENTANE}}{\text{(CH}_2)_5\text{ ring}} + H_2 \xrightarrow[Heat]{Ni} \text{No reaction}$$

The temperatures at which the rings for cyclopropane and cyclobutane open are revealing. The three-membered ring is quite easily opened because its carbon-to-carbon bonds are weak and because the cyclopropane ring is said to be subject to internal strain. The cyclobutane ring is less strained than that of cyclopropane, however, and a higher temperature is required to open it. Cyclopentane has essentially no ring strain and undergoes no useful ring-opening reactions.

The internal angle of a regular hexagon is 114°44′, which is 5°16′ larger than the normal bond angle for tetravalent carbon. This departure might lead us to predict that cyclohexane should be subject to internal strain similar to but a bit less than that in cyclobutane. There is evidence, however, that bonds in cyclohexane are just as strong as those in cyclopentane, certainly far stronger than those in cyclobutane or cyclopropane. The explanation is that the cyclohexane ring is not flat as in a normal hexagon. It is twisted out of the plane, and two conformations having normal bond angles are shown in Figure 2.6. Important structural features of cyclohexane are discussed in the legend for this figure. In similar fashion, the rings with seven carbons or more are also nonplanar and they are subject to no internal strain. Therefore they all have the normal alkanelike chemical and physical properties.

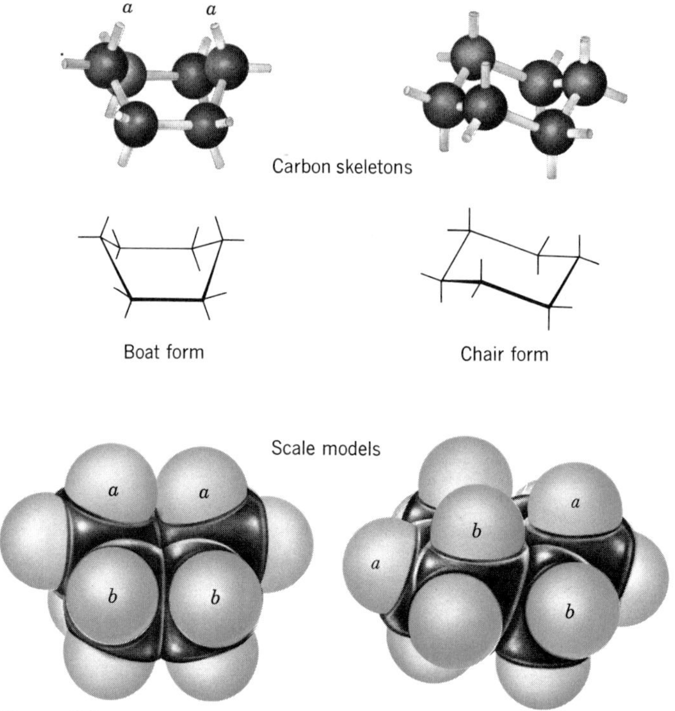

Figure 2.6
Conformations of the cyclohexane ring. In both the boat and the chair form, bond angles in the ring are normal (109°28′). The chair form has less internal energy than the boat form, however, and is therefore much more abundant in a sample of cyclohexane. The boat form brings hydrogens from opposite sides of the ring quite close to each other (see those marked *a*), and their electron clouds tend to repel each other. Moreover, the electron clouds marked *b* are near each other. These interactions involving groups not bonded directly to each other are closer together in the boat form than in the chair form.

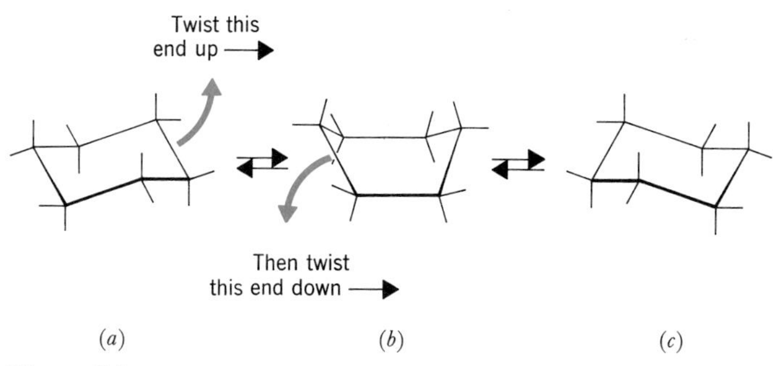

(*a*) (*b*) (*c*)

Figure 2.7
The chair form of cyclohexane (*a*) can "flip-flop" to another chair form (*c*) and back again by means of the boat form (*b*). A dynamic equilibrium exists. Interconversion occurs rapidly enough at ordinary temperatures to make it expedient simply to treat the cyclohexane system as a flat hexagon.

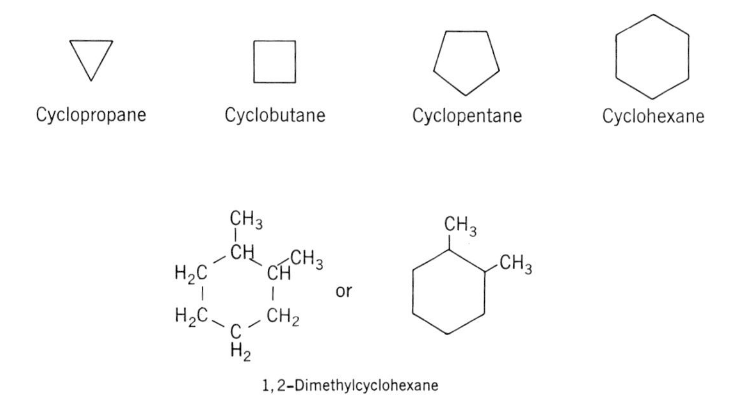

Figure 2.8
Symbolism and nomenclature in cycloalkane systems. At each corner in the geometric symbols a carbon is understood to be present. The lines from corner to corner are carbon-carbon bonds. Unless otherwise indicated, hydrogens are assumed to be present at the corner carbons in sufficient number to fill out carbon's tetravalence. When substituents occupy corners on the ring, the ring positions are numbered in a direction and from a beginning point that together yield the set of smallest numbers possible. Alkyl groups are usually given smaller numbers than halogens.

2.21 Interconversion of Chair Forms

In cyclic systems rotations about the bonds of the ring itself are severely limited. Because ball-and-stick models illustrate this restriction best, the student should examine any that are available to see this restricted rotation for himself.

In cyclohexane a small amount of movement is possible; one chair form can interconvert via a boat form into another equivalent but not identical chair form, as illustrated in Figure 2.7. This ready interconvertibility makes it possible to use simple flat hexagons to represent cyclohexane, as though it were a planar molecule, and chemists usually represent cycloalkane systems with simple geometric forms such as those shown in Figure 2.8. Their symbolism is described in the legend.

2.22 Geometrical Isomerism. "Cis-Trans" Isomerism

Two different 1,2-dimethylcyclopropanes (Figure 2.9) are known. It is the lack of free rotation about single bonds in ring systems that makes their formation possible.

This kind of isomerism is known as *geometric isomerism*. The difference between these isomers lies not so much *where* on a chain or skeleton substituents are located but instead in their geometric orientation. Because two different compounds cannot have the same name, chemists call the isomer in which the substituents project in generally opposite directions the *trans* isomer. In the *cis* isomer the substituents generally protrude in the same direction. Other examples of *cis-trans* isomerism are

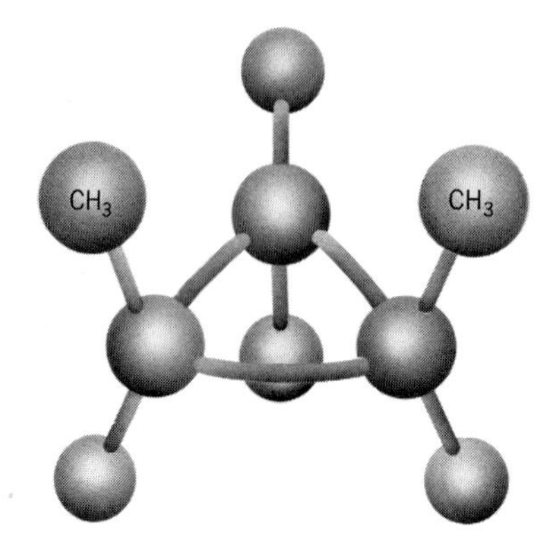

cis-1 2, Dimethylcyclopropane
bp 37°C
mp −141°C
density (20°C) 0.694 g/cc

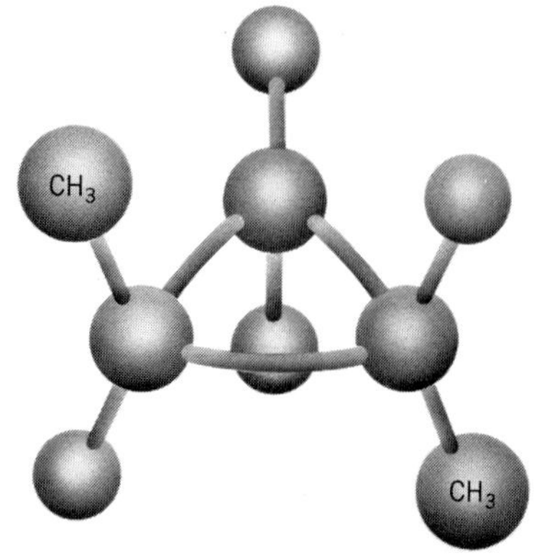

trans-1 2, Dimethylcyclopropane
bp 28°C
mp −150°C
density (20°C) 0.670 g/cc

Figure 2.9
Geometrical isomerism. The differences between the two 1,2-dimethylcyclopropanes mean differences in properties.

OH OH

or

OH HO

cis-1,2-CYCLOHEXANEDIOL

OH OH

or

OH OH

trans-1,2-CYCLOHEXANEDIOL

Br Br

cis-1,2-DIBROMOCYCLOPENTANE

Br Br

trans-1,2-DIBROMOCYCLOPENTANE

Geometrical isomerism is but one type of a general kind of isomerism, *stereoisomerism* (Greek *stereos,* solid, meaning here three-dimensional shape). One other type, optical isomerism, we shall discuss later. In general, stereoisomerism has rather profound implications for the chemical events in living organisms.

Brief Summary: Chemical and Physical Properties of Alkanes

1. Molecules of alkanes, cycloalkanes, and alkanelike parts of the molecules of other substances are nearly nonpolar.

2. All hydrocarbons are insoluble in water.

3. Alkanes and cycloalkanes (except for cyclopropane and cyclobutane) do not react

chemically, especially at room temperature, with water, acids and bases, oxidizing agents and reducing agents, and active metals.

4. The alkanes and cycloalkanes undergo attack by fluorine, chlorine, bromine, and hot nitric acid.

5. Alkanes and cycloalkanes can be "cracked" to alkenes and hydrogen by heat and a catalyst. (This is largely an industrial process.)

Brief Summary: Important Concepts and Theories

1. Free radical chain reaction.
2. *Cis-trans* isomerism.
3. Chair-and-boat interconversions of cyclohexane.
4. Ring strain.

Problems and Exercises

Where you are asked to write structures, provide condensed structures.

1. Write the name and structure of the compound that is the next higher homolog of hexane.
2. Write the structures and the IUPAC names of all of the isomers of C_7H_{16}.
3. Write the structure and the IUPAC name for isononane.
4. Write structures of all the isomeric monochloro derivatives of 2,2,3-trimethylpentane.
5. Ethyl chloride may be prepared by the following reaction:

$$Cl_2 + CH_3{-}CH_3 \xrightarrow[\text{ultraviolet light}]{\text{Heat or}} \underset{\text{ETHYL CHLORIDE}}{CH_3{-}CH_2{-}Cl} + H{-}Cl$$

   The reaction occurs by a free radical chain mechanism. Write the equations for the steps. (Include at least two possible chain-terminating steps.)

6. If the common name of chloromethane is *methyl chloride* and if the common name for chloroethane is *ethyl chloride,* what are the common names for the following?

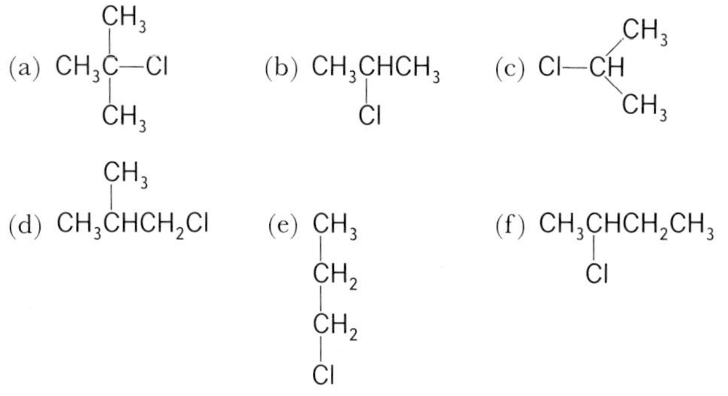

(a) $CH_3C(CH_3)_2{-}Cl$ (central C bearing CH₃ above and CH₃ below)

(b) CH_3CHCH_3 with Cl on the central C

(c) $Cl{-}CH(CH_3)_2$

(d) CH_3CHCH_2Cl with CH_3 on the second C

(e) $CH_3{-}CH_2{-}CH_2{-}Cl$ (written vertically)

(f) $CH_3CHCH_2CH_3$ with Cl on the second C

7. Write "flow schemes" for the synthesis of each of the following:

(a) $CH_3-C(CH_3)_2-CH_2-CH_2-CH(CH_3)-CH_3$ — Show two ways, one from compounds of five or fewer carbons; one from a C_9 compound.

(b) $CH_3CH_2CH_2CH_2CH_3$ — Show two ways, one from compounds of three or fewer carbons; one from a C_5 compound.

(c) $CH_3CH(CH_3)CH_2CH_2CH(CH_3)CH_3$ — Use the Wurtz reaction.

8. Could the compounds of 8(a) and 8(b) be made in good yield by a Wurtz reaction? Explain.

9. If CH_3I and CH_3CH_2I, in a 1:1 mole ratio were heated with sodium, what would be the principal organic products? In what molar proportions would they likely form?

10. Which would be the more soluble in water, **1** or **2**? Explain.

$HOCH_2CH_2OH$ (**1**) $CH_3CH_2CH_2CH_3$ (**2**)

chapter three

Unsaturated Aliphatic Hydrocarbons

3.1 Types

Molecules in butter have some. Those in olive oil have more, corn oil still more. They are in molecules of rubber and many plastics. Some of the hormones and vitamins have them. They are attacked by ozone and other oxidizing agents, by chlorine, bromine, and hydrogen. Even water can be made to react with them. They are carbon-carbon double bonds, and this double bond represents the first functional group of our study. We shall start with the simplest molecules having them, those in alkenes or "olefins."

The carbon-carbon double bond is an "ene" function. A *diene* (di-ene) is a substance whose molecules have two double bonds. (There are trienes, tetraenes . . . up to polyenes, as well.) We'll look briefly at some of these in this chapter, too.

Because of some similarities to alkenes, we take up the alkynes here, also compounds whose molecules have carbon-carbon triple bonds.

$H_2C{=}CH_2$	(cyclohexene ring)	$H_2C{=}CH{-}CH{=}CH_2$	H—C≡C—H
ETHYLENE (AN ALKENE)	CYCLOHEXENE (A CYCLOALKENE)	1,3-BUTADIENE (A DIENE)	ACETYLENE (AN ALKYNE)

Alkenes

The Carbon-Carbon Double Bond

3.2 sp^2 Hybridization

When carbon is attached to *four* other atoms or groups, as in molecules of the alkanes, it utilizes sp^3 hybrid orbitals. At a double bond, however, each carbon is attached to only *three* other groups. Carbon cannot use sp^3 orbitals here. It needs different orbitals, which we shall study in this section. As illustrated in Figure 3.1 we start with the various atomic orbitals at the first and second energy levels. Then we "mix" or "hybridize" the 2*s* and just *two* of the 2*p* orbitals. We leave the third *p*-orbital at level-2 unchanged—for the moment. Because we "mix" one *s* and two *p* orbitals, we call the new hybrid orbitals sp^2 orbitals.

The sp^2 hybrid orbitals have both *s*-character and *p*-character—*p*-character in that there are two lobes, *s*-character in that part of the orbital includes the nucleus. The shapes and distributions of these orbitals about a carbon atom are shown in Figure 3.2. Let us now see how these orbitals interact to give the network of bonds in ethylene.

3.3 The Pi Bond

Figure 3.3 illustrates two carbon atoms whose atomic orbitals have undergone sp^2 hybridization. To form the C—H bonds each atom has used two sp^2 orbitals to accept overlap with 1*s* orbitals of two hydrogens. This leaves each carbon with an unused sp^2 orbital and an unused p_z orbital. Next, there is the sp^2-sp^2 overlap between the two carbons to give one of the bonds, one of the molecular orbitals of the double bond. Because this molecular orbital is symmetrical about the axis of the bond we call it a sigma bond. It is like any ordinary single, covalent bond. We should expect free rotation about it, for example. However, at one "stop" in such a rotation, the two (so far) unused p_z orbitals will be lined up side by side as seen in Figure 3.4. They can and do overlap, side by side, and this locks the

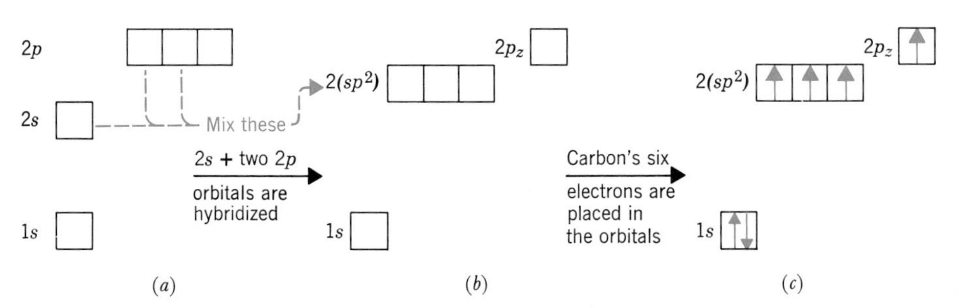

Figure 3.1
A view of how the sp^2 hybrid orbitals arise. (*a*) These boxes represent the orbitals available to electrons at the first and second energy levels. They are, for the moment, empty. (*b*) The result of "mixing" or hybridizing the 2*s* orbital and two of the 2*p* orbitals. The number of places for electrons has not changed, but the shapes and orientations for some have. The boxes are still empty. (We don't hybridize electrons; we hybridize orbitals.) (*c*) The valence state of a carbon at a double bond: $1s^2\ 2(sp^2)\ 2(sp^2)\ 2(sp^2)\ 2p_z$.

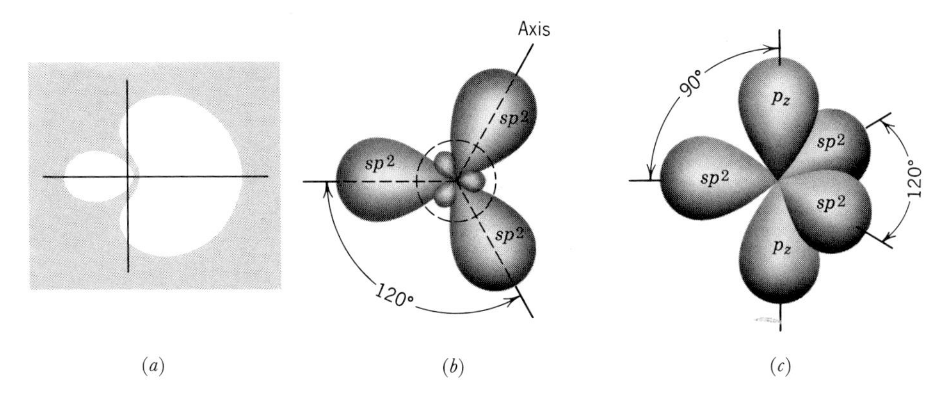

Figure 3.2
The bonding orbitals used by a carbon atom when it holds just three other groups, as in alkenes. (*a*) The cross-section of the sp^2 hybrid orbital. (*b*) Top view of a carbon atom with its three sp^2 orbitals. The axes of these orbitals are in the plane of the page, and they make angles of 120° with each other. (*c*) Perspective view, showing only the larger lobes of the sp^2 orbitals and the full, unhybridized p_z orbital. (In the top view of (*b*), the p_z orbital's location was indicated by the dashed-line circle.

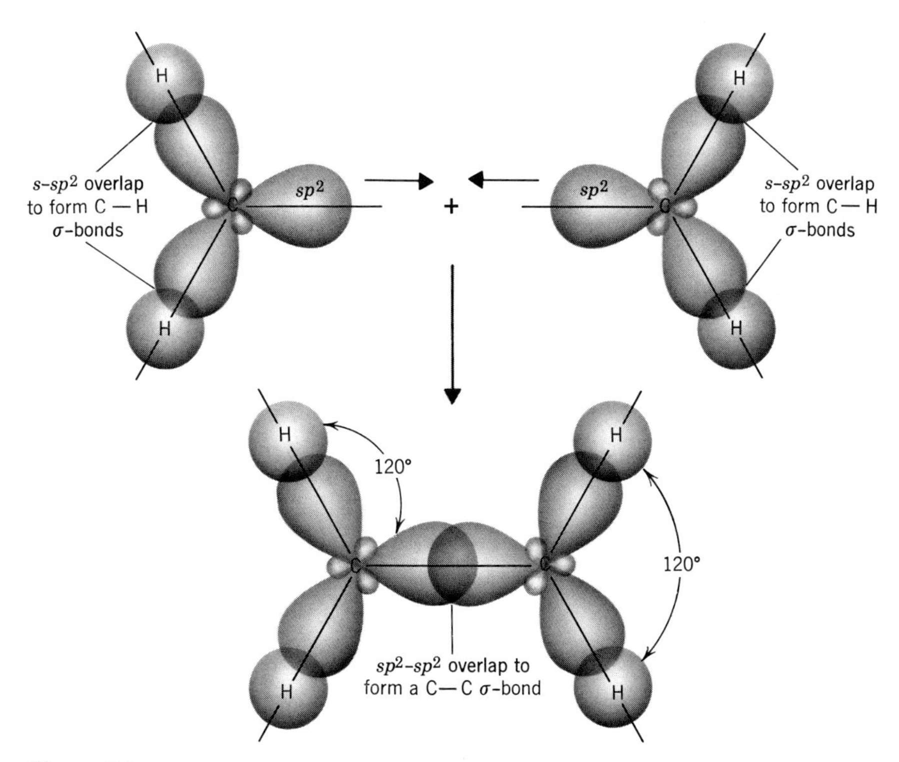

Figure 3.3
The σ-bond network in ethylene consists of one carbon-carbon bond and four carbon-hydrogen bonds. The p_z orbitals are not shown, but their lobes lie above and below the plane of the page with their axes piercing the paper at each carbon.

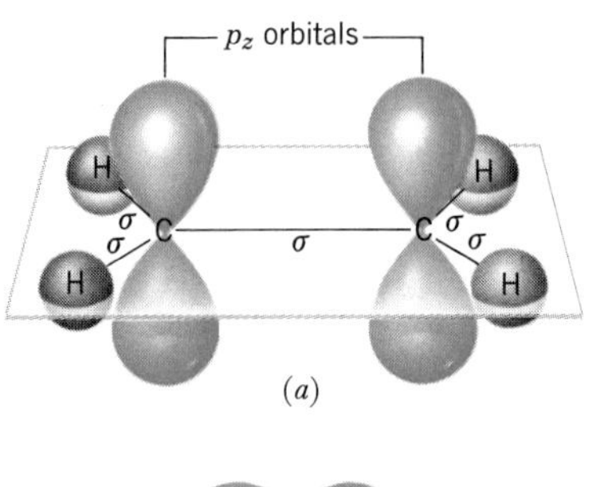

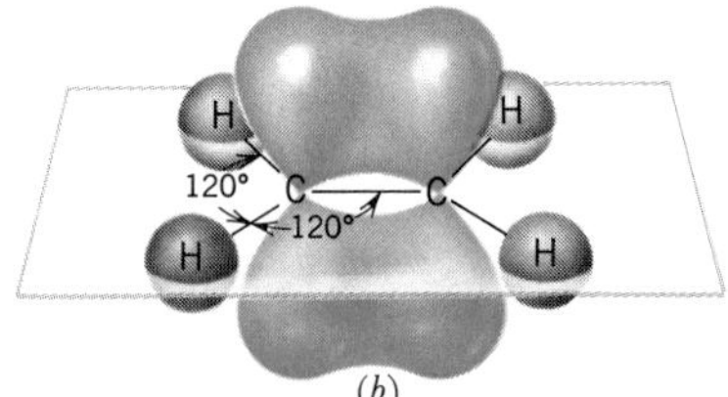

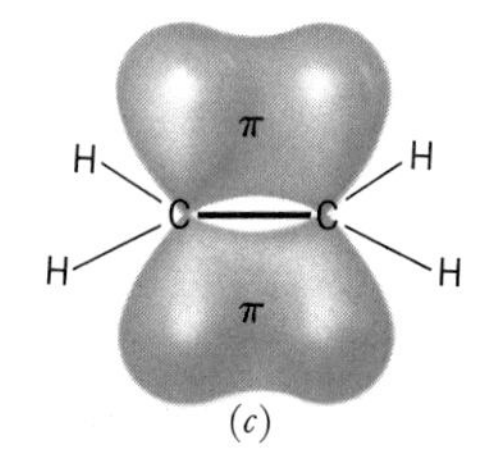

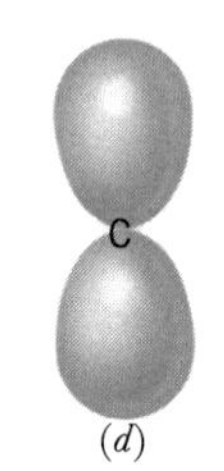

Figure 3.4
The nature of the bonding and the geometry at a carbon-carbon double bond, according to one common theory. The solid lines labeled σ are described in Figure 3.3.
(*a*) This is the hypothetical situation before the p_z orbitals overlap. (*b*) Maximum overlap of the p_z orbitals can occur only when the six nuclei shown here lie in the same plane. (*c*) The double bond in ethylene consists of one σ-bond (the heavy line) and one π-bond. The π-bond has the appearance of two sausages above and below the plane. (*d*) A cross-sectional view of the π-bond looking down the carbon-carbon axis shows its two lobes.

system from free rotation at this point. The side-by-side overlap also is a covalent bond, the second bond of the double bond. It is a new molecular orbital with two electrons, but it does not have cylindrical symmetry. Viewed end-on it has the cross section of a figure eight, and any molecular orbital with this symmetry is called a **pi-bond.** The double bond, therefore, consists of one sigma bond and one pi bond.

3.4 Importance of the Pi Bond

The electrons in the pi bond are called the **pi electrons,** and they are not located as close to the carbon skeleton as the sigma electrons are. The overlap in the pi bond is not as effective as that in the sigma bond. The net result is that the pi electrons are capable of reacting with electron-poor species (e.g., protons). As we shall see, *the key to the chemistry of the alkenes is the double bond and the availability of its pi electrons to reagents seeking electrons.*

3.5 Isomerism Among the Alkenes

Two types of isomerism are possible in the alkene family. One is positional isomerism; the double bond may be located between different carbons. The other is geometrical isomerism; the groups attached to the double bond may have cis-trans relations. Four alkenes of the formula C_4H_8 are known. Their structures are

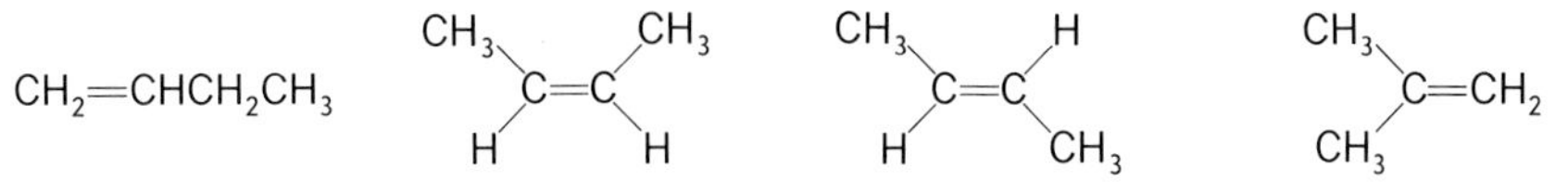

$CH_2{=}CHCH_2CH_3$ — 1-BUTENE (α-BUTYLENE)

$CH_3(H)C{=}C(H)CH_3$ — *cis*-2-BUTENE (*cis*-β-BUTYLENE)

$CH_3(H)C{=}C(CH_3)H$ — *trans*-2-BUTENE (*trans*-β-BUTYLENE)

$(CH_3)_2C{=}CH_2$ — 2-METHYL-1-PROPENE (ISOBUTYLENE)

In 1-butene the double bond is between the first and the second carbon. In both of the 2-butenes it is between the second and third. If there were free rotation about a double bond, 2-butene, like *n*-butane, could have no geometric isomers. But two 2-butenes are known. Their physical properties are listed in Table 3.1. Differences in structure however subtle do mean differences in observable properties. The six isomeric pentenes are also shown in Table 3.1. The word *cis* means that groups are on the same side of the double bonds; *trans* means they are on opposite sides. In complicated molecules this system of terms does not work well, and the E/Z system (Appendix) must be used.

Exercise 3.1 Identify each of the following compounds by name.

(a) 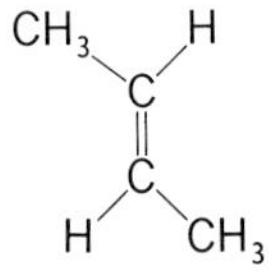 $CH_3(H)C{=}C(H)CH_3$ (CH₃ and H on upper carbon; H and CH₃ on lower carbon, drawn vertically)

(b) $CH_3CH_2CH{=}CH_2$

(c) 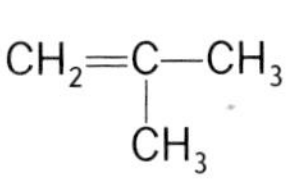 $CH_2{=}C(CH_3){-}CH_3$

(d) 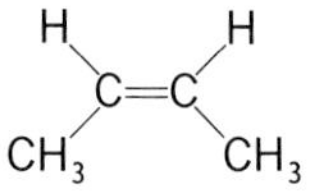 $H(CH_3)C{=}C(H)CH_3$ (H and H on top; CH₃ and CH₃ on bottom)

Exercise 3.2 Examine each of the following structures, write trial structures for geometric isomers, and determine whether or not cis-trans isomerism is possible.

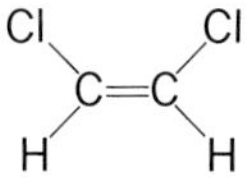

(a) 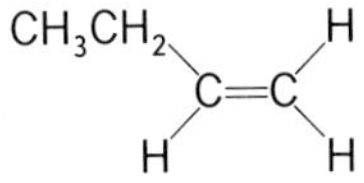 $CH_3CH_2(H)C{=}C(H)H$

(b) $Cl(H)C{=}C(Cl)H$

(c) $Cl(Cl)C{=}C(H)H$

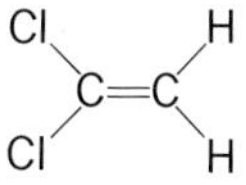

Table 3.1
Physical Properties of the Butenes and Pentenes

	Name	Bp (°C)	Mp (°C)	Density (in g/cc)
Butenes	2-Methyl-1-propene	−6.90	−140.4	0.640 (−20°C)
	1-Butene	−6.26	−185	0.641
	trans-2-Butene	+0.88	−105.6	0.649
	cis-2-Butene	+3.72	−138.9	0.667
Pentenes	1-Pentene	30.0	−165.2	0.641 (+20°C)
	cis-2-Pentene	36.9	−151.4	0.656
	trans-2-Pentene	36.4	−140.2	0.648
	2-Methyl-1-butene	31.2	−137.5	0.650
	3-Methyl-2-butene	20.1	−168.5	0.627
	2-Methyl-2-butene	38.6	−133.8	0.662

(d) Cl(H)C=C(Br)(H)

(e) Cl(Br)C=C(H)(H)

3.6 Writing Structures of Alkenes

While there are some chemical properties of *cis* and *trans* isomers that are remarkably different, most are not. Unless *cis-trans* distinctions are important, when we write structures of alkenes we may write them in a simple manner. For example, 2-butene is usually well enough represented just by $CH_3CH{=}CHCH_3$, which would mean either the *cis* or the *trans* isomer or a mixture of both. What is important is that, unlike single bonds, we never leave a double bond "understood."

Nomenclature of Alkenes

3.7 Common Names

The common names of alkenes have the same ending, "-ylene." An alkene with the carbon skeleton of the three-carbon alkane, propane, is called "propylene." The "prop-" prefix indicates a total of three carbons; the "-ylene" suffix places it in the ethylene or alkene family. The various butylenes are differentiated arbitrarily by the prefixes α, β, and "iso-." In the homologous series the common system loses most of its practicality above the butylenes, and the IUPAC system is used.

3.8 IUPAC Names

The IUPAC rules for naming the alkenes are as follows.

1. The characteristic name ending is "-ene."
2. The longest continuous chain *that contains the double bond* is selected. This chain is named by selecting the alkane with the *identical* chain length and changing the suffix in its name from "-ane" to "-ene."
3. The chain is numbered to give the *first* carbon of the double bond the lowest possible number (e.g., $CH_3CH_2CH{=}CH_2$ is 1-butene, not 1,2-butene and not 3-butene).
4. The locations of the groups attached to the main chain are identified by numbers. The word parts are then assembled in a manner analogous to that in alkane nomenclature.

$CH_3CH_2CH(CH_3)CH_2CH{=}C(CH_3)CH_3$

2,5-DIMETHYL-2-HEPTENE

$CH_3CH(CH_3)CH_2C(CH(CH_3)CH_2CH_3){=}CH_2$

2-ISOBUTYL-3-METHYL-1-PENTENE

Not: 2-sec-BUTYL-4-METHYL-1-PENTENE

The examples in Table 3.2 as well as the following illustrate correct application of these rules. Study them, noting especially the placing of commas and hyphens. Because rule 3 gives numbering precedence to the double bond, rather than to the side chains, positions bearing side chains will sometimes have larger numbers than they would if an alternate numbering were used. Common names are given in parentheses.

Table 3.2
Properties of Some 1-Alkenes

Name (IUPAC)	Structure	Bp (°C)	Mp (°C)	Density (in g/cc at 10°C)
Ethylene[a]	$CH_2{=}CH_2$	−104	−169	—
Propene	$CH_2{=}CHCH_3$	−48	−185	—
1-Butene	$CH_2{=}CHCH_2CH_3$	−6	−185	—
1-Pentene	$CH_2{=}CHCH_2CH_2CH_3$	+30	−165	0.641
1-Hexene	$CH_2{=}CHCH_2CH_2CH_2CH_3$	64	−140	0.673
1-Heptene	$CH_2{=}CHCH_2CH_2CH_2CH_2CH_3$	94	−119	0.697
1-Octene	$CH_2{=}CHCH_2CH_2CH_2CH_2CH_2CH_3$	121	−102	0.715
1-Nonene	$CH_2{=}CHCH_2CH_2CH_2CH_2CH_2CH_2CH_3$	147	−81	0.729
1-Decene	$CH_2{=}CHCH_2CH_2CH_2CH_2CH_2CH_2CH_2CH_3$	171	−66	0.741
Cyclopentene	(structure)	44	−135	0.722
Cyclohexene	(structure)	83	−104	0.811

[a]See footnote 1.

$CH_2{=}CH_2$ ETHYLENE[1]

$CH_3CH{=}CH_2$ PROPENE[2] ("PROPYLENE")

$CH_3CH{=}CHCH_2CH_3$ 2-PENTENE (*cis* or *trans*) ("β-AMYLENE")

Two important unsaturated groups that occur many times in organic systems are:

$CH_2{=}CH{-}$ VINYL

$CH_2{=}CHCH_2{-}$ ALLYL

Thus the compound $CH_2{=}CH{-}Cl$ is called *vinyl chloride,* and the compound $CH_2{=}CHCH_2{-}Br$ is called *allyl bromide.* ("Vinyl" and "allyl" are the IUPAC names for these groups.)

Exercise 3.3 Write condensed structures for each of the following.

(a) 4-methyl-2-pentene
(b) 3-*n*-propyl-1-heptene
(c) 3,3-dimethyl-4-chloro-1-butene
(d) 2,3-dimethyl-2-butene
(e) allyl iodide
(f) vinyl bromide

[1]This nonsystematic ("common") name is retained in the IUPAC system for the *unsubstituted* compound (1957 Rules, paragraph A-3.1). Most chemists, however, do not object to calling it *ethene,* and it must be so named in substituted ethenes.

[2]Whenever the double bond cannot be located differently to make an isomer, its position need not be designated by a number.

Exercise 3.4 Write IUPAC names and the common names if they have already been mentioned for each of the following. Use cis or trans designations if they are applicable.

(a) 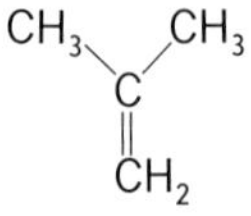 $(CH_3)_2C{=}CH_2$

(b) 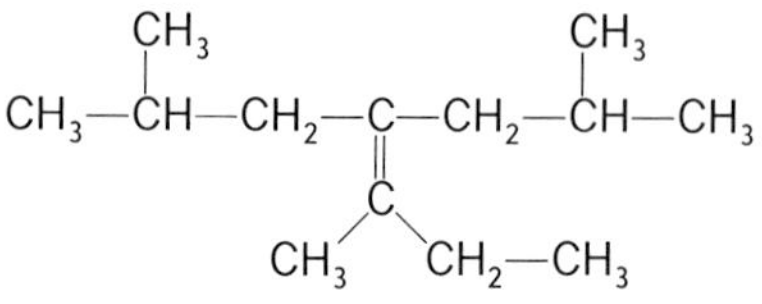 $CH_3{-}CH(CH_3){-}CH_2{-}C({=}C(CH_3){-}CH_2{-}CH_3){-}CH_2{-}CH(CH_3){-}CH_3$

(c) $CH_3{-}CH{=}CH{-}Cl$

(d) $Br{-}CH_2{-}CH{=}CH_2$

(e) 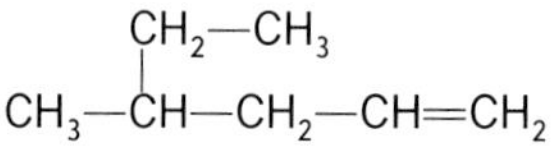 $CH_3{-}CH(CH_2{-}CH_3){-}CH_2{-}CH{=}CH_2$

(f) If [hexagon with one double bond] is called cyclohexene what is its *full* structure? What is the structure of cyclopentene?

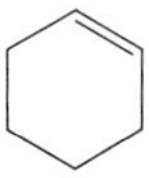

(g) If [cyclohexene ring numbered 1–6 with CH_3 at position 3] is 3-methylcyclohexene,[3] what is [cyclohexene ring bearing two adjacent CH_3 groups] ?

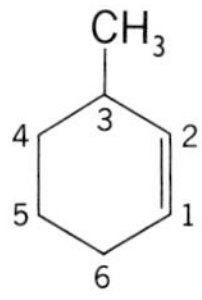

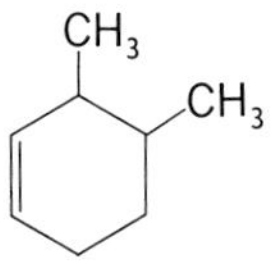

3.9 Syntheses of Alkenes

We shall do no more than mention some of the methods for introducing a double bond into a molecule. Each method constitutes a chemical property of a family of compounds that we have yet to study.

Basically, there are two ways we can put double bonds into molecules. We can begin with a saturated molecule and "desaturate," that is, pull something out of it by an *elimination reaction.* Otherwise, we may start with an acetylene and partially reduce it. The three syntheses of alkenes in this text are:

1. Elimination of H—X from alkyl halides, section 7.9.
2. Elimination of H—OH from alcohols, section 6.25.
3. Alkenes by reduction of a triple bond, section 3.38.

3.10 Physical Properties of Alkenes

The alkenes closely resemble corresponding alkanes in physical properties; this generalization is the most useful one we can carry forward to our study of other families. Alkene molecules, which in the higher homologs are mostly alkanelike

[3] The double bond in a cyclic system is *always* regarded as *starting from* position 1. Numbering then proceeds through the double bond around the ring in the direction that will give the lower set of numbers to the substituents.

anyway, are relatively nonpolar. Alkenes are insoluble in water, but they dissolve in the typical nonpolar solvents such as benzene, ether, carbon tetrachloride, chloroform, and ligroin (a mixture of liquid alkanes).

3.11 Chemical Properties of Alkenes

There are basically two sites in alkenes that are attacked by chemicals. One is the double bond itself, the other the position on the chain directly attached to the double bond, the *allylic position.* In this chapter we study reactions at the double bond itself.

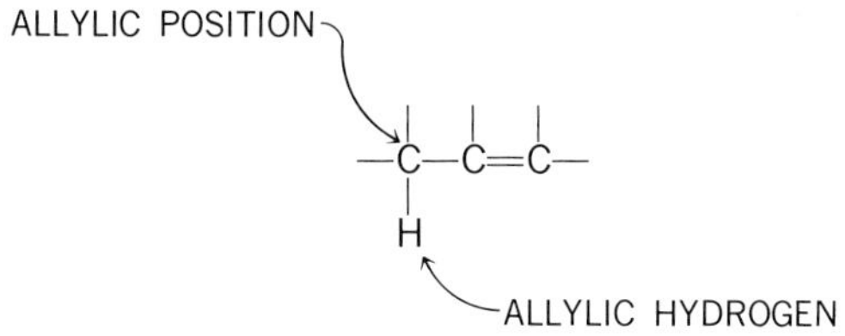

ALLYLIC POSITION → —C(H)—C=C— ← ALLYLIC HYDROGEN

3.12 Addition Reactions

One of the common types of reactions of alkenes is the opening of the pi-electron bond to pick up two groups, one at the carbon at one end of the double bond and the other group at the other carbon. When it is all over, the double bond has changed to a single bond, and its two carbons have become saturated carbons. We call such events addition reactions. We say that "something adds to the double bond."

3.13 Hydrogenation

The addition of the elements of hydrogen, H—H. In general:

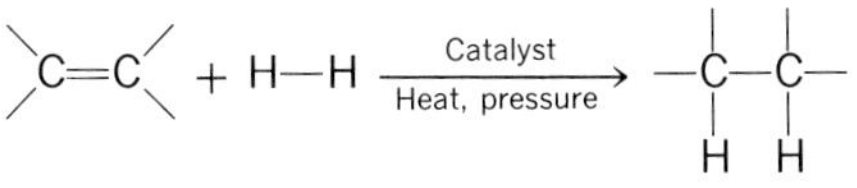

$$>C=C< + H\text{—}H \xrightarrow[\text{Heat, pressure}]{\text{Catalyst}} \text{—}\underset{H}{C}\text{—}\underset{H}{C}\text{—}$$

Specific examples:

$$\underset{\text{ETHYLENE}}{CH_2=CH_2} + H_2 \xrightarrow[\text{Heat, pressure}]{\text{Ni}} \underset{\text{ETHANE}}{CH_3\text{—}CH_3}$$

$$\underset{\text{ISOBUTYLENE}}{(CH_3)_2C=CH_2} + H_2 \xrightarrow[\text{Heat, pressure}]{\text{Ni}} \underset{\text{ISOBUTANE}}{(CH_3)_2CH\text{—}CH_3}$$

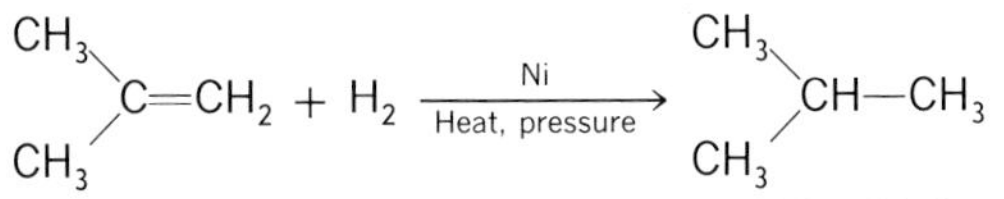

$$\text{3-METHYLCYLOPENTENE} + H_2 \xrightarrow[\text{Heat, pressure}]{\text{Ni}} \text{METHYLCYCLOPENTANE}$$

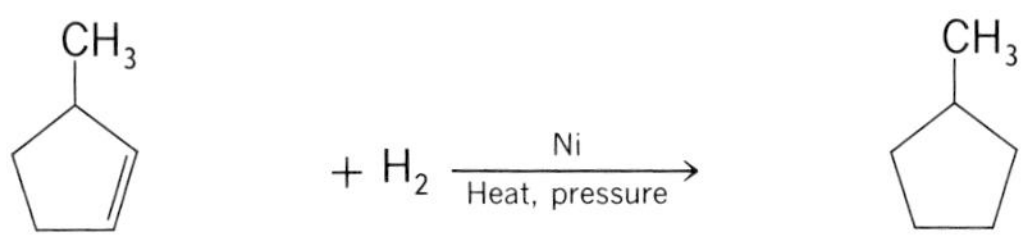

The alkene is placed in a heavy-walled bottle or a steel cylinder together with the catalyst, a metal such as nickel, platinum, or palladium in a specially prepared, powdered form. Hydrogen gas is admitted to the vessel under pressure; the bottle is sometimes heated, and it must be shaken to insure intimate contact of the reactants with the metallic catalyst. Figure 3.5 shows a typical apparatus used when a relatively low hydrogen pressure is sufficient (less than 100 lb/in.^{2}).

3.14 Halogenation

The addition of chlorine or bromine. (Only these two halogens will add to a double bond.) In general:

$$>C{=}C< \; + \; X{-}X \longrightarrow >\underset{X}{C}{-}\underset{X}{C}< \quad (\text{or } X{-}\overset{|}{\underset{|}{C}}{-}\overset{|}{\underset{|}{C}}{-}X)$$

(X = Cl or Br)

Specific examples:

$$CH_3CH{=}CHCH_3 + Cl_2 \longrightarrow CH_3\underset{Cl}{CH}{-}\underset{Cl}{CH}CH_3$$

2-BUTENE → 2,3-DICHLOROBUTANE (81%)[4]

$$CH_3CH{=}CHCH_2CH_3 + Br_2 \longrightarrow CH_3\underset{Br}{CH}{-}\underset{Br}{CH}CH_2CH_3$$

2-PENTENE → 2,3-DIBROMOPENTANE (94%)

$$\text{Cyclohexene} + Br_2 \longrightarrow \text{1,2-dibromocyclohexane}$$

CYCLOHEXENE → 1,2-DIBROMOCYCLOHEXANE (The trans isomer forms)

The dichloro- and dibromo compounds are oily liquids. In this sense, therefore, ethylene was early regarded as an oil-forming gas or olefiant gas. From this came the nickname of all alkenes. Chemists commonly call them *olefins*.

Usually this reaction is carried out at room temperature, or below, and the reagents are dissolved in some inert solvent such as carbon tetrachloride. If chlorine is used, the gas is bubbled into the solution. Bromine, a corrosive liquid that must be handled with great care, is usually added as a dilute solution in carbon tetrachloride.

[4]Throughout the chapters most reactions are shown with a figure in parentheses representing the mole percent yield. Side reactions are almost inevitable in organic synthesis. Moreover, in isolating the product, losses occur. Many reactions do not go to completion; unchanged starting materials remain and must be separated from the product. The finally isolated, pure product is nearly always less, on a mole basis, than would be expected if every mole of starting material forms the desired product. The percentages cited in nearly all examples of this text are yields actually reported and then tabulated in *Synthetic Organic Chemistry* by R. B. Wagner and H. D. Zook (John Wiley and Sons, New York, 1953).

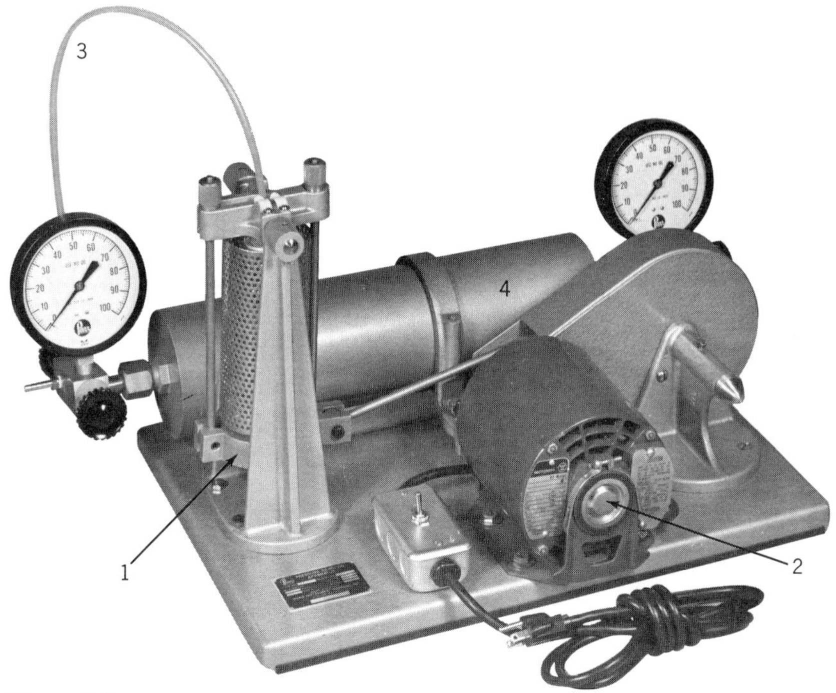

Figure 3.5
Typical low-pressure hydrogenation apparatus. The unsaturated compound together with a catalyst and an inert solvent is placed in a thick-walled glass bottle positioned in a heavy-mesh, protective metal screen and clamped on a rocker, 1. Rocking action, powered by the motor, 2, mixes the contents of the flask while hydrogen gas under pressure is led in by a hose, 3, from a storage tank, 4. (Courtesy of Parr Instrument Company.)

3.15 The Bromine/Carbon Tetrachloride Test for Unsaturation

A dilute solution of bromine in carbon tetrachloride is reddish brown. If you were to add it drop by drop to an "unknown" that you believed to be either an alkene or an alkane, what might you see? If the unknown were an alkene, you would see the color of the bromine solution promptly disappear; dibromo compounds are colorless (or pale yellow). If the unknown were the alkane, however, you would only see a dilution of the color of the bromine. Bromine could react with an alkane only by a free-radical substitution reaction, and those cannot start without some help—heat or ultraviolet light, for example. Of course, stray ultraviolet light might get in and initiate the reaction, and in that case the bromine color would also disappear. How could you tell the difference, then, between the alkene and the alkane? It's quite easy. *If* the alkane does react, the reaction would also produce hydrogen bromide, a gas that is not too soluble in carbon tetrachloride. As it fumes up and out of the test tube and encounters air, the humidity of the air will cause the gas to fume visibly. (Care is needed. The fuming may be so slow that it goes unnoticed.)

In summary, an alkene discharges the bromine color promptly without evolution of hydrogen bromide.

An alkane may or may not react to discharge the color; if it does, hydrogen bromide also forms.

3.16 Addition of Hydrogen Chloride or Hydrogen Bromide

Molecules of hydrogen or of the halogens are symmetrical. Those of hydrogen chloride or hydrogen bromide are not. When a molecule of either HCl or HBr adds to an unsymmetrically substituted double bond, two directions for the addition are possible. Propylene, for example, might react with hydrogen chloride to give either *n*-propyl chloride or isopropyl chloride or a mixture of both:

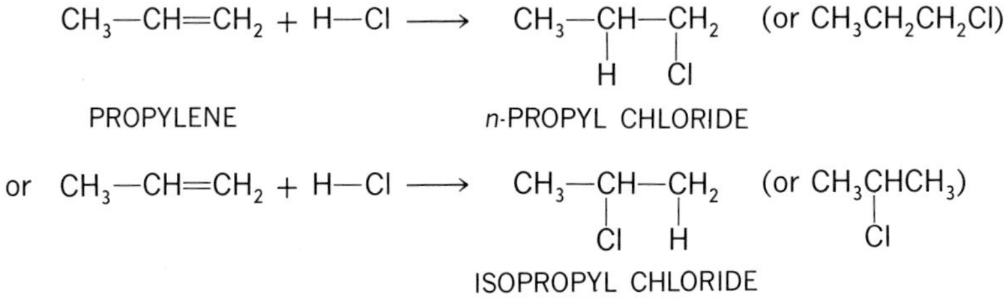

$$CH_3{-}CH{=}CH_2 + H{-}Cl \longrightarrow CH_3{-}\underset{H}{\underset{|}{C}H}{-}\underset{Cl}{\underset{|}{C}H_2} \quad (\text{or } CH_3CH_2CH_2Cl)$$

PROPYLENE → *n*-PROPYL CHLORIDE

$$\text{or}\quad CH_3{-}CH{=}CH_2 + H{-}Cl \longrightarrow CH_3{-}\underset{Cl}{\underset{|}{C}H}{-}\underset{H}{\underset{|}{C}H_2} \quad (\text{or } CH_3\underset{Cl}{\underset{|}{C}H}CH_3)$$

ISOPROPYL CHLORIDE

In either event, the double bond becomes a single bond, and the two parts of the hydrogen chloride molecule become attached to the carbons at each end of the former double bond.

The product that actually forms is largely isopropyl chloride. Essentially no *n*-propyl chloride is produced.

Other examples of this reaction shown below display a consistency first noted by a Russian chemist, Markovnikov.[5] According to Markovnikov's rule, when an unsymmetrical reagent, H—G, adds across an unsymmetrical carbon-carbon double bond of a simple alkene, the hydrogen of H—G attaches to the carbon of the double bond that is already bonded to the greater number of hydrogens. ("Them that has, gets.")

For purposes of this rule, "unsymmetrical carbon-carbon double bond" means one in which the carbons bear unequal numbers of hydrogens. Specific examples:

$$CH_2{=}CH_2 + H{-}Br \longrightarrow CH_3CH_2Br \quad (\text{only possible product})$$

ETHYLENE → ETHYL BROMIDE

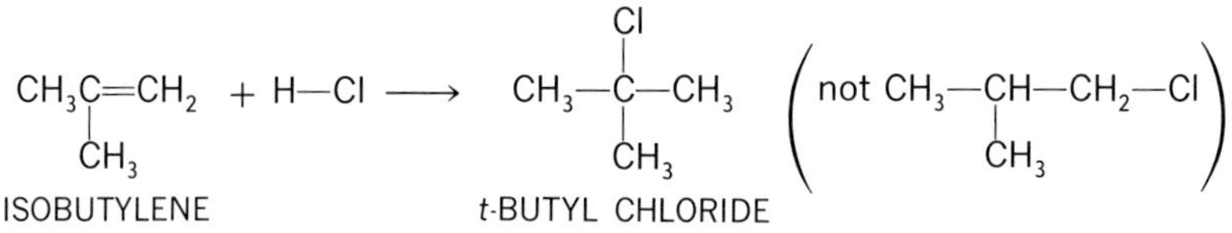

$$CH_3\underset{CH_3}{\underset{|}{C}}{=}CH_2 + H{-}Cl \longrightarrow CH_3{-}\overset{Cl}{\overset{|}{\underset{CH_3}{\underset{|}{C}}}}{-}CH_3 \quad \left(\text{not } CH_3{-}\underset{CH_3}{\underset{|}{C}H}{-}CH_2{-}Cl\right)$$

ISOBUTYLENE → *t*-BUTYL CHLORIDE

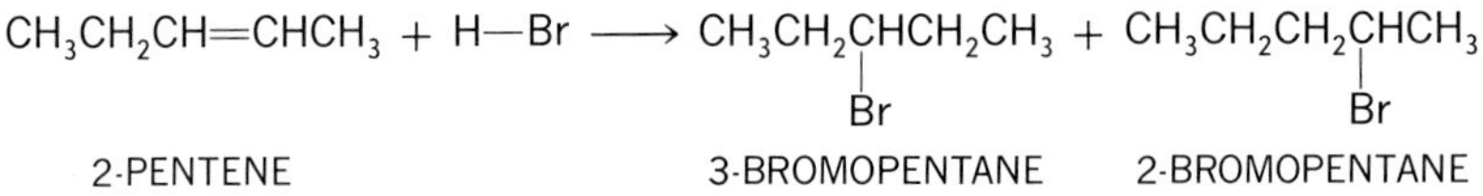

$$CH_3CH_2CH{=}CHCH_3 + H{-}Br \longrightarrow CH_3CH_2\underset{Br}{\underset{|}{C}H}CH_2CH_3 + CH_3CH_2CH_2\underset{Br}{\underset{|}{C}H}CH_3$$

2-PENTENE → 3-BROMOPENTANE + 2-BROMOPENTANE

[5]Vladimir Vasil'evich Markovnikov (1838–1904).

$$\text{1-methylcyclohexene} + H{-}Cl \longrightarrow \text{1-chloro-1-methylcyclohexane} \quad (\text{not } \text{1-chloro-2-methylcyclohexane})$$

1-METHYLCYCLOHEXENE — 1-CHLORO-1-METHYLCYCLOHEXANE

The reaction is carried out by bubbling gaseous hydrogen chloride or hydrogen bromide into the alkene, which may or may not be dissolved in some solvent inert to the reagents (e.g., carbon tetrachloride).[6]

For preparing alkyl chlorides and bromides, this reaction is superior to direct halogenation of an alkane. The latter gives mixtures of mono-, di-, and higher halogenated materials, and the reaction also produces mixtures of isomers.

3.17 Hydration

The addition of water, H—OH. In general:

$$>C{=}C< \; + \; H{-}OH \xrightarrow{H^+} -\underset{H}{\overset{|}{C}}-\underset{OH}{\overset{|}{C}}-$$

AN ALKENE — AN ALCOHOL

Specific examples:

$$CH_2{=}CH_2 + H{-}OH \xrightarrow[\text{240°C (closed vessel)}]{10\%\ H_2SO_4} CH_3{-}CH_2{-}OH$$

ETHYLENE — ETHYL ALCOHOL

$$(CH_3)_2C{=}CH_2 + H{-}OH \xrightarrow[25°C]{10\%\ H_2SO_4} (CH_3)_2C(OH){-}CH_3 \quad (\text{not } (CH_3)_2CH{-}CH_2{-}OH)$$

ISOBUTYLENE — *t*-BUTYL ALCOHOL

Markovnikov's rule applies to this reaction in which H—G is now H—OH. The H in H—OH goes to the carbon of the double bond that has the greater number of hydrogens; the —OH goes to the other carbon.

Acid catalysis is the key condition. In fact, the reaction can be completed in two discrete steps if we use concentrated sulfuric acid, another unsymmetrical reagent whose molecules add to double bonds according to Markovnikov's rule:

$$CH_3{-}CH{=}CH_2 + H{-}O{-}\underset{\downarrow\atop O}{\overset{O\atop\uparrow}{S}}{-}O{-}H \xrightarrow[0°C]{} CH_3{-}\overset{CH_3}{\overset{|}{C}H}{-}O{-}\underset{\downarrow\atop O}{\overset{O\atop\uparrow}{S}}{-}O{-}H$$

PROPYLENE — SULFURIC ACID — ISOPROPYL HYDROGEN SULFATE

[6]When hydrogen bromide is used, it is important that no peroxides, compounds of the type R—O—O—H or R—O—O—R, be present. They catalyze a different mechanism, which we shall not discuss, with the result that in their presence hydrogen bromide (and *only* this hydrogen halide) adds itself to unsymmetrical alkenes against the Markovnikov rule. Traces of peroxides form when organic chemicals remain exposed to atmospheric oxygen for long periods.

If the product, a member of the class of alkyl hydrogen sulfates, is diluted with water, the sulfate group is replaced by —OH and the product is the alcohol:

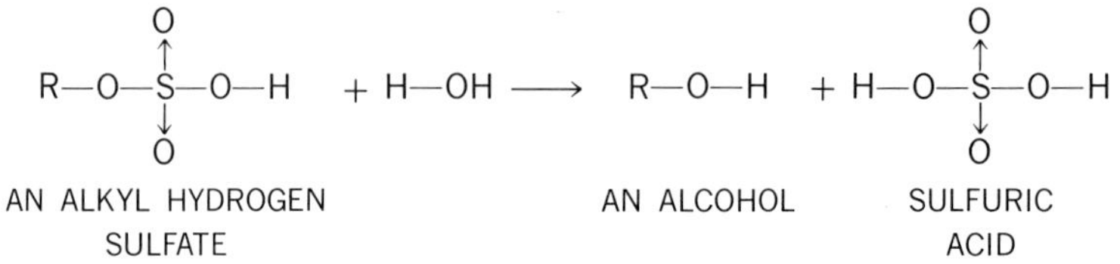

$$R-O-\overset{\overset{O}{\uparrow}}{\underset{\underset{O}{\downarrow}}{S}}-O-H + H-OH \longrightarrow R-O-H + H-O-\overset{\overset{O}{\uparrow}}{\underset{\underset{O}{\downarrow}}{S}}-O-H$$

AN ALKYL HYDROGEN SULFATE — AN ALCOHOL — SULFURIC ACID

Exercise 3.5 Write the structures for the product(s) that would form under the conditions shown. Assume that peroxides are absent.

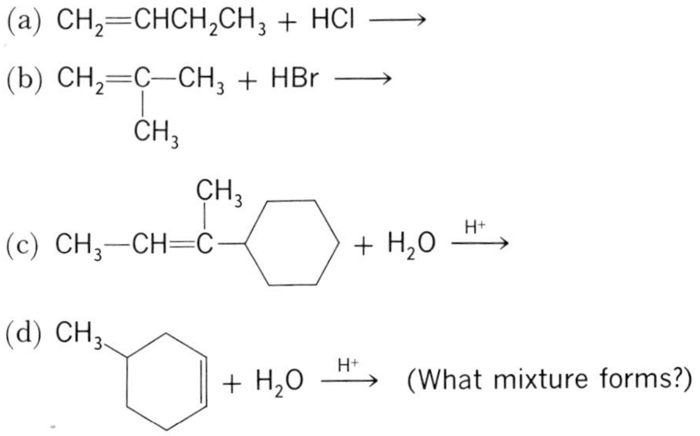

(a) $CH_2{=}CHCH_2CH_3 + HCl \longrightarrow$

(b) $CH_2{=}\underset{\underset{CH_3}{|}}{C}-CH_3 + HBr \longrightarrow$

(c) $CH_3-CH{=}\overset{\overset{CH_3}{|}}{C}-C_6H_{11}$ (cyclohexyl) $+ H_2O \xrightarrow{H^+}$

(d) CH_3-cyclohexene $+ H_2O \xrightarrow{H^+}$ (What mixture forms?)

Exercise 3.6 Write the structures and names of alkenes that are best used to prepare each of the following halides. Avoid selecting those yielding isomeric halides that are difficult to separate. If the halide cannot be prepared by the action of HCl or HBr on an alkene (given the limitations of Markovnikov's rule), state this fact.

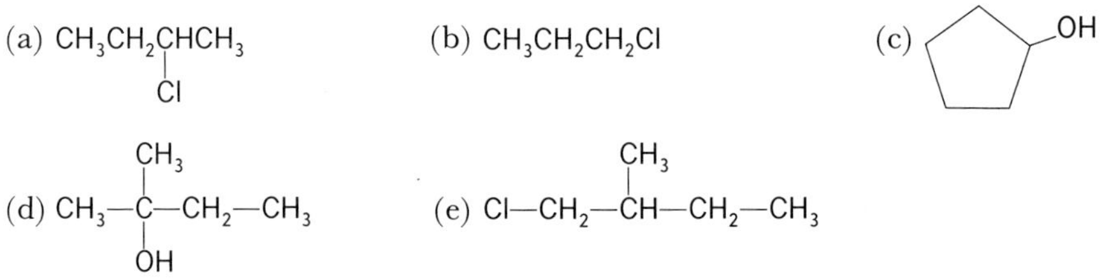

(a) $CH_3CH_2\underset{\underset{Cl}{|}}{C}HCH_3$

(b) $CH_3CH_2CH_2Cl$

(c) cyclopentanol (cyclopentane ring with OH)

(d) $CH_3-\overset{\overset{CH_3}{|}}{\underset{\underset{OH}{|}}{C}}-CH_2-CH_3$

(e) $Cl-CH_2-\overset{\overset{CH_3}{|}}{C}H-CH_2-CH_3$

3.18 Concentrated Sulfuric Acid Test

The alkyl hydrogen sulfates are vastly more polar substances than the alkenes from which they may be made. Being polar, they dissolve readily in concentrated sulfuric acid, which is a polar substance, too. ("Likes dissolve likes.") Hence, we have the basis for another, quick, simple and inexpensive test tube test to tell the difference between an alkane and an alkene. Alkanes, recall, do nothing with concentrated sulfuric acid (at least at ordinary temperatures). If the "unknown," which on other grounds you know is either an alkene or an alkane, reacts with and dissolves in concentrated sulfuric acid, it has to be the alkene. If it simply floats on top of the acid, it has to be the alkane.

3.19 How Addition Reactions Take Place

Having surveyed some common addition reactions of alkenes let us see *how* they happen. We begin as we must begin with devising any theory—with the surest facts we have available. For addition reactions of alkenes we know the following:

1. The reactions happen. (That fact, while obvious, is central.)
2. They follow Markovnikov's rule. (That rule is not a theory; it is a summary of known facts. It *explains* nothing.)
3. Either acids or acid catalysts are involved. (These are proton-donating reagents.)
4. The more substituted alkenes appear more reactive than the less substituted alkenes. (We didn't assemble much data on this point but refer back to the conditions that are milder for the addition of water to isobutylene than to ethylene. Isobutylene is the more branched or substituted olefin.)

Organic chemists have pieced and woven these facts together into the following theory (which we simplify to some extent). The pi electrons of a double bond are not as strongly held as sigma electrons. They can be pivoted out to make a bond to an electron-seeking reagent. We call such reagents, in general, **electrophiles** from "electron loving" (Greek *philos,* loving); or we say they are **electrophilic reagents.** The proton, H^+, is a common example, and chemicals that readily donate protons (strong acids) attack pi electrons of alkenes.

Consider hydrogen chloride, H—Cl, and its reaction with propylene. The first step is the transfer, during a collision, of a proton from H—Cl to the alkene:

Step 1.

$$:\ddot{\underset{\cdot\cdot}{Cl}}\text{—H} + CH_2\text{=CH—}CH_3 \longrightarrow :\ddot{\underset{\cdot\cdot}{Cl}}:^- \quad + \quad \text{H—}CH_2\text{—}\overset{+}{C}\text{H—}CH_3$$

ISOPROPYL CATION
(A SECONDARY CARBONIUM ION)

The product is a carbonium ion, an ion that has a carbon with just a sextet of outer-level electrons, not an octet. Carbonium ions have only the most fleeting existences. They will combine with any particle that can help restore the outer-level octet to the carbon. We have one in this reaction, a chloride ion, also produced in the step.

The second step is simply the joining together of the carbonium ion with the chloride ion:

Step 2.

$$:\ddot{\underset{\cdot\cdot}{Cl}}:^- + CH_3\text{—}\overset{+}{C}\text{H—}CH_3 \longrightarrow CH_3\text{—}\underset{}{\overset{Cl}{\overset{|}{C}}}\text{H—}CH_3$$

On paper, at least, step 1 should just as easily have given a different carbonium ion:

$$:\ddot{\underset{\cdot\cdot}{Cl}}\text{—H} + CH_2\text{=CH—}CH_3 \xrightarrow{?} :\ddot{\underset{\cdot\cdot}{Cl}}:^- \quad + \quad \overset{+}{C}H_2\text{—}CH_2\text{—}CH_3$$

n-PROPYL CATION
(A PRIMARY CARBONIUM ION)

If the *n*-propyl cation formed, then the final product would have to be *n*-propyl chloride, not isopropyl chloride. The reason the *n*-propyl cation does not become the intermediate leading to the product is that it is less stable than the isopropyl cation. The reason it is less stable is that alkyl groups, R-, are electron-releasing. They are slightly electropositive. That is, unlike electronegative groups that attract electron density toward themselves along some bond, alkyl groups tend to pass off electron density away from themselves. If the carbon holding the positive charge in a carbonium ion could have some electron density pushed its way, it would be more stable. It's a simple rule of physics: to stabilize a positive charge, give it some electron density, some negative charge. The isopropyl cation has two alkyl groups attached to the carbon with the positive charge. The *n*-propyl cation has only one alkyl group on that carbon.

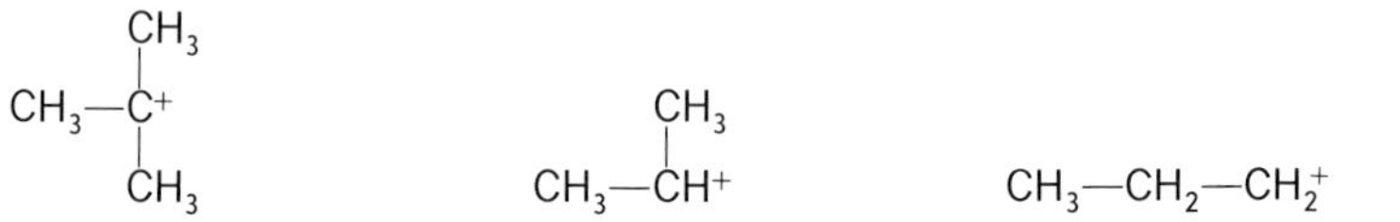

$$CH_3-\overset{CH_3}{\underset{CH_3}{\overset{|}{\underset{|}{C^+}}}} \qquad CH_3-\overset{CH_3}{\overset{|}{CH^+}} \qquad CH_3-CH_2-CH_2^+$$

t-BUTYL CATION (A 3° CARBONIUM ION) — ISOPROPYL CATION (A 2° CARBONIUM ION) — *n*-PROPYL CATION (A 1° CARBONIUM ION)

The *t*-butyl cation, with three alkyl groups on the positively charged carbon, should be more stable than any carbonium ion we have discussed. To generalize, the order of relative stability of carbonium ions is:

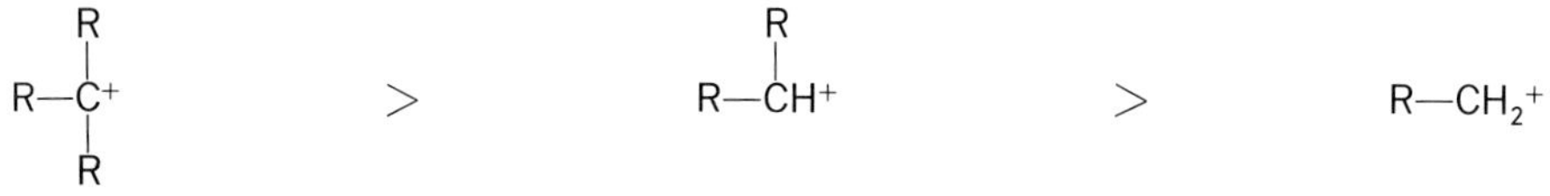

$$R-\overset{R}{\underset{R}{\overset{|}{\underset{|}{C^+}}}} \quad > \quad R-\overset{R}{\overset{|}{CH^+}} \quad > \quad R-CH_2^+$$

TERTIARY . . . MORE STABLE THAN . . . SECONDARY . . . MORE STABLE THAN . . . PRIMARY

The relative stabilities of carbonium ions account for Markovnikov's rule. Propylene reacts with hydrogen chloride or *any* strong proton donor to give a secondary carbonium ion (the isopropyl cation) rather than a primary carbonium ion (the *n*-propyl cation) as the intermediate. The more stable ion forms. It leads naturally to the Markovnikov product.

The acid-catalyzed addition of water proceeds in much the same way as the addition of hydrogen chloride:

Step 1. Catalyst donates proton to alkene; carbonium ion forms.

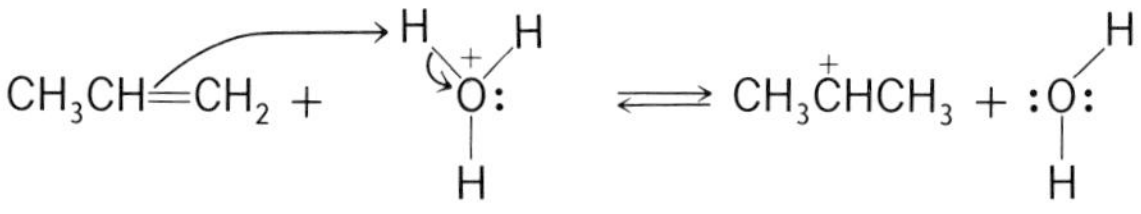

$$CH_3CH{=}CH_2 + H_3\overset{+}{O}: \rightleftharpoons CH_3\overset{+}{C}HCH_3 + :\overset{..}{O}H_2$$

PROPYLENE — CATALYST (HYDRONIUM ION) — ISOPROPYL CATION

Step 2. Electron-dense site on water molecule is attracted to carbonium ion.

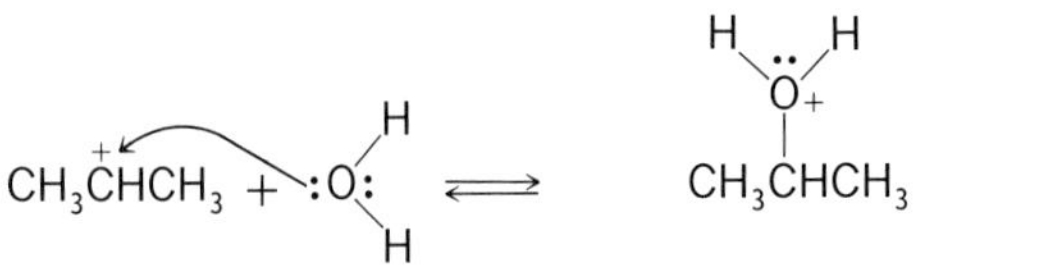

$$CH_3\overset{+}{C}HCH_3 + :\overset{..}{O}H_2 \rightleftharpoons CH_3\underset{\overset{|}{\overset{+}{O}H_2}}{CH}CH_3$$

ISOPROPYL CATION — WATER — ISOPROPYL ALCOHOL IN ITS PROTONATED FORM

Step 3. Proton transfers to water molecule; catalyst is recovered.

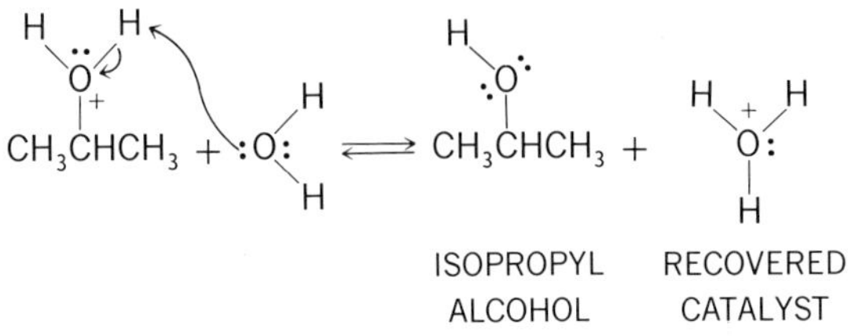

$$CH_3\underset{\underset{+}{O}H_2}{CH}CH_3 + H_2\ddot{O}: \rightleftharpoons CH_3\underset{\underset{\ddot{O}H}{|}}{CH}CH_3 + H_3O^+$$

ISOPROPYL ALCOHOL RECOVERED CATALYST

Exercise 3.7 Write the condensed structures for the two carbonium ions that can conceivably form if a proton becomes attached to each of the following alkenes. Circle the one that is preferred. Where both are reasonable, state that they are. Write the structures of the alkyl chlorides that will form by the addition of hydrogen chloride to each.

(a) $CH_3—CH_2—CH{=}CH_2$ (b) $CH_3—C(CH_3){=}CH_2$ (c) 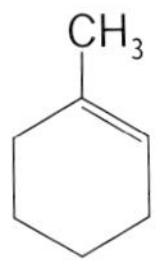 1-methylcyclohexene (ring with CH_3 on the double-bonded carbon)

(d) $CH_3—CH{=}CH—CH_3$ (e) $CH_3—CH{=}CH—CH_2—CH_3$

(f) Addition of water to 2-pentene (part e) gives a mixture of alcohols. What are they? Why is formation of a mixture to be expected here but not from propylene?

3.20 Oxidation of Alkenes

Substances whose molecules have alkene (or alkyne) groups will be attacked by a variety of oxidizing agents. We shall call one kind of attack "vigorous"; it splits an alkene apart at its double bond. Ozone, O_3, does this, and ozone's interaction with substances in smog produces certain very potent eye irritants. Ozone also kills plants partly because it attacks double bonds in molecules of chlorophyll, the plant's green pigment it needs for photosynthesis. Action of *hot* aqueous potassium permanganate, $KMnO_4$, is also a vigorous oxidation.

A gentler oxidation of an alkene does not fracture its molecules. Potassium permanganate acting at room temperature, for example, converts alkenes to dihydroxy compounds called glycols. Let's now study these various oxidations in more detail.

3.21 Ozonolysis—Attack by Ozone

We shall not here study how ozone is generated in smog. It is also produced in nature during electrical storms as lightning bolts strike through the air. In the laboratory it is made by letting oxygen flow through a special ozone generator where an electrical discharge produces an ozone-oxygen mixture with about 5–8% ozone.

One way in which ozone attacks double bonds is as follows:

$$R_2C{=}CR'_2 + O_3 \longrightarrow \underset{\text{AN OZONIDE}}{R_2C(O{-}O)(O)CR'_2} \xrightarrow{H_2O} \underset{\text{KETONES}}{R_2C{=}O + O{=}CR'_2} + H_2O_2$$

The second step, reaction of the ozonide with water, generally takes place spontaneously because enough moisture is generally available.

3.22 Other Vigorous Oxidations

Hot, aqueous potassium permanganate or sodium dichromate also split alkenes apart. The active attacking species are the ions, MnO_4^- or $Cr_2O_7^{2-}$. The equations for the reactions are more complicated than we need for our goals. Let us simply use the symbol (O) to represent any vigorous oxidizing agent, and not try to write balanced equations. We shall be interested only in the products, anyway. The following equations illustrate how various types of alkenes behave.

Terminal alkenes:

$$\begin{matrix}R \\ H\end{matrix}\!\!>\!C{=}CH_2 \xrightarrow{(O)} \underset{\text{A CARBOXYLIC ACID}}{R{-}\overset{O}{\overset{\|}{C}}{-}OH} + CO_2 + H_2O$$

$$\begin{matrix}R \\ R\end{matrix}\!\!>\!C{=}CH_2 \xrightarrow{(O)} \underset{\text{A KETONE}}{\begin{matrix}R \\ R\end{matrix}\!\!>\!C{=}O} + CO_2 + H_2O$$

Internal alkenes:

$$R{-}CH{=}CHR' \xrightarrow{(O)} R\overset{O}{\overset{\|}{C}}OH + R'{-}\overset{O}{\overset{\|}{C}}OH$$

$$R{-}\overset{R}{\overset{|}{C}}{=}CHR' \longrightarrow R{-}\overset{R}{\overset{|}{C}}{=}O + R'{-}\overset{O}{\overset{\|}{C}}{-}OH$$

$$R{-}\overset{R}{\overset{|}{C}}{=}\overset{R'}{\overset{|}{C}}{-}R' \longrightarrow R{-}\overset{R}{\overset{|}{C}}{=}O + O{=}\overset{R'}{\overset{|}{C}}{-}R'$$

The principal organic products depend on the structure of the alkene. You will obtain ketones or carboxylic acids.

Exercise 3.8 Predict the products of the vigorous oxidation of each of the following alkenes.

(a) $CH_3CH{=}CH_2$ (b) $CH_3\overset{CH_3}{\overset{|}{C}}{=}\underset{CH_3}{\underset{|}{C}}HCH_3$ (c) 1,2-dimethylcyclohexene (cyclohexene ring with CH_3 on each carbon of the double bond)

(d) Vigorous oxidation of one mole of an alkene yielded two moles of acetic acid, $CH_3\overset{\overset{O}{\parallel}}{C}OH$, as the sole organic product. What was the alkene?

3.23 Mild Oxidation of Alkenes. Hydroxylation

This reaction looks like the addition of the two —OH groups of hydrogen peroxide, HO—OH, to a double bond. That's why we call it hydroxylation of a double bond. It can be brought about by hydrogen peroxide (in formic acid) or by cold dilute potassium permanganate.

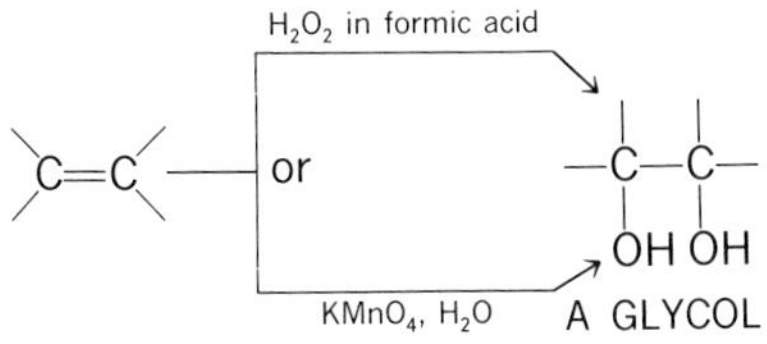

$$>C=C< \xrightarrow[\text{or } KMnO_4,\ H_2O]{H_2O_2 \text{ in formic acid}} -\underset{OH}{\overset{|}{C}}-\underset{OH}{\overset{|}{C}}-$$

A GLYCOL

Specific examples:

$$\underset{\text{1-OCTENE}}{CH_2{=}CH(CH_2)_5CH_3} + \underset{\text{HYDROGEN PEROXIDE}}{H_2O_2} \xrightarrow{\text{Formic acid}} \underset{\text{1,2-OCTANEDIOL (58\%)}}{\underset{OH}{CH_2}-\underset{OH}{CH}(CH_2)_5CH_3}$$

$$\underset{\text{1,2-DIMETHYLCYCLOHEXENE}}{\text{(1,2-dimethylcyclohexene)}} + \underset{\text{POTASSIUM PERMANGANATE}}{KMnO_4} \xrightarrow{H_2O} \underset{\textit{cis}\text{-1,2-DIHYDROXY-1,2-DIMETHYLCYCLOHEXANE (27\%)}}{\text{(cyclohexane with OH, CH}_3\text{ on C1 and C2)}} + MnO_2 + KOH$$

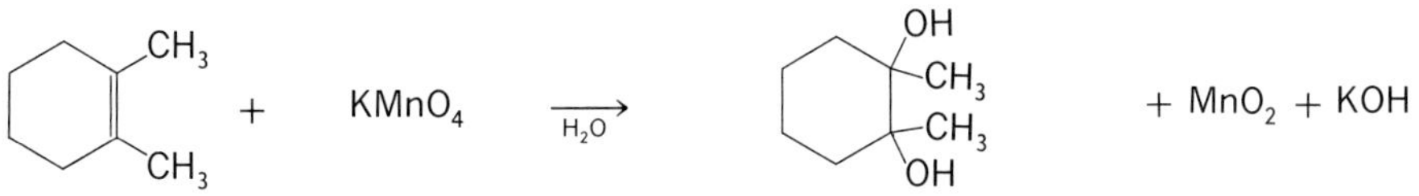

Compounds whose molecules contain two hydroxyl groups on adjacent carbons are called glycols. (See Sections 6.11 and 6.31.)

Exercise 3.9 Write the structures of the alkenes from which the following glycols can be prepared.

(a) $\underset{OH}{CH_2}-\underset{OH}{CH_2}$ (b) $CH_3-\underset{OH}{CH}-\underset{OH}{CH_2}$ (c) cyclopentane-1,2-diol (ring with OH on adjacent carbons)

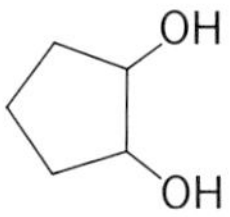

3.24 The Permanganate Test for Easily Oxidized Functional Groups

Potassium permanganate is valuable not so much as a reagent for making glycols from alkenes—the yields are not very high—but rather as a reagent for testing for the presence of an easily oxidized group, such as a carbon-carbon double bond. The inorganic product is manganese dioxide, MnO_2, a dark-brown, water-insoluble solid. The permanganate ion in solution is deep purple in color. Thus if an unknown compound with an easily oxidized group is mixed with permanganate solution, its purple color will give way to a brown sludge. Other families that give this test are the aldehydes, certain types of alcohols, acetylenes, and certain types of aromatic compounds, such as phenols and anilines.

3.25 Polymerization

In general:

$$n \; \rangle C{=}C\langle \xrightarrow{\text{Initiator}} \left(-\overset{|}{\underset{|}{C}}-\overset{|}{\underset{|}{C}}- \right)_n$$

ALKENE — POLYALKENE

Specific examples:

$$nCH_2{=}CH_2 \xrightarrow{\text{Initiator}} (CH_2-CH_2)_n$$

ETHYLENE — POLYETHYLENE (n may be several hundred to several thousand)

$$n\overset{CH_3}{\overset{|}{C}}H{=}CH_2 \xrightarrow{\text{Initiator}} \left(\overset{CH_3}{\overset{|}{C}}H-CH_2\right)_n$$

PROPYLENE — POLYPROPYLENE

Polymerization of alkenes is accomplished through the use of small amounts of chemicals called *initiators*. For years atmospheric oxygen was the initiator for the polymerization of ethylene, but a high temperature, 100°C, and high pressure, 15,000 lb/in.^{2}, were also necessary. Oxygen, a free radical, acts initially to generate a trace of a free radical of the type R—O•. From then on the reaction proceeds by means of a chain mechanism.

Initiation: $R{-}O\cdot + CH_2{=}CH_2 \longrightarrow R{-}O{-}CH_2{-}CH_2\cdot$

Propagation: (a) $R{-}O{-}CH_2{-}CH_2\cdot + CH_2{=}CH_2 \longrightarrow R{-}O{-}CH_2{-}CH_2{-}CH_2{-}CH_2\cdot$

(b) $R{-}O{-}CH_2CH_2CH_2CH_2\cdot + CH_2{=}CH_2 \longrightarrow R{-}O{-}CH_2CH_2CH_2CH_2CH_2CH_2\cdot$

(c) product of (b) + $CH_2{=}CH_2 \longrightarrow$ etc.

Termination: $R{-}O{-}CH_2CH_2(CH_2CH_2)_nCH_2CH_2\cdot + R\cdot \longrightarrow R{-}O{-}CH_2CH_2(CH_2CH_2)_nCH_2CH_2R$

POLYETHYLENE

The starting alkene is called a **monomer,** the product a **polymer** (Greek: *polys,* many; *meros,* part). Polyethylene is largely straight chain throughout, but there are branches (methyl groups) about every nine carbons because a hydrogen shifts, and the odd electron is relocated during the growth of the chain. These branches are like thorns on a stick, inhibiting the close packing of neighboring chains. As a result the low-density polyethylene (0.940 g/cc or less) that is produced is soft and pliable but unsuitable for spinning into fibers.

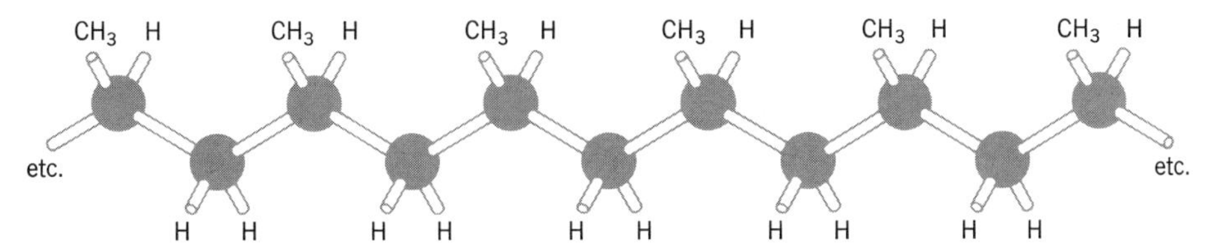

Figure 3.6
Symmetrical (isotactic) polypropylene.

In 1963 the Noble prize in chemistry went to Karl Ziegler of Germany and Giulio Natta of Italy for their work in developing catalysts for the polymerization of ethylene and propylene. Ziegler found that trialkyl derivatives of aluminum, R_3Al, plus titanium chloride catalyzed the formation of a higher-density polyethylene at lower temperatures and pressures. This polyethylene softens at temperatures high enough to permit the sterilization of hospital tubes and bottles made of it.

Natta's work produced the symmetrically polymerized polypropylene illustrated in Figure 3.6. The electron clouds of the side chain methyl groups repel each other, causing the chain to coil into a helix. Polymers whose molecules are very long and very symmetrical give promise of being made into serviceable fibers. Natta's symmetrical polypropylene is used more and more to make carpeting for installation around swimming pools, in open entranceways, basements, bathrooms, and even kitchens. One great advantage of both polyethylene and polypropylene is that they are alkanes and have all the chemical stability of this family. They are also thermoplastic polymers, meaning that when heated, they soften and may be extruded into sheets, molded articles, and tubes. Figures 3.7, 3.8, and 3.9 illustrate some applications of polyethylene and polypropylene. In the early 1970s the United States production of low- and high-density polyethylene was about 6 billion pounds annually.

An interesting relative of polyethylene is Teflon, a polymer in which all the hydrogens in polyethylene have been replaced by fluorines. The monomer is tetrafluoroethylene, $F_2C{=}CF_2$. Teflon is one of the most chemically inert of all

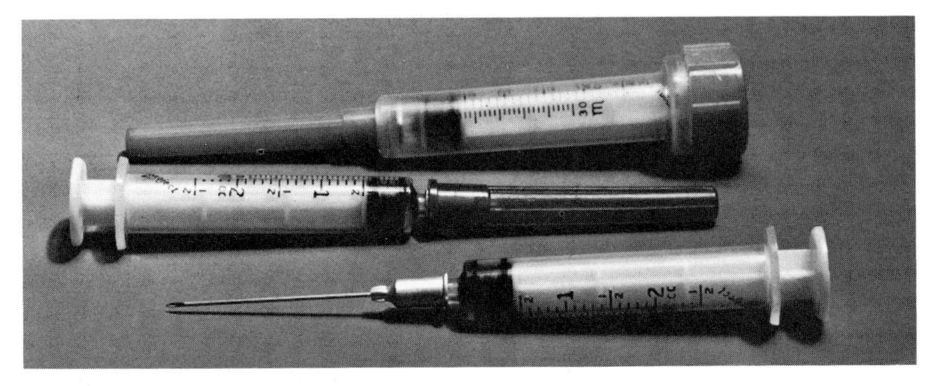

Figure 3.7
Hospitals and clinics cut costs of syringes by using presterilized, disposable plastic barrels and plungers. The energy crunch of 1973, however, cut into these savings, and supplies of such items became very short. (Exxon Chemical Company U.S.A.)

Figure 3.8
Moisture is conserved and weed growth is reduced when rows in vegetable fields are "mulched" with thin polyethylene (0.0015 inch thick). The polyethylene mulch also protects the seedlings from heat and cold. Slits in the plastic let the desired plants grow through. Courtesy of Monsanto Company and Raymond Sheldrake, Jr., New York State College of Agriculture, Cornell University.)

organic substances, for the only chemicals known to attack it are molten sodium and potassium. In addition to the well-known Teflon coating for pots and pans, the polymer is used in electrical insulations and antifriction devices.

Another important monomer related to ethylene is vinyl chloride. In the presence of peroxide catalysts it polymerizes to a brittle resin:

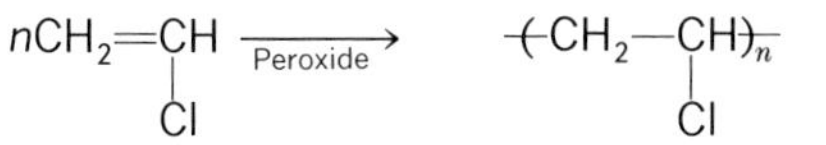

$$nCH_2{=}CH(Cl) \xrightarrow{\text{Peroxide}} {-}(CH_2{-}CH(Cl)){-}_n$$

VINYL CHLORIDE → POLYVINYL CHLORIDE (PVC)

Its brittleness can be overcome and a softer product obtained if during the polymerization we add certain compounds known as *plasticizers;* cresyl phosphate and alkyl phthalates are examples. Polyvinyl chloride is in this way made rubberlike or leatherlike so that it can be widely used in floor tiles, raincoats, tubing, electrical insulation, phonograph records, protective coatings, and bottles.

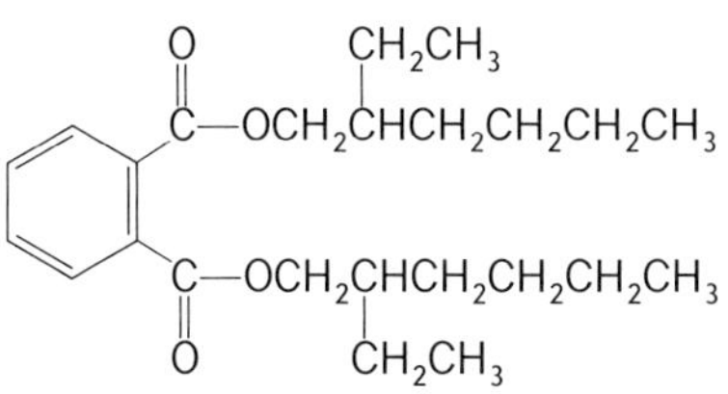

$$C_6H_4[C(=O){-}OCH_2CH(CH_2CH_3)CH_2CH_2CH_2CH_3]_2$$

DI-2-ETHYLHEXYL PHTHALATE (THE MOST COMMONLY USED ALKYL PHTHALATE PLASTICIZER)

$$(CH_3{-}C_6H_4{-}O)_3P{=}O$$

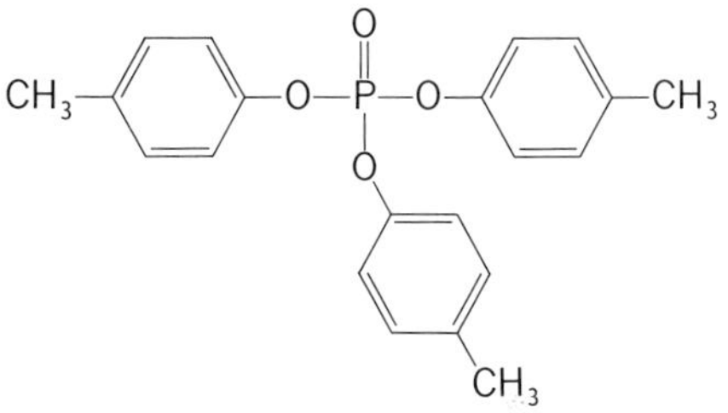

TRICRESYL PHOSPHATE, TCP (ONE ISOMER SHOWN HERE)

As much as 40 percent of the weight of a finished plastic can be the plasticizer whose molecules are scattered throughout the polymer molecules acting essentially as a lubricant. Since plasticizer molecules are not chemically bound, they can and do migrate slowly to the plastic's surface. The stickiness sometimes noted with plastics in hot weather comes at least partly from plasticizers exuded in trace amounts. The phthalate plasticizers are relatively nontoxic in single doses. What might be the long-range effects of human exposure to low concentrations, however, is not yet known.

Polyvinyl chloride is particularly popular for making plastic bottles. Billions of such bottles are made annually in the United States. (Billions are, therefore, also discarded but, as a whole, all of our plastic wastes still account for less than 5 percent of our total solid wastes. Paper is the largest contributor—over 50 percent.)

Another important chlorinated plastic is prepared from vinylidene chloride, the monomer for saran, which is widely used in plastic film for packaging and wrapping.

Orlon, a plastic that can be made into fibers for fabrics, is the polymer of a cyano derivative of ethylene called acrylonitrile.

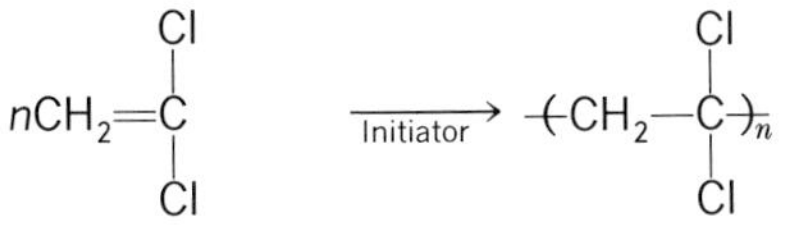

$$nCH_2{=}CCl_2 \xrightarrow{\text{Initiator}} {-}(CH_2{-}CCl_2){-}_n$$

VINYLIDENE CHLORIDE → SARAN

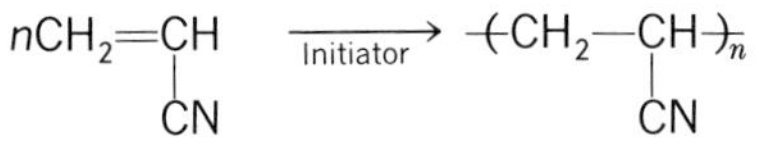

$$nCH_2{=}CH(CN) \xrightarrow{\text{Initiator}} {-}(CH_2{-}CH(CN)){-}_n$$

ACRYLONITRILE → ORLON

Figure 3.9
Pet fish used to go home in glass jars, which could and sometimes did break. Now we bag the fish. While the bag can still break, we don't have all that sharp glass to pick up—another advantage of polyethylene. (Exxon Chemical Company U.S.A.)

Dienes

3.26 Structure and Nomenclature

Alkenes with two carbon-carbon double bonds are called dienes (i.e., two "ene" functions). If the double bonds are widely enough separated, the diene has the properties of an ordinary alkene. Each double bond behaves independently of the other, and the two are said to be *isolated*. Because the allene system shown in Figure 3.10 is rare, we shall consider it no further.

3.27 Naming the Dienes

Table 3.3 shows us that dienes are named by the IUPAC system in essentially the same way as the alkenes are, except that the ending "-diene" is used. Further-

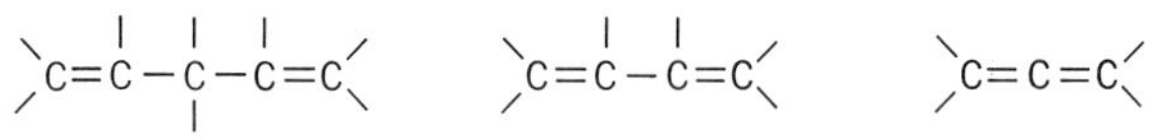

$$\rangle C{=}C{-}\overset{|}{\underset{|}{C}}{-}C{=}C\langle \qquad \rangle C{=}C{-}C{=}C\langle \qquad \rangle C{=}C{=}C\langle$$

(*a*) (*b*) (*c*)

Figure 3.10
(*a*) Nonconjugated or isolated double bonds. (*b*) Conjugated double bonds, meaning that double bonds alternate with single bonds. (*c*) Compounds with adjacent double bonds are allenes.

more, *two* numbers are necessary to specify the locations of the *two* double bonds. The system can obviously be extended to compounds containing three double bonds ("-trienes"), four ("-tetraenes"), etc.

3.28 Conjugated Dienes

Dienes in which the two double bonds are separated by only one single bond, as in 1,3-butadiene, $CH_2{=}CH{-}CH{=}CH_2$, are called *conjugated* dienes.[7] In such compounds the alkene properties are modified. Whenever two or more functional groups are adjacent to each other in organic compounds, the properties of the individual groups are changed in important ways.

3.29 1,4-Addition to Conjugated Dienes

When hydrogen bromide is added to a nonconjugated diene such as 1,4-pentadiene, one molecule of hydrogen bromide will add in a way we would expect from Markovnikov's rule:

$$CH_2{=}CH{-}CH_2{-}CH{=}CH_2 + H{-}Br \longrightarrow CH_3{-}\underset{\underset{Br}{|}}{CH}{-}CH_2{-}CH{=}CH_2$$

[7]In general, a conjugated system is any one consisting of alternating double and single bonds.

Table 3.3
Some Dienes

Name	Structure	Bp (°C)	Mp (°C)	Density (in g/cc at 20°C)
Allene	$CH_2{=}C{=}CH_2$	−34.5	−136	—
1,3-Butadiene	$CH_2{=}CH{-}CH{=}CH_2$	−4.4	−108.9	—
cis-1,3-Pentadiene	$CH_2{=}CH(H)C{=}C(H)CH_3$ (cis)	44.1	−140.8	0.691
trans-1,3-Pentadiene	$CH_2{=}CH(H)C{=}C(CH_3)H$ (trans)	42.0	−87.5	0.676
1,4-Pentadiene	$CH_2{=}CHCH_2CH{=}CH_2$	26.0	−148.3	0.661
2-Methyl-1,3-butadiene ("isoprene")	$CH_2{=}\overset{\overset{CH_3}{|}}{C}{-}CH{=}CH_2$	34.1	−146.0	0.691

However, action of one mole of hydrogen bromide on one mole of the conjugated diene, 1,3-butadiene, gives the following results:

$$CH_2{=}CH{-}CH{=}CH_2 + H{-}Br \xrightarrow{40^\circ C} \underset{\substack{|\\H}}{CH_2}{-}\underset{\substack{|\\Br}}{CH}{-}CH{=}CH_2 \quad (20\%)$$

1,2-ADDITION PRODUCT

+

$$\underset{\substack{|\\H}}{CH_2}{-}CH{=}CH{-}\underset{\substack{|\\Br}}{CH_2} \quad (80\%)$$

1,4-ADDITION PRODUCT

The principal product is not the "expected" 1,2-addition product, the one that would form if the double bonds were isolated. Instead, hydrogen bromide adds at opposite ends of the conjugated system in 1,4-addition. Formation of the 1,2-addition product is not difficult to understand, but to explain the 1,4-product we have to return to our survey of how carbon's atomic orbitals participate in forming bonds. What we have encountered with 1,4-addition is a general phenomenon among organic substances.

> Whenever molecules have two functional groups, especially two unsaturated groups, separated by just one bond, the groups modify each other's chemical properties.

We want to see now why this *must* be so.

3.30 Molecular Orbitals in Conjugated Systems

With isolated double bonds recall that the pi bond arises from side-by-side overlap of p_z orbitals, as was illustrated in Figure 3.4. When two (or more) double bonds are conjugated, when they are separated by only a single bond, this side-by-side overlap can and does take place over all of the carbons involved in the conjugated system. See Figure 3.11. What this means is that *we do not have isolated double bonds in conjugated systems.* We cannot expect such systems to behave "normally," that is, like simple alkenes.

This picture of 1,3-butadiene helps us to see what not to expect in a conjugated system, but it is hard to use it to explain what does happen, at least not at the introductory level of our study of organic chemistry. We cannot go into sophisticated molecular orbital theory. What we can do is what most organic chemists have done since the early 1930s. We adapt to our needs the familiar and the simple for dealing with "unexpected" chemical properties such as 1,4-addition. What is familiar to us now about structure is the use of straight lines to represent shared pairs of electrons. We are familiar with the rules of covalence and the importance of an outer octet. We are familiar with the fact that unlike electrical charges attract each other. We have just seen how the stabilities of 1°, 2°, and 3° carbonium ions vary. We know about isomers and how their molecules differ in the arrangement of their nuclei. What we are going to do is continue to represent 1,3-butadiene as $CH_2{=}CH{-}CH{=}CH_2$. It is somewhat misleading; it seems to indicate two ordinary double bonds. It's not strictly correct, but we shall let our minds adjust as the need arises. We go along with it because it is easier to make corrections from it than to work from a more complicated and sophisticated model. Let us

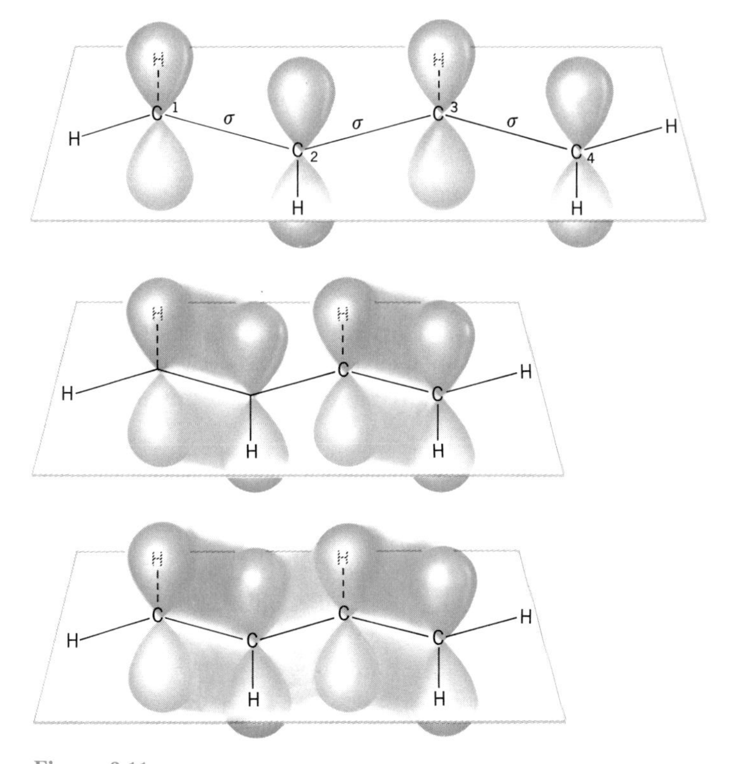

Figure 3.11
Molecular orbitals in 1,3-butadiene, a simple example of a conjugated system. (*a*) All carbons are in the same plane. Their unhybridized p_z orbitals have not yet undergone side-by-side overlap to create pi bonds. (*b*) If 1,3-butadiene were not conjugated, we could not draw overlap between the p_z orbitals on carbons 2 and 3. (*c*) The molecular orbital view of 1,3-butadiene—side-by-side overlap of *p*-orbitals all up and down the carbon chain.

now see how we can use, how organic chemists, in fact, use that structure to account both for 1,2-addition and 1,4-addition, and at the same time account for the facts that we get neither 1,3- nor 2,3-addition to 1,3-butadiene.

3.31 Theory of Resonance —An Introduction

When butadiene is attacked by the strong proton-donor, H—Br, a pair of pi electrons "swings" out to pick up the proton and form a carbonium ion. What we expect (and what nature insists) is that the most stable of all possible carbonium ions forms. Here are the two options, the only two based on our naive structure for 1,3-butadiene:

1. $CH_2{=}CH{-}CH{=}CH_2 + H{-}Br \longrightarrow CH_2{=}CH{-}\overset{+}{C}H{-}CH_3$
   1 — A 2° CARBONIUM ION

or

2. $CH_2{=}CH{-}CH{=}CH_2 + H{-}Br \longrightarrow CH_2{=}CH{-}CH_2{-}CH_2^+$
   A 1° CARBONIUM ION

(Attacks at the other double bond would give these same options.) Option 1 generates, **1**, a 2° carbonium ion and the more stable carbonium ion. It would clearly lead to 1,2-addition, too, in accordance with Markovnikov's rule:

$$\underset{\mathbf{1}}{CH_2{=}CH{-}\overset{+}{C}H{-}CH_3} + Br^- \longrightarrow CH_2{=}CH{-}\underset{\displaystyle Br}{\underset{|}{C}}H{-}CH_3$$

But neither option explains 1,4-addition. For that we introduce the theory of resonance.

The structure we drew for the 2° carbonium ion, **1**, has a pair of pi electrons adjacent to a positive charge. We could well imagine—since unlike charges attract—that these electrons would drift toward the charge. We might represent this as follows:

$$\underset{\mathbf{1}}{CH_2{=}CH{-}\overset{+}{C}H{-}CH_3} \longleftrightarrow \underset{\mathbf{2}}{{}^{+}CH_2{-}CH{=}CH{-}CH_3}$$

In going from **1** to **2** we relocated a pair of electrons and shifted the positive charge from one place to another. Structures **1** and **2** differ only in the location of electrons. (They differ in the location of the + charge, also, but a positive charge is only the absence of an electron.) There is no difference between **1** and **2** in the location of any atomic nuclei; hence, **1** and **2** are not isomers. They are *not* analogous to 1-butene and 2-butene, which are isomers and in which some atomic nuclei are in different locations.

It is a cornerstone of the theory of resonance that whenever two or more structures can be written that differ only in the location of electrons, neither structure will actually represent the system adequately. The real system, it turns out, can be imagined as a hybrid of the structures that have been drawn. (We assume that these structures are otherwise within the rules of valence. Bizarre structures violating these rules are rejected.) The carbonium ion we formed is not **1**. It is not **2**, either; but it does behave as if it were a hybrid of **1** and **2**. Just as a hybrid resembles both parents—compare the mule as a hybrid of a horse and a donkey—so the real carbonium ion resembles both **1** and **2**. We can draw the real carbonium ion as follows, where we use dotted lines to represent a "partial bond."

$$\overset{\delta+}{C}H_2{\overset{\cdots}{-}}CH{\overset{\cdots}{-}}\overset{\delta+}{C}H{-}CH_3$$

A HYBRID CARBONIUM ION
(A HYBRID OF **1** AND **2**)

$$\Big[\underset{\mathbf{1}}{\underset{1\quad\ 2\quad\ 3\quad\ 4}{CH_2{=}CH{-}\overset{+}{C}H{-}CH_3}} \longleftrightarrow \underset{\mathbf{2}}{\underset{1\quad\ 2\quad\ 3\quad\ 4}{\overset{+}{C}H_2{-}CH{=}CH{-}CH_3}}\Big]$$

Structure **1**, in the language of resonance theory, *contributes* to the hybrid—it is one of the *resonance contributors*. Structure **2** is the other contributor. To illustrate what we mean by "contributing to the hybrid" notice that in **2** there is a full + charge on C—1; in **1**, there is zero charge on C—1. Therefore, we predict that in the *real* ion, the hybrid, there is a partial charge at C—1, that is, something between zero and one. In neither contributor is there a charge at C—2; therefore we predict that the hybrid has no charge there either. Given a full + charge on C—3 in **1** and no charge at C—3 in **2**, we predict that the real ion, the hybrid, has a partial + charge at C—3. There is a simple CH_3-group in both. The hybrid must have it, too. The bond from C—1 to C—2 is single in **2** and double in **1**; therefore we predict that in the real ion there is a partial double bond from C—1 to C—2 in the hybrid. For similar reasons we have a partial double bond from C—2 to C—3.

In molecular orbital terms, we could visualize the hybrid as arising from **1** as seen in Figure 3.12. However, we shall find it usually easiest to work with the set of contributors, enclosed in brackets and separated by double-headed arrows. We use the whole set as our representation of the one true structure. It will be easiest this way because we know how to write structures within the rules of valence, structures that show full lines for bonds and full charges, if they are present. We emphasize that contributors are strictly hypothetical; they have no real existence at all. Each is something of a "first approximation." We draw them as a way of working toward an idea of what the true structure is more nearly like. Once they are drawn we let our minds "blend" them into a picture of the hybrid. We do not have to, but sometimes we write down that blend using partial bonds and partial charges as needed. We use double-headed arrows to separate contributors, because we may not even suggest, as we would by using regular arrows, that the contributors are somehow changing back and forth. They are not, not any more than we imagine that a mule changes back and forth between being a horse and a donkey. A mule is a mule, but even an average artist could come close to drawing one if he had seen only a horse and a donkey.

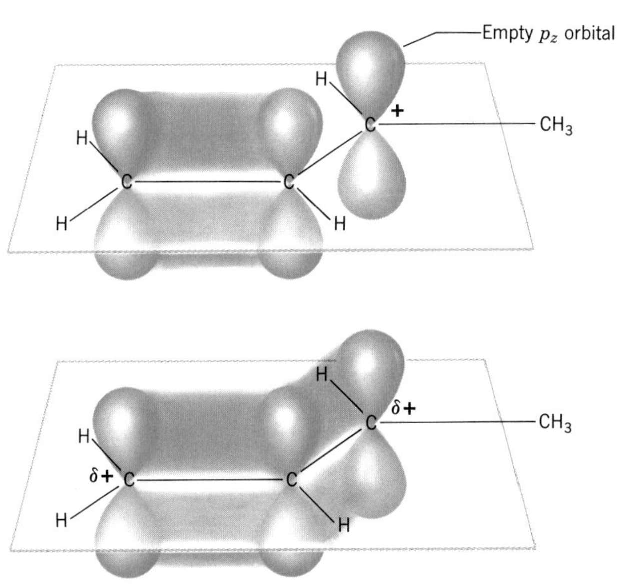

Figure 3.12
The carbonium ion intermediate in the 1,4-addition of HBr to 1,3-butadiene. (*a*) Molecular orbital view of contributing structure **1**, $CH_2{=}CH{-}\overset{+}{C}H{-}CH_3$. (*b*) Molecular orbital view of the actual carbonium ion; the hybrid, in resonance terms, of **1** and **2**. There are two electrons in the long, extended molecular orbital. More importantly, the net positive charge on the carbonium ion is not restricted to one site. It is delocalized over three nuclei.

With the resonance view, **3**, of the carbonium ion it is now easy to see how collisions of some of these ions with bromide ions lead to 1,2-addition, collisions of others lead to 1,4-addition, and why there is no 1,3- or 2,3-addition.

$$\left[\overset{+}{C}H_2{-}CH{=}CH{-}CH_3 \longleftrightarrow CH_2{=}CH{-}\overset{+}{C}H{-}CH_3\right] \begin{cases} \xrightarrow[Br^-]{\text{1,4-addition}} CH_2(Br){-}CH{=}CH{-}CH_3 \\ \xrightarrow[\text{1,2-addition}]{Br^-} CH_2{=}CH{-}CH(Br){-}CH_3 \end{cases}$$

3

3.32 Delocalization Means Stabilization

If routes are at all open, nature will undergo its changes by lowest energy paths. In energy terms, spontaneous events (i.e., natural events) "run down hill." It is this fact that is really behind both 1,4- and 1,2-addition to 1,3-butadiene. The hybrid carbonium ion, as an intermediate in the reaction, is of lower energy than either contributor could be. In either contributor, **1** or **2**, the pi electrons are localized. They are confined to the rather limited region between two carbons. Moreover, the positive charge is also localized in either **1** or **2.** In the real ion, which we visualize as the hybrid, both the electrons and the positive charge are delocalized. The electrons have more room in which to be (Fig. 3.12*c*). Since electrons tend to repel each other anyway, you go to a lower energy status by letting them move into a larger space. At the same time, the positive charge is spread out in the hybrid, too. The more we can "dilute" electrical charge, the lower energy will the system have and the more stable it will be. We shall see how significant delocalization can be when, in the next chapter, we study extensive conjugated systems found in aromatic compounds.

3.33 Polymerization of Dienes

An **elastomer** is any polymeric substance with the properties of rubber. Conjugated dienes can serve as monomers for a variety of elastomers, including natural rubber. Natural rubber is a polymer of isoprene (Table 3.3), and we may imagine its formation as follows:

$$CH_2{=}C(CH_3){-}CH{=}CH_2 + CH_2{=}C(CH_3){-}CH{=}CH_2 + CH_2{=}C(CH_3){-}CH{=}CH_2 + CH_2{=}C(CH_3){-}CH{=}CH_2 + \text{etc.}$$

ISOPRENE

$\downarrow$ Ziegler's catalyst

$$\text{-}CH_2{-}C(CH_3){=}CH{-}CH_2{-}CH_2{-}C(CH_3){=}CH{-}CH_2{-}CH_2{-}C(CH_3){=}CH{-}CH_2{-}CH_2{-}C(CH_3){=}CH{-}CH_2\text{-}$$

NATURAL RUBBER

(Note the isoprene units between the dashed lines. Note also that every double bond is *cis*.)

3.34 Copolymerization

Industrial chemists have extensively studied the mixed polymerization or *copolymerization* of dienes with monoenes (alkenes) to find polymers that have fewer double bonds but are still flexible, elastic, and resistant to solvents and abrasion. One of the most important of these polymers is the copolymer of butadiene and styrene, Buna S rubber. In this substance the polymer molecules are not all identical, but the following illustrates some of their features, without regard to the geometry or the exact proportions of monomers:

$CH_2{=}CHCH{=}CH_2$ (1,3-BUTADIENE) $C_6H_5CH{=}CH_2$ (STYRENE)

$$\underbrace{-CH_2CH{=}CHCH_2-}_{\text{FROM BUTADIENE}}\underbrace{CHCH_2-}_{\substack{|\\C_6H_5\\ \text{FROM STYRENE}}}\underbrace{CH_2CH{=}CHCH_2-}_{\text{FROM BUTADIENE}}\underbrace{CHCH_2-}_{\substack{|\\C_6H_5\\ \text{FROM STYRENE}}}\quad \text{Buna S rubber}$$

This product is also called GRS (Government Rubber Styrene)[8] and "cold rubber."

Another synthetic rubber, *butyl rubber,* is made by polymerizing isobutylene in the presence of a small amount of isoprene.

Alkynes

Molecules of alkynes have carbon-carbon triple bonds. While this functional group does not occur widely in nature, the simplest alkyne, acetylene, is a very important industrial chemical.

3.35 Acetylene

There are principally two ways to make acetylene, H—C≡C—H. One is from limestone, coke, and water.

Preparation of lime:

$$\underset{\text{LIMESTONE}}{CaCO_3} \xrightarrow{\text{Heat}} \underset{\substack{\text{LIME}\\\text{(CALCIUM OXIDE)}}}{CaO} + CO_2$$

Production of calcium carbide:

$$CaO + \underset{\text{COKE}}{3C} \xrightarrow{2500\text{–}3000^\circ} \underset{\substack{\text{CALCIUM}\\\text{CARBIDE}}}{CaC_2} + CO$$

Hydrolysis of calcium carbide:

$$CaC_2 + 2H_2O \longrightarrow H{-}C{\equiv}C{-}H + Ca(OH)_2$$

There is evidence that calcium carbide consists of calcium ions and C_2^{2-} ions

[8] After a government project during World War II.

already having the triple bond: $Ca^{2+}(^{-}{:}C{\equiv}C{:}^{-})$. This means that C_2^{2-} is really an anion of acetylene itself. It acts as a proton-acceptor (a base) toward water molecules:

$$HO{-}H + {}^{-}{:}C{\equiv}C{:}^{-} + H{-}OH \longrightarrow H{-}C{\equiv}C{-}H + 2OH^{-} \xrightarrow{Ca^{2+}} Ca(OH)_2$$

The second industrial synthesis of acetylene is by a very high temperature cracking of methane or other alkanes.

3.36 Nomenclature

The name "acetylene" has been adopted by the IUPAC system. For higher members of the series we use essentially the same kinds of rules we used for alkenes, except that we use "-yne" for the word ending. These examples illustrate application of these rules.

$CH_3{-}C{\equiv}C{-}H$ PROPYNE

$CH_3{-}CH(CH_3){-}C{\equiv}C{-}H$ 3-METHYL-1-BUTYNE

$CH_3{-}C{\equiv}C{-}CH_3$ 2-BUTYNE

3.37 Structure. *sp*-Hybridization

We understand the formation and structure of the carbon-carbon triple bond as we did the double bond. At the triple bond carbon is bound to just two groups, and we invoke still another mode of hybridization: *sp* hybridization. Figure 3.13 illustrates how the hybrid orbitals are formed. Figure 3.14 shows how they participate in forming the bonds in acetylene. The triple bond forces the atoms immediately around it into a linear alignment; the bond angle is 180°. Therefore, there can be no geometric isomerism about this functional group.

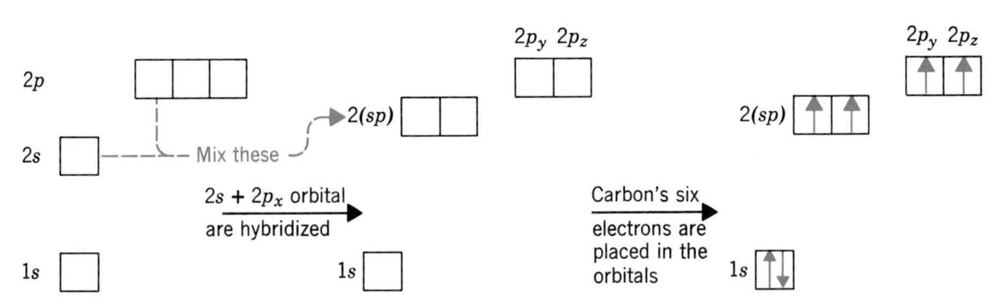

Figure 3.13
A view of how *sp* hybrid orbitals arise. (*a*) These boxes represent the orbitals available to electrons at the first and second energy levels. They are, for the moment, empty. (*b*) The result of "mixing" or hybridizing the 2*s* orbital with just one of the 2*p* orbitals. (*c*) The valence state of a carbon at a triple bond: $1s^2\ 2(sp)\ 2(sp)\ 2p_y\ 2p_z$.

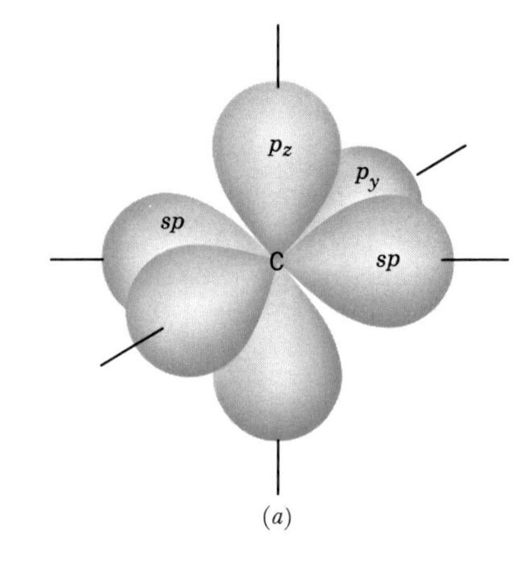

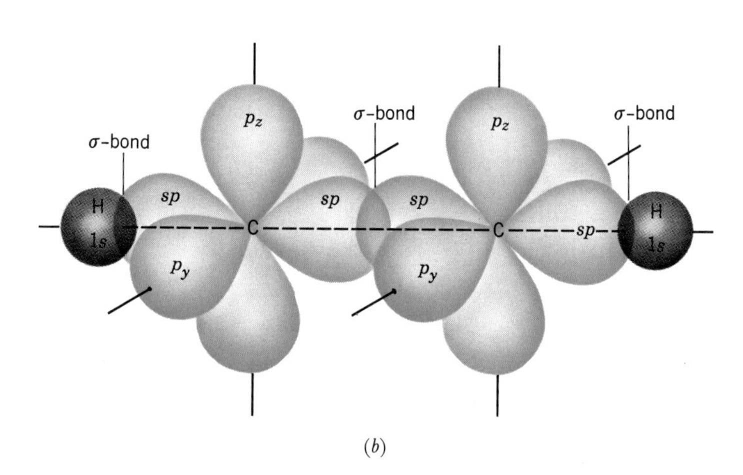

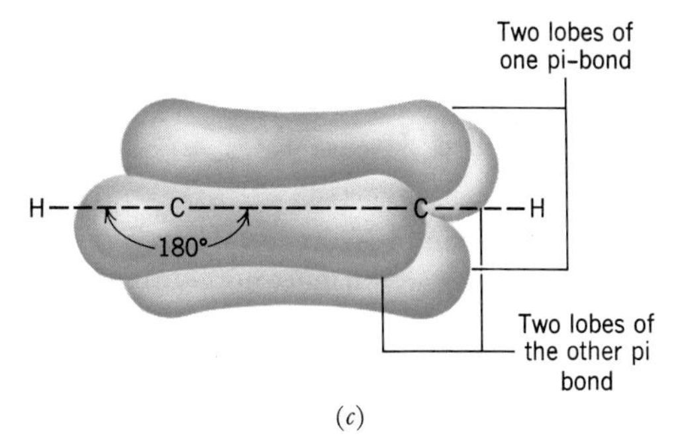

Figure 3.14
The triple bond in acetylene. (*a*) The orbitals of carbon in *sp* hybridization. (*b*) Two overlappings of *s* to *sp* and one of *sp* to *sp* give the sigma bond framework. (*c*) Side-by-side overlap of p_y to p_y and of p_z to p_z give two pi bonds.

Chemical Properties

3.38 Addition Reactions

The following reactions of acetylene illustrate that each pi bond of the triple bond behaves as we might expect of pi bonds. What products are isolated depends on the relative proportions of reactants used. Those resulting from the addition of just one mole of reagent can be isolated without trouble.

Hydrogenation

$$\underset{\text{ACETYLENE}}{H{-}C{\equiv}C{-}H} + H_2 \xrightarrow[\text{Heat, pressure}]{\text{Catalyst}} \underset{\text{ETHYLENE}}{H_2C{=}CH_2} \xrightarrow[\text{(same conditions)}]{H_2} \underset{\text{ETHANE}}{CH_3CH_3}$$

Addition of
Halogen (chlorine or bromine)

$$\underset{\text{PROPYNE}}{CH_3{-}C{\equiv}C{-}H} + Cl_2 \longrightarrow \underset{\text{1,2-DICHLOROPROPENE}}{CH_3{-}\underset{\displaystyle Cl}{\underset{|}{C}}{=}CH{-}Cl} \xrightarrow{Cl_2} \underset{\text{1,1,2,2-TETRACHLOROPROPANE}}{CH_3{-}\overset{\displaystyle Cl}{\overset{|}{\underset{\displaystyle Cl}{\underset{|}{C}}}}{-}CHCl_2}$$

Addition of Hydrogen
Halide (H—Cl, H—Br or H—I)
Notice that Markovnikov's rule operates.

$$\underset{\text{PROPYNE}}{CH_3{-}C{\equiv}C{-}H} + H{-}Cl \longrightarrow \underset{\text{2-CHLOROPROPENE}}{CH_3{-}\underset{\displaystyle Cl}{\underset{|}{C}}{=}CH_2} \xrightarrow{H{-}Cl} \underset{\text{2,2-DICHLOROPROPANE}}{CH_3{-}\overset{\displaystyle Cl}{\overset{|}{\underset{\displaystyle Cl}{\underset{|}{C}}}}{-}CH_3}$$

$$H{-}C{\equiv}C{-}H + H{-}Cl \longrightarrow \underset{\text{VINYL CHLORIDE}}{CH_2{=}CH{-}Cl} \xrightarrow{\text{If polymerized}} \text{Polyvinyl chloride (PVC)}$$

Addition of Water
Both sulfuric acid and mercury (II) sulfate are required.

$$\underset{\text{ACETYLENE}}{H{-}C{\equiv}C{-}H} + H{-}OH \xrightarrow[HgSO_4]{H_2SO_4} \underset{\text{VINYL ALCOHOL (UNSTABLE)}}{[CH_2{=}CH{-}OH]} \xrightarrow{\text{rearranges}} \underset{\text{ACETALDEHYDE}}{CH_3{-}\overset{\displaystyle O}{\overset{\|}{C}}{-}H}$$

Addition of Hydrogen Cyanide
This reaction had no counterpart in our study of the chemistry of alkenes.

$$H{-}C{\equiv}C{-}H + H{-}C{\equiv}N \longrightarrow \underset{\text{ACRYLONITRILE}}{CH_2{=}CH{-}C{\equiv}N} \xrightarrow{\text{If polymerized}} \text{Orlon}$$

Addition of Acetic Acid

$$H{-}C{\equiv}C{-}H + \underset{\text{ACETIC ACID}}{H{-}O\overset{\displaystyle O}{\overset{\|}{C}}CH_3} \longrightarrow \underset{\text{VINYL ACETATE}}{CH_2{=}CH{-}O\overset{\displaystyle O}{\overset{\|}{C}}CH_3} \xrightarrow{\text{If polymerized}} \text{Polyvinyl acetate plastics}$$

The acetic acid can be made by oxidation of acetaldehyde. Since acetaldehyde may be made from acetylene (addition of water, above), acetylene can serve as the sole organic starting material for polyvinyl acetate plastics.

Addition of Acetylene to Acetylene

$$HC{\equiv}CH + H{-}C{\equiv}CH \xrightarrow[H-Cl,\ H_2O]{Cu_2Cl_2,\ NH_4Cl} \underset{\text{VINYL ACETYLENE}}{CH_2{=}CH{-}C{\equiv}CH}$$

The significance of this reaction is that vinylacetylene is an intermediate in the production of chloroprene, a diene that can be polymerized to the very useful polymer, Neoprene.

$$CH_2{=}CH{-}C{\equiv}CH + H{-}Cl \longrightarrow \underset{\text{CHLOROPRENE}}{CH_2{=}CH{-}\overset{\overset{Cl}{|}}{C}{=}CH_2} \xrightarrow{\text{if polymerized}} \underset{\text{NEOPRENE}}{\left(CH_2{-}CH{=}\overset{\overset{Cl}{|}}{C}{-}CH_2\right)_n}$$

These reactions of acetylene serve only to illustrate how commercially important acetylene is. By sequences of reactions not shown here, acetylene can also be changed into 1,3-butadiene, acrylate polymers, and isoprene, all important chemicals in their own right.

3.39 The Acidity of Terminal Acetylenes

A terminal acetylene is one having the triple bond at the end of a chain and, therefore, has a hydrogen on the end carbon: $R{-}C{\equiv}C{-}H$. Provided we use a strong proton-acceptor (e.g., the amide ion, NH_2^-, as in sodium amide, $NaNH_2$) we can remove that terminal proton.

$$R{-}C{\equiv}C{-}H + {:}NH_2^- \longrightarrow \underset{\text{AN ACETYLIDE ION}}{R{-}C{\equiv}C{:}^-} + NH_3$$

We cannot do this to either alkanes or alkenes. With terminal acetylenes we have one of the few examples of hydrocarbons being proton-donors—but then only to very strong bases. Acetylide ions are reactive toward alkyl halides in the following way, illustrated with ethyl bromide:

$$R{-}C{\equiv}C{:}^- + \underset{\underset{CH_3}{|}}{CH_2}{-}Br \longrightarrow R{-}C{\equiv}C{-}CH_2CH_3 + Br^-$$

This type of reaction is one of the principal ways of making higher alkynes from smaller ones.

Brief Summary: Chemical Properties

Alkenes

The carbon-carbon double bond can be oxidized, reduced, and hydrated. It adds halogens (Cl_2 and Br_2) and hydrogen halides. Its characteristic reactions, thus, are addition reactions.

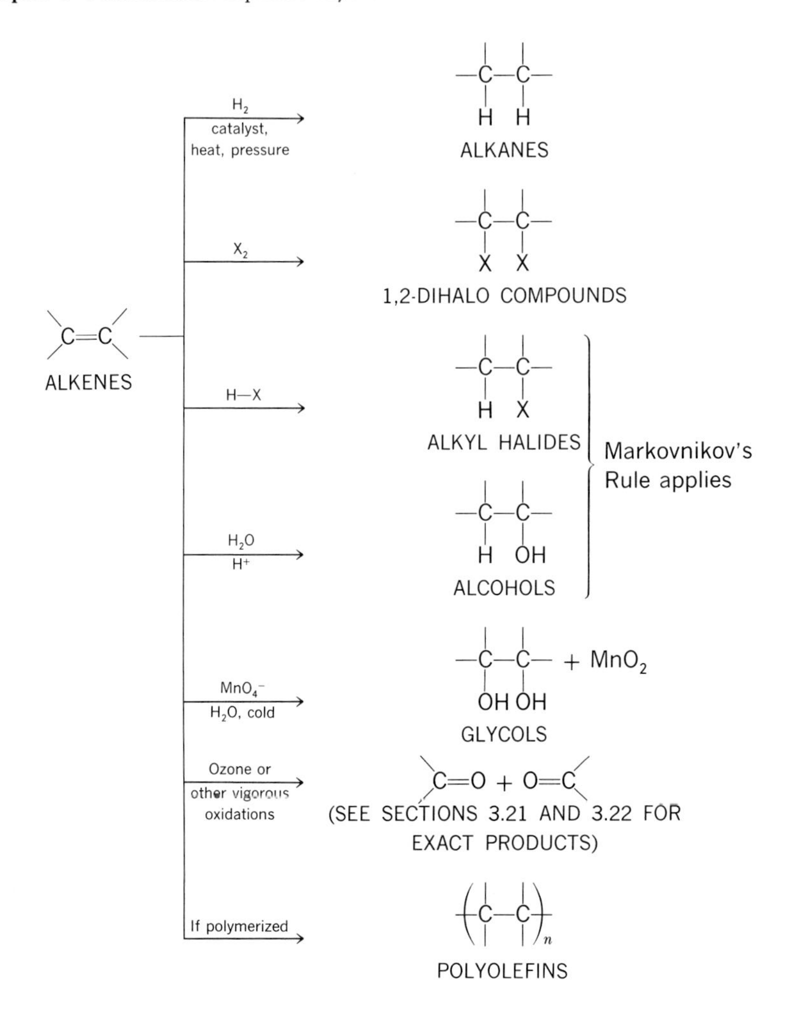

Dienes

Dienes can be polymerized to produce various elastomers.

Dienes also exhibit 1,4-addition, which illustrates an important fact: when two functional groups are adjacent, they modify each other's properties. We introduced the theory of resonance to account both for 1,2- and 1,4-addition to 1,3-butadiene. Central to this theory is the postulate that whenever two or more nonbizarre structures can be written for a substance, structures that differ only in the locations of electrons, then a condition of resonance exists. None of the structures written will adequately explain the properties of the system, and the true structure is really a hybrid of those drawn. Nonbizarre structures are defined as those for which the following statements are true:

1. The rules of valence are obeyed. (For the first row elements we may never have more than eight outer-level electrons, but depending on the exact number of outer electrons and the atom involved we may have a positive or a negative charge.
2. All atomic nuclei are in identical locations.

The postulates and rules of resonance theory enable us to use conventional structures to visualize just how electrons and charges are most probably delocalized in an unsaturated system. When delocalization is possible (i.e., when a condition of resonance is possible) the system will have less energy and be more stable than we could expect if it were any one of the resonance contributors.[9]

Alkynes

Alkynes undergo most of the addition reactions of alkenes and some uniquely their own. Acetylene is at the chemical intersection between the mineral realm of limestone, coke, and water and the organic realms of a large variety of substances.

Simple Chemical Tests

The following were discussed:

1. Bromine in carbon tetrachloride test.
2. Sulfuric acid test.
3. Permanganate test.

Brief Summary: Important Concepts and Theories

1. sp^2 and sp hybridization.
2. Addition reactions.
   (a) Markovnikov's rule.
   (b) Carbonium ion intermediates.
3. Polymerization reactions.
4. Conjugated and nonconjugated systems.
5. Theory of resonance.
   (a) When to expect a condition of resonance (and, therefore, special chemical properties).
   (b) Contributing structures.
   (c) Hybrid structures.
   (d) Delocalization—what it is and what it connotes in terms of stabilization of the system.

Problems and Exercises

1. Write an illustrated discussion of each of the following terms.
   (a) sp^2 hybrid orbital
   (b) pi bond
   (c) cis isomer
   (d) trans isomer
   (e) hydrogenation
   (f) unsymmetrical double bond
   (g) Markovnikov's rule
   (h) glycol
   (i) plasticizer
   (j) *sp* hybridization
   (k) initiator
   (l) polymerization

[9]Concerning resonance theory, the goal for this course is not so much that you develop the skills for writing correct contributing structures. Instead, these will be written for you. Your goal should be to be able to follow arguments based on the use of these structures.

(m) electrophile (n) carbonium ion
(o) ozonolysis (p) conjugated system

2. List the families of organic compounds for which we have learned a method of synthesis in this chapter and write equations illustrating these reactions.
3. Write IUPAC names for each of the following:

(a) $CH_3(CH_2)_7CH{=}CH_2$

(b) 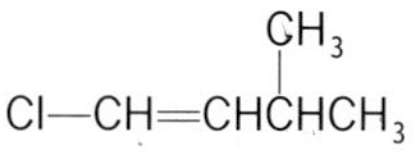 $Cl{-}CH{=}CH\overset{\overset{CH_3}{|}}{C}HCH_3$

(c) $CH_3CH_2CH_2\overset{\overset{CH_2}{\|}}{C}CH_2\overset{\overset{CH_3}{|}}{C}HCH_2CH_3$

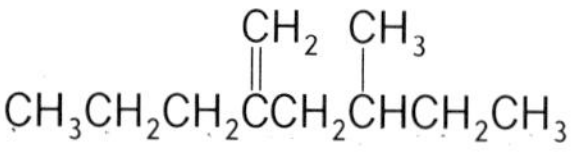

(d) 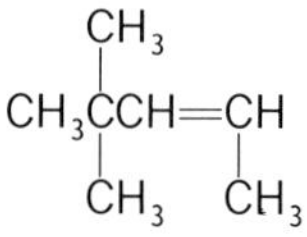 $CH_3\underset{\underset{CH_3}{|}}{\overset{\overset{CH_3}{|}}{C}}CH{=}\underset{\underset{CH_3}{|}}{C}H$

(e) 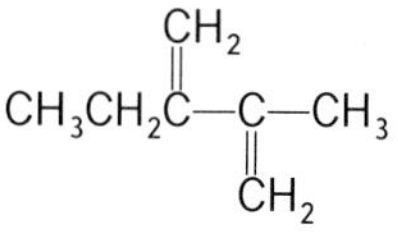 $CH_3CH_2\overset{\overset{CH_2}{\|}}{C}{-}\underset{\underset{CH_2}{\|}}{C}{-}CH_3$

(f) 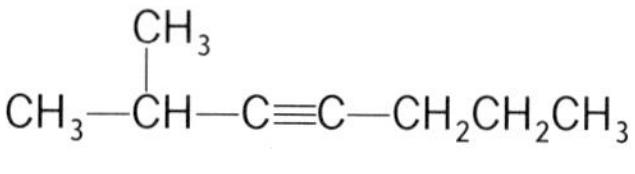 $CH_3{-}\overset{\overset{CH_3}{|}}{C}H{-}C{\equiv}C{-}CH_2CH_2CH_3$

4. Write equations for the reaction of isobutylene with each of the following reagents: (a) cold, concentrated H_2SO_4, (b) H_2 (Ni, heat, pressure), (c) H_2O (H^+ catalysis), (d) H—Cl, (e) H—Br.
5. Write equations for the reaction of 1-methylcyclopentene with each of the reagents of exercise 4.
6. When 1-butene reacts with hydrogen chloride, the product is 2-chlorobutane. The isomer, 1-chlorobutane, does not form. Explain.
7. Ethane is insoluble in concentrated sulfuric acid. Ethylene dissolves readily. Write an equation to show how ethylene is converted into a substance polar enough to dissolve in the highly polar concentrated sulfuric acid.
8. Write out enough of the structures of each of the following substances to show what their molecules are like. Do not use parentheses to condense these structures.
   (a) polyethylene (b) polypropylene (c) Teflon
   (d) PVC (e) polybutadiene (f) Neoprene
9. Describe three simple chemical tests that could be used to distinguish between 2-butene and butane. Describe what you would do and see.
10. Complete each equation by writing the structures of the organic products. If no reaction occurs, write "no reaction."

(a) $CH_3CH{=}CH_2 + H_2O \xrightarrow{H^+}$

(b) $CH_2{=}CHCH_2CH_3 + HCl \longrightarrow$

(c) 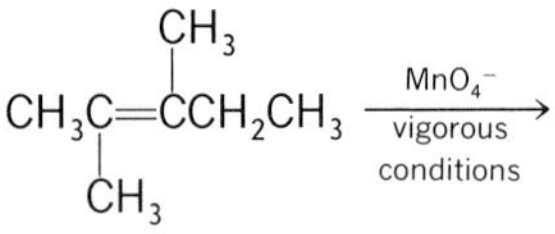 $CH_3\underset{\underset{CH_3}{|}}{C}{=}\overset{\overset{CH_3}{|}}{C}CH_2CH_3 \xrightarrow[\text{vigorous conditions}]{MnO_4^-}$

(d) $(CH_3)_2C{=}C(CH_3)_2 + O_3 \xrightarrow[H_2O^\circ]{}$

(e) $CH_3{-}C{\equiv}C{-}CH_3 + 2HCl \longrightarrow$

(f) $CH_3CH_2CH_3 + H_2O \longrightarrow$

(g) 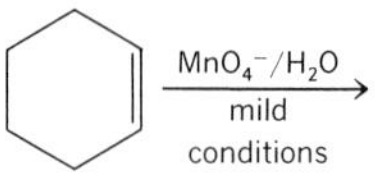 cyclohexene $\xrightarrow[\text{mild conditions}]{MnO_4^-/H_2O}$

chapter four

Aromatic Hydrocarbons

4.1 Aromatic and Aliphatic Substances

Perhaps he was searching for medicinal agents. Perhaps he was just curious. Whatever the reasons, Johann Rudolph Glauber (1604–1668), one of the forerunners of organic chemistry, was interested in what happens to coal when it is strongly heated yet not allowed to burn. He obtained what he called a "pleasant-smelling oil and valuable healing balsams." Working with a variety of plants, Glauber isolated many other fragrant, aromatic oils that people came to prize.

Glauber's "pleasant-smelling oil" might have been benzene or a mixture containing it. His "valuable healing balsams" were undoubtedly members of the class of phenols, known germkillers. Long after, in the nineteenth century, these and many other fragrant oils obtained from plants were found to share a peculiar structural system, a ring of six carbons, and much too few hydrogens to make it a saturated system. Today, regardless of the particular aroma (or lack of it), any compound whose molecules have this system are called *aromatic compounds*. The compounds we have been studying thus far, the alkanes, alkenes, alkynes, and their cyclic relatives and their oxygen or nitrogen derivatives are in the other great class of organic compounds: the *aliphatic compounds*.

4.2 Benzene

The simplest substance in the aromatic family is benzene, C_6H_6. By studying benzene's chemical and physical properties we shall gain a better understanding of the characteristic properties of all aromatic compounds. Equally importantly we shall encounter a major example of the kind of failure of simple structural theory that drove scientists to develop the theory of resonance.

4.3 Structure and Reactions

The following facts about benzene led to what we now call the Kekulé structure of benzene, after August Kekulé (1829–1896). (He was the inventor of ball-and-stick models for molecular structure.)

1. The molecular formula of benzene is C_6H_6. At very high pressure, benzene reacts with hydrogen to form cyclohexane. With six carbons in a six-membered ring holding only six hydrogens, the molecule should be highly unsaturated. Addition reactions such as those described for the alkenes should be the rule. Instead *addition reactions are the rare exception.*
2. *Benzene undergoes substitution reactions* that are catalyzed by acids. When monosubstituted benzenes are made (of the general type C_6H_5—G; G = —Br, —Cl, —NO_2, etc.), no isomers form. This means that all six hydrogens in benzene must be equivalent. The principal aromatic substitution reactions are the following.

Nitration

$$C_6H_6 + \underset{\text{NITRIC ACID}}{HO\text{—}NO_2} \xrightarrow[\text{50–55°C}]{H_2SO_4\ \text{(concd)}} \underset{\substack{\text{NITROBENZENE (85\%)} \\ \text{(bp 211°C)}}}{C_6H_5\text{—}NO_2} + H_2O$$

(Under these conditions alkenes undergo extensive oxidation and decomposition.)

Sulfonation

$$C_6H_6 + SO_3 \xrightarrow[\text{Room temperature}]{H_2SO_4\ \text{(concd)}} \underset{\substack{\text{BENZENESULFONIC ACID (56\%)} \\ \text{(mp 50–51°C)}}}{C_6H_5\text{—}\overset{\overset{O}{\uparrow}}{\underset{\underset{O}{\downarrow}}{S}}\text{—O—H}}$$

(Sulfuric acid *adds* to double bonds in alkenes to form alkyl hydrogen sulfates. See Section 3.17.)

Halogenation

$$C_6H_6 + Cl_2 \xrightarrow[FeCl_3]{\text{Fe or}} \underset{\substack{\text{CHLOROBENZENE (90\%)} \\ \text{(bp 132°C)}}}{C_6H_5\text{—Cl}} + H\text{—Cl}$$

$$C_6H_6 + Br_2 \xrightarrow{\text{Fe}} \underset{\substack{\text{BROMOBENZENE (59\%)} \\ \text{(bp 155°C)}}}{C_6H_5\text{—Br}} + H\text{—Br}$$

Chlorine and bromine add to an alkene with no catalyst required (Section 3.14).

Alkylation, Friedel-Crafts reaction:

$$C_6H_6 + R\text{—Cl} \xrightarrow{AlCl_3} \underset{\text{ALKYLBENZENES}}{C_6H_5\text{—R}} + H\text{—Cl}$$

Acylation, Friedel-Crafts acylation:

$$C_6H_6 + R{-}\overset{\overset{\displaystyle O}{\|}}{C}{-}Cl \xrightarrow{AlCl_3} \underset{\text{AROMATIC KETONES}}{C_6H_5{-}\overset{\overset{\displaystyle O}{\|}}{C}{-}R} + H{-}Cl$$

Oxidation Benzene is unusually stable toward oxidizing agents, which cause extensive changes to alkenes (Sections 3.20–3.24).

3. When a monosubstituted benzene, C_6H_5—G, is made to form a disubstituted product, either $C_6H_4G_2$ or G_6H_4GZ, three isomeric substances can be isolated.

These are principal chemical facts about benzene. Next, we shall study how they are explained.

4.4 The Kekulé Structure of Benzene

Kekulé proposed structure **1** for benzene as being the most consistent with the evidence. All six hydrogens are equivalent, and the structure of a monosubstituted benzene would be of type **2.** Possible disubstituted products would be **3, 4,** and **5.** It was rightly pointed out by Kekulé's critics, however, that if his structure for benzene were correct, the 1,2-disubstituted product **3** should have an isomer **6** that differed from it only in the location of the double bonds. (Compare sites 1-2 in **3** and **6.**) Kekulé therefore modified his theory. He suggested that the double bonds so rapidly shift back and forth that **3** and **6,** while isomers, are in such rapid and mobile equilibrium that they can never be separated. His theory did not solve the problem, however. It failed completely to explain the remarkable fact that benzene is not an alkene.

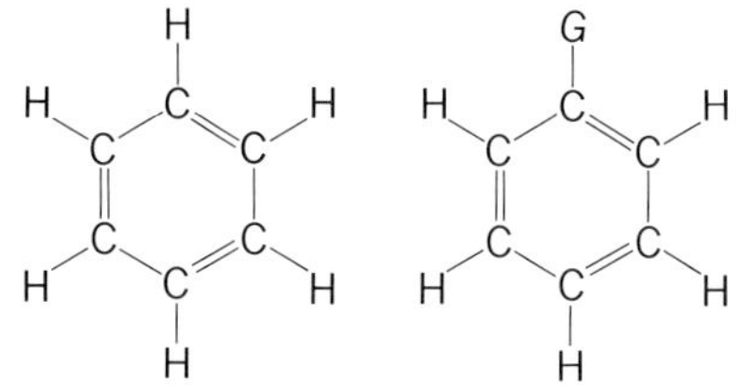

1 **2**

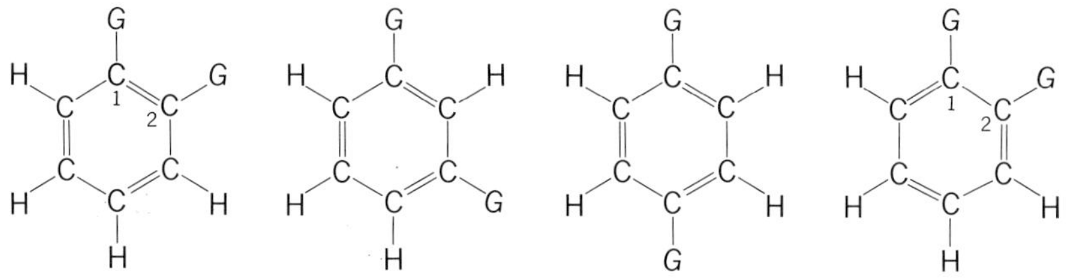

3 **4** **5** **6**

Kekulé's structure is really 1,3,5-cyclohexatriene. Open-chain trienes are known. Like dienes, however, they are quite reactive toward reagents that *add* to a double bond. Cyclic dienes are also known, and they also undergo addition reactions. Moreover they are very vigorously attacked by oxidizing agents.

All efforts to make 1,3,5-cyclohexatriene lead instead to benzene. Although Kekulé's formulation of benzene answered many questions and served very usefully for decades, it left some important questions unresolved. Benzene will undergo one important addition reaction, however: hydrogenation to cyclohexane.[1] We shall next see how data from this reaction shed light on the structure of benzene.

4.5 Hydrogenation of Benzene. Heats of Hydrogenation

Hydrogenation of a double bond is exothermic, and it is possible to measure the heat that is evolved. Between 27 and 32 kcal/mole are evolved when carbon-carbon double bonds in simple alkenes are hydrogenated. For example, the heat of hydrogenation of cyclohexene is −28.6 kcal/mole. We may use that value as the heat of hydrogenation of 1 mole of double bond in a six-membered ring. The compound 1,3-cyclohexadiene has two double bonds. When hydrogenated it should release 2 × 28.6 = 57.2 kcal/mole. The compound "1,3,5-cyclohexatriene," with three ordinary double bonds, should release 3 × 28.6 = 85.8 kcal/mole.

CYCLOHEXENE + H_2 ⟶ CYCLOHEXANE $\Delta H_{obsd} = -28.6$ kcal/mole

1,3-CYCLOHEXADIENE + $2H_2$ ⟶ CYCLOHEXANE $\Delta H_{obsd} = -55.4$ kcal/mole; $\Delta H_{calcd} = -57.2$ kcal/mole

"1,3,5-CYCLOHEXATRIENE" (BENZENE) + $3H_2$ ⟶ CYCLOHEXANE $\Delta H_{obsd} = -49.8$ kcal/mole; $\Delta H_{calcd} = -85.8$ kcal/mole

Cyclohexene is rapidly hydrogenated under mild conditions—at room temperature over nickel at only 20 lb/in.^{2} of pressure. Benzene requires strong conditions for hydrogenation—a pressure of 1500 lb/in.^{2} and a temperature of nearly 200°. The most remarkable result, however, is that the net liberation of heat is much less than we might expect on the basis of benzene's being 1,3,5-cyclohexatriene. Since the products in the three examples are identical, the evolution of 36 kcal/mole less energy for benzene must mean that it has this much less internal energy initially. That is, benzene has less energy per mole than the hypothetical 1,3,5-

[1]In the presence of ultraviolet light, benzene will also add chlorine to form "benzene hexachloride" or BHC. This compound is really 1,2,3,4,5,6-hexachlorocyclohexane and consists of several geometric isomers. One of the isomers is a potent insecticide, and for this reason BHC is an important commerical chemical sold under the names gammexane and lindane.

cyclohexatriene. If it has less internal energy than the expected structural system, it must be more stable than that system. Benzene is obviously not well represented by the Kekulé structures. Its bond lengths heap more coals on the Kekulé structure.

4.6 Bond Lengths in Benzene

If benzene molecules had either Kekulé structure **7** or **8**, with alternating double and single bonds, X-ray diffraction analysis should reveal two different values for carbon-carbon bond distances. Only one is observed, however, 1.39 Å. This result shows that each carbon-carbon bond is equivalent, and that it is intermediate between a single bond (1.54 Å) and a double bond (1.34 Å). Again, the Kekulé structure is inadequate.

4.7 Resonance in Benzene

In Section 3.30 we learned that whenever we can draw two (or more) nonbizarre, conventional structures for a substance, structures that differ only in the location of electrons, then none of these conventional structures will suffice.

Two Kekulé structures of benzene, **7** and **8**, differing only in the locations of electrons can be drawn. Two carbons are numbered to show that although **7** and **8** are equivalent, they are not identical. The C_1—C_2 bond in **7** is single, in **8** double.

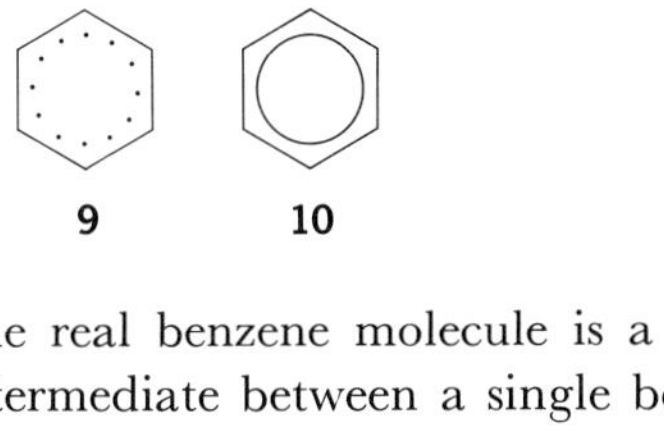

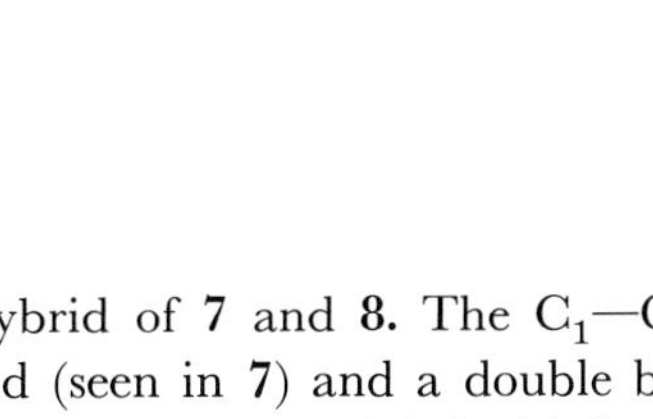

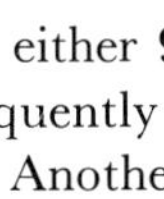

The real benzene molecule is a hybrid of **7** and **8**. The C_1—C_2 bond will be intermediate between a single bond (seen in **7**) and a double bond (seen in **8**). All other carbon-carbon bonds will similarly be partial double bonds, and we may let either **9** or **10** serve as our hybrid structure. (You will generally see **10** more frequently than **9**, and we shall adopt that practice here.)

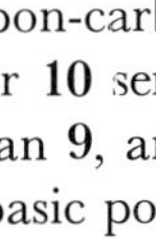

Another basic postulate of resonance theory, one we have not stated yet, says that *resonance is especially important when the contributors are equivalent.* We have this situation with benzene's two contributors, **7** and **8**. From the heats of hydrogenation data, Section 4.5, the 36 kcal/mole that benzene does not have because of resonance is called the resonance energy of benzene.

The chief reason that benzene is so stable is because the six pi electrons of benzene are very delocalized. The molecular orbital model of benzene helps to show how.

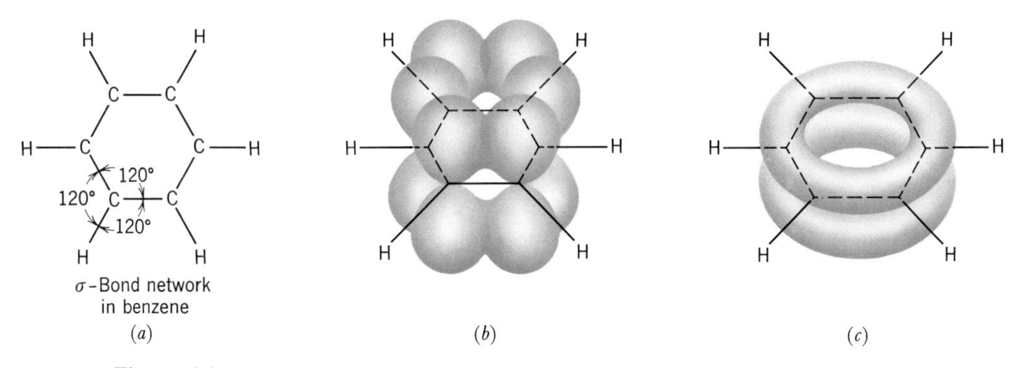

Figure 4.1
Orbital model for benzene. The fundamental framework of the molecule is provided by σ-bonds shown in part *a*. At each carbon is located an unhybridized p_z orbital which can overlap with p_z orbitals on both neighboring carbons as shown in part *b*. The result is the double doughnut-shaped molecular orbital shown in part *c*. This is actually only one of the new molecular orbitals that forms. For our purposes we may treat the shape shown in part *c* as a molecular shell containing three subshells with two electrons in each.

4.8 Molecular Orbital Model for Benzene

Figure 4.1 illustrates by means of the orbital model what delocalization of the pi electrons in benzene means. Since the carbons in benzene are individually attached to three other atoms, sp^2 hybridization may be assumed for each of them. A planar system with bond angles of 120° would then be expected, and this is precisely the internal angle of a regular hexagon. With overlapping of unhybridized p_z orbitals (Figure 4.1), benzene acquires a large circular molecular orbital instead of three isolated, localized double bonds. Instead of the pi electrons being localized as implied by a Kekulé structure, they are considerably delocalized throughout a much larger volume of space. It is believed that this delocalization of electrons into a cyclic molecular orbital largely accounts for the "extra" stability of the benzene molecule. In fact, the term *delocalization energy* is sometimes used synonomously with resonance energy.

The delocalized six electrons in the benzene ring are often called the *aromatic sextet*. This system is not readily broken up. (Witness how hard it is to hydrogenate benzene.) That is why benzene does not easily give addition reactions. Substitution, however, can occur without permanently destroying the aromatic sextet. Keeping track of those electrons is easier if we use a Kekulé structure for benzene. Hence, when we discuss *how* benzene gives substitution reactions, we shall use Kekulé structures. Otherwise, we may follow the practice illustrated often in the next section.

4.9 Naming Benzene Derivatives

Monosubstituted Compounds

For a few compounds the nomenclature has some system. Prefixes naming the substituent are joined to the word "benzene":

Other derivatives have common names that are always used, even though systematic names are possible:

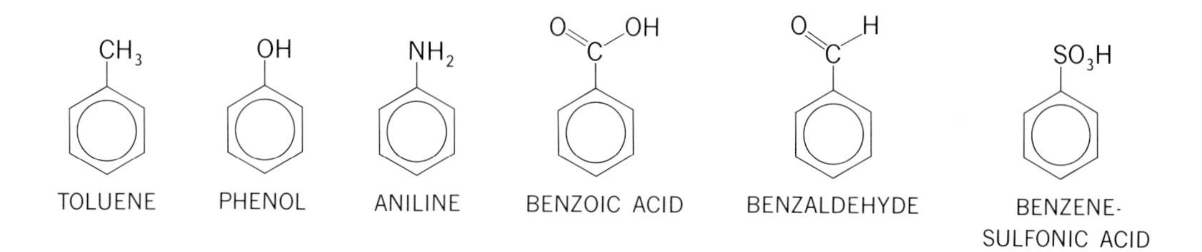

Disubstituted Compounds

When two or more groups are attached to the same benzene ring, both their nature and their relative locations must be specified in the name. The prefixes "ortho," "meta," and "para" are used to distinguish, respectively, the 1,2-, the 1,3-, and the 1,4-relations. These prefixes are usually abbreviated *o*-, *m*-, and *p*-; in the following examples both substituent groups are the same.

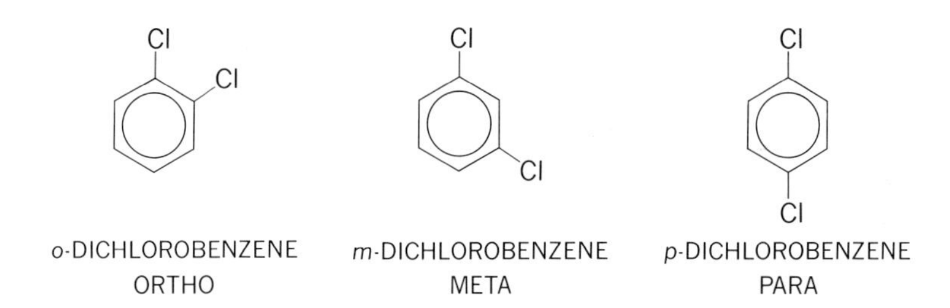

If one of the two groups, alone on the benzene ring, would give it a common name, this name is used, and the second group is designated as a substituent. The following examples illustrate this point.

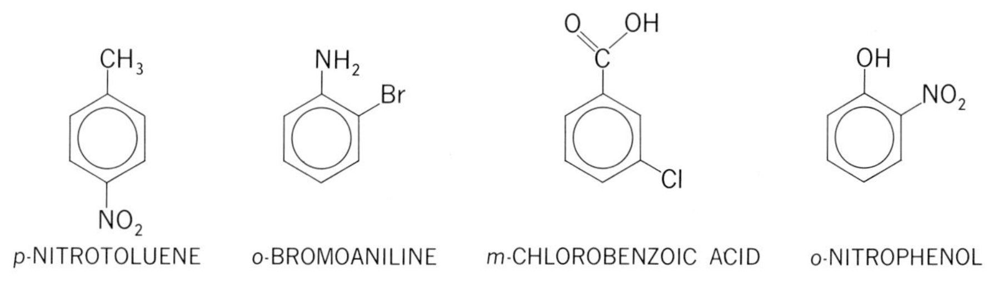

If neither group would be associated with a common name were it alone on the benzene ring, both groups are named and located in the name. For example,

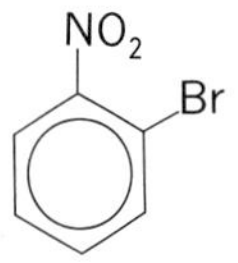

NO_2 Br — *o*-BROMONITROBENZENE

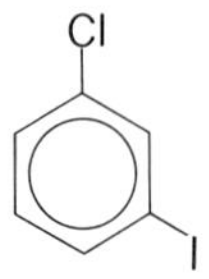

Cl I — *m*-CHLOROIODOBENZENE

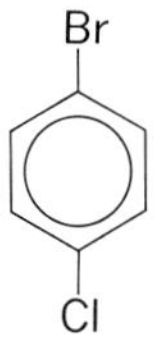

Br Cl — *p*-CHLOROBROMO-BENZENE

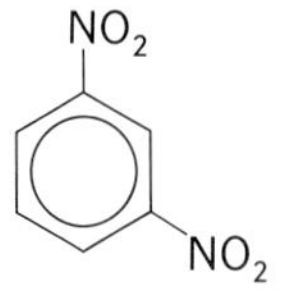

NO_2 NO_2 — *m*-DINITRO-BENZENE

Polysubstituted Benzenes

If three or more groups are on a benzene ring, the ring positions must be numbered. When one group is associated with a common name, its position number is 1. For example,

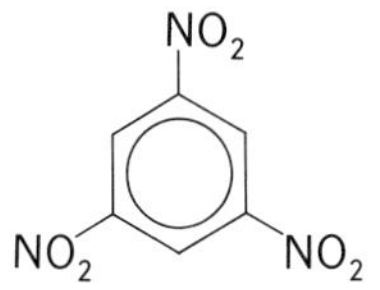

NO_2 NO_2 NO_2 — 1,3,5-TRINITROBENZENE (TNB)

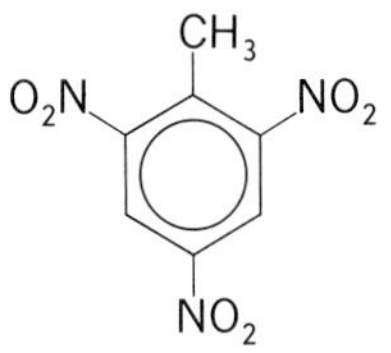

CH_3 O_2N NO_2 NO_2 — 2,4,6-TRINITROTOLUENE (TNT)

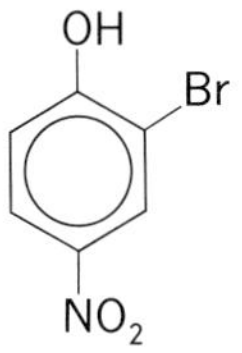

OH Br NO_2 — 2-BROMO-4-NITROPHENOL

4.10 Mechanism of Electrophilic Aromatic Substitution

The commonly accepted mechanism for nearly all electrophilic aromatic substitutions involves three steps.

Step 1. The attacker is made. An electrophilic intermediate is generated.

Step 2. The attack occurs. The electrophile strikes out at the electron-dense benzene ring system. (For a moment during the attack, the circular, pi-electron network is broken. A high-energy carbonium ion intermediate forms, but it happens to receive some stabilization by resonance, too.)

Step 3. The intermediate carbonium ion expels a proton, H^+. Delocalization around the ring is reestablished. On balance, the strong electrophile from step 1 is replaced by the weaker electrophile, H^+.

Let us now see how these steps occur in some specific reactions of benzene.

4.11 The Mechanism of the Nitration of Benzene

When concentrated nitric acid and concentrated sulfuric acid are mixed, freezing point depression studies and other analyses show that the following equilibrium is established. This is step 1 in the nitration of benzene. The nitronium ion which is formed is a powerful electrophilic species.

Step 1.

$$\underset{\text{NITRIC ACID}}{HO{-}NO_2} + \underset{\text{SULFURIC ACID}}{H_2SO_4} \rightleftharpoons \underset{\text{NITRONIUM ION}}{NO_2^+} + H_3O^+ + 2HSO_4^-$$

Step 2. Attack by the electrophilic nitronium ion on a molecule of benzene.

NO_2^+ → [H NO₂ ⟷ H NO₂ ⟷ H NO₂] ≡ H NO₂ (δ+ δ+ δ+)

CONTRIBUTING STRUCTURES FOR RESONANCE-STABILIZED CARBONIUM ION

HYBRID ION (THE THREE δ+ TOTAL 1+.)

Step 3. Loss of a proton, recovery of benzene system, and formation of product.

H NO₂ → NO₂ + H^+

NITROBENZENE

A rough idea of energy changes during this reaction can be visualized in terms of a "Progress-of-Reaction" diagram, Figure 4.2. We show only the last two steps. Step 2 takes the two interacting particles up an energy hill. The energy of their collision is absorbed by their molecules and it goes into breaking up, temporarily, the circular, pi electron network—the aromatic network—of the ring. The carbonium ion is in a high-energy valley. It readily changes over to products. More energy is released in step 3 than needed for step 2, and the reaction is exothermic.

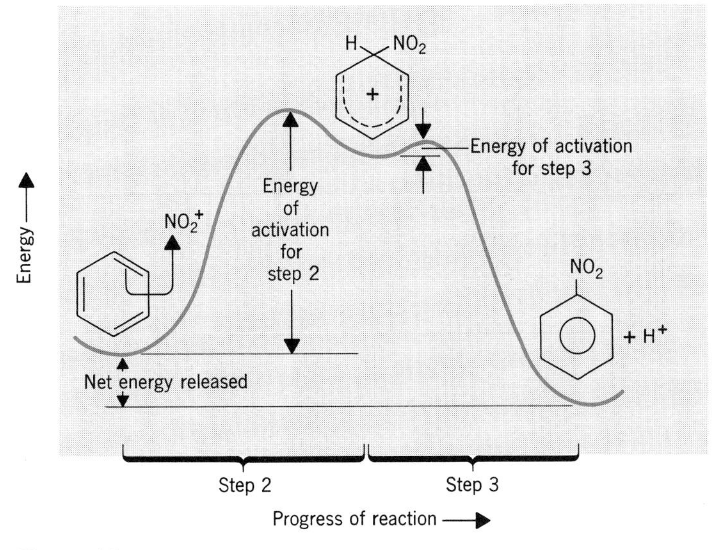

Figure 4.2
Energy changes during the final two steps in the nitration of benzene. Similar diagrams could be drawn for the analogous steps of other substitution reactions of benzene.

4.12 The Mechanism of the Chlorination of Benzene

Iron(III) chloride reacts with chlorine to establish the following equilibrium.

Step 1.

$$Cl{-}Cl + FeCl_3 \rightleftharpoons :\ddot{\underset{..}{Cl}}^+ + FeCl_4^-$$

Because Cl^+ is a strongly electrophilic species, this is the first step in the mechanism. (In all likelihood Cl^+ and $FeCl_4^-$ exist closely associated with each other.)

Step 2. Attack by Cl^+ on the benzene ring.

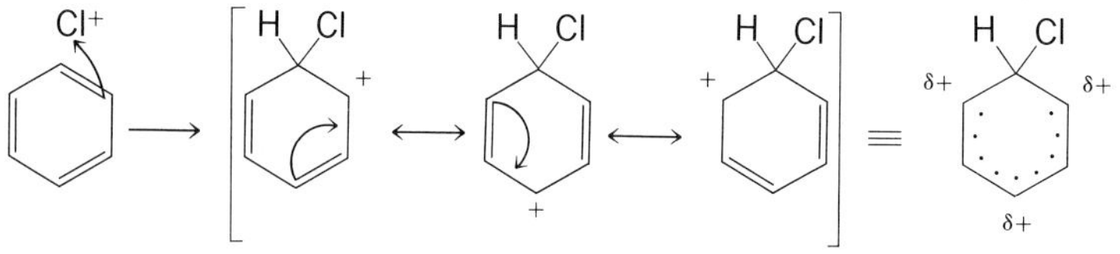

CONTRIBUTING STRUCTURES FOR RESONANCE-STABILIZED CARBONIUM ION

HYBRID ION

Step 3. Loss of a proton from the ring and formation of a resonance stabilized product.

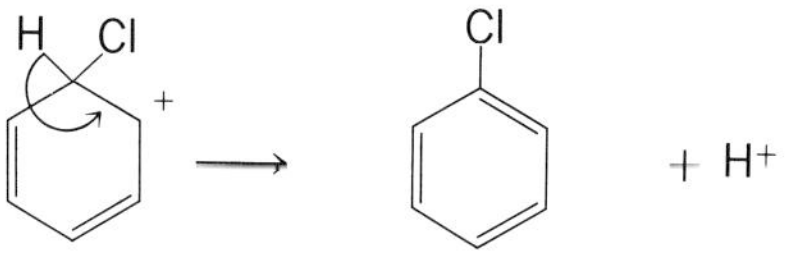

$+ H^+$

CHLOROBENZENE

4.13 The Mechanism of the Friedel-Crafts Alkylation of Benzene[2]

Anhydrous aluminum chloride, a powerful Lewis acid, reacts with alkyl halides in a manner that, with some oversimplification, may be represented as follows. A carbonium ion, a strongly electrophilic species, is generated.

Step 1.

$$R{-}Cl + AlCl_3 \rightleftharpoons R^+AlCl_4^-$$

Step 2. Attack by the carbonium ion, R^+, on a molecule of benzene to form a more stable carbonium ion.

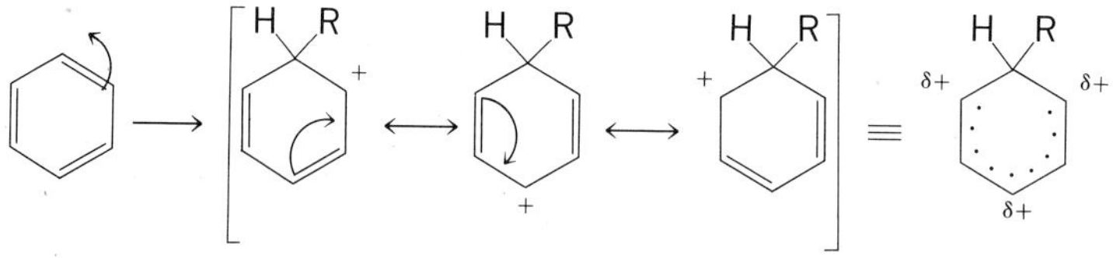

CONTRIBUTING STRUCTURES FOR RESONANCE-STABILIZED CARBONIUM ION

HYBRID ION

[2]Charles Friedel (1832–1899), a French chemist, and James Mason Crafts (1839–1917), an American, discovered this reaction in 1877.

Step 3. Loss of a proton from the benzene ring to form an alkylbenzene.

$$C_6H_6(H)(R)^+ \longrightarrow C_6H_5R + H^+$$

ALKYLBENZENE

4.14 Carbonium-Ion Rearrangements

The following examples illustrate the course of some typical Friedel-Crafts alkylations.

$$C_6H_6 + CH_3Cl \xrightarrow{AlCl_3} \underset{\text{TOLUENE}}{C_6H_5CH_3} + HCl$$

$$C_6H_6 + CH_3CH_2Cl \xrightarrow{AlCl_3} \underset{\text{ETHYLBENZENE}}{C_6H_5CH_2CH_3} + HCl$$

$$C_6H_5 + CH_3CH_2CH_2Cl \text{ or } CH_3CH(Cl)CH_3 \xrightarrow{AlCl_3\ (35\%)} \underset{\text{ISOPROPYLBENZENE (29\%)}}{C_6H_5CH(CH_3)_2} + \underset{n\text{-PROPYLBENZENE (19\%)}}{C_6H_5CH_2CH_2CH_3} + HCl$$

$$C_6H_6 + CH_3CH_2CH_2CH_2Cl \text{ or } CH_3CH_2CH(Cl)CH_3 \xrightarrow[\text{heat}]{AlCl_3} C_6H_5CH(CH_3)CH_2CH_3 + HCl$$

sec-BUTYLBENZENE (50%)
(Only monoalkylbenzene obtained, even from *n*-butyl chloride)

Thus, in the Friedel-Crafts alkylation of benzene with *n*-propyl chloride, the principal product is isopropylbenzene. When *n*-butyl chloride is used, the only monoalkyl product is *sec*-butylbenzene. In both alkylations the carbonium ion we would expect to form in step 1, a primary carbonium ion, cannot be the one that becomes attached to the benzene ring. What happens is that the initially formed primary carbonium ion rearranges to form a more stable ion by a hydride-ion shift. Thus, for the *n*-butyl cation,

$$\underset{\substack{\mathbf{11}\\ n\text{-BUTYL CATION}}}{CH_3CH_2-\overset{H}{\ddot{C}H}-CH_2^+} \longrightarrow \underset{\substack{\mathbf{12}\\ \text{MIGRATION OF “HYDRIDE ION” PARTIALLY COMPLETE}}}{CH_3CH_2-CH\overset{H}{\overset{+}{-}}CH_2} \longrightarrow \underset{\substack{sec\text{-BUTYL}\\ \text{CATION}}}{CH_3CH_2-\overset{+}{C}H-CH_2(H)}$$

Before the hydrogen with its bonding pair of electrons breaks away, it starts to form a new bond at the carbon bearing the initial positive charge (**11** ⟶ **12**). An alkyl group can migrate if there is no available hydrogen at the carbon adjacent to the positive charge.

Carbonium ions rearrange when a more stable ion can be produced. Thus they may go from 1° to 2° or from 2° to 3°.

Reactions of Monosubstituted Benzenes

4.15 Orientation of Further Substitution

Monosubstituted benzenes undergo the same types of substitution reactions as benzene itself. Nitrobenzene can be nitrated to dinitrobenzene. There are three different positions at which the second nitro group could become attached: either of the ortho positions, either of the meta sites, and the para. Statistically, therefore, we might expect the three isomeric dinitrobenzenes, ortho, para, and meta, to form in a ratio of 2:1:2. The experimentally determined ratio, however, is 6.4:0.3:93.5. The meta isomer clearly predominates.

When toluene is nitrated, however, very little meta isomer forms. The isomeric nitrotoluenes, ortho, para, and meta, form in the ratio 58:38:4. A mixture of *o*-nitrotoluene and *p*-nitrotoluene is 96% of the product.

In the two examples given, nitration of nitrobenzene and toluene, the nature of the incoming electrophilic species, nitronium ion, is the same. Only the nature of the group already on the ring is changed. In general, *where* a second substituent becomes attached depends far more on what the first substituent is than on the incoming group.

In Table 4.1 are summarized experimental data for the compositions of the isomer mixtures that form when various monosubstituted benzenes are nitrated. The substituents already on the ring fall into two groups. The first group consists of the *ortho-para directors*, the second of the *meta directors*.

Exercise 4.1 Predict the structures of the principal organic products that would form if further substitution were carried out on the following monosubstituted benzenes. Write the names of the products.

(a) Nitration of toluene.
(b) Chlorination of nitrobenzene.
(c) Sulfonation of phenol.
(d) Bromination of $C_6H_5NH-\overset{\overset{\displaystyle O}{\|}}{C}-CH_3$ (acetanilide).
(e) Chlorination of chlorobenzene.

4.16 Relative Reactivities in Additional Aromatic Electrophilic Substitutions

Having studied *where* the second group enters, we next consider *how rapidly* it goes there, compared with how rapidly it would attack benzene itself. Toluene is nitrated 10 to 20 times faster than benzene under identical conditions. Nitrobenzene, in contrast, is much more slowly nitrated than benzene. What is already on the ring therefore affects the ring's reactivity toward further substitutions. Some

Table 4.1
Orientation in the Nitration of Monosubstituted Benzenes

$$C_6H_5G + HNO_3 \xrightarrow{\text{Catalyst}} C_6H_4GNO_2 + H_2O$$

(Mixture of disubstituted benzenes)

G (Director)	Composition of Mixture of Disubstituted Benzenes: Ortho	Para	Total Ortho plus Para	Meta
Ortho-para directors				
—OH	50–55	45–50	100	Trace
$—NH—C(=O)—CH_3$	19	79	98	2
$—CH_3$	58	38	96	4
—F	12	88	100	Trace
—Cl	30	70	100	Trace
—Br	38	62	100	Trace
—I	41	59	100	Trace
Meta directors				
$—\overset{+}{N}(CH_3)_3$	0	0	0	100
$—NO_2$	6.4	0.3	6.7	93.3
—C≡N	—	—	11.5	88.5
$—SO_3H$	21	7	28	72
—C(=O)—OH	19	1	20	80
—C(=O)—H	—	—	21	79

Data from R. T. Morrison and R. N. Boyd, *Organic Chemistry*, third edition, table 11.1, p. 340. © Copyright 1973 by Allyn and Bacon, Inc., Boston. Used with permission.

Table 4.2
Aromatic Electrophilic Substitution, Activating or Deactivating Influences of Groups Attached to Benzene

Activating groups (all ortho-para directors)	Strong activators $—NH_2$, —NHR, $—NR_2$ —OH	Moderate activators —OR —NH—C(=O)—R	Weak activators $—C_6H_5$ —R (alkyl)
Deactivating Groups	Meta directors: $—\overset{+}{N}R_3$, $—NO_2$, —CN, $—SO_3H$, —C(=O)—OH(R), —C(=O)—H, —C(=O)—R Ortho-para directors: —F, —Cl, —Br, —I		

attached groups *activate* the ring, and others *deactivate* it. The terms "activation" and "deactivation" used in this context mean "relative to benzene."

All the meta directors deactivate the ring in varying degrees. With the exception of one "family," the halogens, all the ortho-para directors activate the ring in varying degrees. The halogens, although ortho-para directors, are deactivators. They form a borderline. Table 4.2 summarizes these effects.

Exercise 4.2 In preparing disubstituted benzenes, starting with benzene, the order in which the two groups are put on the ring is obviously important. The following questions relate to this problem.

1. What will the final product be if benzene is first brominated and then that product is nitrated?
2. What will the final product be if benzene is first nitrated and then the product is brominated?
3. Predict the principal organic products in each of the following.

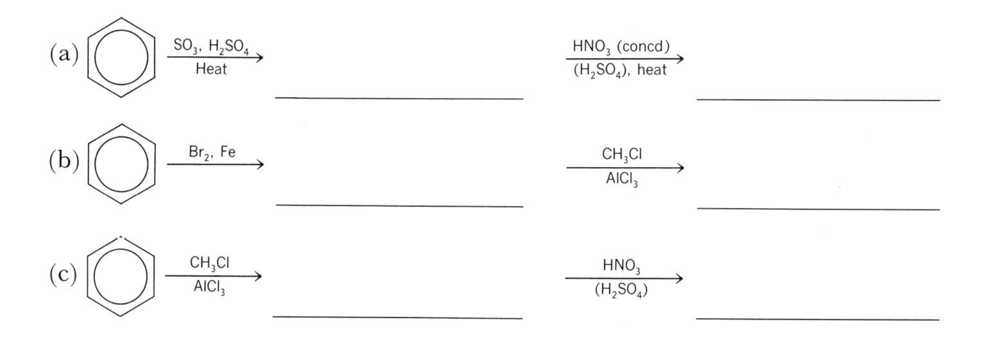

4.17 Theory of Orientation and Reactivity in Aromatic Electrophilic Substitution

Orientation has to do with *where* a group or atom becomes attached if different sites are available. Reactivity has to do with *how rapidly* the group or atom becomes attached as compared with the rate of reaction with a reference compound. Benzene is the reference for aromatic electrophilic substitutions.

Figure 4.2 was for an attack on benzene itself. We now want to see what happens when an electrophile attacks a substituted benzene ring. What will the group on the ring do? How will it affect the reaction? The key is the carbonium ion in the high energy valley. If the group already on the ring can help make that carbonium ion more stable, the whole energy hill is lowered, and the ring will react more readily with some other reagent. The group causing this is called an **activating group.** However, if the group already on the ring makes a carbonium ion less stable, then the whole hill is raised, and the ring is less reactive. Therefore, the group on the ring is a **deactivating group.** A nitro group is an example. It has an electron-withdrawing inductive effect. Other deactivating groups are given in Table 4.2, and they all have electron-withdrawing effects. The reason this effect raises the activation for the next substitution is that the making of the carbonium ion puts a positive charge into the ring system. Electron withdrawal from the ring by the attached group intensifies that charge. You cannot *stabilize* a charged particle by intensifying that charge. To stabilize a charged particle do the opposite—reduce or neutralize the charge. Hence, groups that deactivate the ring do so by destabilizing the intermediate carbonium ion. That is what raises the activation energy for further substitution.

An electron-releasing group on the ring, such as a CH_3— group, is the kind needed to stabilize the positive charge of the intermediate carbonium ion. Let us see how this works in the nitration of toluene, which occurs more rapidly than the nitration of benzene. The methyl group is an activator, but not of all positions. It activates the ortho and the para position, not so much the meta position. The methyl group is an ortho-para director.

4.18 The Methyl Group— An Activator and an Ortho-Para Director

In the nitration of toluene the incoming nitronium ion could become attached at the ortho, the para, or the meta position. In resonance theory the contributing structures for the carbonium ion in each position are as follows:

CH_3 $\overset{+}{N}O_2$ Ortho substitution → (a) ⟷ (b) ⟷ (c) ≡ HYBRID

13

CH_3 NO_2^+ Para substitution → (a) ⟷ (b) ⟷ (c) ≡ HYBRID

14

CH_3 NO_2^+ Meta substitution → (a) ⟷ (b) ⟷ (c) ≡ HYBRID

15

The methyl group has an electron-releasing inductive effect, indicated by the arrowhead drawn on the line representing its bond to the ring. This inductive effect tends to stabilize any partial positive charge that might develop, especially *at the carbon holding the methyl group. Only for ortho or para substitution does such a partial positive charge appear at this point.* (See structures **13**(a) and **14**(b) or the hybrid structures for **13** and **14**.) In meta substitution a partial positive charge does not develop where the methyl joins the ring (see **15**). The carbonium ion for meta substitution is less stable than that for either ortho or para substitution. Hence the energy of activation for making it is higher. The rate of meta substitution therefore is slower, much slower in fact, than the rates of either ortho or para substitution.

4.19 The Nitro Group— A Deactivator and a Meta Director

Both nitrogen and oxygen are more electronegative than carbon, and the nitro group has a strong, electron-withdrawing inductive effect. Let us now consider the options in the chlorination of nitrobenzene.

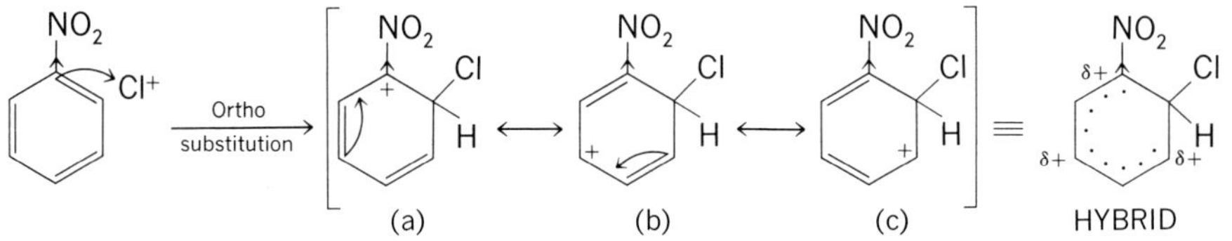

16

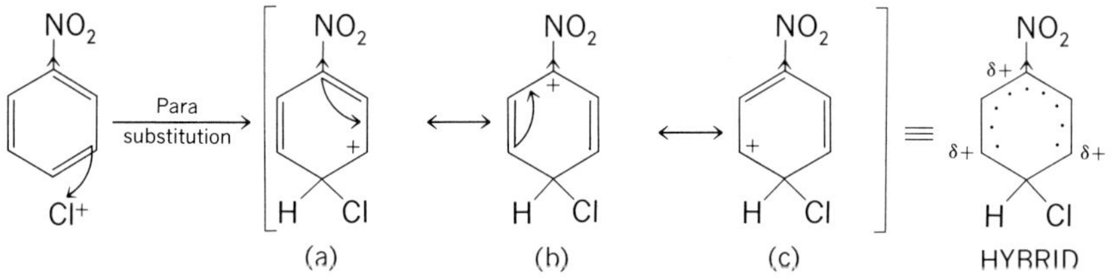

17

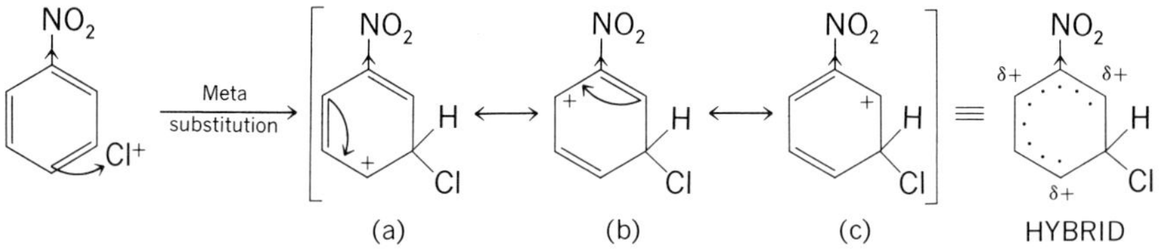

18

In **16** and **17**, according to resonance theory, a partial positive charge develops in the intermediate at the ring carbon holding the nitro group. However, the inductive effect of the nitro group is to make that carbon partially positively charged even *before* Cl^+ attacks. An attack by Cl^+ at either ortho or para positions worsens this. It intensifies the positive charge. Therefore, such attacks require especially violent collisions that do not frequently happen. It is energetically easier for Cl^+ to attack a meta position. That way the incoming positive charge delivered by Cl^+ to the ring need not be delocalized to the carbon holding the nitro group. See structure **18**. The partial positive charge (in the hybrid ion) "skips" this point. Of course, not even **18** is as stable a carbonium ion as would be formed if Cl^+ attacked the benzene ring without the nitro group. Thus the rate of attack by Cl^+ *anywhere* in nitrobenzene will be slower than its rate of attack on benzene itself. This explains the deactivating influence of the nitro group. Because the meta sites of nitrobenzene are *relatively* easier to attack than its ortho or para positions, the nitro group "directs" incoming groups to meta sites.

4.20 The Phenolic Hydroxyl Group—Resonance versus Induction

Oxygen is more electronegative than carbon, and we might predict that the phenolic hydroxyl group is electron withdrawing, somewhat like the nitro group only less strongly so. We would therefore consider it a meta director and a deactivator. Our prediction could not be more incorrect. This group is such a powerful activator (and ortho-para director) that bromination of phenol requires no iron catalyst. Let us see how we can understand this in resonance theory.

The carbonium ion is again the important factor, but with the oxygen attached directly to the ring another pair of electrons is available for writing contributing structures. Depending on whether Br^+ (generated in the first step) makes its attack at the ortho, para, or meta position, one of the following carbonium ions will form. In ortho or para attack contributors (a), (c), and (d) of **19** and **20** are familiar. Contributor (b) is new in both **19** and **20** and shows the positive charge on oxygen

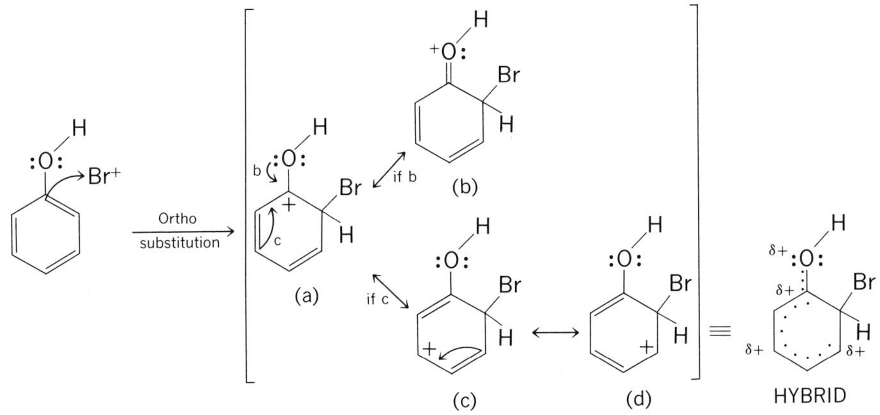

19

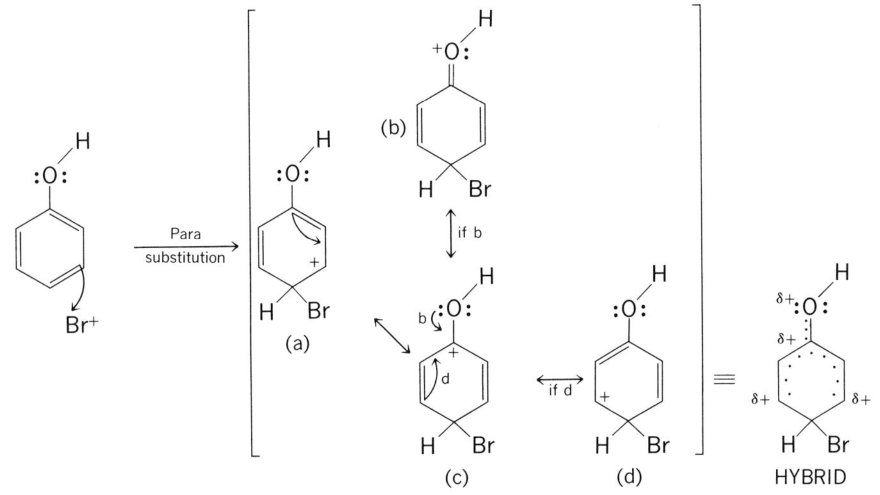

20

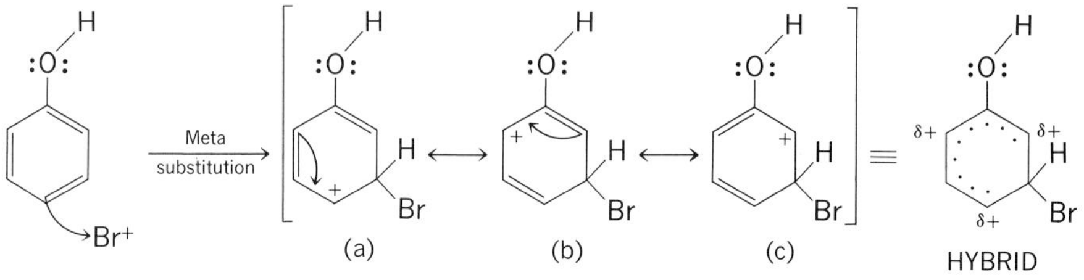

Meta substitution

(a) (b) (c) HYBRID

21

even while the oxygen still has its octet and all the ring atoms have octets. We have seen a positive charge on oxygen before, in the hydronium ion, H_3O^+, and from this ion we learned that if oxygen has an octet it can handle a positive charge. Therefore **19**(b) and **20**(b) should make large contributions. Therefore, we are justified in concluding that the hybrid carbonium ions resulting from ortho and para attacks are more stable than the carbonium ion resulting from the meta attack. In meta attack the positive charge cannot be delocalized onto oxygen. Moreover, substitution into phenol should for these same reasons be easier than substitution into benzene. Thus resonance theory accounts correctly for both the directing and the activating influence of the hydroxyl.

Exercise 4.3 The amino group in aniline, $C_6H_5—\ddot{N}H_2$, is as strong an activator and ortho-para director as the hydroxyl in phenol. The explanation is exactly the same as the explanation for the properties of the hydroxyl group. See whether you can write the steps, paying particular attention to the contributing structures for the intermediate carbonium ions for ortho, para, and meta monobrominations of aniline.

4.21 The Halogens— Resonance versus Induction

The halogens are deactivating groups, as might be predicted from their electron-withdrawing inductive effect. (Compare the effect of the nitro group.) This property should also make them meta directors, however, and they are not. With the halogens we are apparently at a border line. Their inductive effects are not overwhelmed by any resonance effect, as was the case with the inductive affects of the —OH and the —NH_2 groups. The relatively strong inductive effect of a halogen on a benzene ring makes pi electrons *anywhere* on the ring less available than those in benzene to some incoming electrophile. But a resonance effect, illustrated for ortho substitution, must be invoked to explain ortho-para substitution. In the ortho bromination of chlorobenzene, structures **22**(a) to (d) contribute to the hybrid. In **22**(b) the chlorine is shown with a positive charge and an octet, but chlorine is less able to handle a positive charge than either nitrogen or oxygen. Hence, although **22**(b) contributes enough to the hybrid to make the carbonium ions for ortho and para substitutions more stable than the carbonium ion for meta substitution, the ring is not more active than that of benzene itself. Such is the argument from resonance theory for the deactivating but ortho-para-directing influence of a halogen atom on a benzene ring.

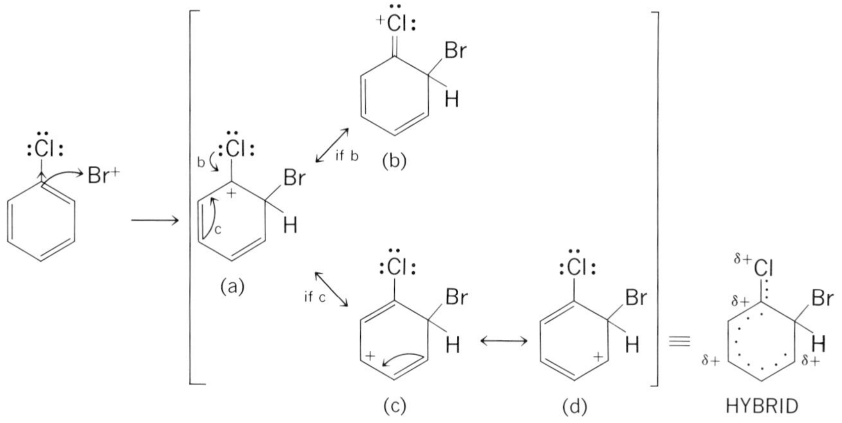

(a) (b) (c) (d) HYBRID

22

4.22 Arenes. Alkylbenzenes

The names and structures of several alkylbenzenes or "arenes" are given in Table 4.3. Petroleum and coal tar are important sources. Because arenes combine an alkanelike group with a benzene system, we might expect them to exhibit properties of both systems. They do.

4.23 Chemical Properties

Halogenation

Under conditions that promote the formation of free radicals (sunlight or high temperature), alkylbenzenes undergo halogenation at the side chain, the alkyl group.

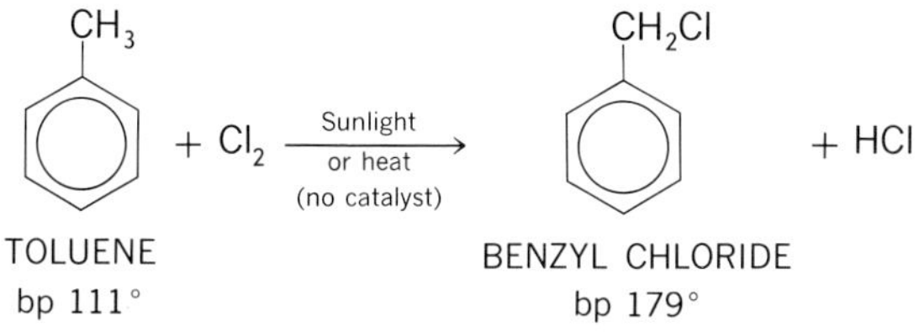

$$C_6H_5CH_3 + Cl_2 \xrightarrow[\text{(no catalyst)}]{\text{Sunlight or heat}} C_6H_5CH_2Cl + HCl$$

TOLUENE bp 111° → BENZYL CHLORIDE bp 179°

We shall study this reaction in greater detail in Section 7.5.

If an iron (or iron halide) catalyst is used, substitution into the ring occurs. Nitration, sulfonation, and Freidel-Crafts reactions also occur at the ring.

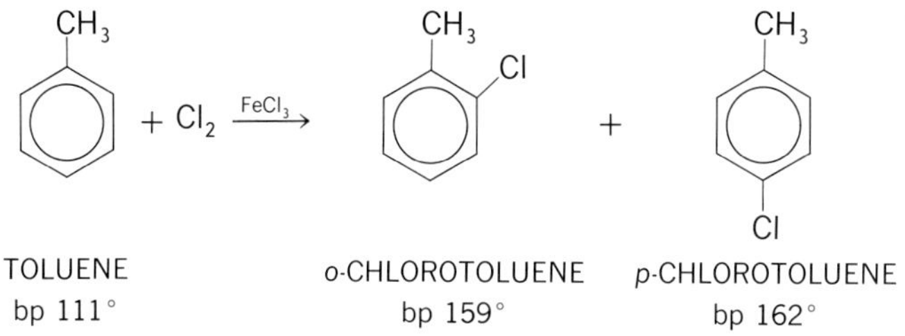

$$C_6H_5CH_3 + Cl_2 \xrightarrow{FeCl_3} o\text{-}ClC_6H_4CH_3 + p\text{-}ClC_6H_4CH_3$$

TOLUENE bp 111° → *o*-CHLOROTOLUENE bp 159° + *p*-CHLOROTOLUENE bp 162°

Table 4.3
Arenes

Name	Structure	Mp (°C)	Bp (°C)	Specific Gravity (at 20°C)
Benzene	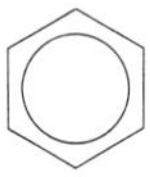	5.5	80	0.879
Toluene	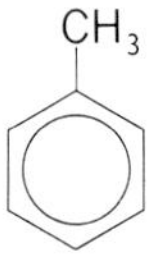 CH_3	−95	111	0.866
o-Xylene	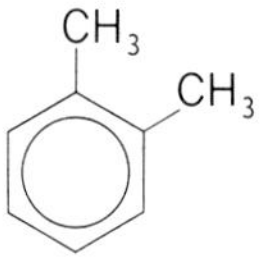 CH_3 CH_3	−25	144	0.897
m-Xylene	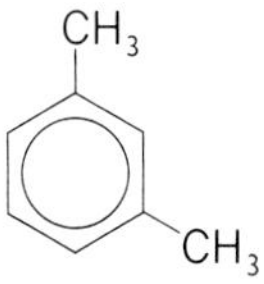 CH_3 CH_3	−48	139	0.881
p-Xylene	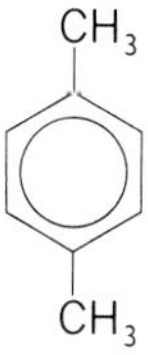 CH_3 CH_3	−13	138	0.854
Ethylbenzene	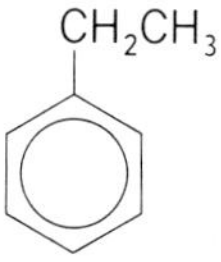 CH_2CH_3	−95	136	0.867
Cumene	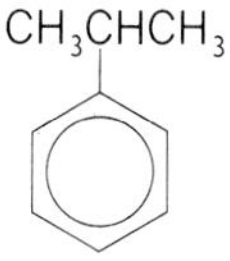 CH_3CHCH_3	−81	152	0.862
p-Cymene	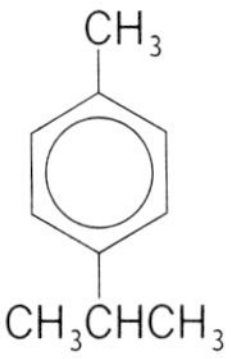 CH_3 CH_3CHCH_3	−70	177	0.857
Biphenyl	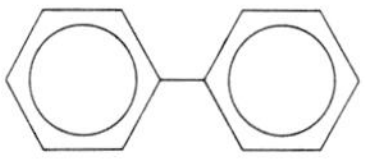	70	255	—

Oxidation

Benzene is exceptionally stable toward strong oxidizing agents and so are alkanes, but strong oxidizing agents will attack the side chain, however long it is, where it joins the ring. For example,

$$C_6H_5CH_2CH_2CH_3 \xrightarrow{\text{hot } KMnO_4} C_6H_5\overset{O}{\overset{\|}{C}}\text{—}OH \quad (+\ CO_2 + H_2O)$$

n-PROPYLBENZENE → BENZOIC ACID

$$CH_3\text{—}C_6H_4\text{—}CH_3 \xrightarrow{\text{hot } KMnO_4} HO\overset{O}{\overset{\|}{C}}\text{—}C_6H_4\text{—}\overset{O}{\overset{\|}{C}}OH \quad (+\ CO_2 + H_2O)$$

p-XYLENE → TEREPHTHALIC ACID

This reaction is an important source of aromatic carboxylic acids. Terephthalic acid is one monomer needed to make Dacron and Mylar (Section 10.19). Reagents such as potassium permanganate, potassium dichromate, or dilute nitric acid may be used for the reaction. In this example the properties of one system are modified by the presence of another. An alkanelike group on a benzene ring is much more susceptible to oxidation than an alkane is.

The benzene ring is not this stable in all situations. If powerful electron-donating groups (for example, amino, hydroxyl) are attached to the ring, it is quite susceptible to oxidation. Atmospheric oxygen slowly attacks anilines and phenols, converting them to deeply colored, complex substances. In human metabolism phenolic substances obtained from proteins are converted into melanins, the dyes responsible for pigmentation of the skin.

4.24 Polynuclear Aromatic Hydrocarbons

Compounds exhibit aromatic properties if their molecules consist of flat, cyclic systems with conjugated multiple bonds when Kekulé structures are written for them. Several hydrocarbons are in this category, and their structures appear to consist of fused or condensed benzene rings. Examples are given in Figure 4.3.

Naphthalene is a common moth repellent, but its use has declined since the introduction of *p*-dichlorobenzene. Some of these condensed aromatic hydrocarbons occur in the tars of tobacco smoke. One of them, benzopyrene, is a known carcinogen (cancer inducer).

Brief Summary: Chemical Properties

1. For a substance to be classified as aromatic its molecules must be cyclic, planar, be quite unsaturated, and yet give substitution reactions—not addition reactions. (The representative reactions of benzene should be learned.) Take special note of overall

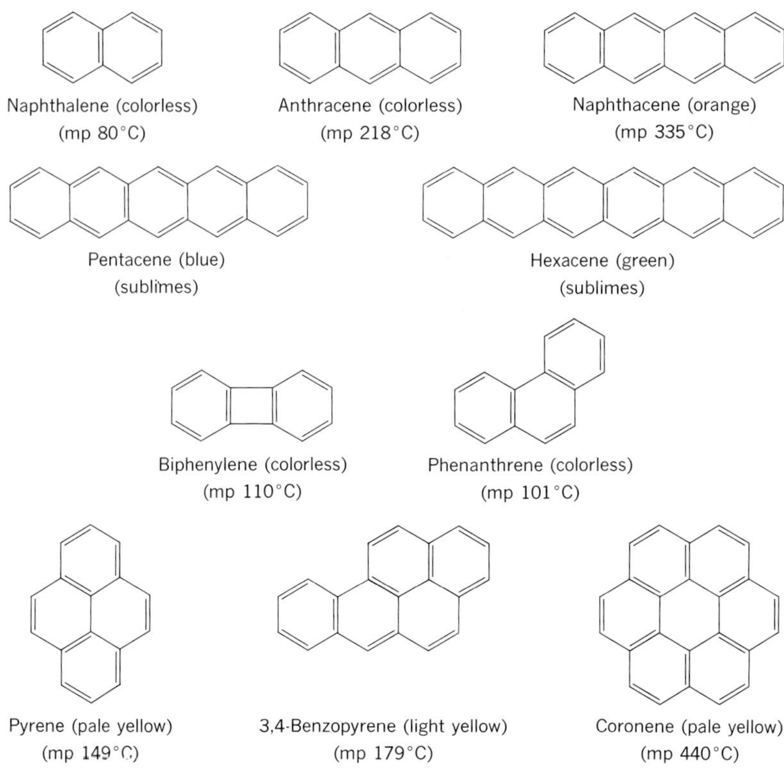

Figure 4.3
Some polynuclear aromatic hydrocarbons.

results—we can put onto the ring a nitro group, a sulfonic acid group, a halogen atom, and an alkyl group (but it may be a rearranged group compared with the starting material).

We cannot directly put onto the ring either —NH_2 (amino) or —OH (phenolic hydroxyl).

2. Monosubstituted benzenes also undergo ring substitution. The group on the ring at the start determines where and how rapidly a second group goes.
The lists of activators/deactivators and of *o,p/m*-directors should be learned.

3. Alkyl benzenes (arenes) may react either at some ring position or at the side chain (and if there, generally at the carbon attached to the ring—the alpha-position).

Brief Summary: Important Concepts and Theories

1. The Kekulé and the hybrid models of the benzene molecule.
2. The use of heats of hydrogenation to estimate resonance energy.
3. The mechanism of aromatic electrophilic substitution. (Be able to write the three steps.)
4. Carbonium ion rearrangements.
5. Theory of activation and deactivation and of directing influences of groups.

Problems and Exercises

1. Write structures of each of the following compounds
   (a) *m*-nitrobenzoic acid (b) 2,4,6-tribromoaniline
   (c) *o*-chlorophenol (d) *p*-toluenesulfonic acid
   (e) *m*-iodotoluene

2. Starting with benzene or toluene and any needed inorganic reagents, write reactions for preparing each of the following compounds. (Assume that mixtures of ortho and para isomers can be separated.)
   (a) *p*-nitrochlorobenzene (b) benzoic acid
   (c) *m*-nitrobenzoic acid (d) *p*-bromobenzenesulfonic acid
   (e) *m*-bromobenzenesulfonic acid

3. Write equations for the steps in the bromination of benzene, catalyzed by $FeBr_3$.

4. Naphthalene has three important contributing structures. One of them is the following:

   Write the structures of the other two.

5. Complete each of the following equations by writing the structure(s) of the principal organic product(s). Where no reaction occurs, state so.

(a) CH_3 (on benzene) $+ Br_2 \xrightarrow{FeBr_3}$

(b) NO_2 (on benzene) $+ H_2SO_4 \xrightarrow[Heat]{}$

(c) $\overset{O}{\overset{\|}{C}}—OH$ (on benzene) $+ Cl_2 \xrightarrow{FeCl_3}$

(d) NO_2, CH_3 (para, on benzene) $+ HNO_3 \xrightarrow[\Delta]{H_2SO_4}$

(e) CH_3 (on benzene) $+ Cl_2 \xrightarrow{Sunlight}$

(f) CH_3, CH_2CH_3 (meta, on benzene) $\xrightarrow[Heat,\ OH^-]{Excess\ KMnO_4}$

(g) $NH—\overset{O}{\overset{\|}{C}}—CH_3$ (on benzene) $+ HNO_3 \xrightarrow{H_2SO_4}$

(h) OCH_3 (on benzene) $+ Cl_2 \xrightarrow{FeCl_3}$

(i) benzene $+ CH_3\overset{Cl}{\overset{|}{C}}HCH_3 \xrightarrow[\Delta]{AlCl_3}$

(j) NO_2, SO_3H (meta, on benzene) $+ HNO_3 \xrightarrow[Heat]{H_2SO_4}$

chapter five

Organic Fuels and Petrochemicals

Our forebears put yokes on oxen and other beasts to make them serve mankind. They thrust paddle wheels into the wind and into flowing water to get help to do the chores. Then the trick of turning heat into useful work was learned. It made the world's resources of wood and coal and oil more significant than as sources of warmth and light. It made possible both the extraordinary benefits and the awesome problems of the industrial-scientific age. It freed us from drudgery, but it made us very dependent on energy. It made possible a population explosion, which couples any hope of going back to a supposedly simpler age with the certainty that the trip means unspeakable human misery and tragedy. It also made possible bright colors, new medicines, sumptuous fabrics, new fuels for engines and new rubber for vehicle tires—all or almost all of which are the "synthetics" made from the chemicals we wring from coal and oil and natural gas. In this chapter we shall survey in a very general, nontechnical way how this is done. We can do little more here, but having just completed our survey of various kinds of hydrocarbons we should pause to learn more of their origins and their several uses.

The Fossil Fuels

5.1 Petroleum

Nature locked up some sunlight long ago. It did it throughout hundreds of thousands of years during the Carboniferous Period (280 to 345 million years ago) while vast areas of the continents, little more than monotonous plains, basked in sunlight near sea level. In the seas countless microscopic plants such as diatoms—tons of them per acre of ocean surface during the early spring—soaked up solar energy to power their fugitive lives. Then they died. According to one theory, death of each such plant released a microscopic droplet of oily material that eventually settled into the bottom muds. The muds grew in thickness and compacted, sometimes into shales, sometimes into limestone and sandstone deposits. The oily material changed to petroleum. In some parts of the world it managed to move through porous rock and collect into pools forming the great petroleum reserves of our planet. In other parts, this movement could not occur, and the oily substances remain to this day locked in enormous deposits of oil shales and oil sands.

5.2 Coal

Plants and animals, of course, grew and died on the lands bordering those ancient seas. In a moist and sunny setting where the land was marshy and boggy, lush vegetation flourished and died. The remains, covered by stagnant, oxygen-poor, often acidic water, decomposed, but this occurred more slowly than the plants died. The rotting mass accumulated, became fibrous, and turned into peat. Where peat layers became unusually thick, they compacted into lignite ("brown coal"). Lignite is over 40% water, but it is still an important fuel. When lignite deposits grew in thickness and became further compacted, sometimes by overlying deposits of sedimentary rock, the water was forced out. It became bituminous coal ("soft coal"), with less than 5% water. Bituminous coal includes roughly 30% volatile,

Figure 5.1
Vast deposits of lignite and soft coal were produced eons ago in what is now the western United States. This strip mining operation in the Four Corners area of northwestern New Mexico fuels one of many gigantic power plants. The plume of smoke from one plant was the only man-made "object" seen on earth from the astronaut's view during the Gemini 12 flight in 1966. (Mimi Forsyth/Monkmeyer)

organic matter. When nature's workings forced that out, the remainder was anthracite ("hard coal") with over 95% fixed carbon.

5.3 Natural Gas

Both in soft coal deposits and in petroleum deposits the most volatile hydrocarbons—methane and ethane—accumulated too. Natural gas is largely methane.

The various coals, petroleum, and natural gas constitute the fossil fuels, a legacy from the past now being consumed so rapidly that for the first time in history worry about running out has become widespread. That we shall run out is quite simply just a matter of time. Almost any imaginable human use of the fossil fuels would use them up much faster than nature can replenish them.

5.4 Estimates of World Supplies of Crude Oil and Coal

M. King Hubbert of the U. S. Geological Survey surveyed and refined estimates of the reserves of crude oil and coal on our planet. Since these materials are not renewable on the time scale of human history, their total production, beginning at zero units per year, must eventually return to zero units per year. The actual production figures tell us about past uses. We have to rely on a variety of sophisticated ways for estimating future supplies. These estimates depend altogether on the assumptions used to make them, but one fact is inescapable. Someday we shall run out. Of what the world does have, over half is either in the Near East or in Latin America.

Figure 5.2 is Hubbert's conclusion concerning the world's complete cycle of crude-oil production. Drawn in the late 1960s, the graph reflects two estimates of total supplies, one conservative and one generous. Each square represents 250 billion barrels; the symbol Q_∞ represents the estimated total supplies, past and future. If $Q_\infty = 1350 \times 10^9$ barrels, then we shall have used up the "heart" 80% of the total supply between 1961 and 2019—58 years! Even with a liberal estimate, $Q_\infty = 2100 \times 10^9$ barrels, this "heart" 80% stretches out only to 64 years. The United States cycle of crude oil production is shown in Figure 5.3.

Since the price of crude oil began to climb sharply, greater efforts will no doubt be made to go after harder-to-recover deposits. Greater attempts will be made to use crude oil with maximum efficiency. All these efforts will do is buy a few years during which other resources can be developed. These include solar energy, geothermal energy, breeder reactor, and possibly fusion power. Petroleum needs will be augmented by removing crude oil from oil shale and oil sand as well as by changing coal into petroleum liquids (coal liquifaction) and into petroleum gases (coal gasification). We shall discuss these topics more fully later.

Hubbert's analysis of the world's coal reserves and the complete cycle of coal production is shown in Figure 5.4. Figure 5.5 shows how this coal is distributed throughout the world. Figure 5.6 is the estimated cycle of coal production in the United States. The principal coal-bearing regions of the United States are shown in the map in Figure 5.7. As these figures indicate, the world's supply of various kinds of coal is still vast. Given the facts that we do not yet have breeder reactors

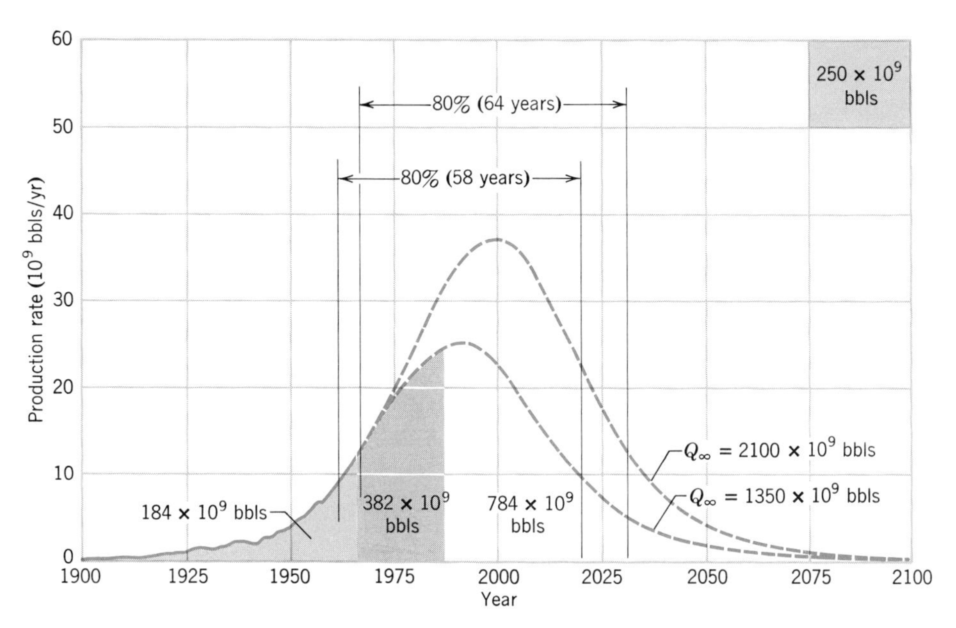

Figure 5.2
Complete cycles of world crude oil production for two different values of Q_{∞}. Values charted included what has been produced (colored area), what is definitely known to be in the ground (gray area), and what will be found additionally, according to scientific estimates. (From *Resources and Man,* Publication 1703, Committee on Resources and Man, National Academy of Sciences–National Research Council, W. H. Freeman, and Company, San Francisco, 1969. Used by permission.)

or commercial solar furnaces, and we do not know if we shall ever have the fusion reactors, coal is our only current source of energy that could meet all of our energy needs for decades to come.[1]

5.5 Oil Shale

The story is told of a gentleman in a western state who built a new home and made the fireplace out of a local shale. When he lit the first fire, the fireplace burned up together with the home. True or not, the Green River shale formations of Colorado, Utah, and Wyoming (Figure 5.8) contain tremendous quantities of "shale oil," something like 2000 billion barrels. (The United States, in the early 1970s, used about 5 billion barrels per year.) Not all is recoverable, but from the highest grade deposits there may be about 80 billion barrels of oil recoverable by early 1970 technology. When this shale is crushed and heated to about 260°C a substance essentially the same as petroleum is released. The rock is called oil shale if it holds an average of 10 gallons per ton. The environmental problems created by working the oil shales are awesome. The waste rock is less dense than the original shale, which means that it cannot simply "be put back." Moreover, its salt content will be attacked by rains and melting snow sending salt into streams and rivers, waters needed downstream for irrigation.

[1]The breeder reactor is a special nuclear reactor, now being developed, that will extend greatly our rapidly disappearing supplies of uranium "fuel."

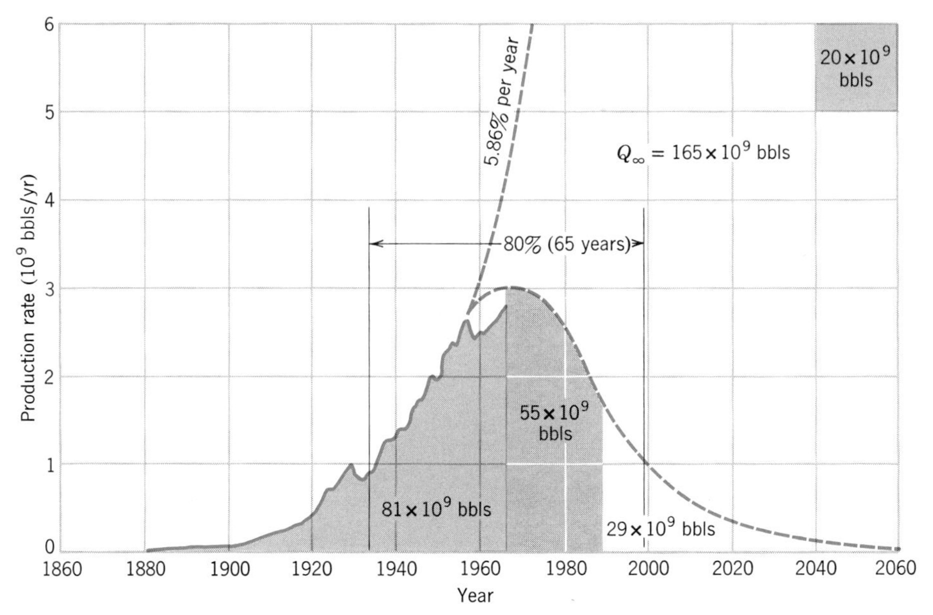

Figure 5.3
Complete cycle of production of crude oil in the United States including offshore wells but excluding Alaska. The dashed line pointing upward is the projection of production based on the rate that prevailed from 1935 to 1955—a rate of increase of 5.86% per year—an obviously impossible development. The cycle is based on a value of Q_∞ of 165 billion barrels. The addition of Alaskan oil would increase Q_∞ to about 190 billion barrels. (From *Resources and Man*, Publication 1703, Committee on Resources and Man, National Academy of Sciences-National Research Council, W. H. Freeman, and Company, San Francisco, 1969. Used by permission.)

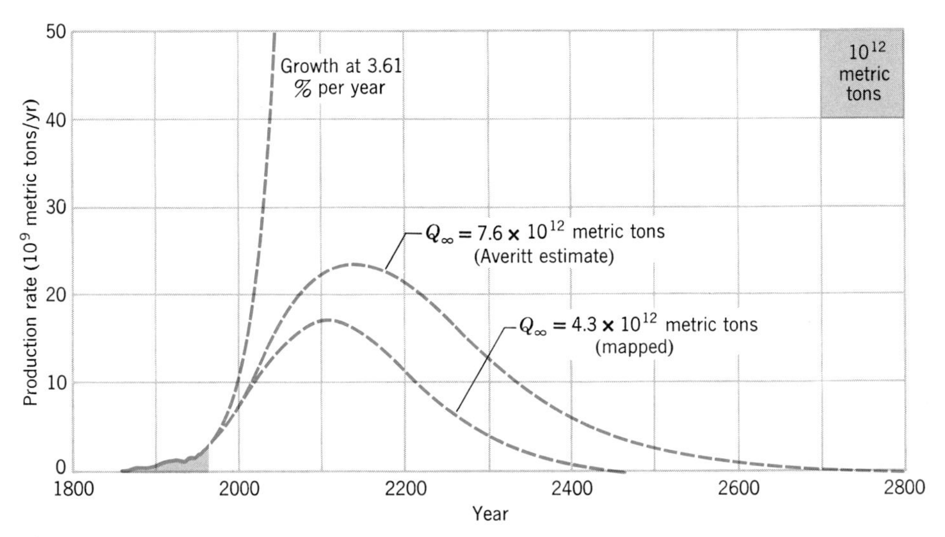

Figure 5.4
Two forecasts of the complete cycle of world coal production based on two values of Q_∞. (From *Resources and Man*, Publication 1703, Committee on Resources and Man, National Academy of Sciences-National Research Council, W. H. Freeman and Company, San Francisco, 1969. Used by permission.)

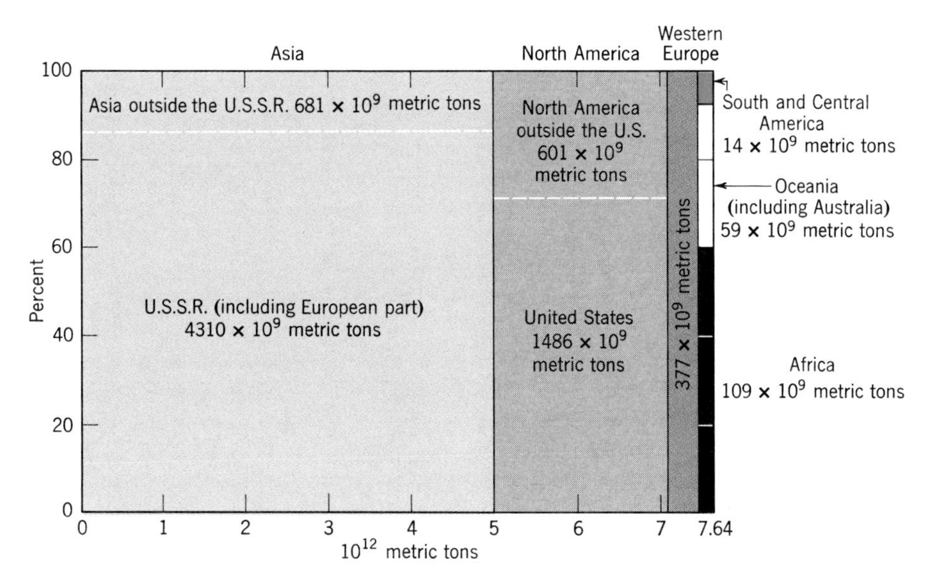

Figure 5.5
Estimates of the world resources of minable coal and lignite, based on data by Paul Averitt, 1969. (From *Resources and Man,* Publication 1703, Committee on Resources and Man, National Academy of Sciences–National Research Council, W. H. Freeman, and Company, San Francisco, 1969. Used by permission.)

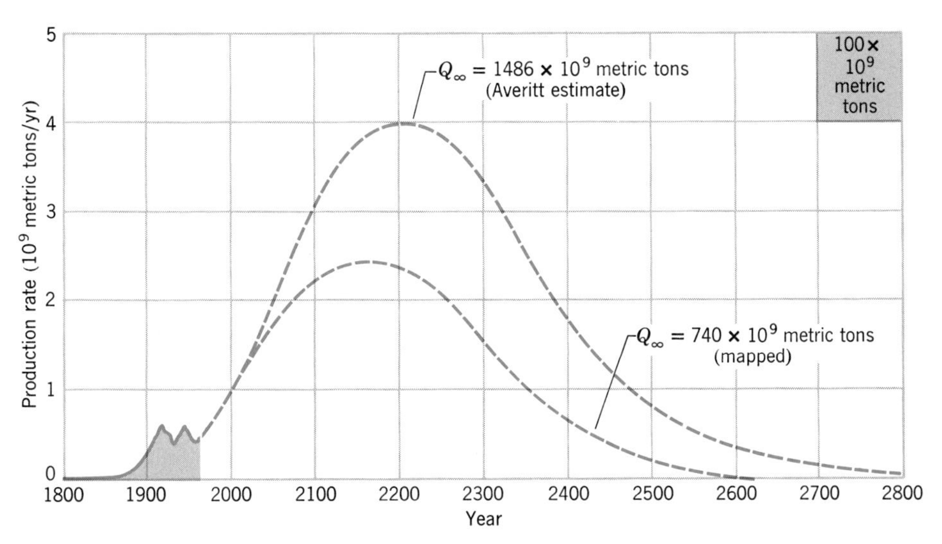

Figure 5.6
Complete cycles of production of United States coal based on two values of Q_∞. At the higher estimate the year of peak production of 2220 A.D. represents an eightfold increase over the present rate of production. The lower estimate, peaking at 2170, represents a fivefold increase over the present rate of production. (From *Resources and Man,* Publication 1703, Committee on Resources and Man, National Academy of Sciences–National Research Council, W. H. Freeman, and Company, San Francisco, 1969. Used by permission.)

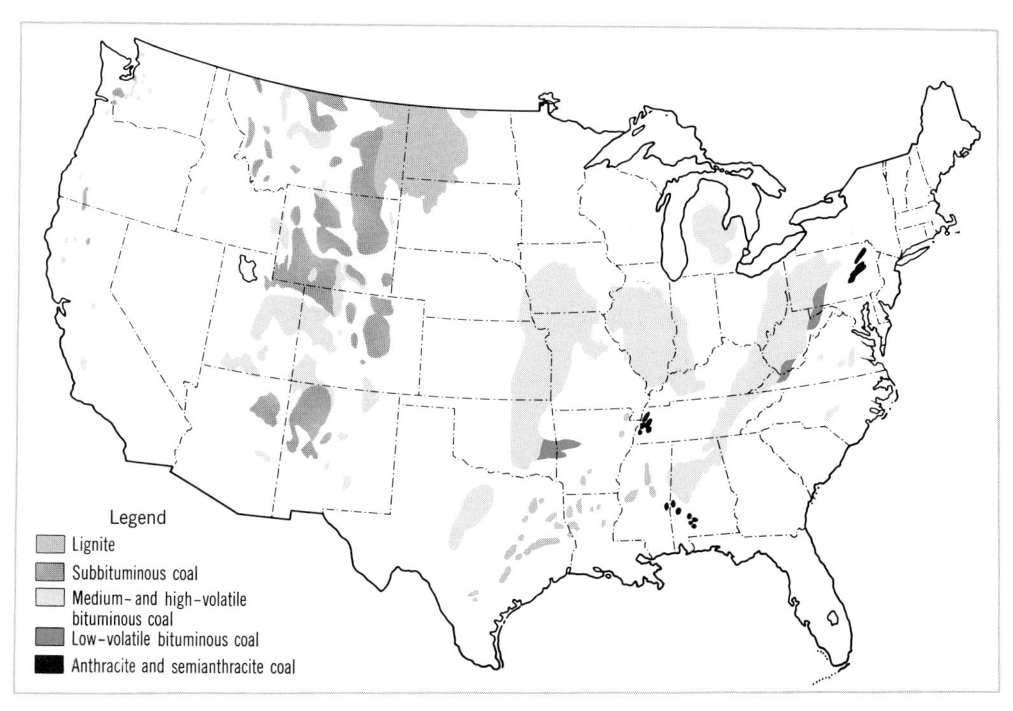

Figure 5.7
Coal fields of the conterminous United States. (From: "Coal Resources of the United States, Jan. 1, 1967." Geological Survey Bulletin 1275, U. S. Government Printing Office, 1969.)

5.6 Oil Sands

In areas of Alberta, Canada, roughly equal in size to Lake Michigan, there are enormous reserves of bitumen mixed with sand, the Athabasca oil sands. When mixed with hot water the sand sinks, and the bitumen rises. Further processing yields a material essentially like crude oil. Each ton of sand yields about two-thirds of a barrel of oil. By present technology about 14 billion barrels can be recovered. Over 600 billion barrels of oil, however, are believed to be present, which is roughly the same as half the entire petroleum reserves of the world.

Only a few years ago virtually no one could bring himself to believe we could ever run out of petroleum or coal. Now almost everyone is painfully aware that we shall. The figures and graphs presented here not only bring this home, they also should prod us into thinking about another important use of the chemicals in the fossil fuels, their use as petrochemicals for drugs, dyes, plastics, and many other things. To set the stage for more study of this, let us go down to the molecular level of the fossil fuels and find out more about their chemical composition.

5.7 Petroleum

Petroleum (*petra,* "rock;" *oleum,* "oil") is a complex mixture of organic substances together with some water. When the water and organic gases are removed, what

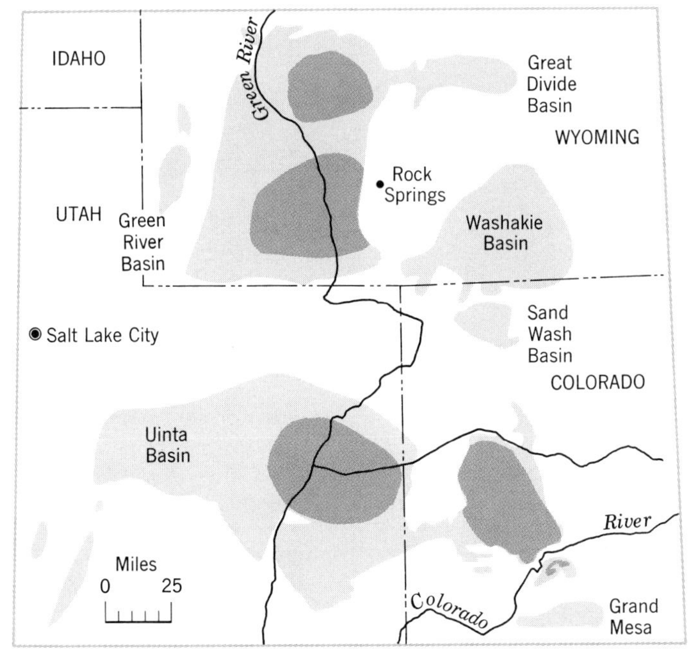

Figure 5.8
Distribution of oil shale in the Green River formation of Wyoming, Utah, and Colorado. The darkest-colored areas hold oil shale yielding 25 or more gallons of oil per ton of shale lying in layers more than 10 feet thick. (Source: U. S. Geological Survey Circular 523, Department of the Interior.)

remains is *crude oil.* Nearly all of the organic compounds are hydrocarbons, but crude oil contains from 1 to 6% nitrogen, oxygen, and sulfur. The object of petroleum refining is to separate it into products of varying uses. Table 5.1 contains a list of the principal fractions. Each fraction is a mixture of hydrocarbons having an overall volatility that makes it useful in certain kinds of engines or furnaces. Residues that do not distill yield some special products, but most are used as residual fuel oil ("resid") or are made into coke and asphalt. An estimated 500 compounds occur in the fractions that range in boiling points up to 200°. In the lower ranges, roughly a third are alkanes, a third are cycloalkanes, and a third are aromatic hydrocarbons.

5.8 Synthetic Gasoline

The gasoline fraction of crude oil is not large enough by far to supply world needs. Molecules in the gasoline fraction have from 5 to 10 or 12 carbons apiece. We can make such molecules by "cracking" larger molecules or by "alkylating" smaller ones. Even though alkanes are very stable, thermally, if they are heated (in the absence of air) well above their boiling points, their molecules will "crack" and sometimes undergo rearrangements. The operation is called **thermal cracking.** When it is done on the mixture in the gas oil range of crude oil, the product is naturally an exceedingly complex mixture of lower boiling alkanes and alkenes. We merely illustrate one of the myriad possibilities:

Table 5.1
Principal Fractions from Distilling Petroleum

Boiling Point Range, in °C (a measure of volatility)	Number of Carbons in Molecules of the Fraction	Uses
Below 20	C_1–C_4	Natural gas; heating and cooking fuel; Bunsen burner fuel; raw material for other chemicals
20–60	C_5–C_6	Petroleum "ether"; nonpolar solvent and cleaning fluid
60–100	C_6–C_7	Ligroin or light naphtha; nonpolar solvent and cleaning fluid
40–200	C_5–C_{10}	Gasoline
175–325	C_{12}–C_{18}	Kerosene; jet fuel; tractor fuel
250–400	C_{12} and higher	Gas oil; fuel oil; diesel oil
Nonvolatile liquids	C_{20} and up	Refined mineral oil; lubricating oil; grease (a dispersion of soap in oil); resid
Nonvolatile solids	C_{20} and up	Paraffin wax (purified solids that crystallize from some oils); asphalt and tar for roads and roofing

$$\underset{\text{OCTADECANE}}{CH_3(CH_2)_{16}CH_3} \xrightarrow{500^\circ} \underset{\text{1-DECENE}}{CH_3(CH_2)_7CH{=}CH_2} + \underset{\text{OCTANE}}{CH_3(CH_2)_6CH_3}$$

$$\underset{\text{(a mixture)}}{\text{"Gas oil"}} \xrightarrow[\Delta]{} \text{Low octane gasoline (octane} \approx 70)$$

It is as if something like the following shuffling of bonds occurred:

$$R{-}\overset{\overset{\displaystyle H}{|}}{C}H{-}\overset{\overset{\displaystyle R'}{|}}{C}H_2 \xrightarrow[\Delta]{} R{-}CH{=}CH_2 + H{-}R'$$

The cracking of octadecane, illustrated above, is true to one fact about thermal cracking. Normal alkanes tend to crack to normal hydrocarbons. Thermal cracking does not tend to increase the population of highly branched hydrocarbons or of aromatics. Therefore the product is not of very high octane rating (Section 5.9), having a rating of about 70 octane. Three very important developments, however, helped solve this problem: catalytic cracking, catalytic reforming, and lead additives.

If cracking is done in the presence of a mixed alumina-silica powder (or granules), the proportions of branched hydrocarbons increases. The product, therefore, has a much higher octane rating. **Catalytic cracking** does not so much permit lower cracking temperatures as it encourages the rearrangements of carbon chains from straight to branched. The advantage is that higher octane gasoline is obtained, gasoline of about 90 octane, without additional reforming. Again, the reactions are numerous, and we can only illustrate.

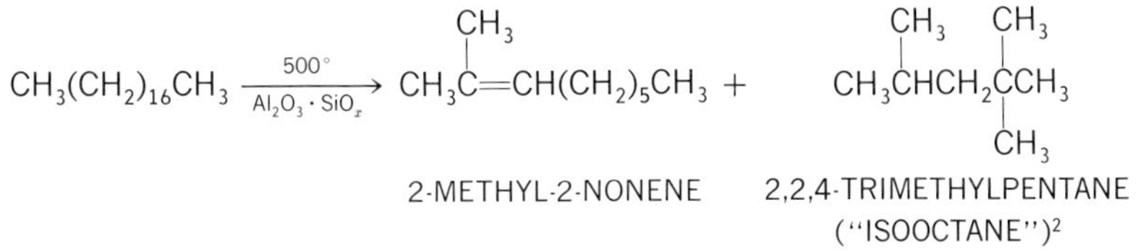

$$CH_3(CH_2)_{16}CH_3 \xrightarrow[Al_2O_3 \cdot SiO_x]{500^\circ} CH_3\overset{CH_3}{\overset{|}{C}}{=}CH(CH_2)_5CH_3 + CH_3\overset{CH_3}{\overset{|}{C}}HCH_2\underset{CH_3}{\underset{|}{\overset{CH_3}{\overset{|}{C}}}}CH_3$$

2-METHYL-2-NONENE 2,2,4-TRIMETHYLPENTANE ("ISOOCTANE")[2]

Reforming is an operation in which lower octane gasoline (richer in *n*-hydrocarbons) is passed over a special catalyst such as platinum or alumina to produce a higher octane gasoline, one richer in branched alkanes. The process also dehydrogenates some hydrocarbons. The following reactions only illustrate:

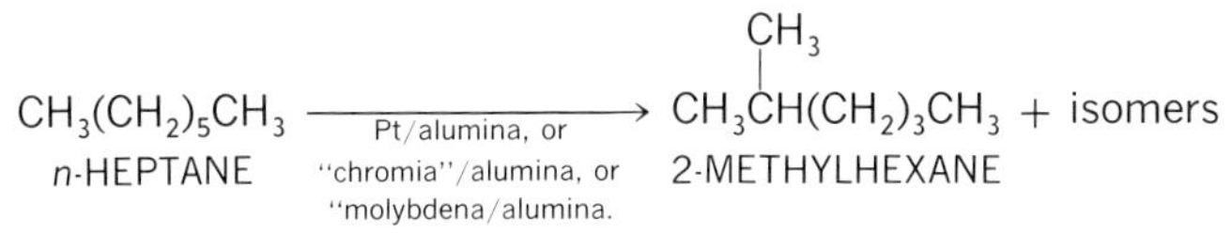

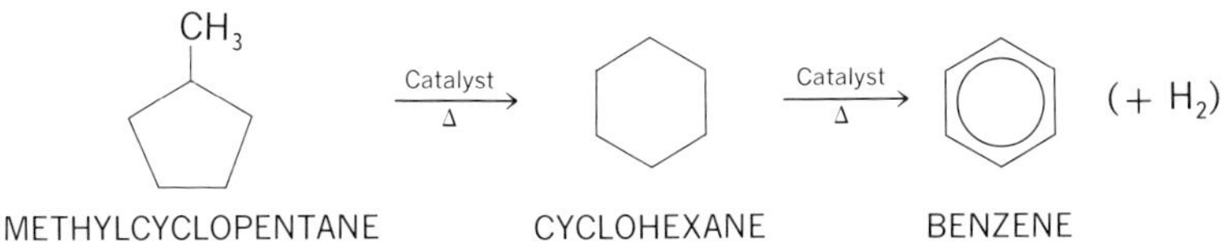

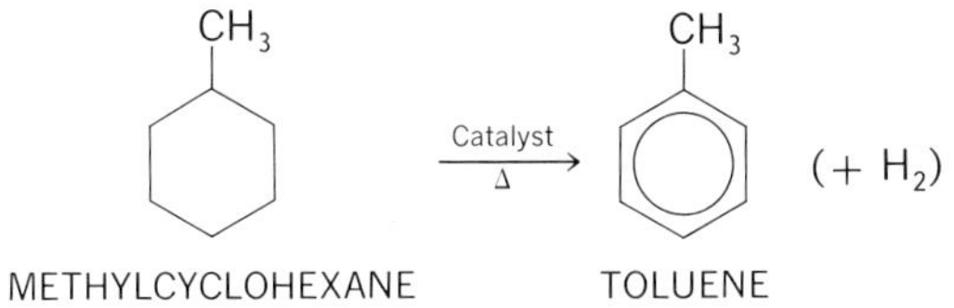

$$\underset{n\text{-HEPTANE}}{CH_3(CH_2)_5CH_3} \xrightarrow[\text{"chromia"/alumina, or "molybdena/alumina.}]{\text{Pt/alumina, or}} \underset{\text{2-METHYLHEXANE}}{CH_3\overset{CH_3}{\overset{|}{C}}H(CH_2)_3CH_3} + \text{isomers}$$

$$\text{METHYLCYCLOPENTANE} \xrightarrow[\Delta]{\text{Catalyst}} \text{CYCLOHEXANE} \xrightarrow[\Delta]{\text{Catalyst}} \text{BENZENE} \quad (+ H_2)$$

$$\text{METHYLCYCLOHEXANE} \xrightarrow[\Delta]{\text{Catalyst}} \text{TOLUENE} \quad (+ H_2)$$

The unsaturated cyclic compounds themselves are present, and some arise during reforming from open chain compounds. Reforming, in other words, can involve increased *branching, cyclizations,* and *aromatizations.*

Benzene and toluene have high octanes, and their formation means that reformed gasoline has a higher octane than prior to the operation. Aside from that, most of the benzene and toluene that are used as industrial chemicals are now made from petroleum by reforming, cyclizations, and aromatizations. (See Figure 5.9.) These compounds, as we shall see, are of major importance in the synthesis of plastics, drugs, dyes, and other substances of commercial value.

Alkylation is still another operation for extending the supply of high quality gasoline. Where cracking makes smaller molecules from the larger, alkylation does the opposite. It makes larger molecules from the smaller. Using concentrated sulfuric acid or hydrofluoric acid as the catalyst, isobutane can be made to add to isobutylene:

[2]The common industrial name, "isooctane," is a normally unacceptable use of "iso."

Figure 5.9
Facilities for producing aromatic chemicals. Ethylbenzene and *o*-xylene units are on the left. On the right is an extraction unit for removing benzene, toluene and xylenes from gasoline fractions of crude oil and reformed products. (Courtesy Tenneco Oil Company. View is of their refinery at Chalmette, Louisiana.)

$$\underset{\text{ISOBUTANE}}{CH_3C(CH_3)_2-H} + \underset{\text{ISOBUTYLENE}}{CH_2{=}C(CH_3)-CH_3} \xrightarrow[\text{HF}]{H_2SO_4 \text{ or}} \underset{\text{"ISOOCTANE" (2,2,4-TRIMETHYLPENTANE)}}{CH_3-C(CH_3)_2-CH_2CH(CH_3)CH_3}$$

The cracking that we discussed earlier is one of the sources of these lower formular weight hydrocarbons. Alkylation is by no means limited to this example. Benzene can be made to add to ethylene or propylene:

$$\underset{\text{BENZENE}}{C_6H_6} + CH_2{=}CH_2 \xrightarrow[95°,\ \text{pressure}]{AlCl_3-HCl} \underset{\text{ETHYLBENZENE}}{C_6H_5CH_2CH_3}$$

$$\underset{\text{BENZENE}}{C_6H_6} + CH_2{=}CH-CH_3 \xrightarrow[H_3PO_4,\ \text{pressure}]{250°} \underset{\text{CUMENE (ISOPROPYLBENZENE)}}{C_6H_5CH(CH_3)_2}$$

Ethylbenzene is a raw material for styrene, monomer for polystyrene. Cumene, among other uses, is the raw material for both phenol and acetone. In the early 1970s ethylbenzene production was over 6 billion pounds per year, placing it among the top 20 of *all* chemicals produced by the chemical industry.

One of the purposes of all these operations is to make a larger supply of high quality gasoline. One of the ways of measuring quality is by the octane rating.

5.9 Octane Number

When gasoline performs poorly the engine makes pinging sounds. We say that it "knocks." Performance is measured in carefully standardized laboratory engines that are fueled by the gasoline being rated. The compound *n*-heptane is taken as the standard of bad performance, and it has an octane rating of zero. An isomer of C_8H_{18}, 2,2,4-trimethylpentane, is taken as the standard of excellence, and it has an octane rating of 100. The special engines are run with varying combinations of *n*-heptane and 2,2,4-trimethylpentane. The test runs give a scale of performance from zero to 100 octane. A particular gasoline fraction is then rated by comparing its performance with the performance of standard combinations of heptane and isooctane.

Since this procedure was developed, fuels that performed even better than isooctane were found. Hence, we now can have fuels rated above 100 octane.

The most common means of improving the octane rating of a gasoline is to add tetraethyllead, $(C_2H_5)_4Pb$, about 2–3 cc per gallon of gasoline. Ethyl Fluid® consists of about 60% tetraethyllead and 38% lead "scavengers," ethylene bromide and ethylene chloride that react with the lead and change it into a form volatile enough to leave the cylinder of the engine via the exhaust. The remainder is dye and kerosene. Tetramethyllead, $(CH_3)_4Pb$, is another useful antiknock compound.

5.10 Coal and Coal By-Products

Most of the coal mined in the world is used to heat buildings or to generate steam for making electricity. However, some bituminous coal is used to make coke and chemicals.

If soft coal is heated in the absence of air to temperatures ranging from 900° to 1200°C it breaks down to coke and a complex mixture of chemicals. The process is called destructive distillation or carbonization. From one ton of coal there is produced 1500 pounds of coke, 11,000 cubic feet of coal gas, 10 gallons of coal tar, and 3.5 gallons of light oil. The gas includes ammonia. If it is removed by reacting with sulfuric acid about 20-30 pounds of ammonium sulfate are made.

Coke is impure carbon, and it is needed by the steel industry to reduce iron oxides to iron.

Coal oil consists mostly of simple aromatic hydrocarbons—benzene, toluene, and the xylenes.

Coal tar is a thick, sticky, black fluid containing a huge number of commercially important substances. Three general types can be separated: the hydrocarbons, tar acids, and tar bases. The hydrocarbons are mostly aromatic. Roughly 10% of coal tar is naphthalene and 4% is phenanthrene. A large number of other polynuclear aromatic hydrocarbons are present, and there are procedures for separating and purifying all of them.

The tar acids are mostly a large variety of phenols (Section 6.32) including phenol itself, all the cresols, the xylenols, the naphthols, and many others. They are raw materials for drugs, dyes, and plastics.

The tar bases are mostly heterocyclic nitrogen compounds such as pyridine and quinoline (Chapter 16) and aromatic amines such as aniline and the toluidines (Chapter 8). These are used to make drugs and dyes and some plastics.

Besides these, coal tar contains small concentrations of sulfur compounds, nitriles, ketones and alcohols.

5.11 Coal Gasification

One of the great advantages of natural gas is that it can be transported by pipelines. Another advantage is that its use is almost pollution free. Our supplies of natural gas, however, are so short that they are more and more being reserved strictly for their best use—heating homes and small buildings. Coal, on the other hand, is in relatively plentiful supply. Shipping coal, however, requires railroads, although for short and intermediate distances it can be transported in pipes as a slurry in water. Coal also has varying amounts of sulfur, which changes to sulfur dioxide and some sulfur trioxide when the coal is burned. For these reasons the Office of Coal Research of the U. S. Department of the Interior has sponsored research aimed at converting coal, *where it is mined,* into the equivalent of natural gas.

A large number of methods are being studied. In one method, the coal, broken up or powdered, is mixed with steam at such a high temperature that water molecules crack, and hydrogen is produced along with carbon monoxide (from the coal).

$$C + H_2O \xrightarrow[\Delta\Delta]{} CO + H_2$$

The hydrogen–carbon monoxide mixture is then subjected to heat, pressure, and a catalyst to change it to methane:

$$3H_2 + CO \longrightarrow CH_4 + H_2O$$

During the various processes, steps are taken to remove sulfur and carbon dioxide. In a variation of this, the extra hydrogen needed is mixed with the steam to change coal directly to methane.

Gasification requires enormous quantities of water. Water is not only a chemical in the process, it is needed as a coolant. Unhappily, the western areas where lignitic coals are abundant (Figure 5.7) are not overly rich in water, given the needs for irrigation, too. We may expect continued controversy in this competition for water as well as in the debate over strip mining and restoration of strip-mined lands.

5.12 Coal Liquifaction

Over 95% of the organic substance in lignite can be changed to gases and liquid products by hydrogenating it under special conditions. Still in the development stage, the method produces both pipeline gas and a liquid rich in the gasoline fractions. In either form, the sulfur content can be made very low, and both can be shipped by pipelines.

5.13 Ethylene and Propylene as Petrochemicals

Besides methane and ethane, natural gas contains varying amounts of propane (3–18%) and butane (2–14%). Since these two gases are easily changed to liquid forms, they are easily removed for other uses. Quantities of them are marketed in cylinders as Skelgas or Philgas for heating and cooking purposes in rural areas (or campers and mobile homes). Perhaps more importantly, these alkanes can be cracked. For example:

$$2CH_3CH_2CH_3 \xrightarrow{500^\circ} CH_2{=}CH_2 + CH_2{=}CHCH_3 + CH_4 + H_2$$

and this makes available a larger supply of ethylene, propylene, and butylenes (from C_4 alkanes). As seen in Figure 5.10, ethylene is the raw material for a large number of commercially important products. Polyethylene (Section 3.25) is just one. By one or a series of reactions ethylene is also used to make Dacron and Mylar. (Section 10.19 has a fuller discussion of the chemistry of these plastics.) The two are chemically the same; however, Dacron is the name for the fiber form and Mylar the name of the film. Mylar is unusually resistant to tearing, making it possible to use very thin films in high-quality magnetic tapes. It is also an excellent insulator. They are made from ethylene glycol, which is important itself as a permanent antifreeze.

The production of ethyl chloride from ethylene leads to two extremely important catalysts. One, tetraethyllead (Section 5.9), has been discussed. The other, triethylaluminum, is one constituent of the famous Ziegler catalyst (Section 3.25).

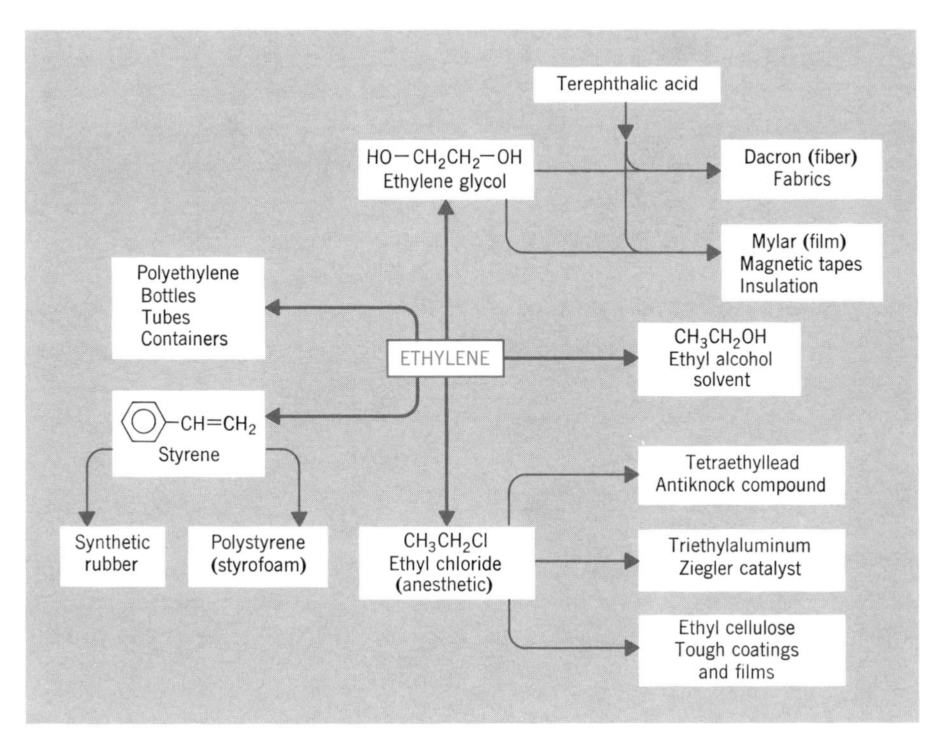

Figure 5.10
Some major uses of ethylene as a petrochemical.

Still another use of ethylene is in the production of styrene, a process also requiring benzene that can be obtained from petroleum (Section 5.8) or another fossil fuel—coal, via coal tar (Section 5.10). Styrene is a raw material for synthetic rubber as well as Styrofoam and other polystyrene plastics.

Ethyl alcohol is made in huge quantities from ethylene, and it is needed as an industrial solvent as well as a solvent for over-the-counter products such as pharmaceuticals.

For most of these products the actual chemistry of the reactions used to make them will be discussed later. Our purpose here is simply to make the case that petrochemicals are as much a part of our daily lives as the more obvious petroleum products: gasoline and oil.

As seen in Figure 5.11, propylene also has many uses. Making polypropylene is only one (Section 3.25).

Propylene is the major source of the glycerol needed to make nitroglycerine. Glycerol is also used to make alkyds, glyptal resins, and coatings and varnishes (Section 10.19). By a series of changes, propylene serves as the basis for the epoxy resins we use in fiberglass products, for the polycarbonates (Merlon and Lexan) so tough they are used to make bulletproof shields, for Bakelite (Section 9.27),

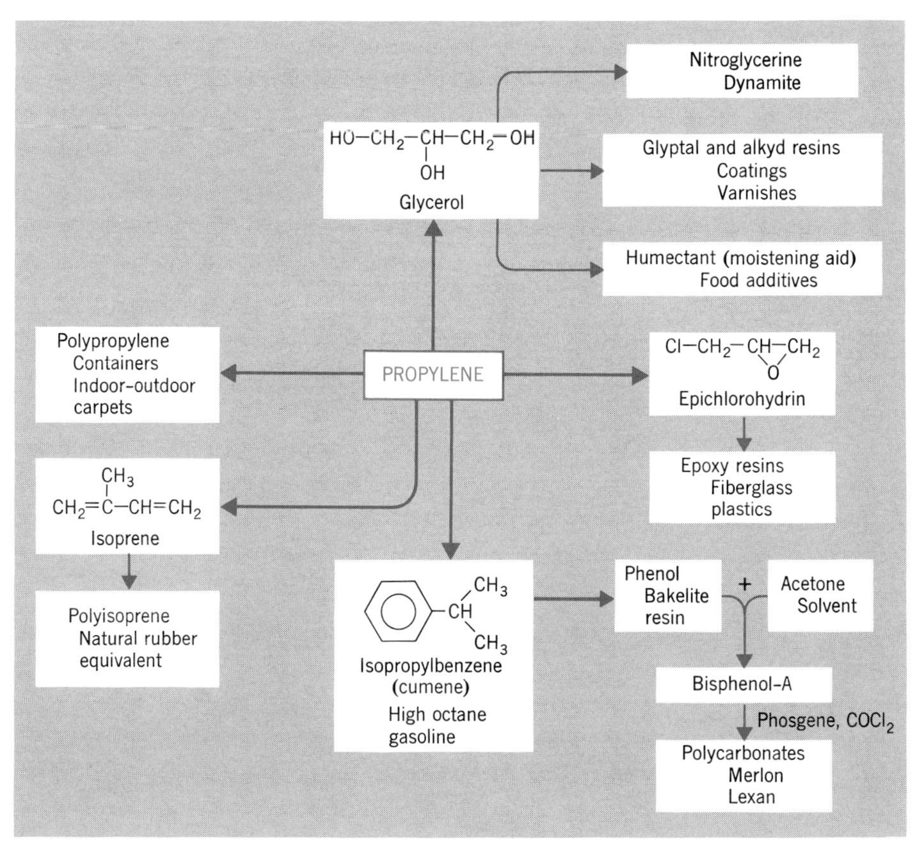

Figure 5.11
Some major uses of propylene as a petrochemical.

and for high-octane fuel. After several changes, propylene yields isoprene from which polyisoprene, essentially identical with natural rubber, is made (Section 3.33).

5.14 The C_4-Hydrocarbons as Petrochemicals

Figure 5.12 outlines several ways by which the various butanes and butylenes are used as raw materials. Two products stand out: high-octane gasolines and rubberlike polymers. Butadiene, after several changes, yields one of the important raw materials for nylon: hexamethylenediamine (Section 10.21).

5.15 Coal Tar Chemicals in Industry

Benzene is one of the important coal tar chemicals. This and the many other aromatic compounds from coal tar make coal a major source of many commercially important aromatic compounds. These range from important dyes (e.g., the aniline dyes) to pharmaceuticals (including aspirin) to pesticides, to plastics. Benzene may be changed to both raw materials for nylon (Section 10.21). The compound *p*-xylene is used to make Dacron, because *p*-xylene is oxidized to terephthalic acid (Sections 4.23, 10.19).

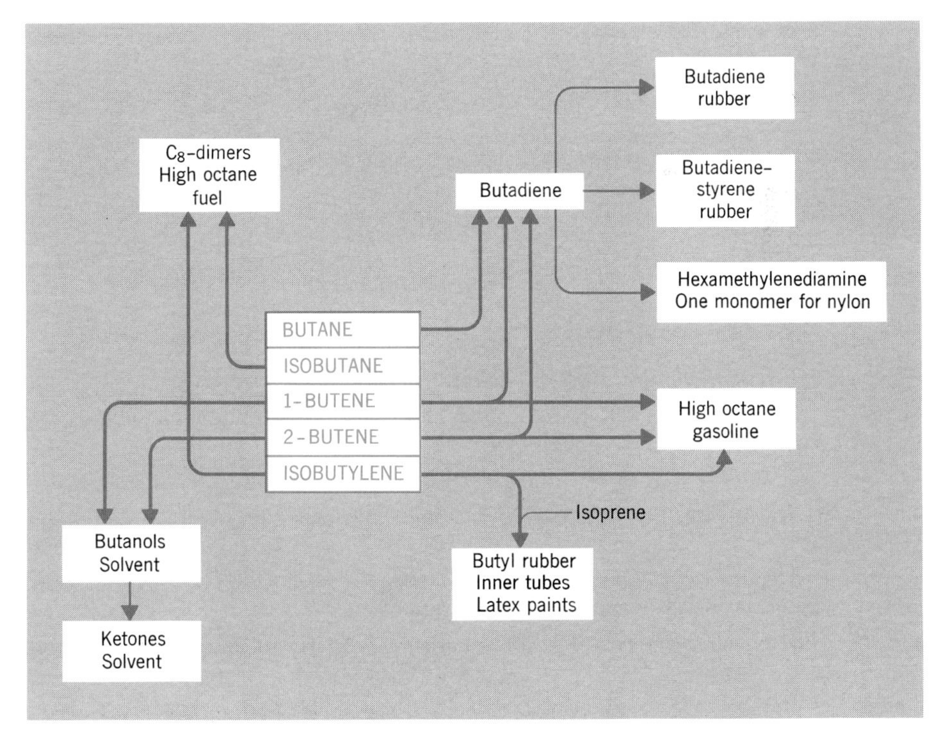

Figure 5.12
Uses for the C_4 petrochemicals.

Selected References

Books (Encyclopedias are good sources of general articles on the various fossil fuels and their technology.)

1. Preston Cloud (committee chairman). *Resources and Man,* 1969. National Academy of Sciences—National Research Council and W. H. Freeman and Company, San Francisco. A study of the adequacy of global resources—food, minerals, and energy—to meet the world's population. It includes a chapter, "Energy Resources," by M. King Hubbert.
2. R. T. Wendland. *Petrochemicals. The New World of Synthetics,* 1969. Doubleday & Company, Inc. This Anchor Book is one of the "Chemistry in Action" Series sponsored by the Manufacturing Chemists Association. It is an excellent introduction to the reactions by which petrochemicals are changed to useful products.
3. R. L. Hershey, M. Harris, and M. Tishler (committee cochairmen). *Chemistry in the Economy,* 1973. American Chemical Society, Washington D. C. Several chapters of this important book deal with materials that we would not have without petrochemicals and substances from coal. These include chapters on "Organic Chemicals," "Plastics and Resins," "Textile Fibers," "Natural and Synthetic Rubber," "Protective Coatings," "Pharmaceuticals," "Soaps and Detergents," and "Pesticides."
4. J. K. Stille. *Industrial Organic Chemistry.* Prentice-Hall, Inc., Englewood Cliffs, New Jersey, 1968. A nice survey, family by family, of industrial organic reactions. Chapter 1 is about industry's ways of changing hydrocarbons into useful fuels or petrochemicals.

Articles

1. A. M. Squires, "Clean Power from Coal." *Science,* August 28, 1970, page 821; T. H. Maugh II, "Gasification: A Rediscovered Source of Clean Fuel," ibid., October 6, 1972, page 44; and A. M. Squires, "Clean Power from Dirty Fuels," *Scientific American,* October 1972, page 26. Three articles about getting natural gas or liquid gasoline from coal.

Brief Summary

1. The three fossil fuels are petroleum, natural gas, and coal.

2. When water and gaseous materials are removed from petroleum the remainder is crude oil.

3. When crude oil is distilled, fractions of progressively higher boiling point ranges are taken for various end uses, mostly as fuels for different kinds of heating devices or engines.

4. Crude oil fractions are predominately saturated hydrocarbons whether open chain or cyclic. Roughly a third are aromatic.

5. A material very much like crude oil can be obtained from heating crushed oil shale to 260° or by mixing oil sand with hot water.

6. The world's reserves of petroleum, both proven and estimated, will probably be used up, for all practical, economic purposes, sometime in the next century.

7. The world's reserves of coal are still enormous, but even they would not last long

if the rate of annual increase in production continues as it was a few years ago.

8. Catalytic cracking of kerosene and reforming of unbranched alkanes are major processes for making gasoline.

9. The octane number of a gasoline is a number on a scale of zero to (now) 115 that discloses how that gasoline's engine performance compares with mixtures of *n*-heptane and "isooctane," 2,2,4-trimethylpentane.

10. The destructive distillation or carbonization of bituminous coal changes it to coke and releases coal oil (mostly benzene, toluene, and the xylenes) and coal tar.

11. Coal tar is processed to yield a variety of polynuclear aromatic hydrocarbons, heterocyclic compounds, aromatic amines, and phenols, all of which are important intermediates in the chemical industry.

12. Coal can be "gasified" by heating it at very high temperatures with water. This makes a mixture of carbon monoxide and hydrogen that can be converted, with additional hydrogen, into methane.

13. Coal can be "liquified" by heating it with hydrogen at high temperatures in the presence of suitable catalysts.

14. Coal gasification and liquification are processes designed to convert coal into a cleaner fuel, environmentally, and to make it easier to ship the energy in coal.

15. Ethylene, propylene, and the various butylenes and butanes are some of the most important petrochemicals. They are raw materials for plastics, synthetic rubber, pharmaceuticals, antifreeze, explosives, catalysts, resins, and high octane gasoline.

Questions

1. In terms of composition, what is the principal difference between anthracite and bituminous coal? Between bituminous coal and lignite?
2. What does the occurrence of vast deposits of oil shale in the mountainous western states suggest about their early geologic history?
3. How is water used as a *chemical* in coal gasification?
4. When catalysts were discovered for cracking high boiling hydrocarbons (kerosene range) to gasoline, all the petroleum refiners switched to the use of catalysts. What was the economic advantage of doing this?
5. In the late 1960s Shell Oil Company advertised that their gasoline contained "Platformate"®. The term is made from the words platinum and reforming. What was "Platformate"® most likely made from and by what process?
6. Which is richer in aromatics, coal tar or crude oil?
7. Name three important polymeric materials that are dependent on ethylene as a raw material.
8. Name six polymeric materials dependent on propylene as a raw material.
9. Write an equation that illustrates each reaction.
   (a) Thermal cracking of hexadecane.
   (b) Catalytic cracking of hexadecane.
   (c) Reforming of *n*-octane.
   (d) Aromatization of 1,4-dimethylcyclohexane.

chapter six

Alcohols, Phenols, and Ethers

Structures of Alcohols, Phenols and Ethers

6.1 Alcohols

Alcohols are compounds whose molecules contain a hydroxyl group, —O—H, attached to a saturated carbon. If the carbon holding an —OH group is unsaturated, as in phenols, the compound is classified in some other way.

—C—O—H (Alcohol group: O—H; Saturated carbon: C)

ESSENTIAL FEATURES OF ALL ALCOHOLS

PHENOL (TWO ACCEPTED SYMBOLS)

6.2 Subclasses of Alcohols

It is sometimes useful to classify an alcohol as primary (1°), secondary (2°), or tertiary (3°) according to the condition of the carbon holding the —OH group. If that carbon has just one other carbon directly attached to it, the alcohol is a primary alcohol. If there are two carbons joined directly to the carbon bearing the —OH group, the alcohol is secondary; if three, tertiary.

R—CH₂—OH: PRIMARY ALCOHOL (1°)

R—CH(R′)—OH: SECONDARY ALCOHOL (2°)

R—C(R′)(R″)—OH: TERTIARY ALCOHOL (3°)

As the primes on the R-groups indicated, these groups need not be identical. Several alcohols are listed in Table 6.1.

6.3 Polyhydric Alcohols

Compounds having two or more hydroxyl groups per molecule are quite common. To be stable, the —OH groups must be on separate carbons. With few exceptions, 1,1-diols are unstable.

Dihydric alcohols—those whose molecules have two —OH groups—are commonly called *glycols*. Glycerol is a common and important trihydric alcohol. Carbohydrates are polyhydric compounds.

$$\underset{OH}{CH_2}\underset{OH}{CH_2}$$
ETHYLENE GLYCOL (1,2-ETHANEDIOL, bp 197°C)

$$CH_3\underset{OH}{CH}-\underset{OH}{CH_2}$$
PROPYLENE GLYCOL (1,2-PROPANEDIOL, bp 189°C)

$$\underset{OH}{CH_2}CH_2\underset{OH}{CH_2}$$
TRIMETHYLENE GLYCOL (1,3-PROPANEDIOL, bp 214°C)

$$\underset{OH}{CH_2}-\underset{OH}{CH}-\underset{OH}{CH_2}$$
GLYCEROL (1,2,3-PROPANETRIOL, bp 290°C)

Table 6.1
Some Monohydric Alcohols

Name	Structure	Mp (°C)	Bp (°C)	Density (in g/cc at 20°C)	Solubility (in g/100 g water, 20°C)
C_1 Methyl alcohol	CH_3OH	−98	64.5	0.791	Soluble
C_2 Ethyl alcohol	CH_3CH_2OH	−115	78.3	0.789	Soluble
C_3 *n*-Propyl alcohol	$CH_3CH_2CH_2OH$	−127	97.2	0.803	Soluble
Isopropyl Alcohol	$(CH_3)_2CHOH$	−86	82.4	0.786	Soluble
C_4 *n*-Butyl alcohol	$CH_3CH_2CH_2CH_2OH$	−90	118	0.810	8.0
Isobutyl alcohol	$(CH_3)_2CHCH_2OH$	−108	108	0.802	10.0
sec-Butyl alcohol	$CH_3CH_2\underset{OH}{CH}CH_3$	−115	99.5	0.806	12.5
t-Butyl alcohol	$(CH_3)_3COH$	25.6	82.5	0.786	Soluble
Higher					
1-Pentanol (*n*-amyl alcohol)	$CH_3(CH_2)_3CH_2OH$	−79	138	0.814	2.2
1-Hexanol	$CH_3(CH_2)_4CH_2OH$	−53	157	0.820	0.6
1-Heptanol	$CH_3(CH_2)_5CH_2OH$	−34	176	0.822	0.1
1-Octanol	$CH_3(CH_2)_6CH_2OH$	−15	195	0.826	0.04
1-Decanol	$CH_3(CH_2)_8CH_2OH$	7	229	0.829	0.004
Cyclopentanol	(cyclopentane ring)—OH		140	0.949	Slightly soluble
Cyclohexanol	(cyclohexane ring)—OH	23	161	0.962	3.6
Allyl alcohol	$CH_2{=}CHCH_2OH$		97	0.852	Soluble
Benzyl alcohol	$C_6H_5CH_2OH$	−15	205	1.046	4

6.4 Phenols

Phenols are compounds whose molecules contain a hydroxyl group attached directly to a carbon in an aromatic ring. While phenols have many of the same properties as alcohols, there are enough differences to make it useful to have them in a separate family. One huge difference is that phenols have highly reactive aromatic ring systems. Another difference is that phenols are much more acidic than alcohols, which are neutral in water.

In Table 6.2 some common phenols are listed.

6.5 Ethers

The ethers are substances whose molecules are of any of the generic formulas:

R—O—R′ R—O—Ar Ar—O—Ar

where R is any aliphatic hydrocarbon group and Ar is any aromatic ring. A few ethers are listed in Table 6.3, which includes some common and important cyclic ethers. Ethylene oxide is a particularly important industrial chemical.

6.6 Occurrence

It is safe to say that we could not begin to understand most natural products without knowing about alcohols and about the particular kinds of reagents to which alcohol groups are vulnerable. The alcohol group is one of nature's most widely occurring functional groups. It occurs in all sugars and starches, cellulose (cotton), proteins, many natural and synthetic drugs, hormones, flavoring agents, and some vitamins.

Many simple alcohols—methyl alcohol ("wood alcohol"), ethyl alcohol ("grain alcohol"), ethylene glycol, propylene glycol, and glycerol, each to be discussed more later in the chapter—are important commercial chemicals.

Phenols are also widespread both in nature and in commerce. One of the amino acid "building blocks" of proteins—tyrosine—contains a phenolic system. The parent compound of this family, phenol itself, is used to manufacture aspirin and Bakelite-type resins that are used, for example, in telephone casings, in laminating agents for wood, in cloth and paper, and in protective coatings.

Other commercial uses of phenol include the synthesis of several bactericides, fungicides, herbicides, detergents, and softening agents (called plasticizers) for certain plastics.[1]

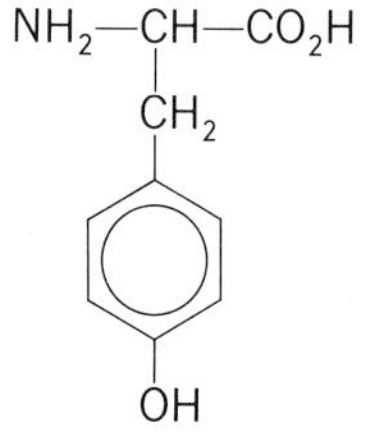

NH_2—CH—CO_2H | CH_2 | (benzene ring) | OH

TYROSINE
(AN AMINO ACID)

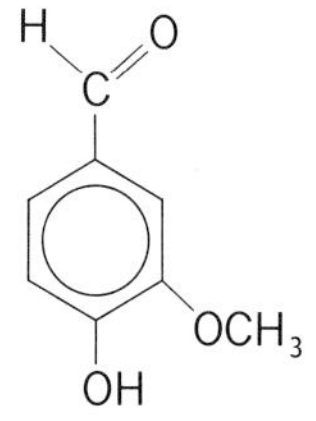

H—C(=O) | (benzene ring) with OCH_3 | OH

VANILLIN
(A FLAVORING AGENT)

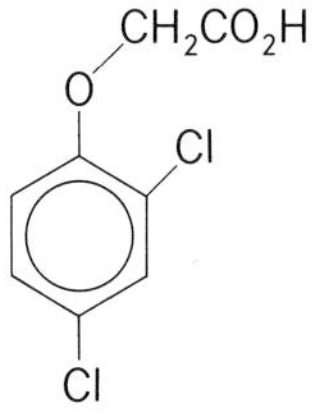

O—CH_2CO_2H | (benzene ring) with Cl | Cl

2,4-D
(A HERBICIDE)

[1]Bactericides are chemicals that kill bacteria.
Fungicides are chemicals used to control fungi, especially on seed grains.
Herbicides are chemicals used to control weeds and brush.

Table 6.2
Some Phenols

Name	Structure	Some Uses
Phenol (Mp, 41°C)	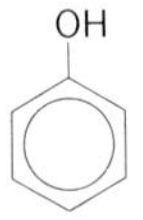 OH	Raw material for plastics, drugs, dyes and other commodities
o-Cresol (Mp 30°C)	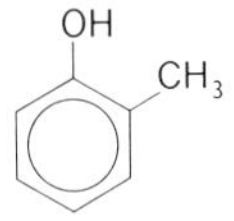 OH, CH_3	The cresols are raw materials for other chemicals and, as a mixture, used in disinfectants
m-Cresol (Mp 11°C)	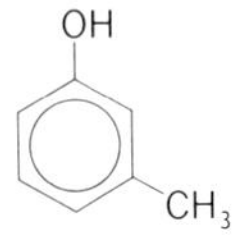 OH, CH_3	
p-Cresol (Mp 36°C)	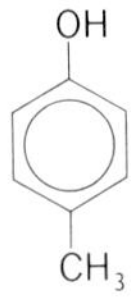 OH, CH_3	
Hydroquinone (Mp 170°C)	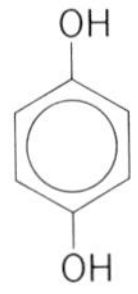 OH, OH	Photographic developer; chemical intermediate
Pyrogallol (Mp 133°C)	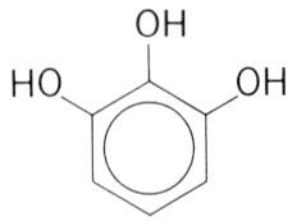 OH, HO, OH	Photographic developer; agent for combining with and removing oxygen from gas mixtures
BHA (Butylated hydroxy anisole)	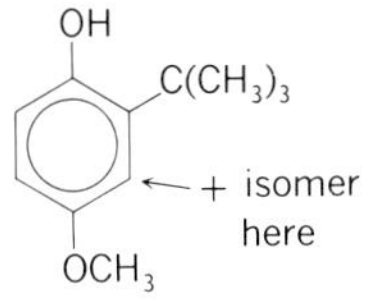 OH, $C(CH_3)_3$, OCH_3, ← + isomer here	BHA and BHT are widely used as antioxidants in gasoline, lubricating oils, rubber, edible fats and oils, and materials used for packaging foods that might turn rancid
BHT (Butylated hydroxyl toluene)	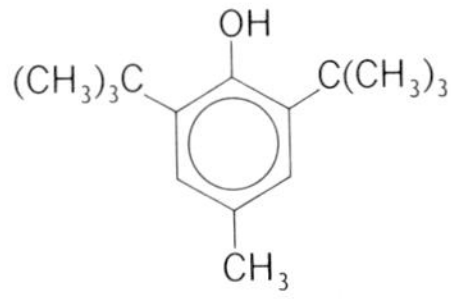 OH, $(CH_3)_3C$, $C(CH_3)_3$, CH_3	
PCP (Pentachloro-phenol)	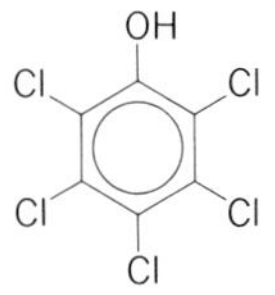 OH, Cl, Cl, Cl, Cl, Cl	One of the widely used fungicides, herbicides and wood preservatives in U.S. agriculture
n-Hexylresorcinol	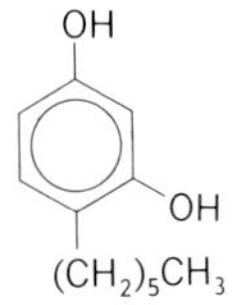 OH, OH, $(CH_2)_5CH_3$	Antiseptic

Table 6.3
Some Ethers

Name	Structure	Mp (°C)	Bp (°C)	Solubility in Water
Dimethyl ether	CH_3OCH_3	−138	−23	37 volumes gas dissolve in 1 volume water at 18°C
Methyl ethyl ether	$CH_3OCH_2CH_3$	−116	11	
Diethyl ether	$CH_3CH_2OCH_2CH_3$	−116	34.5	8 g/100 cc (16°C)
Di-*n*-propyl ether	$CH_3CH_2CH_2OCH_2CH_2CH_3$	−122	91	Slightly soluble
Methyl phenyl ether	$CH_3OC_6H_5$	−38	155	Insoluble
Diphenyl ether	$C_6H_5OC_6H_5$	28	259	Insoluble
Vinethene	$CH_2{=}CHOCH{=}CH_2$		29	
Cyclic Ethers				
Ethylene oxide	CH_2—CH_2 joined through O (three-membered ring)		10.7	Soluble
Tetrahydrofuran	five-membered ring containing O	−109	66	Soluble
Dioxane	six-membered ring with two O atoms (1,4-positions)	12	101	Soluble

The ether group also occurs widely in nature and in commercial products. (Both vanillin and 2,4-D, above, have ether linkages.) One of the very simple ethers, diethyl ether, is the widely used anesthetic commonly called simply "ether."

Nomenclature

6.7 Alcohols. Common Names

Common names are used almost exclusively for simple alcohols. If we have a name for the alkyl group joined to the —OH in an alcohol, we form the name by simply writing the word "alcohol" after the name of the alkyl group, as seen in these examples. Two important groups appear in these names. One is the allyl group, $CH_2{=}CHCH_2—$, and the other is the benzyl group, $C_6H_5CH_2—$. Both should be learned.

CH_3OH	$(CH_3)_3COH$	C_6H_5—CH_2OH
METHYL ALCOHOL	*t*-BUTYL ALCOHOL	BENZYL ALCOHOL

CH_3CH_2OH	$CH_3CH(CH_3)OH$	$CH_3CH(CH_3)CH_2OH$	$CH_2{=}CHCH_2OH$
ETHYL ALCOHOL	ISOPROPYL ALCOHOL	ISOBUTYL ALCOHOL	ALLYL ALCOHOL

6.8 Alcohols. IUPAC Rules for Names

1. For the "parent" structure, select the longest continuous chain of carbons *that includes the —OH group*. Name the alkane that corresponds to that chain. Then drop the final "-e" from that name and replace it by "-ol." For example:

CH_3OH	CH_3CH_2OH	$CH_3CH(CH_3)CH_2CH_2OH$
METHANOL	ETHANOL	-BUTANOL (INCOMPLETE)

2. When the —OH group can be variously located on the chain, specify its location by numbering the carbons of the parent chain from whichever end will give the location of the —OH group the lower number. For example:

$CH_3CH_2CH_2OH$	$CH_3CH(CH_3)OH$	$CH_3CH_2CH(CH_3)OH$	$CH_3CH(CH_3)CH_2CH_2OH$
1-PROPANOL	2-PROPANOL	2-BUTANOL	-1-BUTANOL (INCOMPLETE)

3. Note the names and the locations of any groups (e.g., alkyl, aryl, etc.) on the parent chain. Assemble these names and numbers before the parent name developed thus far. These examples show how it is done. Pay special attention to the use of commas and hyphens to make the final name one word.

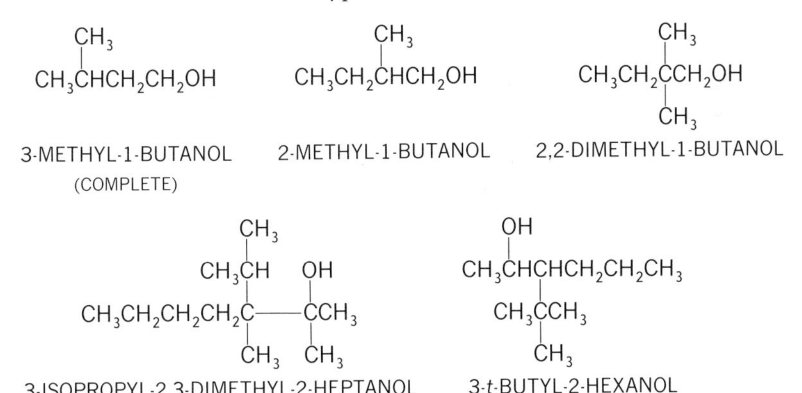

4. If other atoms or groups are attached to or incorporated into the parent chain, first apply rules 1–3 and then work the other groups into the name. For example:

$Br{-}CH_2CH(CH_3)CH_2CH_2OH$	$Cl{-}CH(Cl){-}CH(Cl){-}CH_2OH$
4-BROMO-3-METHYL-1-BUTANOL (*not* 1-BROMO-2-METHYL-4-BUTANOL)	2,3,3-TRICHLORO-1-PROPANOL

5. In a polyfunctional compound having an —OH group where it may be awkward to include the ending, "-ol," the —OH group may be named as a substituent, "hydroxy." For example, *p*-hydroxybenzoic acid.

6.9 Nomenclature of Phenols

"Phenol" is a parent name. Halogen or nitro derivatives of phenol are named with this as the basis, as seen in the examples.

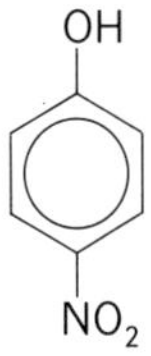

p-NITROPHENOL 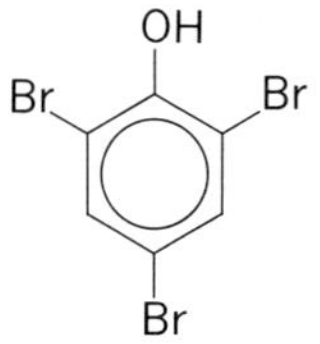   2,4,6-TRIBROMOPHENOL 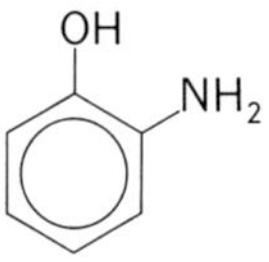   *o*-AMINOPHENOL

Many phenols, however, have common names that are used almost exclusively. The methylphenols, for example, are called *cresols*. See Table 6.2.

6.10 Ethers. Common Names

The two groups joined to oxygen are named, and the word "ether" is placed last. If the two groups are identical the prefix "di-" is used, but it is often omitted by working chemists.

$CH_3—O—CH_3$	$CH_3CH_2—O—CH_2CH_3$	$CH_3—O—CH_2CH_2CH_2CH_3$
DIMETHYL ETHER	DIETHYL ETHER "ETHER"	METHYL *n*-BUTYL ETHER (Note: the parts of the name are not run together.)

6.11 Ethers. IUPAC Names

Provided we can name the "R" in —O—R, we can make a name for —O—R as a substituent. One simply replaces the

$—O—CH_3$	$—O—CH_2CH_3$	$—O—C(CH_3)_2—CH_3$	$—O—C_6H_5$
METHOXY	ETHOXY	*t*-BUTOXY	PHENOXY

"-yl" ending of the R— group's name with "-oxy." You then proceed to name the ether as a substituted hydrocarbon. These examples show how it is done.

$CH_3CH_2—O—CH(CH_3)CH_2CH_3$	$CH_3—O—C(CH_3)=CHCH_2CH_2CH_2—Br$	$CH_3—O—C_6H_5$
2-ETHOXYBUTANE	6-BROMO-2-METHOXY-2-HEXENE	METHOXYBENZENE (ANISOLE)

Physical Properties of Alcohols and Phenols

6.12 Hydrogen Bonds

Ethane and methyl alcohol have similar formula weights (30 versus 32), yet ethane boils 153° below methyl alcohol (−88.6° versus 64.5°). This means that it takes far more energy to separate molecules of the alcohol from each other than to separate those of the alkane. Alcohol molecules cling to each other.

Even more striking is the fact that water, of lower formula weight than either, boils at a higher temperature than both, 100°.

Whenever hydrogen is bonded to a highly electronegative atom—F, N, or O—the bond is very polar. The hydrogen end bears a partial positive charge; the other end, a partial negative charge: $\overset{\delta-}{—O}\overset{\delta+}{—H}$. The molecules in a sample of water or in a sample of an alcohol, therefore, cannot help but attract each other. Moreover, since a hydrogen is a very small substituent, it is easy for it to get very close to a partial negative charge on a neighboring molecule. Since we are talking, of course, about opposite *partial* charges, the resulting force of attraction cannot be as great as we find in a salt crystal, for example, where ions of *full,* opposite charge attract each other very strongly. Nor can it be as great a force as is present when two groups are joined by a covalent bond. Nonetheless, the force *between* molecules in water or in an alcohol sample is enough to make it useful for us to call it a bond. Its name is the *hydrogen bond.*

The hydrogen bond (Figure 6.1) is a force of attraction that establishes a bridge between two highly electronegative atoms (F, O, or N). The hydrogen is in the center. It is *attached* to one of these atoms by a full, covalent bond—a strong bond. It is *attracted* to the other by a partial ionic bond, a much weaker bond, between sites of partial and opposite charge.

6.13 Solubility

Substances in which molecules are attracted to each other by hydrogen bonds are said to be *associated.* They often tend to dissolve in each other. The mutual solubility of water in methyl alcohol is an example.

Hydrocarbons are nonassociated substances. In order for a substance such as methane or any hydrocarbon to dissolve in water, for example, hydrogen bonds between water molecules would have to be broken and reduced in number. The energy cost of preventing water molecules from being nearest neighbors by trying to get hydrocarbon molecules between them cannot be repaid. The formation of new hydrogen bonds to replace the old ones cannot occur, because hydrocarbons do not have their own —OH groups. In contrast, methyl alcohol molecules can slip into the network of hydrogen bonds (or vice versa) in a water solution. See Figure 6.2. This is why low-formula-weight alcohols (and most di- or polyhydric alcohols) are soluble in water in all proportions. (Cf. Table 6.1.)

Straight-chain alcohols having five or more carbons are too hydrocarbonlike, and they are virtually insoluble in water.

Since all alcohols are partly hydrocarbonlike, it is possible for most of them to dissolve in typical "hydrocarbon solvents:" benzene, ether, carbon tetrachloride,

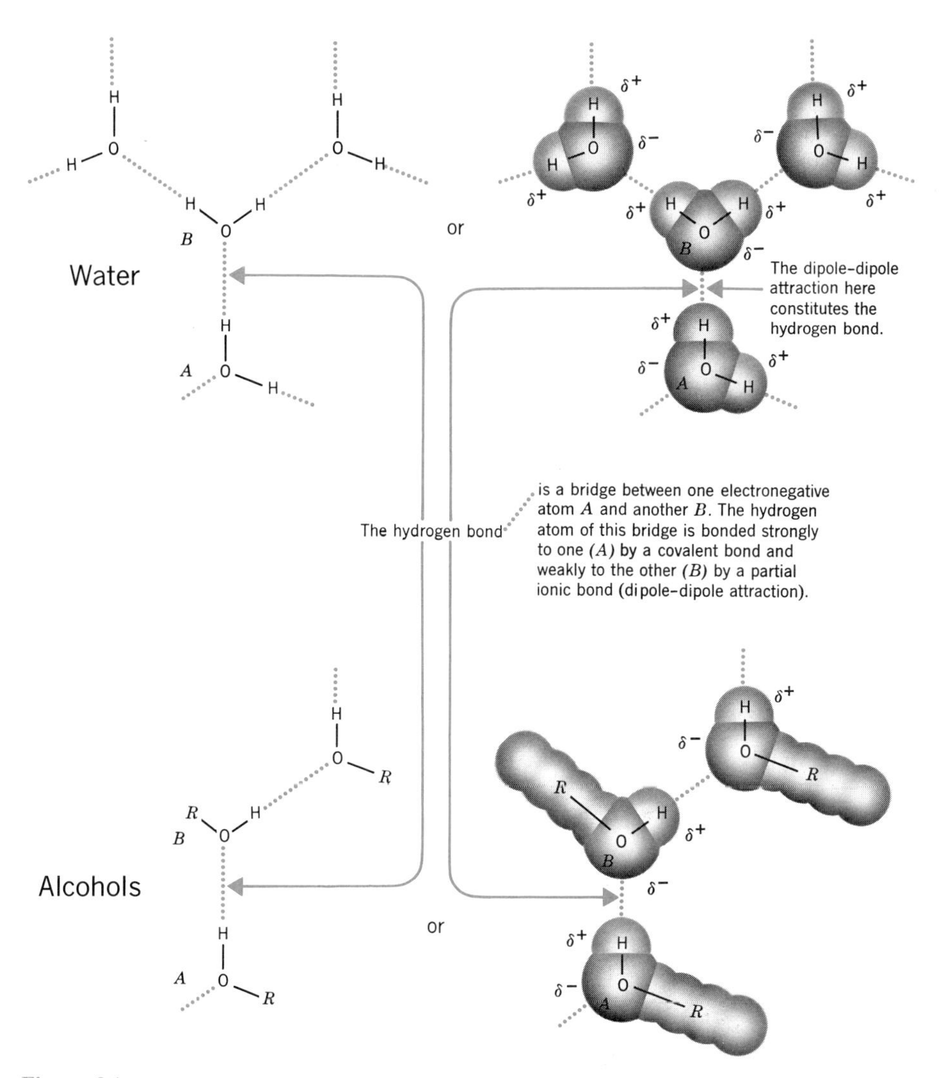

Figure 6.1
Hydrogen bonding in alcohols and water.

chloroform, gasoline, petroleum ether, and methylene chloride. Methanol and ethanol are not very soluble in petroleum ether, a low boiling mixture of alkanes commonly used as a solvent in organic laboratories.

To generalize, water tends to dissolve waterlike molecules; nonpolar solvents tend to dissolve hydrocarbonlike substances. To put it most briefly, "likes dissolve likes." Polar solvents dissolve polar solutes; nonpolar solvents dissolve nonpolar (or moderately polar) solutes.

How Alcohols are Made

Both laboratory and industrial methods for making alcohols involve functional groups either already studied or to be studied. That is why we shall leave the details

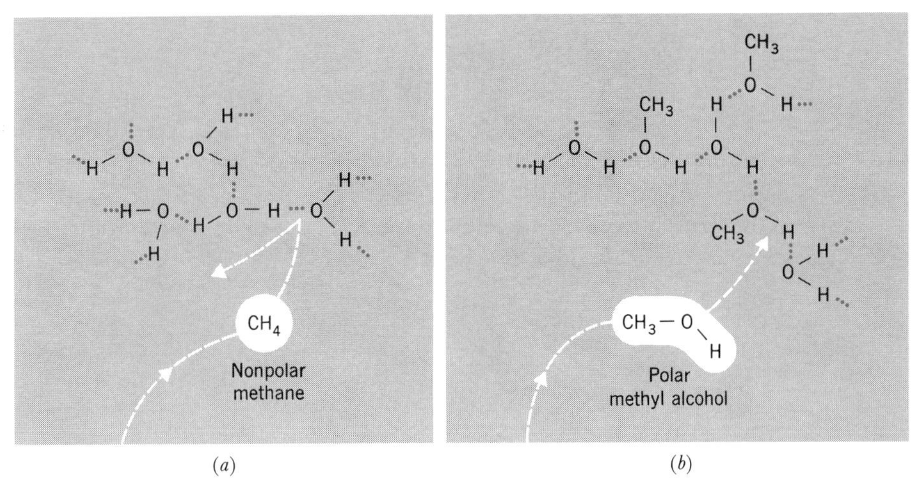

Figure 6.2
Understanding the solubility of short-chain alcohols in water. (*a*) A nonpolar alkane molecule cannot break into the hydrogen-bonded sequence in water. It cannot replace the hydrogen bonds that would have to be broken to let it in. Therefore it is insoluble in water.

(*b*) A short-chain alcohol molecule, capable of hydrogen bonding, can slip into the sequences in water. It can replace at least some of the hydrogen bonds that must be broken to let it in. In any liquid, of course, the molecules shift from place to place. The fixedness of this figure is not meant to indicate any inflexibility. The drawing may be compared to a high-speed flash photograph that has caught the action of the instant.

for some of the ways to make alcohols to other chapters. For the sake of completeness, all of the methods discussed in this text will be listed (see the summary box), and only two will be treated in detail here.

6.14 The Addition of Water to an Alkene

In Section 3.17 we studied the acid-catalyzed hydration of a double bond. Because of Markovnikov's rule this method cannot be used to make any primary alcohol (except ethyl alcohol). Another drawback is that the acid catalyst can cause a rearrangement in the alkene. The method we shall next study solves these problems.

6.15 Hydroboration-Oxidation of an Alkene

The net effect of this two-step synthesis of alcohols from alkenes is an anti-Markovnikov addition of water to a double bond without any complications (rearrangements). We can make primary alcohols by this method. The experimental conditions are mild; the yields are high.

Step 1. Reaction of an alkene with diborane, B_2H_6.

$$6\,RCH{=}CH_2 + B_2H_6 \longrightarrow \underset{\text{A TRIALKYLBORON}}{2\,(RCH_2CH_2)_3B}$$

Step 2. Oxidation and hydrolysis of the trialkylboron.

$$(RCH_2CH_2)_3B \xrightarrow[OH^-;\ H_2O]{H_2O_2} 3RCH_2CH_2OH + H_3BO_3$$

The net result:

$$RCH{=}CH_2 \xrightarrow[2.\ H_2O_2;\ OH^-;\ H_2O]{1.\ B_2H_6} RCH_2CH_2OH$$

Step 1 is the addition not of water but of BH_3 to the double bond. (B_2H_6 may be regarded as a dimer of BH_3.) This is the *addition* step, and it is not actually in violation of Markovnikov's rule. The positive portion of BH_3 is the boron; the negative part is hydrogen. The boron in BH_3 does not have an outer octet; it has a sextet of electrons. That is why it can attract the electrons of the pi bond.

$$R{-}CH{=}CH_2 + BH_3 \longrightarrow R{-}\overset{+}{C}H{-}CH_2{-}BH_3^- \xrightarrow[\text{Transfer}]{H:^-} R{-}CH_2{-}CH_2{-}BH_2$$

(This reacts further, two times, with additional alkene.)

The product of step 1 is not what we isolate. We oxidize and hydrolyze the trialkylboron to yield the alcohol. Thus the overall result appears to be anti-Markovnikov addition of water to a double bond.

SUMMARY—WAYS TO MAKE ALCOHOLS	CROSS REFERENCES
I From Aldehydes and Ketones	
A. Original molecular framework is kept.	
1. Reduction of aldehydes and ketones.	
$>C{=}O + H_2 \longrightarrow -\overset{\vert}{\underset{H}{C}}{-}O{-}H$	
If $R{-}\overset{O}{\overset{\Vert}{C}}{-}H$ (aldehyde) then, $R{-}CH_2{-}OH$ (1° alcohol)	
If $R{-}\overset{O}{\overset{\Vert}{C}}{-}R'$ (ketone) then, $R{-}\underset{R'}{\underset{\vert}{C}}H{-}OH$ (2° alcohol)	
Sources of hydrogen:	
(a) Molecular hydrogen (and special conditions of pressure, temperature and catalyst)	Section 9.17
(b) Metal hydrides (e.g., $LiAlH_4$)	Section 9.17
B. Molecular framework is enlarged. Grignard syntheses of 1°, 2°, and 3° alcohols.	Sections 6.16–6.17
II From Alkenes	
$>C{=}C< + H_2O \longrightarrow -\overset{H}{\overset{\vert}{\underset{\vert}{C}}}{-}\overset{\vert}{\underset{\vert}{C}}{-}OH$	
A. Markovnikov addition of water.	Section 3.17
B. Anti-Markovnikov addition of water. Hydroboration-Oxidation.	Section 6.15
III From Alkyl Halides	
$R{-}X + H_2O \longrightarrow R{-}OH + H{-}X$	Section 7.8
IV Fermentation	Section 14.37

6.16 The Grignard Synthesis of Alcohols

About 1900 Victor Grignard (1871–1935), a French chemist (Nobel Prize, 1912) discovered one of the most versatile reagents in all of organic chemistry. Named after him, the Grignard reagent is commonly written as RMgX.

The Grignard reagent forms from the reaction of magnesium metal with an alkyl halide or an aryl halide (preferably the aryl bromide) in diethyl ether.

$$\underset{\substack{\text{ALKYL HALIDE}\\(\text{X = Cl, Br, I})}}{\text{R—X}} + \text{Mg} \xrightarrow[\text{Diethyl ether}]{} \underset{\text{GRIGNARD REAGENT}}{\text{RMgX}}$$

This reagent can add to a carbon-oxygen double bond in aldehydes and ketones. The R group behaves as if it were $R\colon^-$.

$$\underset{\text{GRIGNARD REAGENT}}{\text{R}\colon^-\text{Mg}^{2+}\text{X}^-} + \underset{\substack{\text{CARBONYL GROUP}\\\text{IN AN ALDEHYDE}\\\text{OR A KETONE}}}{\overset{\delta+\ \ \delta-}{\text{C=O}}} \longrightarrow \underset{\text{A SALT OF AN ALCOHOL}}{\text{R—}\overset{|}{\underset{|}{\text{C}}}\text{—O}^-\text{Mg}^{2+}\text{X}^-}$$

The salt of the alcohol changes to the free alcohol when the mixture is acidified.

$$\text{R—}\overset{|}{\underset{|}{\text{C}}}\text{—O}^-\text{Mg}^{2+}\text{X}^- + \text{H}^+ \longrightarrow \underset{\text{ALCOHOL}}{\text{R—}\overset{|}{\underset{|}{\text{C}}}\text{—O—H}} + \text{Mg}^{2+} + \text{X}^-$$

Primary alcohols form if formaldehyde is the aldehyde used.

$$\underset{\substack{\text{GRIGNARD}\\\text{REAGENT}}}{\text{RMgX}} + \underset{\text{FORMALDEHYDE}}{\text{H—}\overset{\overset{\text{O}}{\|}}{\text{C}}\text{—H}} \xrightarrow[\text{Ether}]{} \underset{\substack{\text{SALT OF ALCOHOL}\\\text{(Not isolated)}}}{\text{RCH}_2\text{O}^-\text{Mg}^{2+}\text{X}^-} \xrightarrow{\text{H}^+} \underset{1^\circ\ \text{ALCOHOL}}{\text{RCH}_2\text{OH}}$$

For example:

$$\underset{n\text{-BUTYLMAGNESIUM BROMIDE}}{\text{CH}_3\text{CH}_2\text{CH}_2\text{CH}_2\text{MgBr}} + \text{H—}\overset{\overset{\text{O}}{\|}}{\text{C}}\text{—H} \xrightarrow[\text{by H}^+]{\text{Followed}} \underset{\text{1-PENTANOL (68\%)}}{\text{CH}_3\text{CH}_2\text{CH}_2\text{CH}_2\text{CH}_2\text{OH}}$$

Primary alcohols form if ethylene oxide is used.

$$\text{RMgX} + \underset{\substack{\text{ETHYLENE}\\\text{OXIDE}}}{\text{CH}_2\text{—CH}_2\ (\text{O bridged})} \longrightarrow \text{R—CH}_2\text{—}\underset{\text{O}^-\text{Mg}^{2+}\text{X}^-}{\underset{|}{\text{CH}_2}} \xrightarrow{\text{H}^+} \underset{\text{A }1^\circ\text{ ALCOHOL}}{\text{R—CH}_2\text{—CH}_2\text{—OH}}$$

Secondary alcohols form if any aldehyde other than formaldehyde is used.

$$\text{RMgX} + \underset{\text{ALDEHYDE}}{\text{R}'\text{—}\overset{\overset{\text{O}}{\|}}{\text{C}}\text{—H}} \xrightarrow{\text{Ether}} \underset{\substack{\text{SALT OF ALCOHOL}\\\text{(Not isolated)}}}{\text{R}'\text{—}\overset{\overset{\text{O}^-\text{Mg}^{2+}\text{X}^-}{|}}{\text{CH}}\text{—R}} \xrightarrow{\text{H}^+} \underset{2^\circ\ \text{ALCOHOL}}{\text{R}'\text{—}\overset{\overset{\text{OH}}{|}}{\text{CH}}\text{—R}}$$

For example:

$$C_6H_5MgBr + C_6H_5\overset{O}{\overset{\|}{C}}H \xrightarrow[H^+]{\text{Followed by}} C_6H_5\overset{OH}{\overset{|}{C}}HC_6H_5$$

PHENYLMAGNESIUM BROMIDE — BENZALDEHYDE — DIPHENYLCARBINOL (70%) ("BENZHYDROL")

Tertiary alcohols form if a ketone is used.

$$RMgX + R'\text{—}\overset{O}{\overset{\|}{C}}\text{—}R'' \xrightarrow[\text{Ether}]{} R'\text{—}\overset{O^-Mg^{2+}X^-}{\overset{|}{\underset{R}{\underset{|}{C}}}}\text{—}R'' \xrightarrow{H^+} R'\text{—}\overset{OH}{\overset{|}{\underset{R}{\underset{|}{C}}}}\text{—}R''$$

KETONE — 3° ALCOHOL

For example:

$$CH_3CH_2CH_2MgBr + CH_3\overset{O}{\overset{\|}{C}}CH_3 \xrightarrow[\text{by } H^+]{\text{Followed}} CH_3\overset{OH}{\overset{|}{\underset{CH_2CH_2CH_3}{\underset{|}{C}}}}CH_3$$

n-PROPYLMAGNESIUM BROMIDE — ACETONE — 2-METHYL-2-PENTANOL (50%)

6.17 Planning a Grignard Synthesis

One of the most common scientific problems handled by the working organic chemist is how to make a certain compound. Suppose it's a simple alcohol. He will mentally run through the list of all the ways he knows can be used. He knows them well, and it does not take long. As he does this, he is thinking all the while about two key problems:

1. What are the limitations of a given method (because he knows that not all methods work for all kinds of alcohols)?
2. What are the needed starting materials (because sometimes these are difficult or impossible to get)?

In planning a Grignard synthesis, you have to recognize that you get part of the skeleton of the desired alcohol from the aldehyde, the ketone, or ethylene oxide. It will always be a part that includes the carbon holding the new —OH group. (The oxygen of this group comes from the carbonyl compound or from ethylene oxide.) The rest of the carbon skeleton must come from some alkyl or aryl halide, because it is from these that we make Grignard reagents.

You start, then, by studying the structure of the desired alcohol. Draw a circle around any one group that is attached to the carbon holding the —OH group. Do not include that carbon; do not break open a cyclic ring, either.

For primary alcohols:

$R\text{—}CH_2\text{—}OH$ The halide for making RMgX is R—X. The carbonyl compound is formaldehyde.

For certain primary alcohols of the type:

$R\text{—}CH_2CH_2\text{—}OH$ The halide could be RCH_2X; the aldehyde, formaldehyde, or
the halide could be RX; the rest, ethylene oxide.

For secondary alcohols:

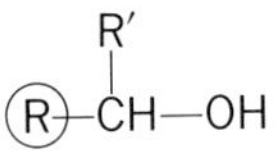

(R)—$\overset{R'}{\underset{}{CH}}$—OH   The halide could be R—X; the aldehyde: $R'—\overset{O}{\overset{\|}{C}}—H$

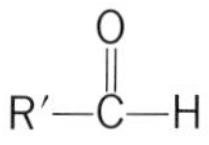

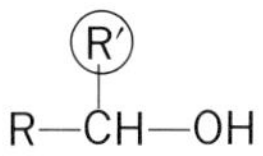

or R—$\overset{(R')}{CH}$—OH   The halide could be R′—X; the aldehyde: $R—\overset{O}{\overset{\|}{C}}—H$

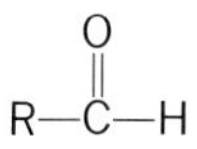

For tertiary alcohols, there are generally three options.

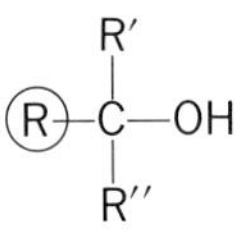

(R)—$\overset{R'}{\underset{R''}{C}}$—OH   The halide could be R—X; the ketone: $R'—\overset{O}{\overset{\|}{C}}—R''$

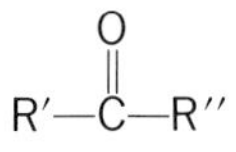

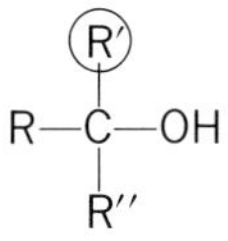

or R—$\overset{(R')}{\underset{R''}{C}}$—OH   The halide could be R′—X; the ketone: $R—\overset{O}{\overset{\|}{C}}—R''$

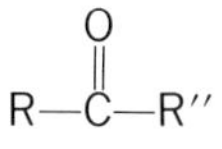

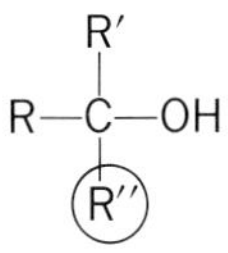

or R—$\overset{R'}{\underset{(R'')}{C}}$—OH   The halide could be R″—X; the ketone: $R—\overset{O}{\overset{\|}{C}}—R'$

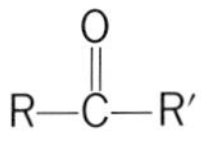

If he has options, what does the chemist do? Generally, one or more of the following. He talks to the stockroom manager. "What have we got in stock?" Or he studies a chemicals catalog. "What can I readily buy?" Nearly always (if he has any sense) he will go to the original chemical literature to find out what other chemists have discovered about the same or a closely related synthesis.

Exercise 6.1 What are the choices open to a chemist faced with making each of the following alcohols by a Grignard synthesis? (Write structures of possible starting materials.)

(a) $CH_3\overset{OH}{\overset{|}{C}}HCH_3$ (b) $C_6H_5\overset{OH}{\overset{|}{C}}HCH_3$ (c) CH_3CH_2OH (d) $CH_3CH_2\overset{OH}{\underset{CH_3}{C}}CH(CH_3)_2$

(e) $C_6H_5CH_2CH_2OH$

6.18 How Phenols are Made

What can we do to attach an —OH group to an aromatic ring? None of the methods described for alcohols can be used. There is no simple, laboratory method for hydrolyzing an aryl halide, although the Dow Chemical Company, by investing in energy and high-pressure apparatus, makes phenol on a large scale from chlorobenzene.

$$\underset{\text{CHLOROBENZENE}}{C_6H_5Cl} + NaOH_{aq} \xrightarrow[\text{200 atmospheres}]{300^\circ C} \underset{\text{PHENOL}}{C_6H_5OH} + NaCl$$

Still another industrial method converts cumene to phenol and acetone by the action of oxygen and a special catalyst. Cumene may be obtained from petroleum or from pine resins or by converting benzene and propylene to this compound by a Friedel-Crafts reaction.

$$\underset{\text{BENZENE}}{C_6H_6} + \underset{\text{PROPYLENE}}{CH_2{=}CHCH_3} \xrightarrow{AlCl_3} \underset{\text{CUMENE}}{C_6H_5CH(CH_3)_2} \xrightarrow[\text{Catalyst}]{O_2} \xrightarrow[H_2O]{H^+} \underset{\text{PHENOL}}{C_6H_5OH} + \underset{\text{ACETONE}}{CH_3\overset{O}{\overset{\|}{C}}CH_3}$$

The most general laboratory synthesis is by the action of nitrous acid and water on an aromatic primary amine, which will be studied in Section 8.15.

6.19 How Ethers are Made

There are two fairly general methods both of which will be studied later.

1. Intermolecular dehydration of alcohols. (Discussed in Section 6.33.)

$$2ROH \xrightarrow{\text{Acid}} R{-}O{-}R + H_2O$$

2. The Williamson-Ether synthesis (Section 7.8).

$$R{-}O^-Na^+ + R'{-}X \longrightarrow R{-}O{-}R' + NaX$$

Chemical Properties of Alcohols and Phenols

6.20 Alcohols as Very Weak Acids

Alcohols cannot neutralize sodium or potassium hydroxide. Alcohols are not acidic in aqueous systems; they are neutral. They cannot release hydroxide ions in water, either. However, if we add sodium metal to an alcohol (water-free, of course), the alcohol changes to a salt, a sodium alkoxide. The reaction is similar to that of sodium with water, but not nearly as violent. Dissolved in the parent alcohol from which it formed, alkoxide ions are reasonably stable, and they are commonly used when chemists want a particularly powerful base.

$$H{-}O{-}H + Na \xrightarrow[\text{Excess water}]{} HO^-Na^+ + \tfrac{1}{2}H_2 \quad \text{(violent reaction)}$$

$$R{-}O{-}H + Na \xrightarrow[\text{Excess ROH}]{} \underset{\substack{\text{A SODIUM} \\ \text{ALKOXIDE,} \\ \text{(Dissolved in ROH)}}}{RO^-Na^+} + \tfrac{1}{2}H_2 \quad \text{(moderate reaction)}$$

6.21 Phenols as Weak Acids

Phenols are stronger acids than alcohols. Phenol itself will neutralize sodium hydroxide, a strong base (but not sodium bicarbonate, a weak base):

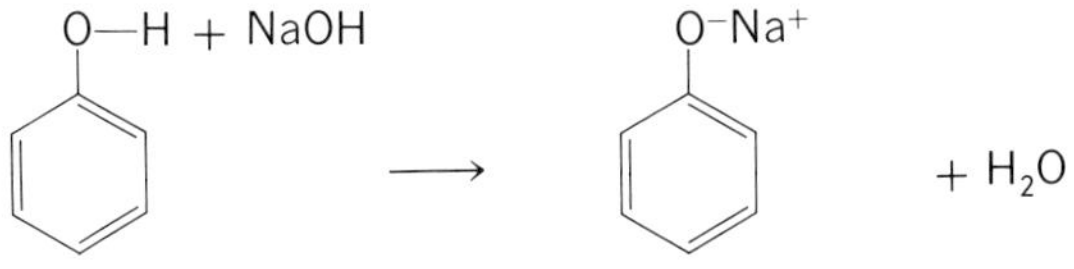

$$C_6H_5OH + NaOH \longrightarrow C_6H_5O^-Na^+ + H_2O$$

PHENOL SODIUM PHENOXIDE

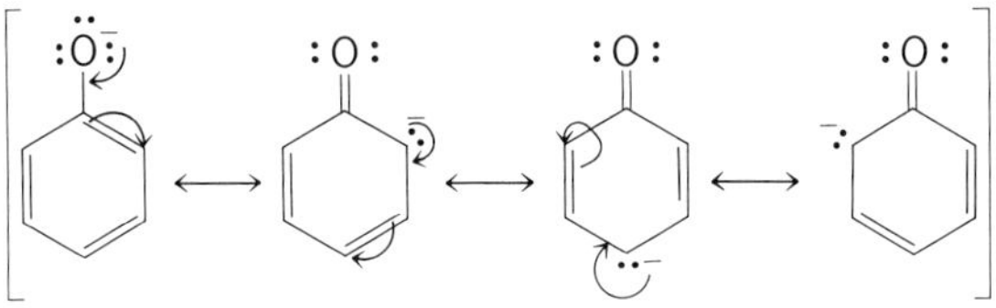

RESONANCE STABILIZATION IN THE PHENOXIDE ION

The alkoxide ion, R—O$^-$, has nowhere to delocalize the negative charge on its oxygen. The phenoxide ion has. The negative charge is delocalized over the whole ring system, and it is this that helps to make this ion somewhat stable. It is this that makes the phenol molecule a better releasor of H$^+$ than an alcohol molecule.

Salts of phenols are ionic compounds, and they are generally quite soluble in water. This property gives us a quick, simple, room-temperature strategy for separating phenols from nonacidic compounds. For example, if you stirred a mixture of phenol and toluene with dilute sodium hydroxide, the toluene would be unchanged (and undissolved). But the phenol would all dissolve in the water as sodium phenoxide. The toluene is left as a separate layer that can be separated.

6.22 Reactions Involving The Carbon-Oxygen Bond of Alcohols

There are two types:

1. The —OH group is replaced: substitution reactions.
2. A molecule of H—OH splits out: elimination reactions.

For either reaction, the carbon-oxygen bond must be weakened first. An acid catalyst usually performs this key service.

When a water molecule is changed to a hydronium ion, all the bonds from oxygen to hydrogen are weak. The hydrogens are all equivalent; they are all equally available for swift transfer to some proton-acceptor—a base. To weaken the *two* oxygen-hydrogen bonds in water put a third hydrogen on the oxygen.

It is like this with alcohols. Put another hydrogen on the —OH group of an alcohol, and all three bonds from the oxygen are weak including the carbon-oxygen bond.

With water:

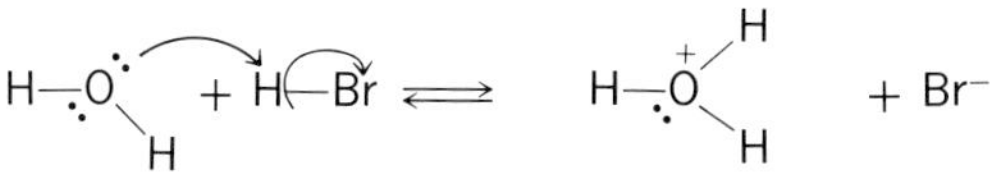

$$H_2O + H{-}Br \rightleftharpoons H_3O^+ + Br^-$$

HYDRONIUM ION

With alcohols:

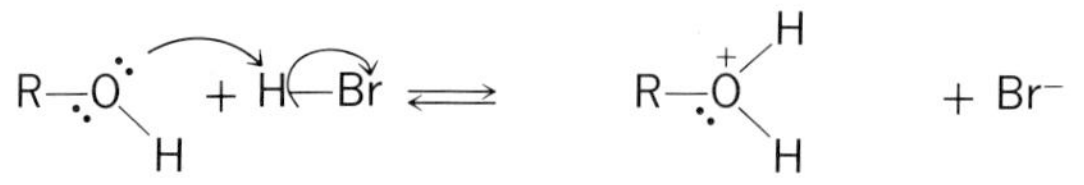

$$R{-}\ddot{O}{-}H + H{-}Br \rightleftharpoons R{-}\overset{+}{\ddot{O}}(H){-}H + Br^-$$

ALKYLOXONIUM ION

We not only change the condition of the carbon-oxygen bond by this, we also change the carbon itself. Oxygen is more electronegative than carbon; in the alcohol the carbon holding an —OH group therefore bears a small, partial positive charge. But when we put a proton on the oxygen—the nearby oxygen—we greatly increase the partial positive charge on the nearby carbon.

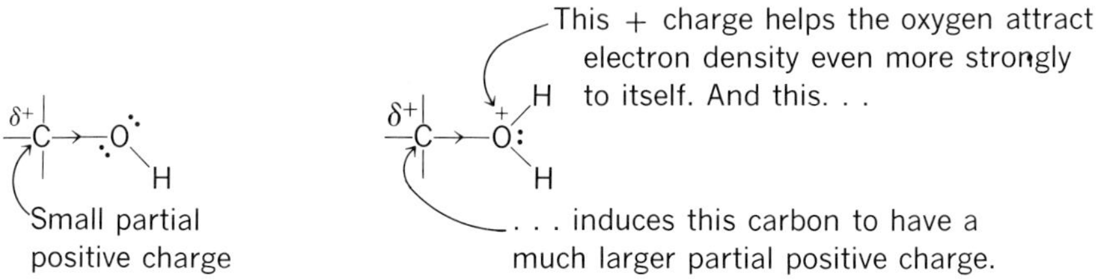

$$\overset{\delta+}{-C-} \rightarrow \ddot{O}{-}H$$

Small partial positive charge

IN THE FREE ALCOHOL

$$\overset{\delta+}{-C-} \rightarrow \overset{+}{\ddot{O}}(H){-}H$$

This + charge helps the oxygen attract electron density even more strongly to itself. And this. . .

. . . induces this carbon to have a much larger partial positive charge.

IN THE PROTONATED ALCOHOL

We have done two things by putting a proton on the oxygen of an alcohol. We have weakened the carbon-oxygen bond. We have made the carbon much more attractive to attack by some electron-rich particle. Let us see what can happen.

6.23 Changing Alcohols to Alkyl Halides. A Substitution Reaction

The overall result is accomplished by mineral acids. In general:

$$R{-}OH + H{-}X \longrightarrow R{-}X + H_2O$$

ALCOHOL (R = ALKYL, *not* ARYL); HYDROGEN HALIDE (X = Cl, Br, I); ALKYL HALIDE

Specific examples:

$$CH_3CH_2CH_2CH_2OH + \text{concd } HBr \xrightarrow{\text{Heat}} CH_3CH_2CH_2CH_2Br + H_2O$$

n-BUTYL ALCOHOL → *n*-BUTYL BROMIDE (95%)

$$CH_3CH_2CH_2CH_2CH_2OH + \text{concd } HBr \xrightarrow{\text{Heat}} CH_3CH_2CH_2CH_2CH_2Br + H_2O$$

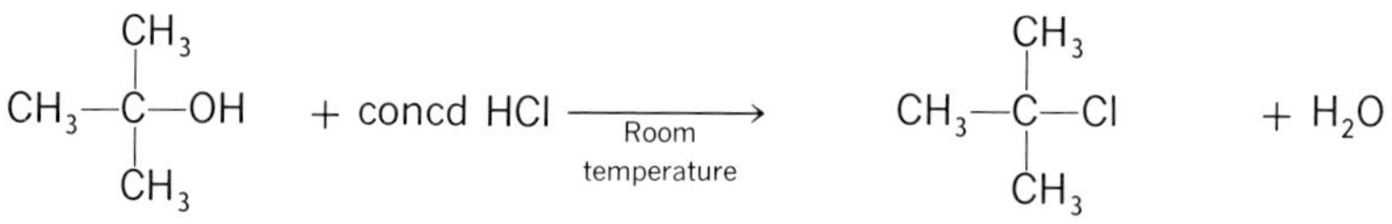

1-PENTANOL → 1-BROMOPENTANE (78%)

$$(CH_3)_3C{-}OH + \text{concd } HCl \xrightarrow[\text{temperature}]{\text{Room}} (CH_3)_3C{-}Cl + H_2O$$

t-BUTYL ALCOHOL → *t*-BUTYL CHLORIDE (88%)

The order of reactivity of the alcohols is

$$3° > 2° > 1° > CH_3OH$$

You cannot do this with salts such as sodium bromide. You must use a strongly acidic medium. Primary alcohols generally react by one mechanism, tertiary alcohols by another. The situation is mixed with secondary alcohols.

With primary alcohols the following steps take place.

Step 1. The acid puts a proton on the alcohol—a relatively rapid step.

$$R{-}CH_2{-}\ddot{O}{-}H + H{-}X \rightleftharpoons R{-}CH_2{-}\overset{+}{O}H_2 + X^- \quad \text{(rapid step)}$$

Step 2. Halide ion, X^-, is attracted to and strikes the backside of the carbon holding the oxygen. This dislodges a water molecule.

$$X^- + RCH_2{-}\overset{+}{O}H_2 \longrightarrow \underset{\text{ALKYL HALIDE}}{X{-}CH_2R} + :\ddot{O}H_2 \quad \text{(slower step)}$$

Chemists call any electron-rich species such as a halide ion that can attack and become attached to carbon a *nucleophile*[2]. This reaction is called a **nucleophilic substitution reaction.**

Step 2 is the "bottleneck" step. Not every collision by X^- dislodges the water molecule. Only a fraction of the collisions do, a fraction that is lower than the fraction of successful proton transfers in Step 1. Step 1 is the more rapid; step 2 is slower. Because *two* particles must hit each other in this slow, second step, we call the overall mechanism *bimolecular.* The reaction is a bimolecular nucleophilic substitution reaction. That is a mouthful to say (or write), and chemists abbreviate it to $\boldsymbol{S_N2}$. The *S* is for substitution, the *N* for nucleophilic, and the 2 for bimolecular.

It is very hard to get tertiary alcohols to react by an S_N2 mechanism. The carbon that has to be hit by the nucleophile is really hedged in by all those groups joined to that carbon, as illustrated in Figure 6.3. We say that there is *steric* interference; or, the S_N2 mechanism shows a **steric effect.** (*Steric* comes from *stereos*, solid. Solid objects exclude each other.) The space-occupying groups in the alcohol do move somewhat out of the way as the nucleophile attacks, as seen in Figure 6.4. But in a tertiary system it is hardest for an attacking nucleophile to get even close. This is partly why tertiary alcohols undergo the change to an alkyl halide by another mechanism. With tertiary alcohols the following steps take place.

Step 1. The alcohol is protonated by the acid.

$$R{-}\ddot{O}{-}H + H{-}X \rightleftharpoons R{-}\overset{+}{O}H_2 + X^- \quad \text{(rapid formation of equilibrium)}$$

Step 2. The C—O bond is now much weaker and breaks, forming a carbonium ion.

[2]Literally, this means "nucleus loving." The nucleus meant here is a carbon's, but a better association of words is with "positive charge." (All nuclei are positively charged.) Thus a nucleophile is something attracted to a positive charge; and it brings its own pair of electrons for the new bond. An electrophile is something attracted to negative charge, as found in an electron-dense area that will donate the electron pair needed for the new bond.

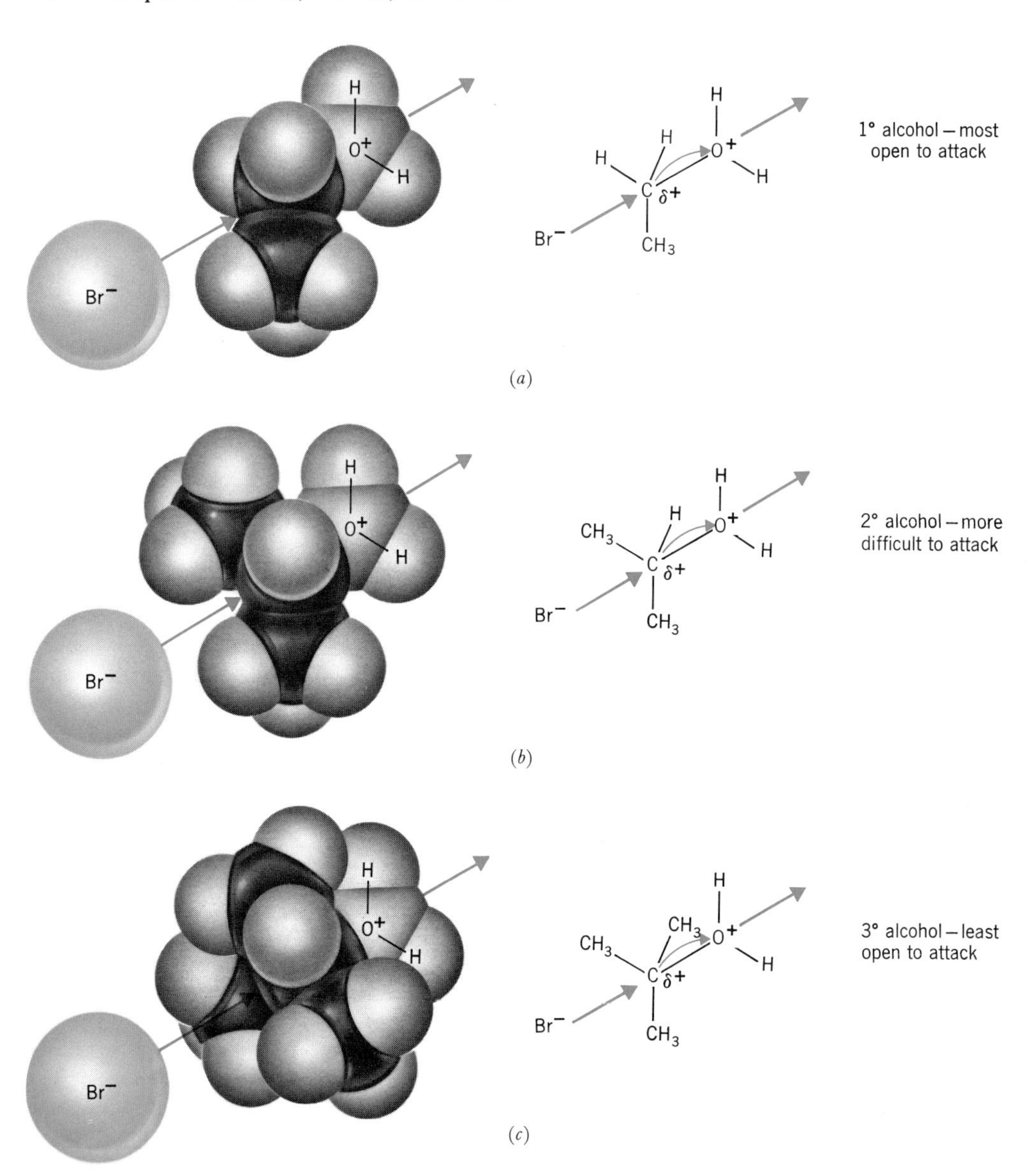

Figure 6.3
The steric effect in S_N2 reactions. The nucleophile, bromide ion in these examples, must move in with much more energy for successful attack at a 3° position (*c*) than at a 1° position (*a*). But the fraction of bromide ions having this high energy of motion at any moment is not large. That is why the rate of an S_N2 reaction at a 3° position is much lower than at a 1° position.

$$R{-}\overset{+}{\ddot{O}}(H)(H): \rightleftharpoons R^+ + H_2O \quad \text{(slow rate of ionization)}$$

CARBONIUM ION

Step 3. The carbonium ion combines with a halide ion.

$$R^+ + X^- \longrightarrow R{-}X \quad \text{(rapid)}$$

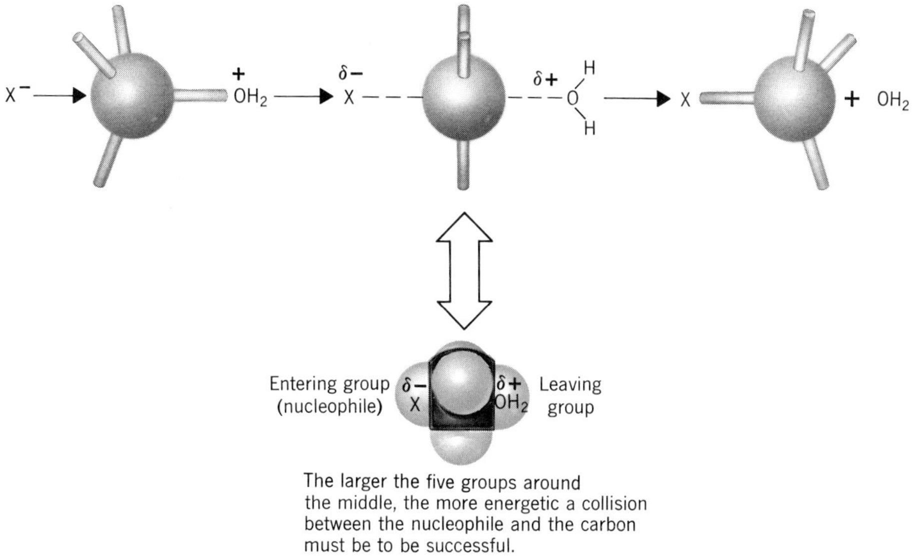

Entering group (nucleophile) $\delta-$ X

$\delta+$ OH_2 Leaving group

The larger the five groups around the middle, the more energetic a collision between the nucleophile and the carbon must be to be successful.

Figure 6.4
Much like an umbrella blowing inside out in a high wind, the groups attached to carbon in the alcohol move from one side to the opposite as an S_N2 attack proceeds. We say that an *inversion* occurs.

The first step, as always, is rapid. It is step 2 that holds up the overall speed of the reaction. The last step cannot take place any faster than carbonium ions, R^+, are produced. In step 2 only one particle is involved, and that is why the entire mechanism is called a *unimolecular* process. Only one molecular species is changed in the bottleneck step. The mechanism is called an S_N1 process: *S* for substitution, *N* for nucleophilic and 1 for unimolecular.

Tertiary alcohols generally react by an S_N1 mechanism. It isn't just that the S_N2 is hard for steric reasons. It is also because tertiary alcohols can produce some of the most stable carbonium ions, ions far more stable than 1° carbonium ions. Step 2, therefore, goes in the direction of relatively stable particles when tertiary alcohols are involved.

Since 2° alcohols would give carbonium ions with stability intermediate between those from 1° and 3° systems, and since steric problems are similarly intermediate, their situation is "mixed." Secondary alcohols undergoing this reaction will send some molecules through the S_N2 track and others through the S_N1.

6.24 Lucas Test

The influence of structure on reactivity in the conversion of alcohols to alkyl halides may be observed in an old chemical test for telling the difference, experimentally, between 1°, 2°, and 3° aliphatic alcohols. The reagent is a solution of zinc chloride in concentrated hydrochloric acid. This highly ionic, strongly acidic medium favors the formation of carbonium ions—even from 1° alcohols. But just as the order of stability of carbonium ions is $3° > 2° > 1° > CH_3OH$, so is the order of their ease of formation.

The test begins with adding the alcohol to be tested to the Lucas reagent. All simple aliphatic alcohols of up to seven carbons will dissolve in the reagent. But tertiary alcohols will then react almost at once to become alkyl chlorides. These are insoluble in the reagent. They come out of solution as they form and produce first a cloudiness and then a definite, separate layer. Tertiary alcohols show this behavior almost at once. Primary alcohols do not react at all, not at least within a few minutes. The test solution remains clear. Secondary alcohols show a cloudiness in a few minutes. Thus the quickness of showing a positive test is in the order of stabilities of carbonium ions: $3° > 2° > 1° > CH_3OH$.

6.25 Changing Alcohols to Alkenes. Dehydration

$$\underset{\text{ALCOHOL}}{-\underset{H}{\overset{|}{\underset{|}{C}}}-\underset{OH}{\overset{|}{\underset{|}{C}}}-} \xrightarrow[\text{heat}]{H^+} \underset{\text{ALKENE}}{>C=C<} + H{-}OH$$

Specific examples:

$$\underset{\text{ETHYL ALCOHOL}}{CH_3CH_2{-}OH} \xrightarrow[170\text{–}180°C]{\text{concd } H_2SO_4} \underset{\text{ETHYLENE}}{CH_2{=}CH_2} + H_2O$$

$$\underset{sec\text{-BUTYL ALCOHOL}}{CH_3CH_2\underset{OH}{\underset{|}{C}}HCH_3} \xrightarrow[100°C]{60\%\ H_2SO_4} \underset{\substack{\text{2-BUTENE}\\ \text{(Principal product)}}}{CH_3CH{=}CHCH_3} + \underset{\text{1-BUTENE}}{CH_3CH_2CH{=}CH_2} + H_2O$$

$$\underset{t\text{-BUTYL ALCOHOL}}{CH_3{-}\underset{CH_3}{\overset{CH_3}{\overset{|}{\underset{|}{C}}}}{-}OH} \xrightarrow[80\text{–}90°C]{20\%\ H_2SO_4} \underset{\text{ISOBUTYLENE}}{CH_2{=}C(CH_3)_2} + H_2O$$

Some of the major facts about the reaction are the following:

1. The acid catalyst is necessary. (We have to weaken the C—O bond.)
2. The ease of dehydration of subclasses of alcohols is
   $3° > 2° > 1°$ (Note the experimental conditions in the examples.)
3. When two alkenes are possible, both are produced but the more highly branched alkene predominates. (See the case of *sec*-butyl alcohol, above.)[3] The reason for this is that the more highly branched alkenes are generally the more stable (a fact we shall have to accept here without reviewing the evidence.)
4. Sometimes carbon skeletons are rearranged during the reaction. (We encountered carbonium-ion rearrangements in Chapter 3, and nothing more will be said about this complication here.)

Tertiary alcohols usually dehydrate via carbonium ions. Primary alcohols react by another mechanism. Secondary alcohols give mixed responses.

[3] "More highly branched" means the greater number of bonds to other carbons going out from the carbons of the double bond. The compound 2-butene has two branches; 1-butene has one.

6.26 The E_1 Mechanism

(Common to 3° alcohols, illustrated with *t*-butyl alcohol.)

Step 1. Proton transfer; a rapid step, and we weaken the C—O bond.

$$(CH_3)_3C{-}\ddot{O}{-}H + H_2SO_4 \rightleftharpoons (CH_3)_3C{-}\overset{+}{O}H_2 + HSO_4^-$$

t-BUTYL ALCOHOL; SULFURIC ACID

Step 2. The C—O bond breaks; a carbonium ion forms.

$$(CH_3)_3C{-}\overset{+}{O}H_2 \longrightarrow (CH_3)_3C^+ + :\ddot{O}H_2$$

t-BUTYL CATION

Step 3. A proton is pulled off from a carbon adjacent to the site of the positive charge. The alkene forms.

$$CH_3{-}\overset{+}{C}(CH_3){-}CH_2{-}H \;\; {}^-O{-}SO_3H \longrightarrow CH_3{-}C(CH_3){=}CH_2 + H_2SO_4$$

ISOBUTYLENE; RECOVERED CATALYST

The slowest step, step 2, is unimolecular. Therefore the whole mechanism is called a unimolecular elimination, or E_1 for short. The E is for elimination; the 1 for unimolecular.

6.27 The E_2 Mechanism

(Common to 1° alcohols; illustrated with ethyl alcohol.)

Step 1. Proton transfer; a rapid step, and we weaken the C—O bond, as before.

$$CH_3CH_2{-}\ddot{O}{-}H + H_2SO_4 \rightleftharpoons CH_3CH_2{-}\overset{+}{O}H_2 + HSO_4^-$$

Step 2. Simultaneous loss of both a water molecule and a proton; the alkene emerges.

$$H{-}CH_2{-}CH_2{-}\overset{+}{O}H_2 \;\; {}^-OSO_3H \longrightarrow CH_2{=}CH_2 + H_2O + HOSO_3H$$

BISULFATE ION; SULFURIC ACID CATALYST IS RECOVERED

6.28 Oxidation of Alcohols

A number of oxidizing agents will convert 1° alcohols to aldehydes or carboxylic acids; 2° alcohols are oxidized to ketones.

$$\underset{\text{ALDEHYDE}}{R-\overset{O}{\overset{\|}{C}}-H} \qquad \underset{\text{CARBOXYLIC ACID}}{R-\overset{O}{\overset{\|}{C}}-OH} \qquad \underset{\text{KETONE}}{R-\overset{O}{\overset{\|}{C}}-R'}$$

When an alcohol molecule is oxidized it suffers the loss of the pieces of the element hydrogen. In other words, *dehydrogenation* occurs. (The word *aldehyde* comes from *al*cohol *dehyd*rogenation.)

$$\underset{\text{1° OR 2° ALCOHOL}}{\text{H}-\overset{|}{\text{C}}-\text{O}-\text{H}} \longrightarrow \underset{\text{ALDEHYDE OR KETONE}}{\text{C}=\text{O}} + \underset{\text{"ELEMENTS" OF HYDROGEN}}{(\text{H}:^- + \text{H}^+)}$$

There are generally three kinds of reagents that can do this.

1. Inorganic oxidizing agents such as potassium permanganate ($KMnO_4$) and sodium dichromate ($Na_2Cr_2O_7$) or other dichromates and chromates.
2. Organic molecules that can accept hydrogen (as in many reactions in living organisms).
3. Hot, finely divided copper metal. (In this case molecular hydrogen is a product.)

6.29 Oxidation of 1° Alcohols. Synthesis of Aldehydes and Carboxylic Acids

Primary alcohols are oxidized to aldehydes. If an inorganic oxidizing agent is used, the aldehyde has to be removed almost as rapidly as it forms. Otherwise, the aldehyde will be oxidized by unchanged oxidizing agent to the carboxylic acid. Aldehydes are even more readily oxidized than alcohols. (Since aldehydes have lower boiling points than their parent alcohols, we can distill them as they form.) In the examples that follow the equations are generally not balanced. Sometimes it is convenient to use a general symbol, (O), to stand for the oxidizing agent—meaning any oxidizing agent that can do the job.

In general, for 1° alcohols:

$$\underset{\text{1° ALCOHOL}}{RCH_2OH} \xrightarrow[200\text{–}300°C]{Cu} \underset{\text{ALDEHYDE}}{R-\overset{O}{\overset{\|}{C}}-H} + H_2$$

$$RCH_2OH + \underset{\text{POTASSIUM PERMANGANATE}}{KMnO_4} \longrightarrow R-\overset{O}{\overset{\|}{C}}-H \xrightarrow{KMnO_4} \underset{\text{SALT OF CARBOXYLIC ACID}}{R-\overset{O}{\overset{\|}{C}}-O^-K^+} + \underset{\text{MANGANESE DIOXIDE (Brown sludge)}}{MnO_2} + KOH$$

$$\underset{\text{SALT OF CARBOXYLIC ACID}}{R-\overset{O}{\overset{\|}{C}}-O^-K^+} \xrightarrow{H_3O^+} \underset{\text{CARBOXYLIC ACID}}{R-\overset{O}{\overset{\|}{C}}-O-H} + H_2O + K^+$$

$$\underset{}{RCH_2OH} + \underset{\substack{\text{SODIUM}\\\text{DICHROMATE}\\\text{(Bright orange)}}}{Na_2Cr_2O_7} \xrightarrow{H^+} R{-}\overset{\overset{O}{\|}}{C}{-}H \xrightarrow{Cr_2O_7^{2-}} R{-}\overset{\overset{O}{\|}}{C}{-}OH + \underset{\substack{\text{CHROMIUM}\\\text{ION}\\\text{(Bright green)}}}{Cr^{3+}}$$

Specific examples:

$$\underset{\substack{n\text{-PROPYL ALCOHOL}\\\text{(bp 97°C)}}}{CH_3CH_2CH_2OH} \xrightarrow{(O)} \underset{\substack{\text{PROPIONALDEHYDE}\\\text{(bp 55°C)}}}{CH_3CH_3\overset{\overset{O}{\|}}{C}H} \qquad \begin{aligned}(O) &= \text{Cu, high temperature; 67\% yield}\\ &= Cr_2O_7^{2-}, H^+\text{; 49\% yield}\end{aligned}$$

$$\underset{\substack{n\text{-BUTYL ALCOHOL}\\\text{(bp 118°C)}}}{CH_3CH_2CH_2CH_2OH} \xrightarrow{(O)} \underset{\substack{n\text{-BUTYRALDEHYDE}\\\text{(bp 82°C)}}}{CH_3CH_2CH_2\overset{\overset{O}{\|}}{C}H} \qquad \begin{aligned}(O) &= \text{Cu, heat; 62\% yield}\\ &= Cr_2O_7^{2-}, H^+\text{; 72\% yield}\end{aligned}$$

Carboxylic acids from 1° alcohols:

$$\underset{n\text{-PROPYL ALCOHOL}}{CH_3CH_2CH_2OH} \xrightarrow{Cr_2O_7^{2-},\ H^+} \underset{\text{PROPIONIC ACID}}{CH_3CH_2\overset{\overset{O}{\|}}{C}OH} \quad \text{(65\% yield)}$$

6.30 Oxidation of 2° Alcohols. Ketones

In sharp contrast to aldehydes, ketones are quite resistant to oxidation, and they need not be removed as they form.

Specific examples:

$$\underset{\text{2-BUTANOL}}{CH_3\overset{\overset{OH}{|}}{C}HCH_2CH_3} \xrightarrow[\text{warm}]{Cr_2O_7^{2-},\ H^+} \underset{\text{2-BUTANONE}}{CH_3\overset{\overset{O}{\|}}{C}CH_2CH_3} \quad \text{(74\% yield)}$$

$$\text{CYCLOHEXANOL} \xrightarrow[\text{warm}]{Cr_2O_7^{2-},\ H^+} \text{CYCLOHEXANONE} \quad \text{(85\% yield)}$$

6.31 Important Individual Alcohols

Methyl Alcohol.
(Wood Alcohol. Methanol)

When selected, dried hardwoods are strongly heated, gases and a liquid condensate separate—leaving a residue of charcoal. The watery part of the liquid contains a small concentration of methyl alcohol (Greek *methu,* wine; *hule,* wood or material).

In 1923 chemists of the Badische Company in Germany found a synthetic route from very simple chemicals, carbon monoxide and hydrogen:

$$2H_2 + CO \xrightarrow[\substack{\text{Temperature} = 300\text{–}400^\circ\text{C} \\ \text{Catalyst} = \text{mixed metal oxides}}]{\text{Pressure} = 200\text{–}300 \text{ atmospheres}} CH_3OH$$

The annual U. S. production is over 5 billion pounds or about 750 million gallons, which would fill enough of the largest railroad tank cars to make 100 trains of 225 tank cars each. Roughly half of this is used to make formaldehyde, a raw material for plastics. It is also a jet fuel, a commercial and laboratory solvent, a temporary antifreeze, a shellac thinner, and a convenience fuel for fondue burners.

Methyl alcohol is a dangerous poison. It blinds or kills if it is taken internally, which can happen to those thinking "alcohol" is "ethyl alcohol."

Ethyl Alcohol. (Grain Alcohol. Ethanol)

This is the alcohol of alcoholic drinks that are made by the fermentation of carbohydrates. Most ethyl alcohol is produced from ethylene, which is obtained by cracking ethane and other hydrocarbons in petroleum. Besides being an important industrial solvent and chemical intermediate, it is used as a solvent in making preparations of drugs, perfumes, lotions, tonics, and rubbing compounds. The government requires that the alcohol used for these be *denatured*. Denatured alcohol is ethyl alcohol to which have been added several substances that cannot be removed and hence make the product utterly unacceptable for drinking.

Alcohol containing 5% water is an interesting mixture in that the vapor in equilibrium with the liquid at atmospheric pressure is of the identical composition. Any mixture having this property is called an *azeotrope,* and azeotropes cannot be separated by fractional distillation. The components of an azeotrope boil off together at a constant boiling point and a constant composition.

Taken internally, ethyl alcohol gives the illusion that it is a stimulant because its first effect is to depress activity in the uppermost level of the brain—the center of judgment, inhibition, and restraint.

Isopropyl Alcohol

This alcohol, twice as toxic as ethyl alcohol, is a common substitute for it as a rubbing compound.

Ethylene Glycol. Propylene Glycol

These two are viscous liquids with high boiling points. They dissolve in water in all proportions, and they are used widely as permanent antifreezes and coolants in refrigerator systems.

Glycerol. (Glycerin)

This colorless, syrupy liquid with a sweet taste is freely soluble in water but not in nonpolar solvents. It is used in the industrial synthesis of glyptal resins, to keep tobacco moist, for use as a softening agent in cellophane, and in the preparation of cosmetics and drugs. The powerful explosive, trinitroglycerine, is made by the reaction of glycerol with nitric acid.

6.32 Chemical Properties of Phenols

With respect to chemical properties the important fact about phenols is that the —OH on a ring acts as a powerful activator and an ortho-para director. We illustrate this by two reactions, bromination and nitration.

Bromination

No iron catalyst is needed. A mixture of bromine in water, shaken with aqueous phenol, promptly tribrominates the phenol.

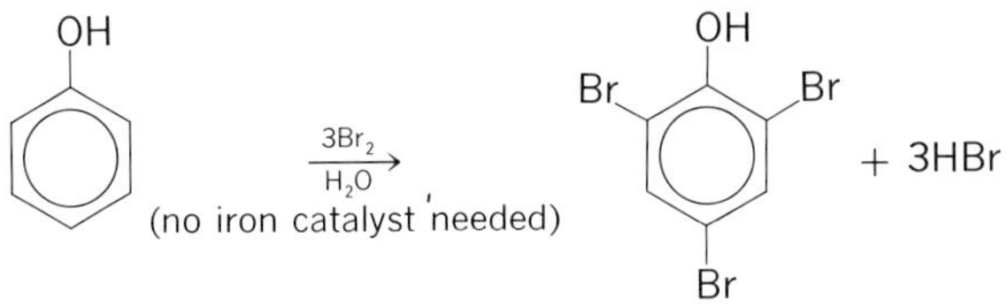

$3Br_2$, H_2O (no iron catalyst needed) + 3HBr

Nitration

Nitration of phenol occurs so rapidly that only dilute nitric acid need be used.

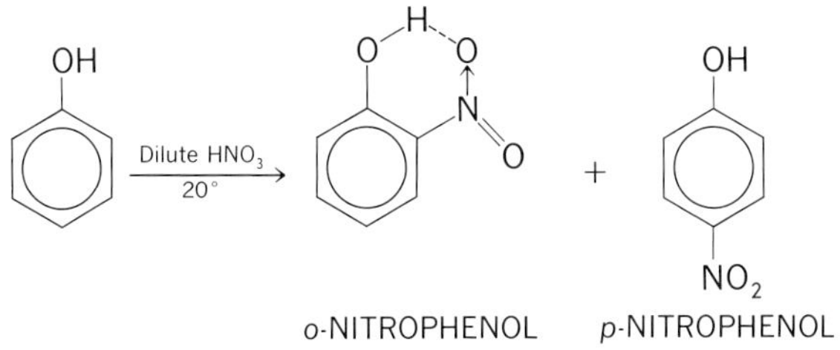

Dilute HNO_3, 20°

o-NITROPHENOL + *p*-NITROPHENOL

The isomers are easily separated. In molecules of the ortho isomer there is an *internal* hydrogen bond. Neighbor molecules of this isomer, therefore, cannot associate (cf. Section 6.13.) The ortho isomer is thus more volatile than the para isomer whose molecules can associate by hydrogen bonds. If live steam is bubbled through a mixture of the two isomers, the ortho isomer is carried out leaving the para isomer behind. This strategy is called *steam distillation,* and it is often used to "distill" compounds whose boiling points are otherwise too high to permit direct distillation. The compound *p*-nitrophenol is used in the manufacture of an important insecticide, parathion.

Preparation of Salicylic Acid

Action of carbon dioxide on the sodium salt of phenol at 150°C changes it to the sodium salt of salicylic acid, the parent compound for aspirin.

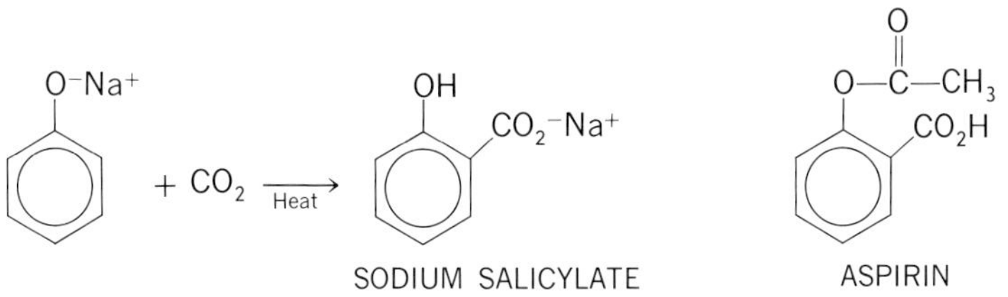

O^-Na^+ + CO_2 (Heat) → SODIUM SALICYLATE

ASPIRIN

Ethers

6.33 Making Ethers from Alcohols

When acid catalysts are used to change alcohols to alkenes, one of the by-products is the corresponding ether. Thus, at 170°C concentrated sulfuric acid changes ethyl alcohol principally to ethylene, but some diethyl ether is produced, also. By lowering the operating temperature to 140°C, the ether becomes the main product

$$CH_3CH_2{-}O{-}H + H{-}O{-}CH_2CH_3 \xrightarrow[140°C]{H_2SO_4} CH_3CH_2{-}O{-}CH_2CH_3 + H_2O$$

In principal, other symmetrical ethers, those with identical R— groups, R—O—R, can be made, but diethyl ether is the only one of importance prepared this way.

6.34 Physical Properties of Ethers

Ether molecules are slightly polar. Having no —OH groups they cannot hydrogen bond or associate with each other. Hence, the ether of formula weight 74—diethyl ether—boils at almost the identical temperature, 34.6°C, as the alkane of nearly the same formula weight, 72, *n*-pentane, boiling point 36°C. In marked contrast the alcohol of comparable formula weight, 74, *n*-butyl alcohol boils much higher at 117°C. Hydrogen bonding in the alcohol makes the huge difference.

In many respects, both physical and chemical, ethers are very similar to alkanes of similar formula weights. Ethers, however, are more soluble in water. Ether molecules can *accept* hydrogen bonds from water; alkanes cannot. See Figure 6.5. Both *n*-butyl alcohol and diethyl ether dissolve to the same extent in water—about 8 g in 100 cc water. The butanes and pentanes, in contrast, are insoluble in water.

6.35 Chemical Properties of Ethers

These are easy to learn—there are almost none. They resemble alkanes. The one reaction of any importance is cleavage by concentrated hydrobromic or hydriodic acid.

$$R{-}O{-}R + 2HBr \xrightarrow[Heat]{} 2R{-}Br + H_2O$$

Otherwise, concentrated aqueous alkalies and acids, even metallic sodium, do not attack simple ethers at room temperature.

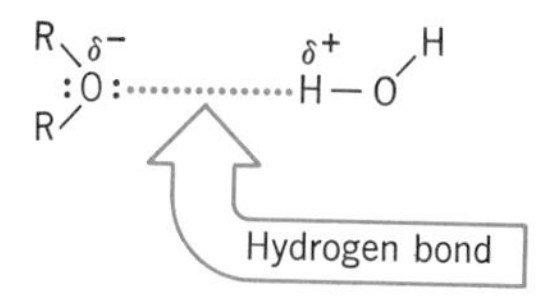

Figure 6.5
Hydrogen bond between molecules of water and ether. We say that the δ^- end of a hydrogen bond is the "acceptor" of the hydrogen bond.

Peroxides in Ethers. Warning!

When a container of an ether is opened and re-closed, some air is trapped inside. Liquid, aliphatic ethers react very slowly with oxygen in the air. Substances called peroxides are produced in very small concentrations. At the time of using the next portion of ether from the container the chemist must be careful. If he lets that ether evaporate to dryness (or, worse, distills it to dryness), the residue may have a much higher concentration of these peroxides. They are dangerous; they may explode violently.

To check ether for peroxides shake a sample of the ether with an aqueous solution of iron(II) ammonium sulfate plus potassium thiocyanate. Peroxides will produce a red color.

6.36 Important Individual Ethers

Diethyl Ether ("Ether")

This very volatile, extremely flammable liquid is widely used as a general anesthetic. It acts to depress the central nervous system and stimulate somewhat the sympathetic nervous system.

Divinyl Ether ("Vinethene")

This is another anesthetic, being more rapid in its action than diethyl ether.

Cyclic Ethers. Epoxides

Epoxides are cyclic ethers in which three atoms make up the ring. Ethylene oxide is the simplest member of the family.

6.37 Ethylene Oxide

This cyclic ether is, at ordinary temperatures and pressures, a colorless flammable gas. Its three-membered ring, like that of cyclopropane, is highly strained and it is readily opened by chemical attack. (We saw one example in its use in a Grignard synthesis of certain 1° alcohols.) Industrially, it is made two ways, both starting with ethylene.

The more important method (in terms of production volume) is direct oxidation of ethylene in the presence of a silver catalyst.

$$2CH_2{=}CH_2 + O_2 \xrightarrow[\text{Pressure; 250°C}]{\text{Silver}} 2\,\underset{\diagdown\;O\;\diagup}{CH_2{-}CH_2}$$

ETHYLENE — ETHYLENE OXIDE

The other method uses a mixture of chlorine and water, which may be considered as supplying HOCl, hypochlorous acid.

$$CH_2{=}CH_2 + HOCl \longrightarrow \underset{Cl\qquad OH}{CH_2{-}CH_2} \xrightarrow{\text{NaOH}} \underset{\diagdown\;O\;\diagup}{CH_2{-}CH_2}$$

ETHYLENE CHLOROHYDRIN

6.38 Ring-Opening Reactions of Ethylene Oxide

Almost any nucleophile can open the ring: H_2O or OH^-, NH_3 or NH_2^-, alcohols (ROH) or RO^- or ArO^-, and of course $R:^-$ from a Grignard reagent. For example, the reaction with water, which must be catalyzed by an acid, is a major route, industrially, to ethylene glycol.

$$H_2\ddot{O}: + CH_2\text{—}CH_2(\text{—}\ddot{O}^+\text{H—}) \longrightarrow H_2\overset{+}{O}\text{—}CH_2\text{—}CH_2\text{—}\ddot{O}\text{—H} \xrightarrow{-H^+} HO\text{—}CH_2CH_2\text{—}OH$$

ETHYLENE GLYCOL

$2CH_2\text{—}CH_2$ (epoxide, O) + H^+ (catalyst)

When alcohols are used instead of water, the commercially important cellosolves are produced.

$$CH_3OH + CH_2\text{—}CH_2(\text{—O—}) \xrightarrow{H^+} CH_3\text{—}O\text{—}CH_2\text{—}CH_2\text{—}OH$$

METHYL ALCOHOL; METHYL CELLOSOLVE

$$CH_3CH_2OH + CH_2\text{—}CH_2(\text{—O—}) \xrightarrow{H^+} CH_3CH_2\text{—}O\text{—}CH_2\text{—}CH_2\text{—}OH$$

ETHYL ALCOHOL; ETHYL CELLOSOLVE ("CELLOSOLVE")

The cellosolves are used as solvents in lacquers and varnishes. Methyl cellosolve is a jet fuel additive where it inhibits the formation of ice crystals. Ethyl cellosolve is a solvent for cellulose nitrate (hence, the name "cellosolve").

6.39 Epichlorohydrin

This epoxide has two functional groups, the ring itself and the chloride unit. As such it is an important raw material for making the epoxy resins, a family of protective coatings, casting resins, and bonding agents (adhesives).

$$Cl\text{—}CH_2\text{—}CH\text{—}CH_2(\text{—O—})$$

EPICHLOROHYDRIN

6.40 Other Epoxides

In principle any alkene can have a corresponding epoxide, and all epoxides display the kinds of ring-opening reactions we have just studied. The functional groups that attack and open the ring of an epoxide are commonly found in nucleic acids, the chemicals of genes and heredity. Both alcohol groups and amino groups abound. That is why epoxides must be handled with great care. If they get inside a cell they will interact with cellular chemicals including genes.

Brief Summary

Structure

1. Alcohols: R—OH (The carbon holding the —OH must be saturated.)
2. Phenols: Ar—OH (The —OH is directly attached to the carbon of an aromatic ring.)
3. Ethers: R—O—R′ or R—O—Ar or Ar—O—Ar

Names

1. Alcohols
   (a) Common: write the name of the R— group and then, as a separate word, the word "alcohol."
   (b) IUPAC: Pick the longest chain that includes the —OH group and substitute -ol for the -e ending of the corresponding alkane. Number the chain from whichever end gives the location of the —OH the lower number. Then proceed by the rules for naming alkanes.
2. Phenols
   (a) Common: names for some simple phenols simply have to be learned. We use ortho, meta, or para relations for naming some.
   (b) IUPAC: in complicated cases where the ring positions are numbered, the phenolic —OH may be named "hydroxy."
3. Ethers
   (a) Common: name each group joined to oxygen and write the word "ether" as a separate part of the name.
   (b) IUPAC: formulate the name as an alkoxyalkane. The alkane portion is the longer of the two hydrocarbon chains that are attached to the oxygen.

Physical Properties

1. Opportunities for associating—forming hydrogen bonds—dominate physical properties of alcohols and phenols. These substances boil much higher than would be predicted on the basis of their formula weights alone.

2. The —OH group is one of nature's most important aids to making substances soluble in water. Even the oxygen of ether molecules helps to get them somewhat soluble in water (compared with corresponding alkanes).

Chemical Properties

(What can we make from alcohols and phenols?)

1. We can make alkyl halides. The —OH group in alcohols can be replaced by halogen atoms.

$$ROH + HX \rightarrow RX + H_2O$$

2. We can make alkenes. The —OH in alcohols can be pulled out, together with H on an adjacent carbon, to introduce a double bond. By changing conditions, we may make *ethers.*

3. From 1° and 2° alcohols we can pull out the pieces of the element hydrogen, leave behind a carbon-oxygen double bond, and thereby make *aldehydes* or *ketones.* From 1° alcohols we can carry oxidation further to *carboxylic acids.*

4. Alcohols are neither acidic nor basic in the ordinary sense; phenols are weak acids.

5. The ring system of phenols is highly reactive toward reagents that can give aromatic substitution reactions.

6. Epoxides undergo reactions of a ring-opening type very readily to produce *glycols, cellosolves,* or other compounds.

7. Ordinary ethers are resistant to chemical change by most reagents under ordinary conditions. Hot hydrobromic or hydriodic acid will split ethers into alkyl bromides (or iodides).

Preparation

1. Alcohols. See the Summary Chart, page 136.
2. Phenols.
   (a) Industrially, phenol is made by action of base on chlorobenzene.
   (b) Action of nitrous acid on an aromatic amine.
3. Ethers.
   (a) Action of an acid on an alcohol under carefully studied conditions.
   (b) Williamson synthesis (Chapter 7.)

Problems and Exercises

1. Write structures for each
   (a) allyl alcohol
   (b) *sec*-butyl alcohol
   (c) glycerol
   (d) isopropyl alcohol
   (e) benzyl alcohol
   (f) propylene glycol
   (g) *t*-butyl alcohol
   (h) *n*-heptyl alcohol
   (i) *o*-nitrophenol
   (j) *m*-cresol
   (k) propylene oxide
   (l) phenyl ethyl ether
   (m) diisopropyl ether
   (n) ethyl isobutyl ether
   (o) *n*-propyl cellosolve
   (p) sodium isopropoxide
   (q) 2-methoxypentane
2. Write common names for each.

(a) $(CH_3)_2CHOH$

(b) $CH_3\underset{\displaystyle OH}{\underset{|}{C}}HCH_2CH_3$

(c) $CH_3{-}O{-}\underset{\displaystyle CH_3}{\underset{|}{C}}HCH_2CH_3$

(d) $C_6H_5{-}O{-}CH_2CH_2CH_3$

(e) $CH_3{-}C_6H_4{-}OH$ (para)

(f) $C_6H_4(OH)(NO_2)$ (OH and NO_2 meta on ring)

(g) $CH_2{=}CHCH_2{-}OH$

(h) $CH_3CH_2CH_2OH$

3. Write IUPAC names for the structures given, in question 2, at (a), (b), (g), and (h). Write the IUPAC name also for the following compound:

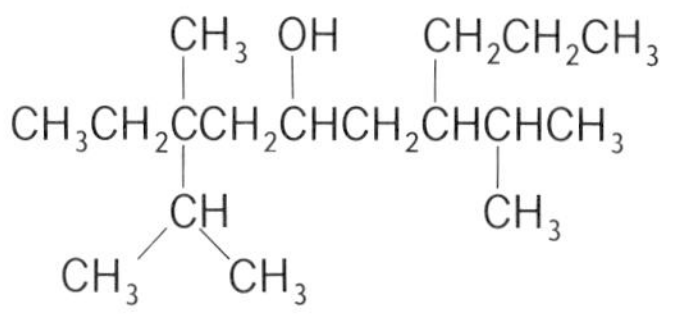

$$\begin{array}{l} \quad\quad\;\; CH_3\;\; OH \quad\; CH_2CH_2CH_3 \\ CH_3CH_2\overset{|}{\underset{|}{C}}CH_2\overset{|}{C}HCH_2\overset{|}{C}H\underset{|}{C}HCH_3 \\ \quad\quad\;\; CH \qquad\qquad\quad CH_3 \\ \quad CH_3 \;\; CH_3 \end{array}$$

4. What do the following expressions mean? Whenever appropriate, illustrate your answer with structures or reactions.
   (a) hydrogen bond
   (b) associated liquid
   (c) nucleophile
   (d) S_N1
   (e) alkoxide ion
   (f) S_N2
   (g) steric factor
   (h) Grignard reagent
   (i) E_2
   (j) epoxide
   (k) E_1
   (l) glycol

5. Write an equation (not necessarily balanced) for the reaction of *n*-propyl alcohol with each of the following reagents.
   (a) sodium metal
   (b) concentrated hydrochloric acid
   (c) potassium permanganate (excess)
   (d) hot copper metal
   (e) concentrated sulfuric acid, heat (Show two kinds of reactions.)

6. Write an equation (not necessarily balanced) for the reaction, if any, of *p*-cresol with each of the following reagents.
   (a) sodium hydroxide
   (b) sodium bicarbonate
   (c) dilute nitric acid
   (d) dilute hydrochloric acid
   (e) hot copper

7. You are given an unknown liquid and told that it is either *t*-butyl alcohol or *n*-butyl alcohol. Moreover, you are told that when a few drops of the unknown are shaken with dilute, potassium permanganate, the purple permanganate color disappears and a brown precipitate forms. Which alcohol is it? Explain in detail.

8. Complete (but do not balance) each of the following equations by writing the structure(s) of the principal organic product(s). If no reaction occurs, write "none." Some examples are a review.

(a) $(CH_3)_2CHOH \xrightarrow[H^+]{Cr_2O_7^{2-}}$

(b) $CH_3CH_2CH_2CH_2OH + HBr_{concd} \xrightarrow[heat]{}$

(c) $CH_3{-}O{-}CH_2CH_3 + HI_{concd} \xrightarrow[heat]{}$

(d) $CH_3CH_2CH_2CH_3 + H_2SO_4 \longrightarrow$

(e) $CH_3CH_2CH_2\underset{\displaystyle OH}{\underset{|}{C}}(CH_3)_2 \xrightarrow{MnO_4^-}$

(f) $C_6H_5Br + Mg \xrightarrow{ether}$

(g) product of (f) $+ CH_3\overset{\displaystyle O}{\overset{\|}{C}}H \xrightarrow{then\ H^+}$

(h) $CH_3CH_2CH_2OH + NaOH_{aqueous} \longrightarrow$

(i) $C_6H_5OH + CH_2\!-\!CH_2$ (epoxide, O bridging) $\xrightarrow{H^+}$

9. What alcohols would be needed to make the following compounds by oxidation? Write their structures.

(a) $CH_3C(=O)CH_2CH_3$

(b) 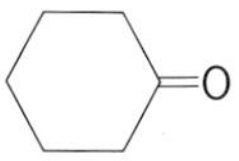 cyclohexanone (cyclohexane ring =O)

(c) $CH_3CH_2CO_2H$

(d) $H{-}C(=O){-}CH_2CH(CH_3)_2$

(e) $C_6H_5C(=O){-}H$

10. What alkenes would be needed to make the following by hydroboration-oxidation?

(a) $CH_3CH_2CH_2OH$

(b) $CH_3CH(CH_3){-}CH(OH)CH_3$

(c) cyclohexyl$-CH_2OH$

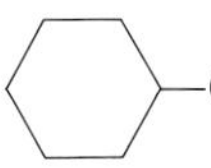

(d) $(CH_3)_2CHCH_2OH$

11. Outline steps for making these compounds by a Grignard synthesis.

(a) $CH_3CH_2CH_2OH$

(b) $C_6H_5C(OH)(CH_3)CH_3$

(c) $CH_3CH(CH_3)CH_2OH$

12. Suppose the only organic compounds available to you were alcohols of four carbons or fewer and benzene, plus needed solvents and inorganic reagents. How would it be possible to make the starting materials required for the Grignard syntheses of exercise 11? Write the equations.

13. What is the chief factor that affects relative rates in an S_N2 reaction?

14. Why are tertiary alcohols more reactive than primary alcohols in the Lucas test?

chapter seven

Organohalogen and Organometallic Compounds

7.1 Monohalogen Compounds

These are the simple aliphatic and aromatic halides. Examples of types and names are in Table 7.1. They are chiefly important as intermediates in making other compounds. We have already seen how certain of them can be converted to hydrocarbons (section 2.16) and to Grignard reagents (6.16). Our survey will be very brief, and we make it largely to review old principles and introduce others.

7.2 Types and Names

There are several ways of classifying monohalogen compounds.

1. According to the halogen: fluoro-, chloro-, bromo- or iodo-compounds.
2. According to the hydrocarbon group:
   Alkyl halides. General symbol: R—X, where X may be F—, Cl—, Br— or I— for purposes of this section. Of these, there are the subclasses of primary (1°), secondary (2°), and tertiary (3°) halides.
   Aryl halides. General symbol: Ar—X.
   Vinyl halides. No general symbol; the key structural feature is shown by **1.** The halogen is attached directly to a carbon that is involved in an alkene link.

VINYL POSITION
>C=C—X
VINYL HALIDES
1

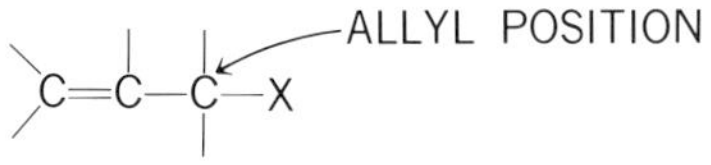

ALLYL POSITION
>C=C—C—X
ALLYL HALIDES
2

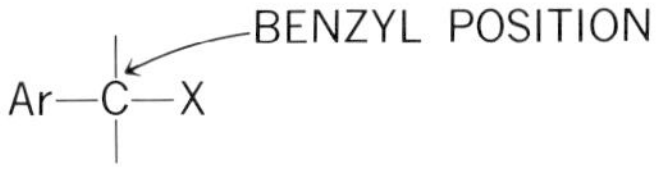

BENZYL POSITION
Ar—C—X
BENZYL HALIDES
3

Table 7.1
Some Common Monohalogen Compounds

Type	Boiling Points (in °C) X =			Densities g/cc (at 20 °C) X =		
	Cl	Br	I	Cl	Br	I
Alkyl halides						
R—X						
Methyl	−24	5	43	(gas)	(gas)	2.28
Ethyl	13	38	72	(gas)	1.44	1.93
Propyl	47	71	102	0.89	1.34	1.75
Butyl	79	102	130	0.88	1.22	1.52
Cyclohexyl	143	165	180 (dec)	1.00	1.32 (15°)	1.62 (15°)
Vinyl halides						
CH_2=CH—X	−14	16	56	—	1.51 (14°)	2.04
CH_3CBr=CH_2	23	48	—	0.92 (9°)	1.36	—
Allyl halides						
CH_2=$CHCH_2$—X (allyl halide)	45	71	103	0.94	1.40	1.85 (22°)
CH_3CH=$CHCH_2$—X (crotyl halide)	84	—	132	—	—	—
Benzyl halides						
C_6H_5—CH_2—X (benzyl halide)	179	201	mp 25°	1.10	1.44 (22°)	—
Aryl halides						
C_6H_5—X	132	156	189	—	—	—
CH_3—C_6H_4—X	162	184	211	—	—	—
Cl—C_6H_4—X	175	196	226	—	—	—

Allyl halides. No general symbol; the key structural feature is given by **2.** The halogen is attached at what is called an allylic carbon or the allyl position. An allylic carbon is defined as one attached to the carbon of a double bond. This puts the attached halogen one carbon removed from the double bond. *Benzyl halides.* Again, there is no general symbol. The key structural feature is shown by **3**. The halogen is attached to what is called a benzylic carbon, defined as a carbon that itself is joined to a carbon in a benzene ring. The halogen is then one carbon removed from an aromatic ring.

These classes are useful because they organize halides into groups whose members display similar chemical properties. For example, we shall see that vinyl halides and aryl halides are exceptionally unreactive toward reagents that rapidly displace halide ions from allyl halides or benzyl halides.

There are no special rules for naming these compounds. The IUPAC rules have been presented. We make common names in the expected way. We use the common name of the hydrocarbon group and follow it, as a separate word, with fluoride, or chloride, or bromide, or iodide, as the case may be. (Cf. Table 7.1.)

The three important unsaturated hydrocarbon groups with common names are:

Vinyl	$CH_2{=}CH{-}$	as in vinyl bromide:	$CH_2{=}CH{-}Br$
Allyl	$CH_2{=}CH{-}CH_2{-}$	as in allyl iodide:	$CH_2{=}CH{-}CH_2{-}I$
Benzyl	$C_6H_5{-}CH_2{-}$	as in benzyl chloride:	$C_6H_5CH_2Cl$

7.3 Physical Properties of Monohalogen Compounds

All of these compounds are insoluble in water and soluble in common organic solvents (e.g., ether, benzene, alcohols, etc.). In general, the halogen derivatives of any of the hydrocarbons are still hydrocarbonlike in solubility. So, too, are the fats of fatty tissue. That is why organohalogen pesticides tend to accumulate in such tissue.

The simple alkyl chlorides are all less dense than water whereas the corresponding bromides and iodides are all more dense than water.

Variations in boiling points for the alkyl halides occur very regularly with formula weights, as we should expect for substances whose molecules cannot associate by hydrogen bonds and are otherwise very similar to each other. Figure 7.1 is a plot of boiling points versus numbers of carbons in their molecules for the homologous series of simple, straight-chain alkyl chlorides, bromides, and iodides. (Two specific branched-chain chlorides are included to illustrate a general rule that isomers having the more "bunched" or compact molecules are the lower boiling isomers.)

7.4 Preparation of Monohalogen Compounds

The methods described briefly below do not apply equally well to all types of halides, whether we look at them as fluorides, chlorides, bromides, and iodides or in other ways of classifying them. To cite just one example, direct halogenation of hydrocarbons works well only for certain chlorides and bromides and not at all for iodides. Direct fluorination does not work for fluorides either, but for a different reason: the unusual reactivity of fluorine with hydrocarbons. It is an uncontrollable reactivity giving polyfluorohydrocarbons. In fact, the study of organofluorine compounds is generally separated from that of the other organo-

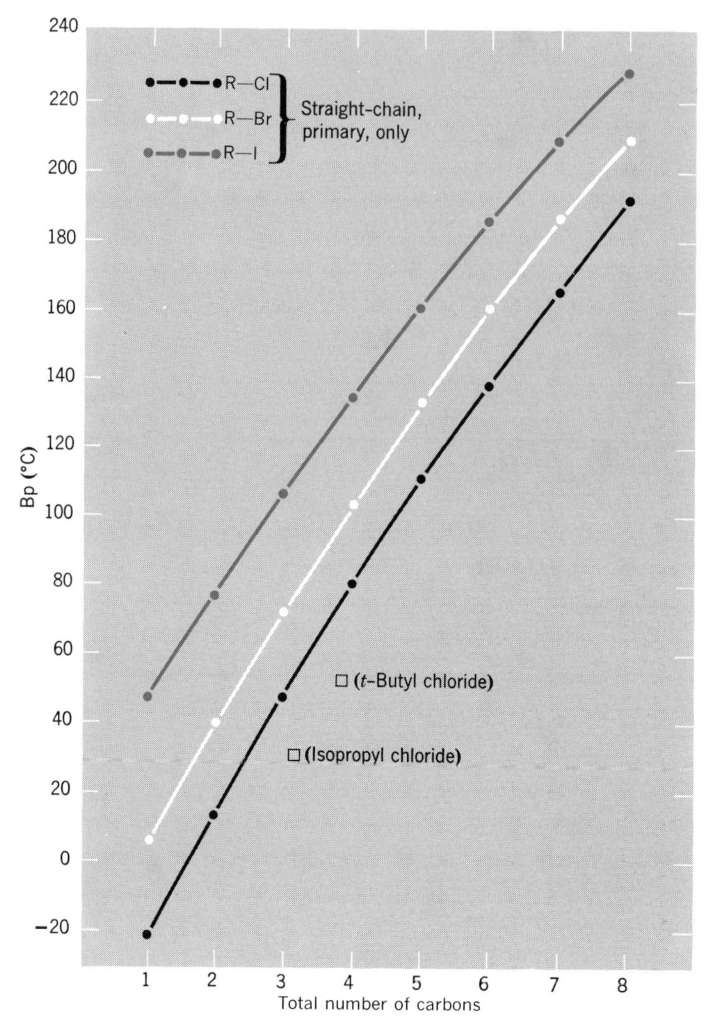

Figure 7.1
Boiling point versus carbon content for straight-chain, primary alkyl halides: $CH_3(CH_2)_nX$.

halogen compounds, as we shall do. We shall limit our study here to the ways to make the chlorides, bromides, and iodides.

7.5 Direct Halogenation of Selected Hydrocarbons

This was studied in some detail in Sections 2.13–2.15.

$$R{-}H + X_2 \xrightarrow[\text{UV light}]{\text{Heat or}} R{-}X + H{-}X$$

(X = Cl or Br; not I; not F)

Recall that one problem is the formation of a mixture of isomers. Therefore, this method of synthesis is usually used with "selected" hydrocarbons, those that cannot give isomers.

$$CH_4 + Cl_2 \longrightarrow CH_3Cl + HCl$$

METHANE → METHYL CHLORIDE

$$CH_3C(CH_3)_2CH_3 + Cl_2 \longrightarrow CH_3C(CH_3)_2CH_2Cl + HCl$$

NEOPENTANE → NEOPENTYL CHLORIDE

CYCLOHEXANE + Cl_2 → CYCLOHEXYL CHLORIDE (C₆H₁₁—Cl) + HCl

Benzylic halides are often made by direct halogenation because the benzylic position is more readily attacked than a position in the benzene ring itself, provided that an iron or iron salt catalyst is not present. We have here the option of halogenating either the ring itself or a benzylic position, and we pick our choice by picking the experimental conditions.

Iron catalyst (or iron salt)—ring halogenation.
Heat or sunlight (and no iron catalyst)—benzylic halogenation.

TOLUENE (C_6H_5—CH_3) + Cl_2 —(Fe or $FeCl_3$)→ Cl—C_6H_4—CH_3 (p-CHLOROTOLUENE) + o-CHLOROTOLUENE

TOLUENE + Cl_2 —(Heat or UV)→ C_6H_5—CH_2—Cl (BENZYL CHLORIDE)

The reason for this selectivity is that iron catalysts induce the reaction to occur by an ionic mechanism in which the catalyst generates Cl^+ that seeks electrons where they are—on the ring, not a side chain. The methyl group is naturally an ortho-para director and an activator. Without the iron catalyst the reaction must occur by a free radical mechanism (Section 2.13) in which Cl· attacks a benzyl hydrogen. The key intermediate is a benzyl free radical, **4**, and the one significant fact about such radicals is that the "odd" (unpaired) electron can be delocalized (resonance). This stabilizes the whole benzyl free radical system. It stabilizes it in a way not matched in any other hydrocarbon system except the allyl system. That is why an attack by Cl· is so much more frequently successful at the benzyl position than anywhere else.

(1) Initiation of a chain:

$$Cl{-}Cl \xrightarrow[\text{or UV}]{\text{Heat}} 2Cl\cdot$$

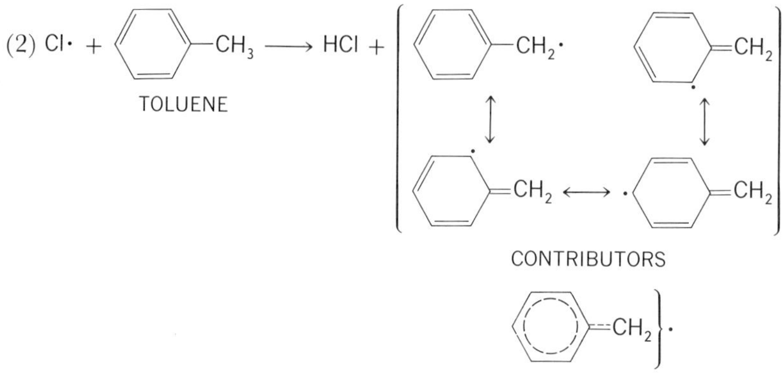

(2) Cl· + C_6H_5—CH_3 ⟶ HCl + [C_6H_5—CH_2· ⟷ ... =CH_2 ⟷ ·... =CH_2 ⟷ ... =CH_2]

TOLUENE

CONTRIBUTORS

HYBRID

4

RESONANCE-STABILIZED
BENZYL RADICAL

Then:

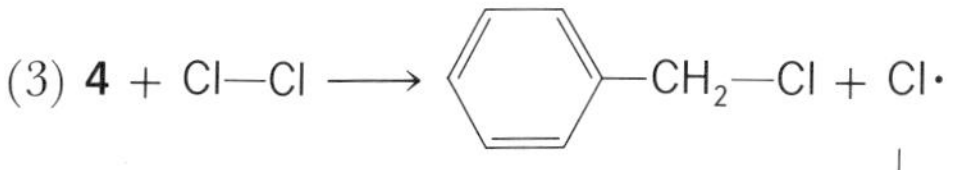

(3) **4** + Cl—Cl ⟶ C_6H_5—CH_2—Cl + Cl·

BENZYL CHLORIDE ⟶ then, repeat (2)-(3), (2)-(3), etc.

If we use ethylbenzene instead of toluene, the attack is still almost exclusively at the benzylic position even though it offers fewer hydrogens to be attacked than the methyl group. The attack goes through the free radical, **5**,

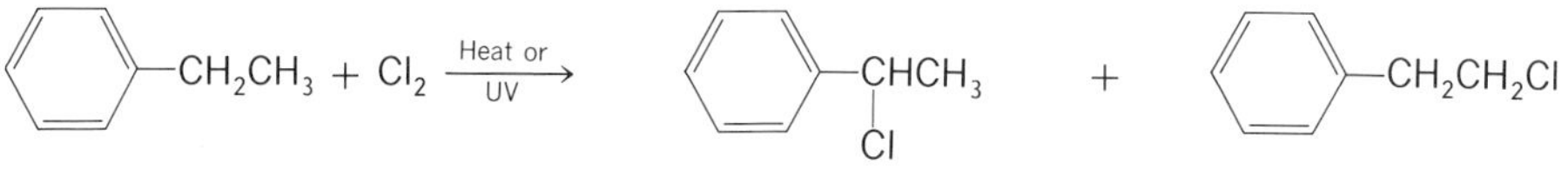

C_6H_5—CH_2CH_3 + Cl_2 $\xrightarrow[\text{UV}]{\text{Heat or}}$ C_6H_5—CH(Cl)CH_3 + C_6H_5—CH_2CH_2Cl

ETHYLBENZENE — 1-CHLORO-1-PHENYLETHANE (90%) — 2-CHLORO-1-PHENYLETHANE (10%)

a benzylic free radical, instead of **6**, in which no resonance stabilization of the odd electron can occur.

C_6H_5—$\dot{C}H$—CH_3 ⟷ Resonance as in **4**

5

C_6H_5—CH_2—CH_2· ⟷ (crossed) No resonance

6

Allylic halides are sometimes made by direct free-radical halogenation of alkenes, but the conditions are critical. At room temperature, alkenes very readily undergo *addition* of chlorine or bromine. However, at high temperatures, 500–600°, and at low concentration of halogen, the reaction of addition is reversed, and the alkene then undergoes free radical attack at an allylic position. For example, with propylene:

CH_2=CH—CH_3 (Allylic position: CH_3) + Cl_2 (Or Br_2) $\xrightarrow{500\text{–}600°}$ CH_2=CH—CH_2—Cl (Or —Br) + HCl (Or HBr)

PROPYLENE — ALLYL CHLORIDE (Or BROMIDE)

The important intermediate is the allyl free radical, **7**, the only one obtainable from propylene that can be stabilized by resonance.

$$[CH_2{=}CH{-}CH_2\cdot \longleftrightarrow \cdot CH_2{-}CH{=}CH_2] \equiv [CH_2{\cdots}CH{\cdots}CH_2]\cdot$$

7 HYBRID

7.6 The Addition of H—X to Alkenes

This was studied in Section 3.16. Recall that HCl, HBr, and HI add to the double bond in accordance with Markovnikov's rule.

7.7 The Conversion of Alcohols to Alkyl Halides

Concentrated hydrochloric, hydrobromic, or hydriodic acid act on alcohols to make alkyl halides as studied in Section 6.23. Other reagents that convert alcohols to halides are certain inorganic compounds such as thionyl chloride, phosphorus tribromide, or a mixture of potassium iodide and phosphoric acid.

$$R{-}OH + \underset{\text{THIONYL CHLORIDE}}{SOCl_2} \longrightarrow R{-}Cl + SO_2 + HCl$$

$$3ROH + \underset{\text{PHOSPHORUS TRIBROMIDE}}{PBr_3} \longrightarrow 3RBr + H_3PO_3$$

$$ROH + KI + H_3PO_4 \longrightarrow RI + KH_2PO_4 + H_2O$$

7.8 Chemical Properties of Monohalogen Compounds

The important reactions of alkyl halides are those in which the halogen is displaced by some other group. They are, in other words, nucleophilic substitution reactions. Table 7.2 shows just a few of the possibilities, and these demonstrate how alkyl halides are important intermediates in organic syntheses. See also Figure 7.2. They often serve as bridges between substances readily available in nature or easily made from some natural substance and other compounds needed for some application or for testing—testing as pharmaceuticals, for example.

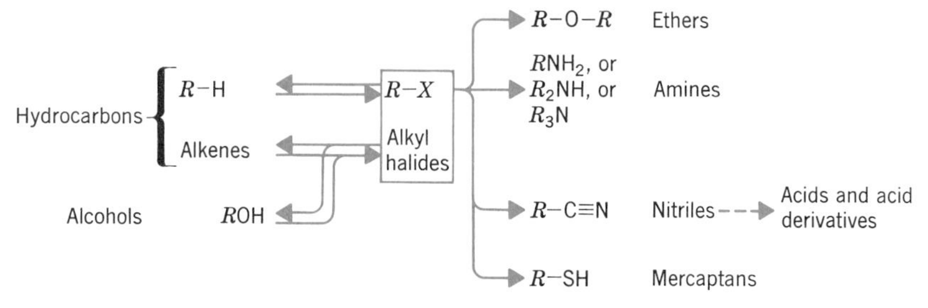

Figure 7.2
The alkyl halide "crossroads" in organic syntheses. The double arrows in this display signify only that interconversions are possible one way or the other depending on conditions and reagents. They are not meant to signify chemical equilibria.

Table 7.2
Some Nucleophilic Substitution Reactions of Alkyl Halides[a]

	Nucleophilic Agent	R—X + Y: ⟶ R—Y + X:
Negatively charged nucleophiles	Hydroxide ion, OH^-	R—X + HO^- ⟶ R—X + HO^- Alcohols
	Alkoxide ion, $R'O^-$ (Williamson ether synthesis)	+ $R'O^-$ ⟶ R—O—R′ + X^- Ethers
	Cyanide ion, ^-CN	+ ^-CN ⟶ R—CN + X^- Nitriles
	Amide ion, $^-NH_2$	+ $^-NH_2$ ⟶ R—NH_2 + X^- Amines
	Hydrogen sulfide ion, ^-SH	+ ^-SH ⟶ R—SH + X^- Mercaptans
Neutral nucleophiles	Ammonia, $:NH_3$	+ $:NH_3$ ⟶ R—$\ddot{N}H_2$ + (HX) Amines
	Water, $:OH_2$	+ $H_2\ddot{O}:$ ⟶ R—$\ddot{O}H$ + (HX) Alcohols

[a]The alkyl halide may be 1° or 2°, alkyl or cycloalkyl, allylic or benzylic. It may not be vinyl or aryl, which are usually inert toward nucleophilic substitution.

One kind of raw material in Figure 7.2, the hydrocarbons, are abundant in petroleum. Alcohols also are often obtained directly from natural sources or they may be made from hydrocarbons. Neither family, however, can be converted directly into amines, nitriles (and thence, organic acids), or mercaptans. These can be made from halides, however. That is why the alkyl halides are important in organic chemistry. The Williamson reaction is but one illustration.

The **Williamson ether synthesis** (Table 7.2) is the most versatile method for making ethers. It is essential for making mixed ethers, R-O-R′ or R-O-Ar, which cannot be made by simple dehydration of alcohols (Section 6.33). An alkoxide ion, RO^-, or a phenoxide ion, ArO^-, is first made, and then the alkyl halide furnishing the other half of the ether is added.

Examples:

$$\underset{\textit{sec}\text{-BUTYL ALCOHOL}}{CH_3CH_2\underset{|}{\overset{CH_3}{C}}H{-}OH} + \underset{\text{SODIUM}}{Na} \longrightarrow \tfrac{1}{2}H_2 + \underset{\text{SODIUM } \textit{sec}\text{-BUTOXIDE}}{CH_3CH_2\overset{CH_3}{C}H{-}O^-Na^+} \xrightarrow[CH_3CH_2I]{\text{then add}}$$

$$\underset{\text{ETHYL } \textit{sec}\text{-BUTYL ETHER (70\%)}}{CH_3CH_2\overset{CH_3}{C}H{-}O{-}CH_2CH_3} + NaI$$

$$C_6H_5OH + NaOH \longrightarrow H_2O + C_6H_5O^-Na^+ \xrightarrow[CH_3CH_2CH_2Br]{\text{then add}} C_6H_5{-}O{-}CH_2CH_2CH_3 + NaBr$$

PHENOL — SODIUM PHENOXIDE — n-PROPYL PHENYL ETHER (73%)

What was said above about alkyl halides applies also to allylic and benzylic halides. In fact, of all the classes of halides we have considered, these two types are the most reactive of all in nucleophilic substitution reactions. They react even with weak nucleophiles such as water and alcohol because these halides can ionize to give resonance stabilized allyl and benzyl cations, as structures **8** and **9** show.

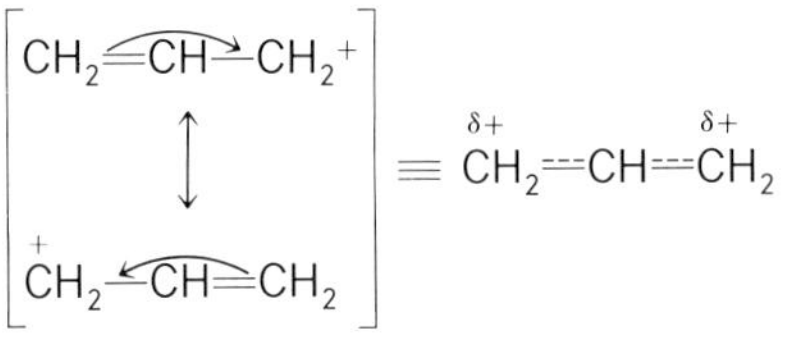

$$\left[CH_2{=}CH{-}CH_2^+ \longleftrightarrow {}^+CH_2{-}CH{=}CH_2 \right] \equiv \overset{\delta+}{C}H_2{\cdots}CH{\cdots}\overset{\delta+}{C}H_2$$

CONTRIBUTORS — HYBRID

8
ALLYL CATION

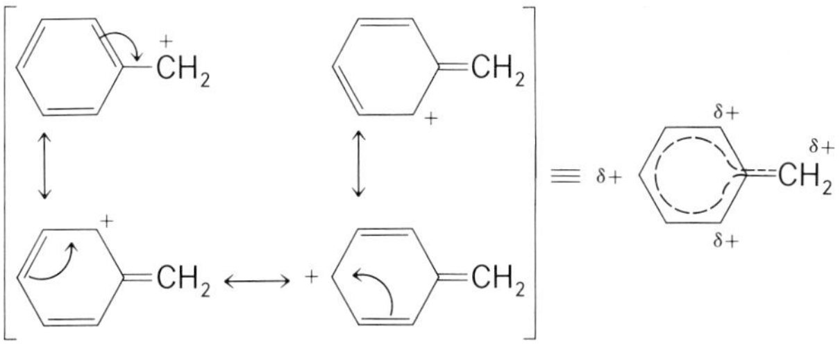

CONTRIBUTORS — HYBRID

9
BENZYL CATION

In sharp contrast, vinyl and aryl halides are quite inert to nucleophilic substitutions. Molecules of neither type are able to give stable carbonium ions. This prevents S_N1 reactions. They resist S_N2 changes partly because nucleophiles (being electron-rich themselves) cannot easily get into the carbon holding the halogen. That carbon itself has an "extra" electron density because of the pi electrons next to it. Direct backside attack is hindered. In fact this route of attack is impossible in aryl halides or in cyclic halides having cagelike and rigid structures.

These are chemical properties of these families. We are touching here on matters not only of considerable importance in organic chemistry but also significant in environmental affairs. Anything that is relatively unreactive with chemicals in air, water or living systems will be persistent in the environment. (cf. Exercise 5.)

7.9 Elimination versus Substitution

It is an often annoying fact that when you want to carry out a particular substitution reaction a side reaction occurs to cut down the yield. An elimination reaction is nearly always a side reaction to substitution and *vice versa.* The problem varies from one type of halide to another. For example, an extreme example, if a tertiary halide is mixed with a nucleophile that is also a strong base, the halide will react almost exclusively by elimination. No substitution will occur. An alkene will be produced instead, as illustrated by the behavior of *t*-butyl chloride.

$$CH_3-\underset{CH_3}{\overset{CH_3}{\underset{|}{\overset{|}{C}}}}-Cl + OH^- \longrightarrow CH_2{=}C(CH_3)_2 + H_2O + Cl^-$$

t-BUTYL CHLORIDE (A 3° halide) — ISOBUTYLENE (And no *t*-butyl alcohol)

Primary halides, on the other hand, are the easiest to make undergo nucleophilic substitution, and the hardest to get to undergo elimination.

The reasons for the difference are complex. The biggest factor seems to be the fact that from tertiary halides you get more highly branched alkenes than you can get from primary halides. The more branched an alkene the more stable it is. "Nature goes in the direction of increasing stability," is the rule. Hence, tertiary halides, in the presence of a strong base, have a lower energy path for changing to alkenes than to some substitution product. Primary halides, moreover, are much more "open" in a steric sense than are tertiary halides to bimolecular attack.

7.10 Organic Compounds of Fluorine

So irregular seem the behavior and properties of some organofluorine compounds that when we enter their realm we might think we have come to one of nature's surprise parties. They include some of the least toxic and most toxic of organic compounds. They also include some of the most reactive and least reactive of substances. In the area of their physical properties, carbon tetrafluoride, CF_4 (bp $-129°$), boils 51° *lower* than methyl fluoride, CH_3F (bp $-78°$), in spite of having nearly three times the formula weight.

7.11 Perfluorocarbons

The methods of preparing the other organohalogen compounds generally do not work for the organofluorines although hydrogen fluoride does add normally to the double bond. If you mix fluorine and an alkane—any alkane—the only product you will get in any respectable yield, after the violent reaction has subsided, is carbon tetrafluoride. Using special methods for moderating the reaction, industrial firms use direct fluorination to make perfluorocarbons, compounds containing only carbon and fluorine. (The "per-" prefix means that all hydrogens of the corresponding hydrocarbon are replaced by fluorines.) Heptane, C_7H_{16}, can be converted to perfluoroheptane, C_7F_{16}, for example. Cobalt(III) fluoride, CoF_3, is also used as a fluorinating agent, one that is easier to control than fluorine:

$$C_7H_{16} + 32CoF_3 \longrightarrow C_7F_{16} + 32CoF_2 + 16HF$$

The perfluorocarbons are very stable to heat; virtually no other chemical attacks them; they do not conduct electricity; and they conduct heat well. These properties make the liquid perfluorocarbons valuable as liquid heat exchangers and hydraulic fluids. Their disadvantage is that they dissolve gases very well. Ironically, this fact has aroused the interests of biologists, biochemists, and specialists in respiratory conditions. Emulsions of perfluorocarbons in water have a high capacity for dissolving both oxygen and carbon dioxide. Because they dissolve essentially nothing else, and because they do not react chemically with much, these emulsions have been used for perfusion of animal organs, and even whole animals. (See Figure 7.3.) The emulsions have even successfully replaced the entire blood volume of experimental animals.

7.12 Teflon

Polymerization of tetrafluoroethylene gives a high-formula-weight polymer called Teflon.

$$CF_2{=}CF_2 \xrightarrow[700\ \text{lb/in.}^2]{\text{Peroxides}} \underset{\text{TEFLON}}{(-CF_2-CF_2-)_n}$$

This remarkable substance is not attacked by hot alkalies, hot acids, oxidizing agents, or organic solvents. Molten sodium and potassium are about the only materials to affect it. It does not conduct electricity. Besides its well-known use as a coating for cooking utensils and irons, Teflon's heat-resistant, nonstick qualities make it useful for bearings in some applications and for supports for printed circuits

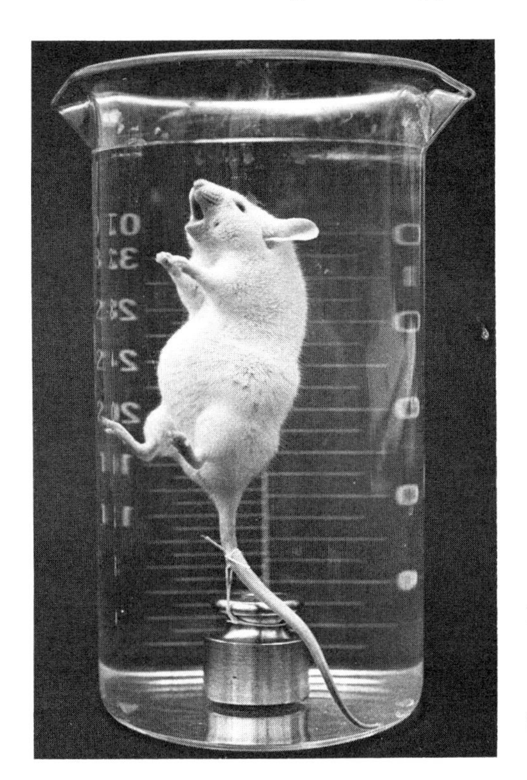

Figure 7.3
A mouse submerged in perfluorobutyl tetrahydrofuran. Removed one hour later the mouse was still alive and in a healthy condition. The oxygen it needs is dissolved in this solvent. The carbon dioxide it must "exhale" also dissolves in the solvent. (Courtesy Dr. Leland C. Clark, Jr.)

or for insulation in electrical systems. Tubes made of Teflon fabric have been used to replace faulty segments of blood vessels.

Other industrially important fluoropolymers are **Teflon-100** (from copolymerization of tetrafluoroethylene and hexafluoropropylene), **Kel-F** or **fluorothene** (monomer: $CFCl{=}CF_2$), **Fluorolubes** (low-formula-weight polymer of $CFCl{=}CF_2$), **Kynar** (monomer: $CH_2{=}CF_2$), and **Tedlar** (monomer: $CH_2{=}CHF$).

7.13 Freons

Freons are compounds whose molecules have very few carbons and that consist also of fluorine, chlorine, and sometimes hydrogen. They generally are quite nontoxic, nonflammable, without odors, and chemically unreactive in an ordinary environment even at relatively high temperatures. A few of the important examples and their uses are given in Table 7.3.

Freons are often given numbers, and what these numbers mean is seen in the following:

Freon-*cba*.

a = the right-hand, outside number = number of fluorines
b = the next number in from a = (1 + number of hydrogens)
c = the third number in from a = (number of carbons − 1)

When $c = 0$, i.e., when c is absent, the Freon has just one carbon per molecule. Examples:

Freon-112	Freon-13
$a = 2$ = two fluorines	$a = 3$ = three fluorines
$b = 1 = 1$ + zero hydrogens	$b = 1 = 1$ + zero hydrogens
$c = 1$ = two carbons − 1	$c = 0$; therefore, 1 carbon

Whatever is needed to make a saturated molecule is assumed to be made up of

Table 7.3
Some Important Organofluorine Compounds

Name	Structure	Uses
Freon 11	CCl_3F (bp 24°C)	Propellant; refrigerant
Freon 12	CCl_2F_2 (bp −30°C)	First fluorocarbon aerosol propellant; refrigerant
Freon 114	$CClF_2CClF_2$ (bp 4°C)	Refrigerant (very stable, odorless); propellant for shaving creams, colognes
Bromotrifluoromethane	$CBrF_3$ (bp −58°C)	Fire extinguishers (low toxicity)
Perfluorocyclobutane	$F_2C{-}CF_2$ / $F_2C{-}CF_2$ (four-membered ring) (bp −4°C)	Propellant for aerosol cans of food products (e.g., whipped cream)

chlorines. When using this system the fluorines are assumed to be distributed on the carbons as symmetrically as possible. Otherwise other names are needed (and we shall not pursue that further). Hence Freon-112 with two carbons, two fluorines, and no hydrogen is $CCl_2F—CCl_2F$. Similarly, Freon-13 would be $CClF_3$.

The Freons are not totally harmless. If aerosol vapors, where a Freon is the propellant, are deeply inhaled, death by cardiac arrest can occur. Dozens of deaths from this bizarre drug abuse have been reported. An entirely different possible danger of the Freons reached the "need-for-more-study" stage in late 1974. Freons, being inert in the environment, drift upward into the stratosphere, where they have been detected. (We made 800 million pounds of Freons in 1974.) There was widening concern that they might reduce the stratospheric ozone shield, and the National Academy of Sciences began an investigation. (cf., *Science News* October 5, 1974, page 212.)

Organometallic Compounds

7.14 Some Types and How They Are Named

For a substance to be classified as an organometallic compound its molecules must include at least one direct carbon-to-metal bond. That bond can be of any type—completely ionic, completely covalent, a special coordination bond involving orbitals above *s* and *p*, or a blend of any or all of these.

Besides at least one bond to carbon, the metal may also have a bond to some other element such as —OH, $—NH_2$, or halogen. A large number of organic derivatives of metals are not organometallics, however. In sodium acetate, $Na^{+-}O_2CCH_3$ the metal ion is joined (ionic bond) to oxygen, not carbon. The following are examples of relatively simple types.

$(CH_3)_2Hg$	$(CH_3CH_2)_3Al$	C_6H_5Li
DIMETHYLMERCURY	TRIETHYLALUMINUM	PHENYLLITHIUM

The name of the metal is last and the whole name is written as one word. If another group is present, one bonded to the metal by something other than carbon, as in the following examples, that other group is named separately.

CH_3MgBr	CH_3CH_2HgCl	$C_6H_5HgO_2CCH$
METHYLMAGNESIUM BROMIDE	ETHYLMERCURY CHLORIDE	PHENYLMERCURY ACETATE

The practice for naming these substances has not, however, become fully standardized, and it is common to see variations of these names used (e.g., phenylmercuric acetate or phenylmercury(II) acetate).

Names for simple organometallic compounds of the Group IV elements have the "-ane" ending found among alkanes. Since methane is CH_4, the other hydrides have similar names:

SiH_4	Silane	SnH_4	Stannane	SeH_4	Germane	PbH_4	Plumbane

Therefore, and we attempt only to illustrate and not to treat exhaustively, a compound such as $Cl_2Si(CH_3)_2$ is called dimethyldichlorosilane.

7.15 Some Methods for Preparing Organometallic Compounds

From Grignard Reagents

The preparation of Grignard reagents, themselves organometallic compounds, was discussed in Section 6.16. Many other organometallics may be made from them. For examples:

$$2RMgCl + CdCl_2 \longrightarrow \underset{\text{ALKYLCADMIUM COMPOUNDS}}{R_2Cd} + 2MgCl_2$$

$$2RMgCl + HgCl_2 \longrightarrow \underset{\text{ALKYLMERCURY COMPOUNDS}}{R_2Hg} + 2MgCl_2$$

This approach is useful also to prepare organoderivatives of other metals, such as lead, zinc, and silicon.

From Other Organometallics

Using active metals:

$$2Na + (C_6H_5)_2Hg \longrightarrow 2C_6H_5Na + Hg$$

A MORE REACTIVE METAL..........REPLACES..........A LESS REACTIVE METAL—NEVER THE OTHER WAY AROUND.

Using metal salts:

$$2HgCl_2 + \underset{\text{TETRAETHYLLEAD}}{Pb(CH_2CH_3)_4} \longrightarrow \underset{\text{ETHYLMERCURY CHLORIDE}}{2CH_3CH_2HgCl} + (CH_3CH_2)_2PbCl_2$$

Direct Synthesis

One synthesis of the very important chemical, dimethyldichlorosilane, is accomplished directly:

$$2CH_3Cl + Si \xrightarrow[300°C]{\text{copper powder}} (CH_3)_2SiCl_2$$

Addition Reaction

The synthesis of another important industrial chemical, triethylaluminum, is carried out as follows (Ziegler reaction):

$$Al + 6CH_2{=}CH_2 + 3H_2 \xrightarrow[\text{High pressure}]{} 2(CH_3CH_2)_3Al$$

7.16 Some Important Properties of Organometallic Compounds

From this enormous field of chemistry we shall examine only those reactions that involve chemicals in the environment: water and air. Moreover, we shall limit the study to those reactions that might bear on consumer products. We shall bring in others only to illustrate some general trends.

As we should expect, properties reflect bond types. Ionic substances tend to dissolve in water; nonpolar substances do not. Molecules with very polar bonds might be expected to be more reactive toward water, a polar compound. Where metals themselves react very exothermically with oxygen, the corresponding organometallic compounds might be expected to be unstable in air.

Carbon-metal bonds are generally quite polar with those to metals in Groups I and II being the most polar. As we go across the second row in the periodic chart we have this trend:

C—Na C—Mg C—Al C—Si C—P C—S C—Cl
INCREASING POLARITY OF CARBON-METAL BOND—

Table 7.4 summarizes how simple alkylmetals of the elements in the second row behave toward water and air.

In general, the methyl and ethylmetals of Groups I and II all react extremely vigorously with water to release methane or ethane and the metal hydroxides. Those of the less reactive metals (those not so readily oxidized), such as mercury, silicon, lead, and iron, do not react with water. It is not surprising, therefore, that where organometallics are consumer products, the metals are the less reactive ones. Their stability in air bears on this, too. The methyl or ethylmetals of mercury, lead or

Table 7.4
Some Alkylmetals of Second-Row Elements

Substance	Some Properties
Ethylsodium	White solid that decomposes in air or water; insoluble in benzene
Dimethylmagnesium	White solid that burns in air; stable to 240° in absence of air; slightly soluble in ether
Triethylaluminum	Colorless liquid that ignites in air. Bp 194°; explodes in water
Tetraethylsilane	Bp 153°. Soluble in ether; insoluble in concentrated sulfuric acid
Triethylphosphorus	Bp 128°; insoluble in water
Dimethyl sulfide	Bp 92°; insoluble in water
Ethyl chloride	Bp 13°; insoluble in water

silicon, for example, are not attacked by air. Many others must be carefully protected from air.

7.17 Some Consumer Uses of Organometallics

In Medicine

Two important bacteriocides are organomercury compounds, Mercurochrome® and Merthiolate®.

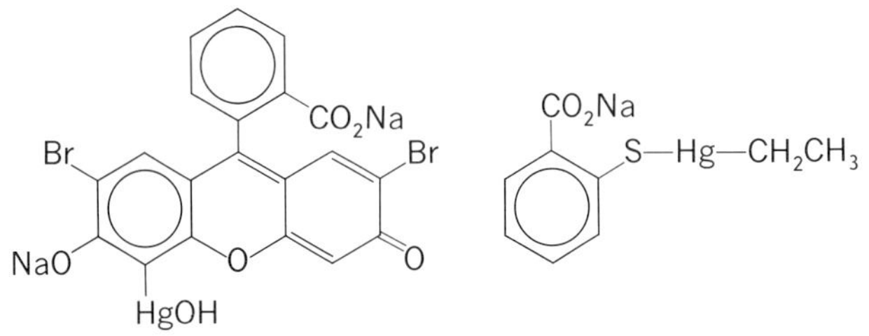

MERCUROCHROME® MERTHIOLATE®

As Pesticides

Organomercury compounds have been widely used to prevent fungi on seed grains, to retard the onset of slime and molds on submerged lumber (wharves, sawmills, sluiceways), and to combat certain weeds. Stringently regulated now, this use contributed in the past to the general problem of mercury pollution in the environment. Certain organotin compounds such as ditributyltin oxide, $[(C_4H_9)_3Sn]_2O$, have successfully been used for treating fabrics to prevent their rotting and to keep them relatively free of bacteria. This compound is also used to treat submerged lumber installations.

In Transportation and Industry

Tetraethyllead and tetramethyllead are antiknock additives for boosting octane-ratings of gasoline. (See Section 5.9.) Triethylaluminum is used in the catalyst system for the polymerization of alkenes such as ethylene and propylene in a way that insures a regular geometry. (See Section 3.25.)

7.18 The Silicones

These are polymers. Depending on the recipe used as well as the conditions, industry makes silicon rubber, silicone resins or silicone oils. In all cases, the products have the following properties:

1. They are thermally stable and resist combustion. Silicon rubber can be heated with a blowtorch without melting or suffering serious change. See Figure 7.4.
2. They not only do not react with water, they repel water. When a gauze or other fabric surface is sprayed with a silicone, the pores and air spaces in the fabric are not closed. It will "breathe." But water beads so strongly on the surface that it is impossible for it to go through the fabric's pores and holes. Many

Figure 7.4
The intense heat of the blow torch flame does not penetrate this silicone board. (Courtesy General Electric Corporation.)

other substances do not stick to silicones either. When auto tires are made, for example, one problem was that the newly made tire would stick—almost as if it had been welded—to the steel mold. When the mold is sprayed with a silicone, however, the tire releases readily. Silicones are commonly used as "release agents."

Because of these properties, silicones are used to seal joints and seams in jet planes, as insulators (against either heat or electricity), as seals and gaskets, as fluids in hydraulic devices, as lubricants in machines operating between wide ranges of temperatures, as water-repellent films on concrete and other surfaces, and in protective hand creams.

To make an elastic silicone—a silicone elastomer or rubber—the polymer is formed as a linear molecule of great chain length and little cross linking. The silicone oils have longer chains and special end groups. The resins have considerable cross linking.

$$(CH_3)_2SiCl_2 \xrightarrow[HCl,\ H_2O]{(CH_3)_3SiCl} \longrightarrow CH_3-\underset{CH_3}{\overset{CH_3}{Si}}-O\left(-\underset{CH_3}{\overset{CH_3}{Si}}-O\right)_x-\underset{CH_3}{\overset{CH_3}{Si}}-CH_3$$

SILICONE OIL

$$(CH_3)_2SiCl_2 \xrightarrow{H_2O} [(CH_3)_2SiO]_4$$

A tetramer

$$[(CH_3)_2SiO]_4 \xrightarrow{\Delta\ \text{Trace of KOH}} \text{etc}-\underset{CH_3}{\overset{CH_3}{Si}}-O\left(-\underset{CH_3}{\overset{CH_3}{Si}}-O\right)_n-\underset{CH_3}{\overset{CH_3}{Si}}-O-\text{etc.}$$

SILICONE GUM
(When vulcanized: Silicone rubber)

Brief Summary

1. The most common ways to prepare halides (where the halogen is Cl, Br, or I) are:

From alcohols	by displacement of the —OH group using HX or $SOCl_2$ or PBr_3.
From alkenes	by addition of HX to a double bond.
From hydrocarbons	by substitution of —H using Cl_2 or Br_2 (and either heat or UV light).

2. Alkenes and alkyl-substituted benzenes undergo free radical chlorination or bromination at allylic or benzylic sites where odd electrons in free radical intermediates are stabilized by resonance.

3. Alkyl, allyl, and benzyl halides undergo nucleophilic substitution reactions and may be changed to compounds in—

—This family	When the Nucleophile is
Alcohols	H_2O or OH^-
Amines	NH_3 (or RNH_2 or R_2NH)
Mercaptans	HS^-
Ethers	RO^-
Nitriles	CN^-

4. When the nucleophile happens also to be a strong base (e.g., OH^-, NH_2^- RO^- and CN^-), a halide is susceptible to losing HX and changing into an alkene. Tertiary halides are most prone to this.

5. Properties of the simpler organometallics depend on the metal involved. Where the metal is from Group I or II (or of comparable reactivity), the organometallic can be expected to be unstable both in air and water. Where the metal is less reactive (e.g., silicon, lead, mercury), the organometallic is generally stable in air and water.

Problems and Exercises

1. Write the structure of each compound.
   (a) isobutyl bromide
   (b) allyl iodide
   (c) *p*-nitrobenzyl chloride
   (d) vinyl bromide
   (e) Freon-113
   (f) methyltrichlorosilane
   (g) *t*-butyl chloride
   (h) methylmercury chloride
2. Name each compound by directions given in this chapter.
   (a) $CCl_3—CCl_2F$
   (b) $(CH_3)_3Al$
   (c) CH_3CH_2HgCN
   (d) $\left(—\overset{CH_3}{\underset{CH_3}{\overset{|}{\underset{|}{Si}}}}—O— \right)_n$

(e) (cyclopentyl)—Cl (common and IUPAC)

3. Write the structure(s) of the organic product(s) in each of the following reactions. If no reaction occurs, write "None."

(a) Cl—C_6H_4—CH_3 + Cl_2 $\xrightarrow[\text{or heat}]{\text{UV}}$

(b) $CH_3OH \xrightarrow{SOCl_2}$

(c) $CH_3CH{=}CH_2 + HCl_{(g)} \longrightarrow$

(d) $(CH_3)_3COH + HCl_{(aq)} \longrightarrow$

(e) $CH_3I + CN^- \longrightarrow$

(f) $CH_3CH_2Br + OH^- \xrightarrow[\text{Alcohol/ water}]{}$

(g) $(CH_3)_3CCl + KOH \xrightarrow[\text{Alcohol}]{}$

(h) $CH_2{=}CHCl + NH_3 \longrightarrow$

(i) $CH_2{=}CHCH_2Br + SH^- \longrightarrow$

(j) $CH_3CH_2MgBr + CdCl_2 \longrightarrow$

4. We can make halides from hydrocarbons, alcohols, or alkenes. Study each of the following halides and consider which specific compound from each of the three families of starting materials one would use to make the halide. One or more of the options may be unwise or impossible. Identify which would be the best starting material to use and write the equation for the synthesis of the halide from it. (Assume that any desired starting material is available.)

(a) $CH_3CH_2CH_2Cl$

(b) C_6H_5—Br

(c) $ClCH_2CH_2$—C_6H_4—CH_3

(d) $CH_3CHCH_2CH_3$ (Br on C-2)

(e) (cyclohexyl)—Cl

5. *Part A*. Where a halogen is located at a "bridgehead" of a bicyclic compound, such as **10**, the substance is unusually stable. Neither S_N1 nor S_N2 conditions will induce displacement of the halide group. Before putting the question, we note that facts such as these led chemists to conclude that one important condition for the relative stability of a 3° carbonium ion is the opportunity for the bonding network and atomic nuclei around the carbon with the plus charge to go from a tetrahedral to a planar shape. The carbonium ion **11** is obviously prevented from being planar. The question here is this. Why is **10** unreactive in S_N2 reactions? Another question: why is **10** unreactive in elimination reactions?

Part B. Aldrin, a persistent, organochlorine pesticide, has the structure **12**. The soil is rich in nucleophiles—moisture, ammonialike substances, and others. Why is aldrin relatively stable in this environment?

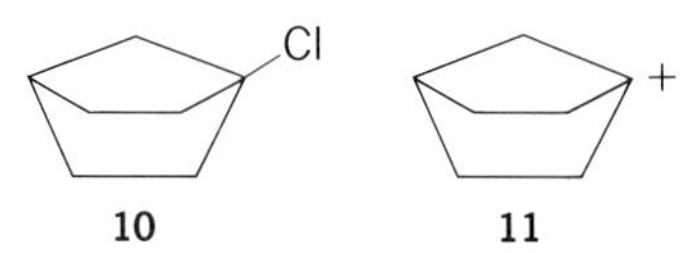

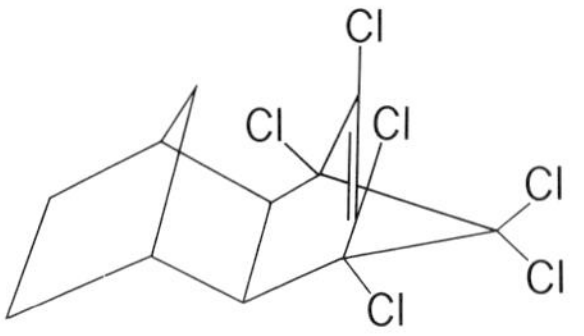

ALDRIN

12

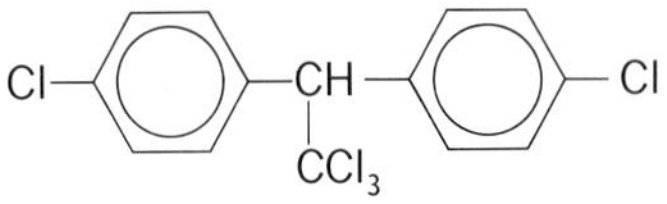

$Cl-C_6H_4-CH(CCl_3)-C_6H_4-Cl$

DDT

13

6. DDT has structure **13.** In the environment one change it undergoes is the loss of HCl. What forms? (Write the structure.) (This product is also dangerous to certain wildlife.)

chapter eight

Amines and Other Organonitrogen Compounds

8.1 Structural Features and Types

Amines are alkyl or aryl derivatives of ammonia. One, two, or all three of the hydrogens on a molecule of ammonia may be replaced by such groups. For example:

CH_3NH_2	$(CH_3)_2NH$	$(CH_3)_3N$	$CH_3NHCH_2CH_3$
METHYLAMINE	DIMETHYLAMINE	TRIMETHYLAMINE	METHYLETHYLAMINE

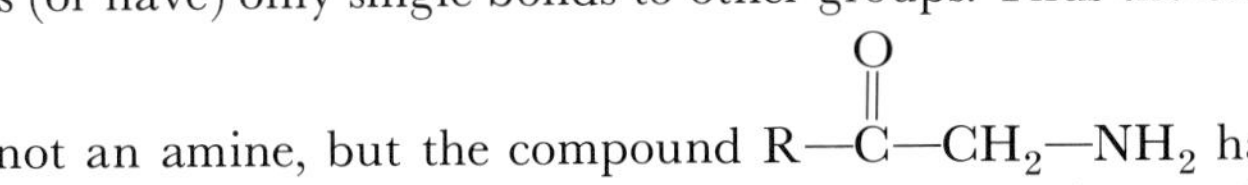

$NH_2—C_6H_5$	$CH_3NH—C_6H_5$	$(CH_3)_2N—C_6H_5$
ANILINE	*N*-METHYLANILINE	*N,N*-DIMETHYLANILINE

Aliphatic amines are those in which the carbon(s) attached directly to the nitrogen has (or have) only single bonds to other groups. Thus the compound $R—\overset{O}{\overset{\|}{C}}—NH_2$ is not an amine, but the compound $R—\overset{O}{\overset{\|}{C}}—CH_2—NH_2$ has an amine function. If even one of the groups attached directly to the nitrogen is aromatic, the amine is classified as an *aromatic amine.*

Another way of subclassifying amines is by the designations 1° (primary), 2° (secondary), and 3° (tertiary) amines. The numerals denote the number of groups attached to the *nitrogen:*

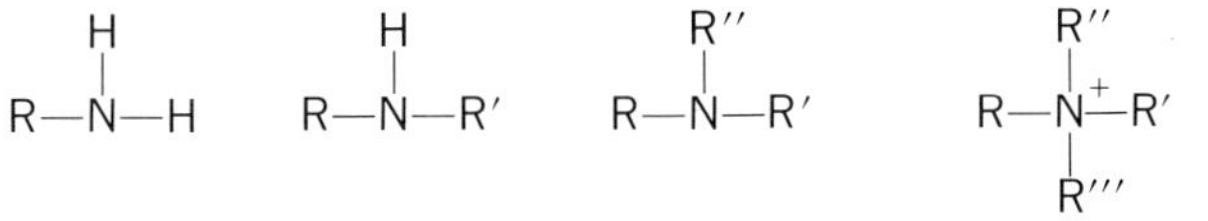

$R—NH—H$ (N bearing H)	$R—N(H)—R'$	$R—N(R'')—R'$	$R—N^+(R'')(R''')—R'$
PRIMARY AMINES 1°	SECONDARY AMINES 2°	TERTIARY AMINES 3°	QUATERNARY AMMONIUM IONS 4°

Several amines are listed in Table 8.1.

Because of structural similarity to amines, tetraalkyl derivatives of the ammonium ion, NH_4^+, called *quaternary ammonium ions,* R_4N^+, are also included here.

8.2 Heterocyclic Amines

Many compounds of biochemical importance consist, at least in part, of ring systems in which one of the ring atoms is a nitrogen. Two of the important saturated heterocyclic amines are pyrrolidine and piperidine. They behave as 2° amines. The important unsaturated heterocyclic amines are discussed in chapter 16.

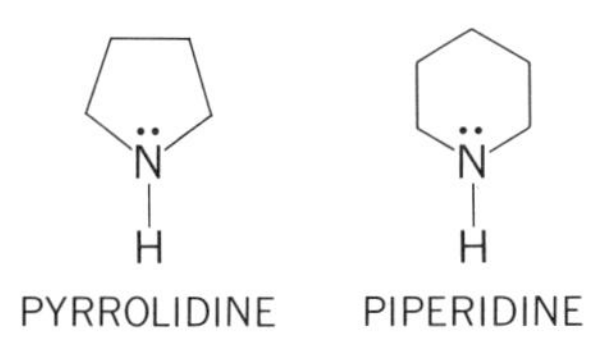

PYRROLIDINE PIPERIDINE

8.3 Naming the Amines

In the *common system* of nomenclature, aliphatic amines are named by designating the alkyl groups attached to the nitrogen and following this series by the word part "-amine."

$(CH_3)_2CHCH_2—NH_2$ ISOBUTYLAMINE

$(CH_3)_2CH—NH—CH_2CH_3$ ETHYLISOPROPYLAMINE

$CH_3CH_2CH_2—N(CH_3)—CH_2CH_2CH_2CH_3$ METHYL-*n*-PROPYL-*n*-BUTYLAMINE

Where this system breaks down, the IUPAC system can be employed. In this system the $—NH_2$ group is named the *amino* group, or *N*-alkyl or *N,N*-dialkylamino, and the specific groups are, of course, named.

$CH_3—CH(NH_2)—CH_2—OH$ 2-AMINO-1-PROPANOL

$CH_3CH_2NH—CH_2CH_2CH_2CH_2CH_2CH_2CH_2CH_3$ 1-(*N*-ETHYLAMINO)OCTANE (That is, at the 1-position in octane an *N*-ethylamino group is attached.)

The capital *N* designation means that the group immediately following it is attached to nitrogen. (See also the earlier examples of *N*-methylaniline and *N,N*-dimethylaniline.)

The simplest aromatic amine is always called *aniline.*

Exercise 8.1 Write structures for the following compounds.

(a) triethylamine
(b) dimethylethylamine
(c) *t*-butyl-*sec*-butyl-*n*-butylamine
(d) *p*-nitroaniline
(e) *p*-aminophenol
(f) tetraethylammonium chloride
(g) cyclohexylamine
(h) diphenylamine
(i) 2-(*N*-methyl-*N*-ethylamino)butane
(j) benzylmethylamine

Table 8.1 **Amines**

Name	Structure	Mp (°C)	Bp (°C)	Solubility in H_2O	K_b (at 25°C)
Ammonia	NH_3	−78	−33	Very soluble	1.8×10^{-5}
Methylamine	CH_3NH_2	−94	−6	Very soluble	4.4×10^{-4}
Dimethylamine	$(CH_3)_2NH$	−96	7	Very soluble	5.2×10^{-4}
Trimethylamine	$(CH_3)_3N$	−177	3.5	Very soluble	0.5×10^{-4}
Ethylamine	$CH_3CH_2NH_2$	−84	17	Very soluble	5.6×10^{-4}
Diethylamine	$(CH_3CH_2)_2NH$	−48	56	Very soluble	9.6×10^{-4}
Triethylamine	$(CH_3CH_2)_3N$	−115	90	14 g/100 cc	5.7×10^{-4}
n-Propylamine	$CH_3CH_2CH_2NH_2$	−83	49	Very soluble	4.7×10^{-4}
Di-*n*-propylamine	$(C_3H_7)_2NH$	−40	110	Very soluble	9.5×10^{-4}
Tri-*n*-propylamine	$(C_3H_7)_3N$	−90	156	Slightly soluble	4.4×10^{-4}
Isopropylamine	$(CH_3)_2CHNH_2$	−101	33	Very soluble	5.3×10^{-4}
n-Butylamine	$CH_3CH_2CH_2CH_2NH_2$	−51	78	Very soluble	4.1×10^{-4}
Isobutylamine	$(CH_3)_2CHCH_2NH_2$	−86	68	Very soluble	3.1×10^{-4}
sec-Butylamine	$CH_3CH_2CH(CH_3)NH_2$	< -72	63	Very soluble	3.6×10^{-4}
t-Butylamine	$(CH_3)_3CNH_2$	−68	45	Very soluble	2.8×10^{-4}
Ethylenediamine	$NH_2CH_2CH_2NH_2$	9	117	Very soluble	8.5×10^{-4}
Tetramethylammonium hydroxide	$(CH_3)_4N^+OH^-$	130–135 (decomposes)	—	Very soluble	Very strong base
Aromatic Amines					
Aniline	C_6H_5–NH_2	−6	184	3.7 g/100 cc	3.8×10^{-10}
N-Methylaniline	C_6H_5–$NHCH_3$	−57	196	Slightly soluble	5×10^{-10}
N,*N*-Dimethylaniline	C_6H_5–$N(CH_3)_2$	3	194	Slightly soluble	11.5×10^{-10}

Exercise 8.2 Classify the amines of the preceding exercise according to (a) aromatic versus aliphatic subclasses, (b) 1°, 2°, or 3° subclasses.

Exercise 8.3 Write names for each of the following according to the common system whenever possible, otherwise according to the IUPAC system.

(a) $CH_3CH_2N(CH_2CH_2CH_2CH_3)_2$ (b) $NH_2CH_2CH_2NH_2$

(c) $C_6H_5-N(CH_2CH_3)_2$ (d) cyclopentyl–NH_2 (e) $(CH_3)_4N^+Br^-$

8.4 Synthesis of Amines by Ammonolysis

Among the few dozen ways to make amines most of them fall into one or another of two broad types: ammonolysis of some organohalogen compound or reduction of some other nitrogen compound.

Ammonolysis

This is a reaction between ammonia and an alkyl halide. (Cf. Section 7.8.) Let us study a specific example. The heart of the reaction is a simple nucleophilic substitution:

$$CH_3I + NH_3 \longrightarrow CH_3{-}\overset{+}{N}H_3I^-$$

The product is methylammonium iodide, a salt. To liberate methylamine, we simply add a base:

$$CH_3{-}\overset{+}{N}H_3 + OH^- \longrightarrow CH_3{-}NH_2 + H_2O$$

We might have used excess ammonia instead:

$$CH_3{-}\overset{+}{N}H_3 + \underset{\text{(excess)}}{:NH_3} \rightleftharpoons CH_3{-}\ddot{N}H_2 + NH_4^+$$

In fact, if we do not use excess ammonia, the methylamine inevitably present in the equilibrium will, itself, begin to attack unchanged halide. We shall return to this problem in Section 8.11. Suffice it to say no more here than that an excess of ammonia is needed to maximize formation of the 1° amine.

Allyl and benzyl halides as well as 1° and 2° halides may be used in ammonolysis. Tertiary halides react with ammonia but suffer a loss of HX instead; alkenes form. Vinyl and aryl halides generally are unreactive (except under stringent conditions of temperature and pressure).

8.5 Syntheses of Amines by Reduction

Reduction of Nitro Compounds

This is the most common method for making aromatic amines. Some desired aromatic nitro compound is first made; then it is reduced usually by a metal such as iron or tin in the presence of hydrochloric acid.

$$Ar—NO_2 \xrightarrow[HCl_{aq}]{\text{Fe or Sn}} Ar—NH_2 \quad \text{(after the acid is neutralized)}$$

For example:

$$C_6H_5NO_2 \xrightarrow{\text{Fe; HCl}} \underset{\text{ANILINE (86\%)}}{C_6H_5NH_2}$$

Reduction of Nitriles

Catalytic hydrogenation of nitriles (alkyl or aryl cyanides, RCN or ArCN) gives amines.

$$RC{\equiv}N + 2H_2 \xrightarrow[\text{Heat, pressure}]{\text{Ni}} RCH_2NH_2$$

For example:

$$CH_3(CH_2)_5CN + 2H_2 \xrightarrow[\text{Heat, pressure}]{\text{Ni}} \underset{n\text{-HEPTYLAMINE (95\%)}}{CH_3(CH_2)_5CH_2NH_2}$$

Because nitriles are generally made from halides (Section 7.8) what this method amounts to is the conversion of an alkyl halide to an amine of one more carbon atom.

$$CH_3(CH_2)_4CH_2Br + NaCN \longrightarrow CH_3(CH_2)_5CN + NaBr$$

$$CH_3(CH_2)_5CN \xrightarrow[2H_2]{\text{Ni}} CH_3(CH_2)_6NH_2$$

8.6 Physical Properties

What we are interested in is how an amino group affects such properties as boiling points and solubilities. The key is the fact that an amino group is ammonialike, that is, moderately polar and capable of giving and receiving hydrogen bonds. Hydrogens on nitrogen in amines can form hydrogen bonds to nitrogens on neighboring amines or to oxygens on neighboring molecules of water if that is present. As a result, when compounds of similar formula weights are compared, amines have higher boiling points than alkanes but lower boiling points than alcohols.

	F.Wt.	BP (°C)
CH_3CH_3	30	−89
CH_3NH_2	31	−6
CH_3OH	32	65

Like the —OH group in alcohols, the $—NH_2$ group in amines helps make them soluble in water (Figure 8.1). Amines are also soluble in less polar solvents.

The lower-formula-weight amines (e.g., methylamine, ethylamine) have odors

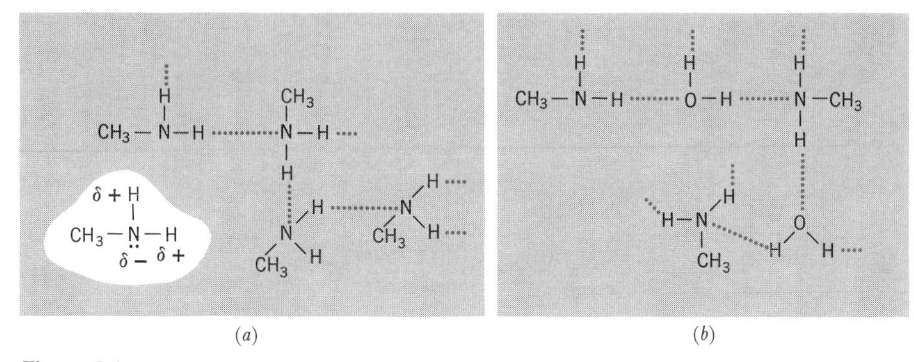

Figure 8.1
Amines as hydrogen bond donors and acceptors. (*a*) Hydrogen bonds (dotted lines) can exist between molecules of an amine. (*b*) Low-formula-weight amines are soluble in water because they can slip into the hydrogen-bonding network of water.

very much like that of ammonia. At higher formula weights the odors become "fishy." Aromatic amines have moderately pleasant, pungent odors, but they are at the same time very toxic. They are absorbed directly through the skin with sometimes fatal results.

8.7 Basicity of Amines

Ammonia is a stronger base than water. Nitrogen sustains a positive charge better than oxygen. The following transfer of a proton, H^+, transfers the positive charge from oxygen to nitrogen (Section 1.14):

$$\underset{\text{AMMONIA (STRONGER BASE)}}{H_3N\!:} + \underset{\text{HYDRONIUM ION (STRONGER ACID)}}{H\overset{+}{O}H_2\!:} \rightleftarrows \underset{\text{AMMONIUM ION (WEAKER ACID)}}{NH_4^+} + \underset{\text{WATER (WEAKER BASE)}}{H\!-\!\ddot{O}\!-\!H}$$

The ammonium ion will give up its extra proton to hydroxide ion, ^-OH, a stronger base than either water or ammonia.

$$\underset{\text{STRONGER ACID}}{H_3\overset{+}{N}\!-\!H} + \underset{\text{STRONGER BASE}}{^-OH} \rightleftarrows \underset{\text{WEAKER BASE}}{:NH_3} + \underset{\text{WEAKER ACID}}{H_2O}$$

Much of the chemistry of amines can be understood by comparing them to ammonia and to ammonium ions. Amines are simply close relatives of ammonia.

$$\underset{\text{1° AMINE STRONGER BASE}}{R\!-\!\ddot{N}H_2} + \underset{\text{STRONGER ACID}}{H_3O^+} \rightleftarrows \underset{\text{WEAKER ACID}}{R\!-\!NH_3^+} + \underset{\text{WEAKER BASE}}{H_2O}$$

$R_2\ddot{N}H + H_3O^+ \rightleftarrows R_2NH_2^+ + H_2O$
2° AMINE

$R_3N: + H_3O^+ \rightleftarrows R_3NH^+ + H_2O$
3° AMINE

$R_4N^+ + H_3O^+ \not\rightarrow$ no reaction (The 4° ion has no unshared electrons.)
4° AMMONIUM ION

We used arrows of unequal length here to say in a simple qualitative way that chemical equilibria are involved.

8.8 Structure and Basicity

We learn about the relative basicities of amines by comparing their basicity constants, or K_b values, for their reaction with water.

$$RNH_2 + H_2O \rightleftharpoons RNH_3^+ + OH^-$$

By definition:

$$K_b = \frac{[RNH_3^+][OH^-]}{[RNH_2]}$$

To have a high K_b value we have to have the numerator, which includes $[OH^-]$, relatively greater than the denominator. Hence, *the higher the K_b, the stronger the basicity of the amine.* The K_b values for several amines are listed in Table 8.1, and it is apparent that the aliphatic amines for the most part are *stronger* bases than ammonia. We shall next see why.

The difference, structurally, between ammonia and an aliphatic amine such as methylamine is the presence of the alkyl group. We learned (Section 3.19) that alkyl groups have an electron-releasing effect. Its electron release is directed toward nitrogen in methylamine. This makes the unshared pair of electrons on nitrogen more available for sharing with and binding a proton. Once the proton is accepted, the same electron-releasing effect of the methyl group releases electron density toward a positively charged site, stabilizing the charge.

$$CH_3 \rightarrow \ddot{N}H_2 + H^+ \rightleftharpoons CH_3{-}\overset{+}{N}H_3$$

Methyl releases electron density, making the unshared pair of electrons more available. Hence methylamine is a stronger base than ammonia.

Methyl releases electron density, tending to stabilize the positive charge. Hence methylamine is a stronger base than ammonia.

According to the K_b values in Table 8.1, aromatic amines are weaker bases than ammonia or aliphatic amines by factors of about one million (10^{-4} versus 10^{-10}). Apparently the unshared pair on nitrogen in an aniline is much less available than in an aliphatic amine. The resonance effect helps explain why. The principal contributing resonance forms for aniline are structures **1** through **5**. Of these, **2, 3,** and **4** involve both a separation of unlike charges (energy demanding)

1 2 3 4 5

and a disruption of the aromatic sextet of the benzene ring (also energy demanding). Therefore, these three structures cannot contribute significantly to the true state of the aniline molecule. Forms **1** and **5** must make the principal contributions. However, *to the extent that* ***2, 3,*** *and* ***4*** *make any contribution,* the unshared pair of electrons on the nitrogen is *at least somewhat delocalized into the ring.* The pair is, to that extent, less available for accepting a proton, and aniline is, therefore, a weaker base than ammonia or aliphatic amines.

This theory also accounts for the high reactivity of the benzene ring in aniline toward electron-seeking reagents (NO_2^+ in nitration, Br^+ in bromination, for example), or oxidizing agents, also electron seekers.

Such attacking species find a higher electron density in the ring in aniline than they find in benzene. The ring in benzene cannot benefit from an electron "feeding" from an attached group. Moreover, the theory also accounts for the ortho-para directing ability of the amino group. Resonance contributors **2, 3,** and **4** enable us to predict that the increased electron density of the ring will be greatest at the ortho and para positions.

8.9 Amine Salts

Table 8.2 contains a list of a few representative salts of amines. These are salts in the true sense; that is, they are typical ionic compounds. All are solids, all have relatively high melting points, especially compared with the melting points of their

Table 8.2
Amine Salts

Name	Structure	Mp (°C)
Methylammonium chloride	$CH_3NH_3^+Cl^-$	232
Dimethylammonium chloride	$(CH_3)_2NH_2^+Cl^-$	171
Dimethylammonium bromide	$(CH_3)_2NH_2^+Br^-$	134
Dimethylammonium iodide	$(CH_3)_2NH_2^+I^-$	155
Tetramethylammonium hydroxide (base as strong as KOH)	$(CH_3)_4N^+OH^-$	130–135 (decomposes)

corresponding amines, and, like ammonium salts, all are generally soluble in water and insoluble in such nonpolar compounds as carbon tetrachloride.

In amine salts the positive ion can be called a *protonated amine*. Such positively charged systems are of considerable importance as proton donors in proteins. Hence in cells they can act to neutralize bases and help control the pH of body fluids.

Chemical Properties of Amines

8.10 Amines as Nucleophiles

Several of the reactions of amines with organic compounds will be studied in later sections on these materials. Anticipating these future discussions, we call attention to the most important way of regarding an amine molecule. The unshared pair of electrons makes it not only a basic substance (that is, a proton-seeking molecule) but also a nucleophilic substance. Whenever an organic molecule contains a partial positive charge on carbon, it must be regarded as a likely site for attack by an amine.

8.11 Reaction of Amines with Alkyl Halides

This reaction is an example of an amine attacking an organic molecule with a partial positive charge on carbon. It closely resembles ammonolysis of an alkyl halide (Section 8.4). In general:

$$\underset{1^\circ \text{ AMINE}}{R{-}\ddot{N}H_2} + R'{-}X \longrightarrow R{-}\overset{+}{N}H_2{-}R' + X^- \xrightarrow{OH^-} \underset{2^\circ \text{ AMINE}}{R{-}\ddot{N}H{-}R'} + H_2O$$

Ordinarily, sodium hydroxide is added after the initial reaction is over to liberate the amine.

$$\underset{2^\circ \text{ AMINE}}{R{-}\underset{\displaystyle R'}{\underset{|}{\ddot{N}}}{-}H} + R''{-}X \xrightarrow[\text{hydroxide ion}]{\text{Followed by}} \underset{3^\circ \text{ AMINE}}{R{-}\underset{\displaystyle R'}{\underset{|}{\ddot{N}}}{-}R''}$$

$$\underset{3^\circ \text{ AMINE}}{R{-}\underset{\displaystyle R'}{\underset{|}{\ddot{N}}}{-}R''} + R'''{-}X \longrightarrow \underset{4^\circ \text{ AMMONIUM SALT}}{R{-}\overset{\displaystyle R'''}{\overset{|}{\underset{\displaystyle R'}{\underset{|}{N^+}}}}{-}R'' \quad X^-}$$

When a 1° amine is allowed to react with an alkyl halide, a mixture consisting principally of 2° and 3° amines plus smaller amounts of 4° ammonium salt and unchanged 1° amine results. The relative amounts of these products can be controlled by adjusting the proportions of reactants. A large excess of 1° amine over alkyl halide, for example, will favor the 2° amine with only traces of higher alkylated amines. (We discussed this strategy in Section 8.4 when we noted that a large excess of ammonia is needed to get the most 1° amine out of ammonolysis of an alkyl halide.)

$$C_6H_5-NH_2 + Cl-CH_2-C_6H_5 \longrightarrow C_6H_5-NH-CH_2-C_6H_5$$

ANILINE (4 PARTS) : BENZYL CHLORIDE (1 PART) → PHENYLBENZYLAMINE (96%)

$$(CH_3CH_2)_2NH + Br-CH(CH_3)_2 \longrightarrow (CH_3CH_2)_2NCH(CH_3)_2$$

DIETHYLAMINE (1.3 PARTS) : ISOPROPYL BROMIDE (1 PART) → DIETHYLISOPROPYLAMINE (60%)

Mechanism

In the alkyl halide the bond is polarized as shown. The carbon holding the halogen has a partial positive charge, a likely site for an electron-rich reagent (the amine) to attack. The attack is therefore usually of the S_N2 type:

$$R-\underset{H}{\overset{H}{N}}: + \overset{\delta+}{\underset{R'}{C}H_2}-\overset{\delta-}{X} \longrightarrow R-\underset{H}{\overset{H}{N^+}}-CH_2-R' + X^- \xrightarrow[\text{(added later)}]{^-OH} R-\underset{\cdot\cdot}{\overset{H}{N}}-CH_2-R' + H-OH$$

8.12 Reactions of Amines with Nitrous Acid

Generated when and where needed by mixing sodium nitrite and hydrochloric acid, nitrous acid behaves differently toward the various classes of amines. Some of the reactions are very important in organic synthesis.

1° Amines

1° aliphatic amines:

$$R-\ddot{N}H_2 \xrightarrow[HX]{HONO} [R-\overset{+}{N}\equiv N]X^- \xrightarrow[\text{prevented)}]{\text{(cannot be}} R^+ + :N\equiv N:\uparrow + X^-$$

ALKYL DIAZONIUM ION (Unstable even at 0°C); CARBONIUM ION; NITROGEN GAS

$$R^+ \xrightarrow{H_2O} R-OH + \text{by-products}$$

ALCOHOL

1° aromatic amines:

$$Ar-\ddot{N}H_2 \xrightarrow{HONO} Ar-\overset{+}{N}\equiv\ddot{N} \xrightarrow[\text{Warm the solution}]{H_2O} Ar-OH + N_2\uparrow + H^+$$

ARYL DIAZONIUM ION (Stable at 0°C); PHENOL

2° Amines

(both aliphatic and aromatic amines)

Here is where we get nitrosamines:

$$\text{(R)Ar—}\overset{\text{R}}{\overset{|}{\ddot{\text{N}}}}\text{H} \xrightarrow{\text{HONO}} \text{(R)Ar—}\overset{\text{R}}{\overset{|}{\ddot{\text{N}}}}\text{—}\ddot{\text{N}}\text{=O} + H_2O \quad \text{(no evolution of } N_2\text{)}$$

N-NITROSAMINE
(Neutral compounds, insoluble in acid; usually yellow or orange-colored oils or solids)

3° Amines

3° aliphatic amines: these dissolve in the acidic medium containing the nitrous acid. No nitrogen evolves. Products of varying complexity may or may not be generated.

3° aromatic amines:

$$C_6H_5\text{—}N(CH_3)_2 \xrightarrow{\text{HONO}} \text{O=N—}C_6H_4\text{—}N(CH_3)_2$$

p-NITROSO-*N,N*-DIMETHYLANILINE

In this reaction the benzene ring is so activated that ring substitution occurs. No nitrogen evolves.

The variety of responses of the subclasses of amines to nitrous acid makes this reagent useful in placing an "unknown" amine in its appropriate category.

—If reaction occurs at 0° and nitrogen evolves, the unknown is a 1° aliphatic amine.

—If reaction occurs at 0° but nitrogen does not evolve until the solution is warmed, a 1° aromatic amine is indicated.

—If reaction occurs without evolution of nitrogen and a yellow or orange oil (or solid) separates, it is a 2° aliphatic or aromatic amine.

—If the amine dissolves without evolution of nitrogen and without separation of a yellow or orange oil (or solid), the product is a 3° aliphatic or aromatic amine.

Although these tests work, it is usually simpler and more certain to use spectroscopy, especially infrared spectroscopy. (Cf. Chapter 17.)

Nitrosamines are among the most potent known carcinogens (cancer-causers). Many scientists are worried that nitrosamines might be formed in the stomach or might be produced during the cooking of certain foods. Their concern stems from the fact that sodium nitrite is legally used in the United States as a food additive for many meat products (e.g., bacon, sausages and frankfurters, bologna, "corned" meats, beef jerky). Its presence in bacon is known to give rise to traces of nitrosamines when bacon is fried. (Virtually all foods include molecules having amino groups that can be nitrosated.) Whether nitrosamines actually can form in the stomach is not known. The possibility is suspected, because the stomach's gastric juice contains hydrochloric acid, and we have just seen how that acid changes nitrite ion to nitrous acid, a nitrosating agent.

The use of sodium nitrite as a food additive began long ago and for worthwhile reasons. It inhibits the growth of botulism-causing bacteria. There are alternatives, and many countries have banned the use of sodium nitrite in meat products.

Resistance to such prohibition is based partly on cost and partly on the "cosmetic effect" of sodium nitrite. Its presence in meat products helps them look redder and, therefore, presumably "meatier."

8.13 The Van Slyke Determination of Amino Groups

Proteins usually have several free amino groups ($—NH_2$) per molecule. Their relative number can be estimated because they react quantitatively with nitrous acid to yield one mole of nitrogen gas per mole of $—NH_2$ groups. The nitrogen can be trapped in a gas buret and its volume measured and then converted by simple gas law calculations into moles of nitrogen. The determination is named after its developer, Donald Van Slyke.

Exercise 8.4 Write the structures of the principal products that can be expected to form in the following reactions. For the purposes of this exercise, assume that 1° aliphatic amines react with nitrous acid to produce an alcohol plus nitrogen gas:

$$R—NH_2 \xrightarrow{HNO_2} R—OH + N_2\uparrow$$

(a) $(CH_3)_2NH \xrightarrow{NaNO_2,\ HCl}$

(b) 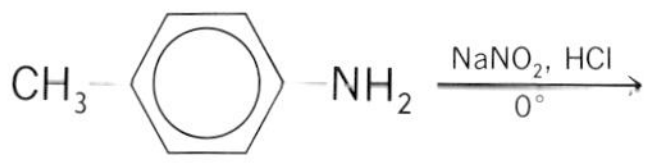 $CH_3—C_6H_4—NH_2 \xrightarrow[0^\circ]{NaNO_2,\ HCl}$

(c) $C_6H_5—CH_2—NH_2 \xrightarrow{NaNO_2,\ HCl}$

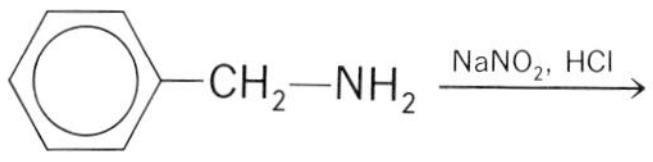

(d) 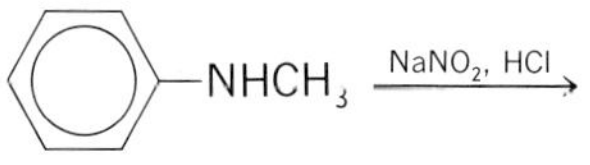 $C_6H_5—NHCH_3 \xrightarrow{NaNO_2,\ HCl}$

(e) 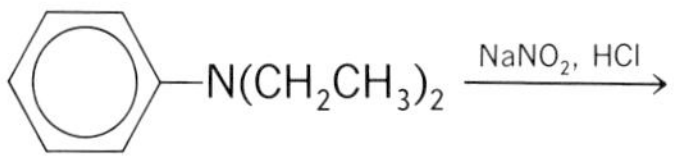 $C_6H_5—N(CH_2CH_3)_2 \xrightarrow{NaNO_2,\ HCl}$

8.14 Reactions of Aromatic Amines

We have learned that aromatic amines are generally much less basic than aliphatic amines. Otherwise, they give similar reactions: they diazotize, and they are nucleophiles toward alkyl halides. In the next section we shall study them as intermediates—through their diazonium salts—in preparing many other aromatic compounds. In this section we briefly survey some of their reactions that involve the aromatic ring. An amino group on a benzene ring activates the ring powerfully, and it directs incoming groups to ortho and para positions. The reaction of aniline with bromine (or chlorine) illustrates these facts. Note particularly the mild condition that can be (indeed, must be) used.

$$C_6H_5NH_2 \xrightarrow[\text{(no catalyst)}]{Br_2 \text{ in } H_2O} \text{2,4,6-}Br_3C_6H_2NH_2 + HBr$$

2,4,6-TRIBROMOANILINE

Attempts to nitrate aniline directly lead largely to complex mixtures and highly colored compounds. This is because nitric acid is also an oxidizing agent and because another important property of aromatic amines is their susceptibility to oxidation. Aniline, when pure, is colorless. In contact with air, however, it slowly turns black because it is being oxidized.

The sulfonation of aniline is also not directly attempted. Instead, anilinium hydrogen sulfate is made and then heated strongly:

$$C_6H_5NH_2 + H_2SO_4 \longrightarrow C_6H_5\overset{+}{N}H_3HSO_4^- \xrightarrow{\text{Heat}} p\text{-}H_2NC_6H_4SO_3H \rightleftarrows p\text{-}H_3N^+C_6H_4SO_3^-$$

ANILINIUM HYDROGEN SULFATE

SULFANILIC ACID

The product, sulfanilic acid, is interesting because it exists as a dipolar ion. Its proton-donating group, $—SO_3H$, has yielded its proton to its proton-accepting group, $—NH_2$. It is a true salt; its melting point, for example is so high that it decomposes at roughly 280 to 300° without melting. It is insoluble in organic solvents, and it happens to be a salt that is also insoluble in water. Sulfanilic acid can be diazotized, which makes it an important intermediate in the synthesis of certain dyes. (Cf. Section 8.15.) The amides of sulfanilic acid are the sulfa drugs that we shall study further in Section 10.25.

8.15 Reactions of Aromatic Diazonium Salts

Aromatic diazonium salts are the workhorses of aromatic chemistry. Through them we can convert an amino group to almost any other substituent; and, of course, the amino group generally comes from reduction of a nitro group. The rule of thumb, then, is "if you can get a nitro group where you want it you can exchange it for almost anything else."

$$Ar—NO_2 \xrightarrow{\text{Reduce}} Ar—NH_2 \xrightarrow{\text{HONO}} Ar—N_2^+$$

The diazonium salt is easily generated from the amine. Depending on the selection of the next reagent to be added, the following transformations occur.

Reactions in Which Nitrogen Is Lost

ArN_2^+ (From Ar—H via $ArNO_2$ and $ArNH_2$)

- $\xrightarrow[\text{Heat}]{Cu_2Cl_2}$ Ar—Cl + $N_2\uparrow$ (Sandmeyer reaction)
- $\xrightarrow[\text{Heat}]{Cu_2Br_2}$ Ar—Br + $N_2\uparrow$ (Sandmeyer reaction)
- $\xrightarrow[\text{Heat}]{Cu_2(CN)_2}$ Ar—CN + $N_2\uparrow$ (synthesis of cyano compounds or nitriles)
- $\xrightarrow[\text{Heat}]{KI}$ Ar—I + $N_2\uparrow$
- $\xrightarrow[\text{Heat}]{HBF_4}$ Ar—F + $N_2\uparrow$
- $\xrightarrow[\text{Warm}]{H_2O}$ Ar—OH + $N_2\uparrow$ (synthesis of phenols)
- $\xrightarrow{R—OH}$ Ar—O—R + $N_2\uparrow$ (synthesis of phenolic ethers)
- $\xrightarrow{H_3PO_2}$ Ar—H + $N_2\uparrow$ The important result is that an aromatic amino group can be removed by hypophosphorus acid.

Reactions in which Nitrogen Is Retained.

Coupling

With proper control of experimental conditions, aryl diazonium ions will react with phenols or aromatic amines to give compounds of the general type Ar—N=N—Ar′, called *azo compounds.* The following are examples.

$C_6H_5N_2^+$ + C_6H_5OH ⟶ C_6H_5—N=N—C_6H_4—OH + H^+

BENZENE-DIAZONIUM ION | PHENOL | *p*-HYDROXYAZOBENZENE (ORANGE)

$Na^{+}\ {}^{-}O_3S$—C_6H_4—N_2^+ + C_6H_5—$N(CH_3)_2$ ⟶

DIAZOTIZED SULFANILIC ACID | *N,N*-DIMETHYLANILINE

$Na^{+}\ {}^{-}O_3S$—C_6H_4—N=N—C_6H_4—$N(CH_3)_2$

METHYL ORANGE
(An acid-base indicator: above pH 4, yellow; below pH 3, red)

The reaction is little more than an aromatic substitution into the highly activated benzene ring of the phenol or the aromatic amine. The diazonium ion acts as the electron-seeking (electrophilic) reagent.

These aromatic azo compounds are all highly colored, and many are used as dyes. Close to half of all commercial dyes are azo compounds. In biochemistry the coupling reaction is the basis of the **van der Bergh test,** a method of identifying tissues that are rich in enzymes capable of catalyzing the hydrolysis of esters. An ester of phenol is added to the tissue, together with a diazonium salt. The ester will be hydrolyzed by action of the enzyme, if it is present. This liberates phenol, and, as soon as it forms, it couples with the diazonium ion to give a highly colored product.

$$CH_3-\overset{O}{\overset{\|}{C}}-O-C_6H_5 + H_2O \xrightarrow[\text{is present}]{\text{If the enzyme}} CH_3\overset{O}{\overset{\|}{C}}-OH + \underset{\text{Phenol}}{H-O-C_6H_5}$$

$$\text{Phenol} + \underset{\text{DIAZOTIZED } p\text{-NITROANILINE}}{\overset{+}{N_2}-C_6H_4-NO_2} \longrightarrow \underset{\text{COLORED AZO COMPOUND}}{HO-C_6H_4-N{=}N-C_6H_4-NO_2}$$

If the tissue does not contain the enzyme, no phenol is produced, and no colored spot develops.

8.16 Important Individual Amines

Dimethylamine. $(CH_3)_2NH$

One of the constituents that gives herring brine its distinctive odor, dimethylamine is synthesized and used industrially in the manufacture of fungicides and accelerators for the vulcanization of rubber.

Several aliphatic amines (e.g., ethyl-, isobutyl-, isopentyl-, and β-phenylethylamine) are synthesized in animal cells. They stimulate the central nervous system.

Ethylenediamine. $NH_2CH_2CH_2NH_2$

The action of ammonia on 1,2-dichloroethane (from ethylene and chlorine) produces this amine. If it is allowed to react with the sodium salt of chloroacetic acid, one of the most versatile known complexing (chelating) agents, EDTA, forms:

$$NH_2CH_2CH_2NH_2 + 4Cl-CH_2\overset{O}{\overset{\|}{C}}-O^-Na^+ \longrightarrow$$

$$\left[\begin{matrix} {}^-O-\overset{O}{\overset{\|}{C}}CH_2 \\ {}^-O-\underset{O}{\underset{\|}{C}}CH_2 \end{matrix} \!\!>\! \overset{+}{N}H-CH_2CH_2-\overset{+}{N}H \!<\!\! \begin{matrix} CH_2\overset{O}{\overset{\|}{C}}-O^- \\ CH_2\underset{O}{\underset{\|}{C}}-O^- \end{matrix} \right] 2Na^+$$

ETHYLENEDIAMINETETRAACETIC ACID
(EDTA, as its sodium salt)

Invert Soaps.
$CH_3(CH_2)_n\overset{+}{N}(CH_3)_3Cl^-$

Quaternary ammonium salts in which one of the alkyl groups is long (e.g., $n = 15$) have detergent and germicidal properties. In ordinary soaps the detergent action is associated with a negatively charged organic ion. In an invert soap a positive ion has the detergent property. (Detergent action is discussed in Section 11.11.)

Aniline. $C_6H_5NH_2$

The annual production of aniline in the United States is well over 100,000 tons. Virtually all of it is used in the manufacture of aniline dyes, pharmaceuticals, and chemicals for the plastics industry. Two methods of synthesizing aniline are used industrially. In one benzene is first converted to nitrobenzene, which is in turn reduced by the action of iron and hydrochloric acid.

$$C_6H_6 \xrightarrow[H_2SO_4]{HNO_3} C_6H_5{-}NO_2 \xrightarrow[\text{HCl, heat}]{Fe} C_6H_5{-}NH_3^+Cl^- \xrightarrow{Na_2CO_3} C_6H_5{-}NH_2$$

BENZENE NITROBENZENE ANILINIUM CHLORIDE ANILINE

The other method begins with benzene, which is chlorinated to produce chlorobenzene. Then, ammonia in the presence of copper(II) oxide, high pressure, and an elevated temperature converts the chlorobenzene into aniline:

$$C_6H_6 \xrightarrow[Fe]{Cl_2} C_6H_5{-}Cl \xrightarrow[\text{CuO, 900 lb/in., 200°C}]{NH_3} C_6H_5{-}NH_2$$

Benzene is required for both syntheses. The cost and availability of benzene, therefore, affects the prices of all the dyes, drugs, and plastics that are made indirectly from it.

8.17 Nitro Compounds

Nitro compounds are those whose molecules contain the nitro group: Ar—NO_2 or R—NO_2. In the details of the structure given by **6**, we see the nitro group as one in which resonance gives it symmetry.

$$\left[-\overset{|}{\underset{|}{C}}-\overset{+}{N}\begin{smallmatrix} \ddot{O}:^- \\ \ddot{O}: \end{smallmatrix} \longleftrightarrow -\overset{|}{\underset{|}{C}}-\overset{+}{N}\begin{smallmatrix} \ddot{O}: \\ :\ddot{O}:^- \end{smallmatrix} \right] \equiv -\overset{|}{\underset{|}{C}}-\overset{+}{N}\begin{smallmatrix} O^{\delta-} \\ O^{\delta-} \end{smallmatrix}$$

CONTRIBUTORS HYBRID

6
NITRO GROUP

Aromatic nitro compounds are made by liquid-phase nitration in the presence of sulfuric acid (Section 4.3). They are used to make aromatic amines and, through their diazonium salts, many other aromatic compounds (Section 8.15).

Nitro compounds are generally quite toxic substances, and they are quickly absorbed through the skin. The symptoms of poisoning include headaches, dizziness, and irregular pulse. With time, cyanosis develops because nitro compounds can render hemoglobin incapable of carrying oxygen. (The same thing happens in cyanide poisoning. In cyanosis the finger tips and lips become bluish.)

Many di- and tri-nitro compounds are explosives and must be handled extremely carefully. (Mononitro compounds should not be distilled to dryness because they often contain the higher nitrated substances as impurities.) TNT, 2,4,6-trinitrotoluene, is the most famous of the polynitro explosives. Alfred Nobel discovered how to handle TNT safely, and the fortune he accumulated now provides the money for the Nobel prizes.

Brief Summary: Chemical and Physical Properties

1. We make amines from:
   (a) alkyl halides (1°, 2°, allyl, benzyl) by action of ammonia or a 1° or 2° amine
   (b) nitro compounds by reduction

2. The amino group (at least as much as $-\overset{|}{\underset{\cdot\cdot}{N}}-H$ must be present) can donate and receive hydrogen bonds.

3. Because of the unshared pair of electrons on the nitrogen, amines are basic. Aliphatic amines are slightly more basic than ammonia; aromatic amines are much less basic than ammonia.

4. Amines are nucleophiles and can displace halide ions from alkyl halides (but not from vinyl or aryl halides unless extreme conditions are used).

5. Nitrite ion in acid (e.g., nitrous acid) reacts with amines.
   (a) 1° amines undergo diazotization:
   —if aliphatic, the diazonium ion, RN_2^+, cannot be kept from decomposing to N_2 and R^+ (from which a mixture of products emerges).
   —if aromatic, ArN_2^+ can be kept from decomposing at icebath temperature.
   (b) 2° amines give nitrosamines (potent carcinogens); no nitrogen evolution.
   (c) 3° amines do not react unless they are aromatic, in which case nitrosation of the ring occurs (no nitrogen evolution).

6. Aryl diazonium salts can be converted to aryl halides, cyanides, and phenols (or their ethers); or they will couple with phenols or aryl amines to give azo dyes.

7. Aromatic amines have rings that are very susceptible to oxidation and electrophilic substitution.

Problems and Exercises

1. Give common names to each of the following compounds.

(a) $(CH_3)_2NCH_2CH_3$

(b) $(CH_3)_2CHCH_2NH\underset{}{\overset{CH_3}{\overset{|}{C}}}HCH_3$

(c) 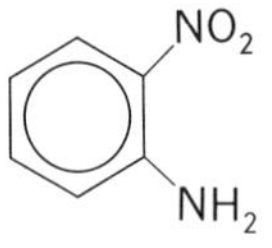 (benzene ring with NO_2 and NH_2 on adjacent carbons)

(e) 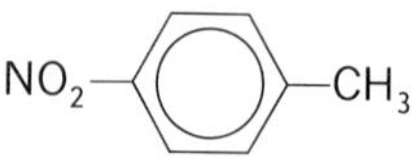 $NO_2-C_6H_4-CH_3$ (para)

(d) 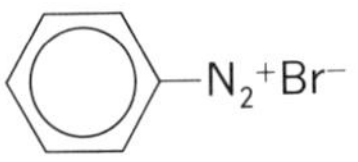 $C_6H_5-N_2^+Br^-$

(f) $(CH_3)_3CNH_2$

2. Complete the following equations by supplying the structures of the principal organic products. Do not balance the equations. If no reaction is to be expected, write "None." (These equations involve not just this chapter but preceding ones, too.)

(a) $CH_3CH_2Br + \underset{\text{(Excess)}}{NH_3} \xrightarrow[\text{(then OH}^-)]{}$

(b) 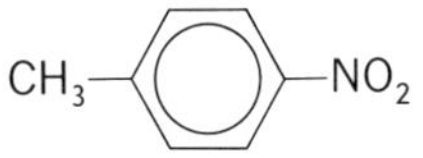 $CH_3-C_6H_4-NO_2 \xrightarrow[HCl_{aq}]{Fe}$

(c) $\underset{\text{(Excess)}}{CH_3NH_2} + CH_3I \xrightarrow[\text{(then OH}^-)]{}$

(d) $CH_3CH_2OH + SOCl_2 \longrightarrow$

(e) 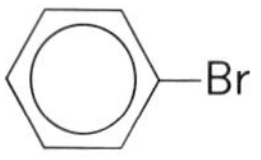 $C_6H_5-Br + NH_{3(aq)} \longrightarrow$

(f) $CH_3CH_2NH_2 + HNO_2 \longrightarrow$

(g) $(CH_3)_4N^+I^- + CH_3I \longrightarrow$

(h) $CH_3NHCH_2CH_3 + HNO_2 \longrightarrow$

(i) 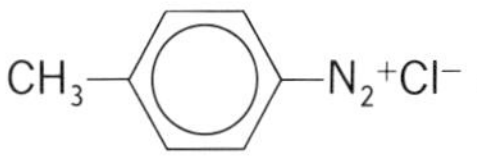 $CH_3-C_6H_4-N_2^+Cl^- + Cu_2(CN)_2 \longrightarrow$

(j) $CH_3CH_2CH_2OH + PBr_3 \longrightarrow$

(k) $CH_3NH_3^+Cl^- + NaOH_{aq} \longrightarrow$

(l) $(CH_3)_3N + HNO_2 \longrightarrow$

(m) $CH_3Br + \underset{\text{(Excess)}}{CH_3NHCH_3} \longrightarrow$

(n) 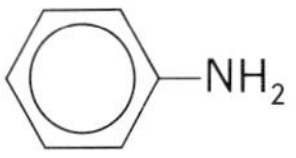 $C_6H_5-NH_2 + NaOH_{aq}$

(o) 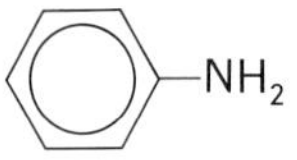 $C_6H_5-NH_2 + HCl_{aq}$

3. Amides have the general structure: $R-\overset{O}{\overset{\|}{C}}-\ddot{N}H_2$. On the basis of relative electronegativities, would the $-\overset{|}{C}=O$ (the carbonyl group) be an electron-withdrawing or an electron-releasing group? Therefore, would amides be stronger bases or weaker bases than aliphatic amines?

4. Arrange the following in their order of increasing solubility in water. (You should not have to refer to chapter tables to do this.)

$CH_3(CH_2)_6CH_2NH_2$ (a) $CH_3(CH_2)_6CH_3$ (b) $CH_3(CH_2)_5CH(NH_2)-CH_2NH_2$ (c)

5. In Table 8.2 we learn that tetramethylammonium hydroxide is as strong a base as potassium hydroxide. First, what does it mean to be "a strong base?" Second, why is tetramethylammonium hydroxide such a strong base?

6. Arrange the following in their order of increasing basicity. (You should not have to refer to tables to answer this.)

$CH_3-C_6H_4-NH_2$ (a) $NO_2-C_6H_4-NH_2$ (b) $Cl-C_6H_4-NH_2$ (c) $C_6H_5-NH_2$ (d)

7. Outline by means of reaction sequences the steps that will accomplish the following overall transformations.

(a) $CH_3-C_6H_4-NH_2$ to $CH_3-C_6H_4-I$

(b) $NO_2-C_6H_3(NH_2)$ to $NO_2-C_6H_3(Br)$

(c) $NO_2-C_6H_3(NH_2)$ to $Cl-C_6H_3(Br)$

(d) $CH_3-C_6H_4-NO_2$ to $CH_3-C_6H_4-OH$

(e) $CH_3-C_6H_4-NO_2$ to $CH_3-C_6H_3(Br)-OH$

(f) $CH_3-C_6H_4-NO_2$ to $CH_3-C_6H_3Br_2$

(g) CH_3NH_2 to $(CH_3)_2NH$

(h) $C_6H_5-NHCH_3$ to $C_6H_5-\overset{+}{N}(CH_3)_3I^-$

(i) $CH_3CH_2NH_2$ to CH_3CH_2OH

(j) $C_6H_5-NH_2$ to $C_6H_5-N{=}N-C_6H_4-OH$

chapter nine

Aldehydes and Ketones

Carbonyl Compounds

9.1 The Carbonyl Group

The carbonyl group consists of a carbon-to-oxygen double bond. The oxygen atom of this group, being more electronegative than carbon, carries a partial negative charge, which leaves the carbonyl carbon with a partial positive charge. These two simple considerations will be of great help to us in understanding several of the reactions of this group as it occurs in several families of compounds.

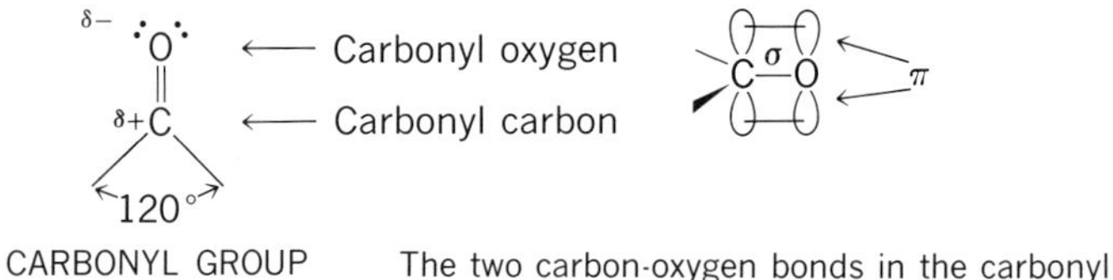

CARBONYL GROUP

The two carbon-oxygen bonds in the carbonyl group consist of one σ-bond and one π-bond.

9.2 Families of Organic Compounds with Carbonyl Groups

Several important families of organic compounds consist of molecules containing the carbonyl group, each family characterized by the nature of the two groups attached to the carbonyl carbon. The essential structural features that define these families, outlined in Table 9.1 and discussed in the following paragraphs,[1] should be memorized.

Aldehydes. $(H)R-\overset{O}{\overset{\|}{C}}-H$

To be classified as an aldehyde, the molecules of the substance must have a *hydrogen* attached to the carbonyl carbon. The other group must involve a bond from the carbonyl carbon to another *carbon* (or a hydrogen in formaldehyde).

[1] In the examples given, R— is not limited just to alkyl groups. It may also be aryl, or any other group in which the bond to the carbonyl group is from a *carbon*, saturated or otherwise.

Table 9.1
The Carbonyl Group in Families of Organic Compounds

Family Name	Generic Family Structure[a]
Aldehydes	$(H)R-\overset{O}{\overset{\|}{C}}-H$
Ketones	$R-\overset{O}{\overset{\|}{C}}-R'$
Carboxylic acids	$(H)R-\overset{O}{\overset{\|}{C}}-OH$
Derivatives of carboxylic acids	
Acid chlorides	$R-\overset{O}{\overset{\|}{C}}-Cl$
Anhydrides	$R-\overset{O}{\overset{\|}{C}}-O-\overset{O}{\overset{\|}{C}}-R'$
Esters	$(H)R-\overset{O}{\overset{\|}{C}}-O-R'$
Amides	$(H)R-\overset{O}{\overset{\|}{C}}-NH_2$ (simple amides)
	$(H)R-\overset{O}{\overset{\|}{C}}-NH-R'$ (*N*-alkyl amides)
	$(H)R-\overset{O}{\overset{\|}{C}}-\underset{R''}{\underset{\|}{N}}-R'$ (*N,N*-dialkyl amides)

[a] When the symbol for hydrogen is placed in parentheses before (or after) R [e.g., (H)R— or —R(H)], the specific R— group may be either an alkyl group or hydrogen. In these compounds, R— may also be aryl. When R— is primed (e.g., R′— or R″—), the two or more R— groups in a given symbol may be different.

The aldehyde group, then, is $-\overset{\overset{\displaystyle O}{\|}}{C}-H$, often written as —CHO. Specific examples are listed in Table 9.3.

Ketones. $R-\overset{\overset{\displaystyle O}{\|}}{C}-R'$

The molecules in ketones must have a carbonyl group flanked on both sides of the carbonyl carbon by bonds to other *carbons*. (Specific examples are listed in Table 9.4, page 211.) When it occurs in ketones, the carbonyl group is frequently called the *keto group*.

Carboxylic Acids. $(H)R-\overset{\overset{\displaystyle O}{\|}}{C}-OH$

The molecules in substances classified as carboxylic acids must have a *hydroxyl group* attached to the carbonyl carbon. The other bond from the carbonyl carbon must be to another *carbon* (or to hydrogen in formic acid, $H-\overset{\overset{\displaystyle O}{\|}}{C}-OH$). The group, $-\overset{\overset{\displaystyle O}{\|}}{C}-OH$, is called either the *carboxylic acid group* or *carboxyl group*. It is often condensed to —COOH or to $-CO_2H$, but when you see it this way remember the presence of the carbon-oxygen double bond and the hydroxyl group. Although their molecules possess hydroxyl groups, carboxylic acids are definitely not alcohols. The carbonyl group changes the properties of the —OH attached to it (and vice versa) too much. (Specific examples of carboxylic acids are listed in Table 10.1, page 238.)

Carboxylic Acid Chlorides,
Acid Chlorides. $R-\overset{\overset{\displaystyle O}{\|}}{C}-Cl$

The molecules of these compounds have carbonyl-to-chlorine bonds, and the group, $-\overset{\overset{\displaystyle O}{\|}}{C}-Cl$, is called the *acid chloride function*. (Specific examples are given in Table 10.5, page 251.) Some of the corresponding acid fluorides, bromides, and iodides are known but they are seldom used. We shall ignore them.)

Carboxylic Acid Anhydrides,
"Anhydrides." $R-\overset{\overset{\displaystyle O}{\|}}{C}-O-\overset{\overset{\displaystyle O}{\|}}{C}-R$

In the molecules of these compounds an oxygen atom is flanked on both sides by carbonyl groups. They are called anhydrides ("not hydrated") because they can be considered to result from the loss of water between two molecules of a carboxylic acid:

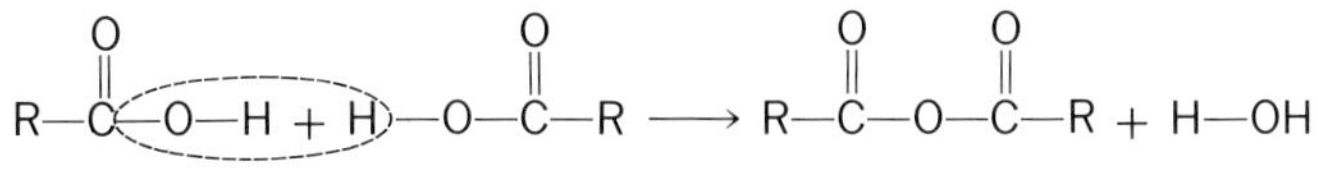

$$R{-}\overset{\displaystyle O}{\overset{\|}{C}}{-}O{-}H + H{-}O{-}\overset{\displaystyle O}{\overset{\|}{C}}{-}R \longrightarrow R{-}\overset{\displaystyle O}{\overset{\|}{C}}{-}O{-}\overset{\displaystyle O}{\overset{\|}{C}}{-}R + H{-}OH$$

The R— groups may not be replaced by hydrogens in either acid chlorides or anhydrides. Such substances (e.g., $H{-}\overset{\displaystyle O}{\overset{\|}{C}}{-}Cl$) are not known. They are too unstable to exist. (A few examples of anhydrides are given in Table 10.5, page 251.)

Carboxylic Acid Esters, Esters. $(H)R{-}\overset{\displaystyle O}{\overset{\|}{C}}{-}O{-}R'$ or $(H)RCO_2R'$

The *ester group* has the features shown below. The bond drawn with a heavier line is called the *ester linkage.* To be an ester the molecule must have a carbonyl-oxygen-carbon network. (Several are listed in Tables 10.6 and 10.7, pages 254 and 255.)

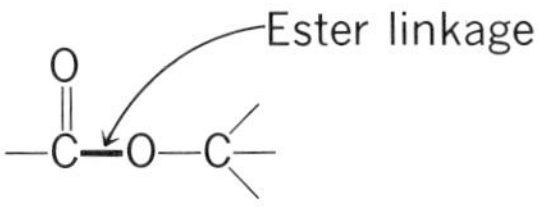

$$-\overset{\displaystyle O}{\overset{\|}{C}}{-}O{-}\overset{/}{\underset{\backslash}{C}}{-}$$

Ester linkage

ESTER GROUP

Carboxylic Acid Amides, "Amides." $(H)R{-}\overset{\displaystyle O}{\overset{\|}{C}}{-}\overset{\displaystyle R''(H)}{\overset{|}{N}}{-}R'(H)$

So-called "simple amides" are ammonia derivatives of carboxylic acids. They may be regarded as having formed by the splitting out of water between an acid and ammonia:

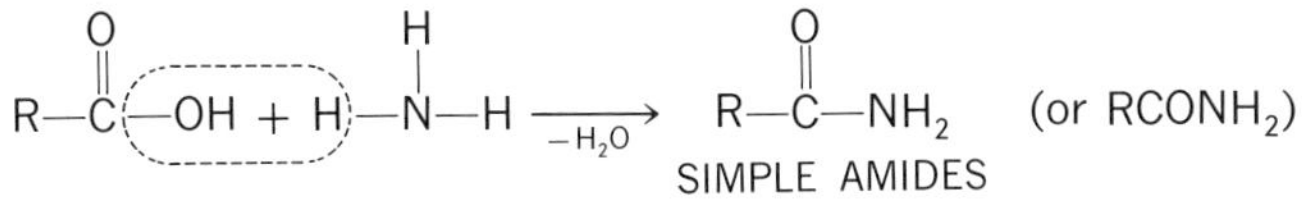

$$R{-}\overset{\displaystyle O}{\overset{\|}{C}}{-}OH + H{-}\overset{\displaystyle H}{\overset{|}{N}}{-}H \xrightarrow[-H_2O]{} R{-}\overset{\displaystyle O}{\overset{\|}{C}}{-}NH_2 \quad (\text{or } RCONH_2)$$

SIMPLE AMIDES

Other amides (the so-called *N*-alkyl or *N,N*-dialkyl amides, where *N* denotes attachment to the nitrogen of a simple amide) are also well-known types of compounds:

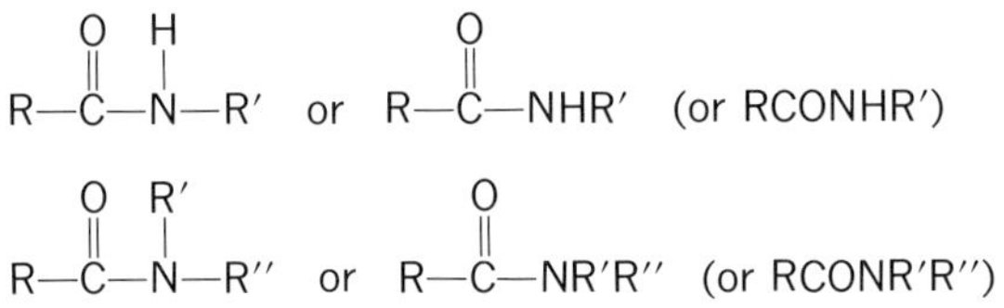

$$R{-}\overset{\displaystyle O}{\overset{\|}{C}}{-}\overset{\displaystyle H}{\overset{|}{N}}{-}R' \quad \text{or} \quad R{-}\overset{\displaystyle O}{\overset{\|}{C}}{-}NHR' \quad (\text{or } RCONHR')$$

$$R{-}\overset{\displaystyle O}{\overset{\|}{C}}{-}\overset{\displaystyle R'}{\overset{|}{N}}{-}R'' \quad \text{or} \quad R{-}\overset{\displaystyle O}{\overset{\|}{C}}{-}NR'R'' \quad (\text{or } RCONR'R'')$$

Any or all of the R— groups may be aryl, too.

To be classified as an amide, the molecule must have a carbonyl-to-nitrogen bond, called the *amide linkage.* In proteins the name peptide bond is synonymous with amide linkage. (Several examples of amides are given in Table 10.8, page 261.)

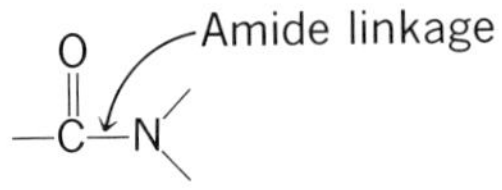

$$-\overset{O}{\overset{\|}{C}}-N\langle \quad \text{(Amide linkage)}$$

THE AMIDE GROUP

Exercise 9.1 Before you go any farther, if you can master right now the skill of stating quickly the family (or families) to which a structure belongs, you will make the rest of the study much easier. Study each of these structures and assign them to their right families.

Examples:

$CH_3\overset{O}{\overset{\|}{C}}OH$ Carboxylic acid

$CH_3\overset{O}{\overset{\|}{C}}CH_2\overset{O}{\overset{\|}{C}}CH_3$ Two keto groups

$C_6H_5\overset{O}{\overset{\|}{C}}H$ Aldehyde

$CH_3\overset{O}{\overset{\|}{C}}O\overset{O}{\overset{\|}{C}}CH_3$ Anhydride

$CH_3\underset{OH}{\underset{|}{C}}H\overset{O}{\overset{\|}{C}}OH$ Carboxylic acid and alcohol

$H\overset{O}{\overset{\|}{C}}OH$ Carboxylic acid (This acid is not in the aldehyde class, but it does have some aldehyde properties.)

(a) $HO\overset{O}{\overset{\|}{C}}CH_3$

(b) $H\overset{O}{\overset{\|}{C}}C_6H_5$

(c) $CH_3\overset{O}{\overset{\|}{C}}Cl$

(d) $H\overset{O}{\overset{\|}{C}}NH_2$

(e) $CH_3O\overset{O}{\overset{\|}{C}}CH_3$

(f) $CH_3\overset{O}{\overset{\|}{C}}CH_3$

(g) CH_3CH_2OH

(h) $CH_3\overset{O}{\overset{\|}{C}}NHCH_3$

(i) $H\overset{O}{\overset{\|}{C}}H$

(j) 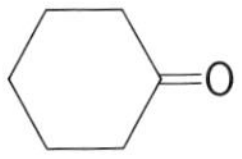 cyclohexanone ring $=O$

(k) $C_6H_5\overset{O}{\overset{\|}{C}}OH$

(l) $C_6H_5O\overset{O}{\overset{\|}{C}}H$

(m) $HOCH_2CH_2\overset{O}{\overset{\|}{C}}CH_3$

(n) $CH_3\underset{OH}{\underset{|}{C}}H\overset{O}{\overset{\|}{C}}CH_3$

(o) $(CH_3)_2CHCOOH$

(p) $H\overset{O}{\overset{\|}{C}}CH_2CH_2\overset{O}{\overset{\|}{C}}CH_3$

(q) $CH_3O\overset{O}{\overset{\|}{C}}CH_2CH_3$

(r) $CH_3OCH_2\overset{O}{\overset{\|}{C}}CH_3$

(s) $CH_3O\overset{O}{\overset{\|}{C}}CH_2\overset{O}{\overset{\|}{C}}H$

(t) $NH_2\overset{O}{\overset{\|}{C}}CH_2CH_3$

(u) $CH_3CH_2CO_2H$

(v) $CH_3CH_2NH\overset{O}{\overset{\parallel}{C}}CH_3$ (w) $NH_2\overset{O}{\overset{\parallel}{C}}CH_2\overset{O}{\overset{\parallel}{C}}NH_2$ (x) $C_6H_5\overset{O}{\overset{\parallel}{C}}O\overset{O}{\overset{\parallel}{C}}C_6H_5$

(y) $C_6H_5\overset{O}{\overset{\parallel}{C}}Cl$ (z) $HO-C_6H_4-\overset{O}{\overset{\parallel}{C}}H$ (aa) $CH_2{=}CHO\overset{O}{\overset{\parallel}{C}}CH_3$

(bb) $CH_3CH_2O\overset{O}{\overset{\parallel}{C}}CH{=}CH_2$ (cc) $CH_3\overset{O}{\overset{\parallel}{C}}OCH_2CH_2O\overset{O}{\overset{\parallel}{C}}CH_3$ (dd) $CH_3\overset{O}{\overset{\parallel}{C}}NHCH_2\overset{O}{\overset{\parallel}{C}}OH$

9.3 Nomenclature of Organic Compounds with Carbonyl Groups

Although discussions of families with carbonyl groups will be spread out over more than one chapter, this study will be easier if we assemble in one place the rules for naming them. All families have both a common and an IUPAC system of nomenclature.

Common Names

Common names for aldehydes and for the carboxylic acids and their derivatives share common prefix portions. These are summarized in Table 9.2. Considering first the carboxylic acids and their derivatives, we have to learn to recognize first what is the carbonyl portion or acid portion of the compound. It is

$$\underbrace{Cl-\overset{O}{\overset{\parallel}{C}}-CH_2CH_3}_{\text{CARBONYL PORTION}} \qquad \underbrace{CH_3\overset{O}{\overset{\parallel}{C}}}_{\text{CARBONYL PORTION}}-O-CH_2CH_3 \qquad CH_3CH_2O\underbrace{\overset{O}{\overset{\parallel}{C}}CH_2CH_3}_{\text{CARBONYL PORTION}}$$

that part of the *carbon* skeleton having the carbonyl group. It is for this part of the molecule that the prefix portions given in Table 9.2 are characteristic. In learning these word parts, be certain to associate the *total* carbon content *of the acid portion* with the prefixes. In counting carbon content, the carbonyl carbon has to be included.

Exercise 9.2 Circle the carbonyl portion in each structure and then write the prefix associated with it.

Examples:

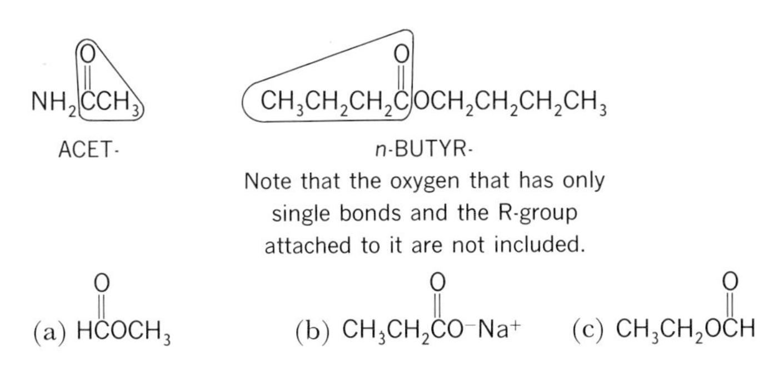

(a) $H\overset{O}{\overset{\parallel}{C}}OCH_3$ (b) $CH_3CH_2\overset{O}{\overset{\parallel}{C}}O^-Na^+$ (c) $CH_3CH_2O\overset{O}{\overset{\parallel}{C}}H$

Table 9.2 **Common Names for Aldehydes, Acids, and Acid Derivatives**

Class	Characteristic Suffix for the Name	Characteristic Prefix in the Carbonyl Portion of the Name			
		C_1 form-	C_2 acet-	C_3 propion-	C_4 butyr-*
Aldehydes	-aldehyde	$HC(=O)H$ Formaldehyde	$CH_3C(=O)H$ Acetaldehyde	$CH_3CH_2C(=O)H$ Propionaldehyde	$CH_3CH_2CH_2C(=O)H$ Butyraldehyde
Carboxylic acids	-ic acid	HCO_2H Formic acid	CH_3CO_2H Acetic acid	$CH_3CH_2CO_2H$ Propionic acid	$CH_3CH_2CH_2CO_2H$ Butyric acid
Carboxylic acid salts	-ate (with name of positive ion written first)	$HCO_2^-Na^+$ Sodium formate	$CH_3CO_2^-Na^+$ Sodium Acetate	$CH_3CH_2CO_2^-K^+$ Potassium propionate	$CH_3CH_2CH_2CO^-NH_4^+$ Ammonium butyrate
Acid chlorides	-yl chloride	(unstable)	$CH_3C(=O)Cl$ Acetyl chloride	$CH_3CH_2C(=O)Cl$ Propionyl chloride	$CH_3CH_2CH_2C(=O)Cl$ Butyryl chloride
Anhydrides	-ic anhydride	(unstable)	$CH_3C(=O)OC(=O)CH_3$ Acetic anhydride	$CH_3CH_2C(=O)OC(=O)CH_2CH_3$ Propionic anhydride	$(CH_3CH_2CH_2C(=O))_2O$ Butyric anhydride
Esters	-ate (with name of alkyl group on oxygen written first)	$HC(=O)OCH_2CH_3$ Ethyl formate	$CH_3C(=O)OCH(CH_3)_2$ Isopropyl acetate	$CH_3CH_2C(=O)OCH_2CH_3$ Ethyl propionate	$CH_3CH_2CH_2C(=O)OCH_3$ Methyl butyrate
Simple amides	-amide	$HC(=O)NH_2$ Formamide	$CH_3C(=O)NH_2$ Acetamide	$CH_3CH_2C(=O)NH_2$ Propionamide	$CH_3CH_2CH_2C(=O)NH_2$ Butyramide

* It is also common practice to write "*n*-" before "butyr-" in all these names. Thus, "*n*-butyraldehyde" or "*n*-butyric acid," meaning the *normal* isomers. This practice, however, is not necessary.

(d) $CH_3CH_2OC(=O)CH_3$ (e) $CH_3CH_2C(=O)OCH_3$ (f) $(CH_3)_2CHC(=O)OCH(CH_3)_2$

In naming esters, first figure out the name of the group attached by a *single* bond to oxygen. Then name the carbonyl portion, attaching the suffix "-ate" to the appropriate prefix.

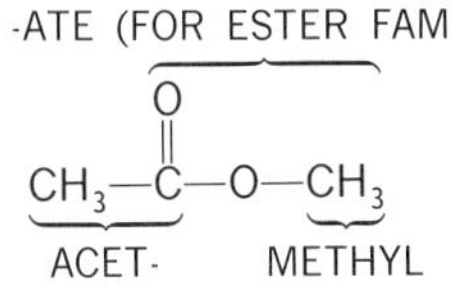

-ATE (FOR ESTER FAMILY)

$CH_3—C(=O)—O—CH_3$ Methyl acetate

ACET- ($CH_3—C$) METHYL (CH_3)

Common names for aldehydes are the easiest. Simply take the prefix for their corresponding carboxylic acids and add the suffix "-aldehyde."

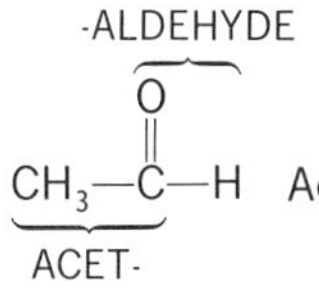

-ALDEHYDE

$CH_3—C(=O)—H$ Acetaldehyde

ACET- ($CH_3—C$)

Exercise 9.3 Complete the following table according to the example given.

Structure	*Family*	*Common Name*
$CH_3C(=O)OCH_3$	ester	methyl acetate
(a) $(CH_3)_2CHOC(=O)CH_3$		
(b) $CH_3CH_2C(=O)NH_2$		
(c) $CH_3OC(=O)CH_2CH_2CH_3$		
(d) $(CH_3)_2CHCH_2OC(=O)CH_2CH_2CH_3$		
(e) $H—C(=O)CH_2CH_2CH_3$		

To give common names for slightly more complicated carbonyl compounds we build on what we have. We use Greek letters to identify the positions in the carbon skeleton as it extends away from the carbonyl group:

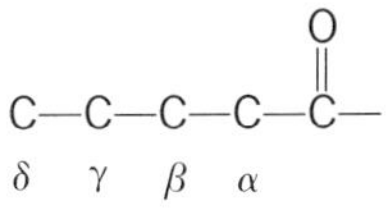

$$\underset{\delta}{C}—\underset{\gamma}{C}—\underset{\beta}{C}—\underset{\alpha}{C}—C(=O)—$$

The α-position is always the carbon attached directly to the carbonyl. The carbonyl carbon is *not* lettered. (The IUPAC system, discussed later, will be different at this point.) The following examples illustrate how the common system can be extended this way.

$CH_3CH(Br)CHO$ (structure: CH_3CHCH with C=O and Br substituent)
α-BROMOPROPIONALDEHYDE

$CH_3CH(Br)—CH(Br)—COOH$
α,β-DIBROMOBUTYRIC ACID

$CH_3CH_2OCOCH_2CH_2OH$ ($CH_3CH_2O\overset{O}{\overset{\|}{C}}CH_2CH_2OH$)
ETHYL β-HYDROXYPROPIONATE

$CH_3CH(CH_3)—CH(CH_3)—CO^-Na^+$ ($CH_3CH—CH—\overset{O}{\overset{\|}{C}}O^-Na^+$ with CH_3 groups)
SODIUM α,β-DIMETHYLBUTYRATE

We make up common names for ketones by giving the name *ketone* to the carbonyl group and then by naming the groups attached to it. The following examples illustrate the method. The simplest ketone possible, dimethyl ketone, however, is always called by its trivial name, acetone.

$CH_3\overset{O}{\overset{\|}{C}}CH_3$
DIMETHYL KETONE
(ACETONE)

$CH_3CH_2\overset{O}{\overset{\|}{C}}CH_3$
ETHYL METHYL KETONE

$CH_3CH_2\overset{O}{\overset{\|}{C}}CH_2CH_3$
DIETHYL KETONE

Some phenyl ketones are often specially named. We contract "phenyl ketone" to "-phenone," and we use it strictly as a suffix. With a little phonetic altering, the rest of the name comes from the prefix we have used for acid portions of esters, aldehydes, etc. A few examples will illustrate.

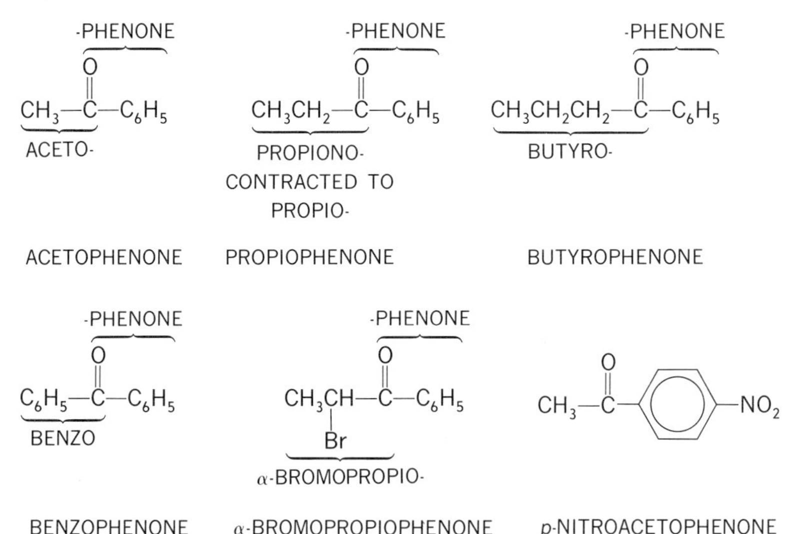

Exercise 9.4 Write the condensed structural formulas for each of the following:

(a) methyl acetate
(b) α,β-dimethylbutyraldehyde

(c) propionic anhydride
(d) isobutyrophenone
(e) *sec*-butyl acetate
(f) α-methylbutyryl chloride
(g) sodium β-chloropropionate
(h) dibenzyl ketone

Exercise 9.5 Write suitable common names for each of the following:

(a) $(CH_3)_2CH\overset{O}{\overset{\|}{C}}CH_2CH_3$ (b) $CH_3\underset{OH}{\underset{|}{C}}H\overset{O}{\overset{\|}{C}}NH_2$ (c) $H\overset{O}{\overset{\|}{C}}OCH_2CH_2CH_2CH_3$

(d) $CH_3CH_2CH_2\overset{O}{\overset{\|}{C}}CH_2CH_2CH_3$ (e) $NH_2\overset{O}{\overset{\|}{C}}CH_2\underset{Br}{\underset{|}{C}}HCH_3$ (f) $CH_3\underset{Cl}{\underset{|}{C}}H-\underset{Cl}{\underset{|}{C}}H\overset{O}{\overset{\|}{C}}H$

(g) $CH_3\underset{Br}{\underset{|}{C}}HCH_2\overset{O}{\overset{\|}{C}}C_6H_5$ (h) $CH_3O\overset{O}{\overset{\|}{C}}CH_2CH_3$ (i) $CH_3\overset{O}{\overset{\|}{C}}OCH_2CH_3$

9.4 IUPAC Nomenclature

For aldehydes, ketones, and carboxylic acids, we take as the base name the longest continuous sequence of carbons, including the carbonyl carbon. We number this chain from the end that will give the carbonyl carbon the lower number, regardless of the other groups that may be present.

In molecules having two or more groups the numbering of the chain is done according to the following order of precedence:

acids > acid halides > amides > aldehydes > ketones > alcohols > alkenes

For aliphatic or cycloaliphatic systems, we select the name of the saturated hydrocarbon corresponding to the longest chain or largest ring. We then alter this name in a way characteristic for the kind of carbonyl compound involved.

*For **aldehydes,** we change the "-e" at the end of the alkane name to "-al."* The "-al" function requires no number to designate its location because it can occur only at the end of a chain and the 1-position for the aldehyde carbon is understood.

$H-\overset{O}{\overset{\|}{C}}-H$ METHANAL (FORMALDEHYDE)

$CH_3\overset{O}{\overset{\|}{C}}-H$ ETHANAL (ACETALDEHYDE)

$CH_3CH_2\overset{O}{\overset{\|}{C}}-H$ PROPANAL (PROPIONALDEHYDE)

*For **ketones,** we change the "-e" ending of the alkane corresponding to the longest chain in the ketone to "-one."* Whenever it is possible for the keto group to be located in different positions, we have to specify the number of its carbon.

$CH_3{-}\overset{\overset{\large O}{\|}}{C}{-}CH_3$ PROPANONE (ACETONE)

$CH_3{-}\overset{\overset{\large O}{\|}}{C}{-}CH_2CH_3$ BUTANONE (ETHYL METHYL KETONE)

$CH_3CH_2{-}\overset{\overset{\large O}{\|}}{C}{-}CH_2CH_3$ 3-PENTANONE (DIETHYL KETONE)

For ***carboxylic acids,*** *we change the "-e" ending of the alkane corresponding to the longest chain in the acid to "-oic acid."* As with the aldehydes, the location of the carbonyl carbon is not specified because it can occur only at the end of the chain and its 1-position is understood. We work into the final name in the usual way all of the substituents on the basic chain, as illustrated by 2-methylpropanoic acid.

$H{-}\overset{\overset{\large O}{\|}}{C}{-}OH$ METHANOIC ACID (FORMIC ACID)

$CH_3{-}\overset{\overset{\large O}{\|}}{C}{-}OH$ ETHANOIC ACID (ACETIC ACID)

$CH_3{-}\overset{\overset{\large CH_3}{|}}{CH}{-}\overset{\overset{\large O}{\|}}{C}{-}OH$ 2-METHYLPROPANOIC ACID (ISOBUTYRIC ACID)

We base the IUPAC names for **acid derivatives** (salts, acid chlorides, anhydrides, esters, and amides) on the name of the acid itself. To name any of these derivatives, first write the name of the parent acid. Then drop the "-ic acid" suffix portion and replace it by the suffix specified for the derivative in the second column of Table 9.2. These word parts are the same in both the common and the IUPAC systems. For example:

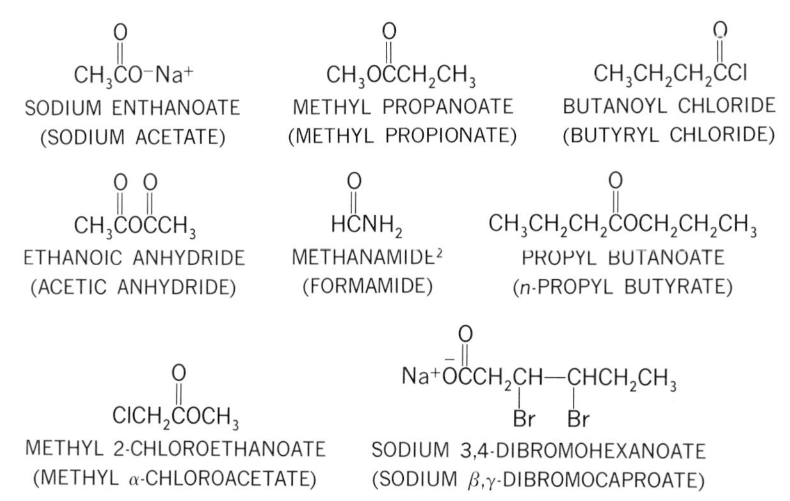

Always figure out the name of the acid first and then operate on that name to write a name for the acid derivative.

Exercise 9.6 Write IUPAC and common names for each of the following.

(a) $CH_3CH_2\overset{\overset{\large O}{\|}}{C}H$ (b) $Na^+\,{}^-O\overset{\overset{\large O}{\|}}{C}CH_2CH_3$ (c) $CH_3CH_2O\overset{\overset{\large O}{\|}}{C}CH_3$

[2]Note that the "o" is also dropped for phonetic reasons. The name is not methanoamide.

(d) $CH_3CH_2C(=O)CH_2CH_2CH_3$ (e) $CH_3C(=O)OC(=O)CH_3$ (f) $HC(=O)OH$

(g) $NH_2C(=O)CH(OH)CH_2CH_3$ (h) $(CH_3)_2CHC(=O)OCH(CH_3)_2$ (i) $CH_3CH_2CH_2OC(=O)CH_3$

(j) $(CH_3)_2CHC(=O)H$

Aldehydes and Ketones

9.5 Physical Properties of Aldehydes and Ketones

Aldehydes and ketones are moderately polar compounds. We can see how their polarity relates to that of other families from the following boiling point data for compounds of closely similar formula weights.

		Formula Weight	BP
Propane	$CH_3CH_2CH_3$	44	−45°C
Dimethyl ether	CH_3OCH_3	46	−25
Methyl chloride	CH_3Cl	50	−24
Ethylamine	$CH_3CH_2NH_2$	45	17
Acetaldehyde	$CH_3CH{=}O$	44	21
Ethyl alcohol	CH_3CH_2OH	46	78.5

Low-formula-weight aldehydes and ketones are soluble in water, but by the time the molecule has five carbons, this solubility has become quite low. The carbonyl group cannot act as a hydrogen bond donor; it has no —H on oxygen (or nitrogen). But the carbonyl oxygen is, of course, a hydrogen bond acceptor. Trends in typical properties are indicated by the data in Tables 9.3 and 9.4.

Syntheses of Aldehydes and Ketones

Aldehydes and ketones may be prepared in dozens of ways. One reference[3] discusses thirty-seven ways for making aldehydes and thirty-seven for making ketones. We shall look briefly at just a few.

[3] R. B. Wagner and H. D. Zook, *Synthetic Organic Chemistry*. John Wiley & Sons, New York, 1953.

Table 9.3
Aldehydes

Common Name	Structure	Mp (°C)	Bp (°C)	Solubility in Water
Formaldehyde	$CH_2{=}O$	−92	−21	Very soluble
Acetaldehyde	$CH_3CH{=}O$	−125	21	Very soluble
Propionaldehyde	$CH_3CH_2CH{=}O$	−81	49	16 g/100 cc (25°)
n-Butyraldehyde	$CH_3CH_2CH_2CH{=}O$	−99	76	4 g/100 cc
Isobutyraldehyde	$(CH_3)_2CHCH{=}O$	−66	65	9 g/100 cc
Valeraldehyde	$CH_3CH_2CH_2CH_2CH{=}O$	−92	102	Slightly soluble
Caproaldehyde	$CH_3CH_2CH_2CH_2CH_2CH{=}O$	−56	128	Very slightly soluble
Acrolein	$CH_2{=}CHCH{=}O$	−87	53	40 g/100 cc
Crotonaldehyde	$CH_3CH{=}CHCH{=}O$	−77	104	Moderately soluble
Benzaldehyde	C_6H_5—CH=O (benzene ring—CH=O)	−56	178	0.3 g/100 cc
Salicylaldehyde	benzene ring with OH (ortho)—CH=O	−10	197	Slightly soluble
Vanillin	HO—benzene ring (with CH_3O)—CH=O	81	285	1 g/100 cc
Cinnamaldehyde (*trans*)	benzene ring—CH=CHCH=O	−8	253	Insoluble
Furfural	furan ring (O)—CH=O	−31	162	9 g/100 cc
Glucose[a] (dextrose)	$CH_2(OH)—CH(OH)—CH(OH)—CH(OH)—CH(OH)—CH{=}O$	146 (decomposes)	—	Very soluble

[a]Only the general structural features of the open-chain form of glucose are shown. Additional details are discussed in Chapter 14.

Table 9.4
Ketones

Name	Structure	Mp (°C)	Bp (°C)	Solubility in Water
Acetone	$CH_3\overset{O}{\overset{\|}{C}}CH_3$	−95	56	Very soluble
Methyl ethyl ketone	$CH_3\overset{O}{\overset{\|}{C}}CH_2CH_3$	−87	80	33 g/100 cc (25°)
2-Pentanone	$CH_3\overset{O}{\overset{\|}{C}}CH_2CH_2CH_3$	−84	102	6 g/100 cc
3-Pentanone	$CH_3CH_2\overset{O}{\overset{\|}{C}}CH_2CH_3$	−40	102	5 g/100 cc
2-Hexanone	$CH_3\overset{O}{\overset{\|}{C}}CH_2CH_2CH_2CH_3$	−57	128	1.6 g/100 cc
3-Hexanone	$CH_3CH_2\overset{O}{\overset{\|}{C}}CH_2CH_2CH_3$	—	124	1.5 g/100 cc
Cyclopentanone	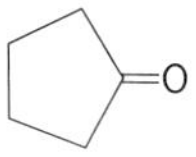 (cyclopentane ring)=O	−53	129	Slightly soluble
Cyclohexanone	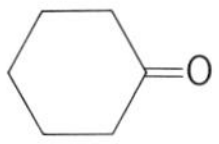 (cyclohexane ring)=O	—	156	Slightly soluble
Camphor	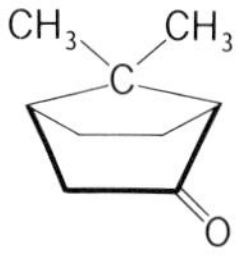 CH_3, CH_3 on C bridge; ring with =O	176	209	Insoluble
Acetophenone	$C_6H_5-\overset{O}{\overset{\|}{C}}CH_3$	−23	137	Insoluble

9.6 Oxidation of 1° and 2° Alcohols

This was discussed in Sections 6.28 to 6.30. For example (and to review):

$$CH_3\underset{}{\overset{OH}{\overset{|}{C}}}H(CH_2)_nCH_3 \xrightarrow[H_2SO_4]{CrO_3} CH_3\overset{O}{\overset{\|}{C}}(CH_2)_nCH_3$$

$n = 2$, 74% $n = 4$, 83%
$= 3$, 80% $= 5$, 96%

$$CH_3(CH_2)_nCH_2OH \xrightarrow[\Delta\Delta]{Cu} CH_3(CH_2)_n\overset{O}{\overset{\|}{C}}H$$

$n = 3$, 72%
$= 7$, 90%

The experimental conditions employed here cannot exist in living systems, of course. Yet, dehydrogenation (oxidation) of 1° and 2° alcohols is a routine event in the chemistry of life. What living cells have are enzymes whose molecules include double bonds that can accept the pieces of hydrogen ($H:^-$ and H^+) from 1° and 2° alcohols. The hydrogen acceptors are often pieces of molecules of the B vitamins.

9.7 Friedel-Crafts Acylation

In Section 4.13 we learned about the alkylation of aromatic rings by alkyl halides in the presence of aluminum chloride. Acyl halides will give the same kind of reaction except that molar proportions, not catalytic portions of aluminum chloride are needed. (See Section 10.9 for the synthesis of acyl halides from carboxylic acids.)

In general:

$$\underset{\text{AROMATIC HYDROCARBON}}{Ar{-}H} + \underset{\text{ACID CHLORIDE}}{Cl{-}\overset{O}{\overset{\|}{C}}{-}R} \xrightarrow{AlCl_3} \underset{\text{AROMATIC KETONE}}{Ar{-}\overset{O}{\overset{\|}{C}}{-}R} + HCl$$

Specific examples:

$$C_6H_6 + Cl{-}\overset{O}{\overset{\|}{C}}{-}CH_2C_6H_5 \xrightarrow{AlCl_3} \underset{\text{BENZYL PHENYL KETONE (83\%)}}{C_6H_5{-}\overset{O}{\overset{\|}{C}}{-}CH_2C_6H_5}$$

$$CH_3{-}C_6H_5 + Cl{-}\overset{O}{\overset{\|}{C}}{-}CH_2CH_3 \xrightarrow{AlCl_3} \underset{\text{ETHYL }p\text{-TOLYL KETONE (86\%)}}{CH_3{-}C_6H_4{-}\overset{O}{\overset{\|}{C}}{-}CH_2CH_3}$$

Two important limitations with this reaction are:

1. The aromatic ring cannot be deactivated (e.g., by a nitro group). If so deactivated, the Friedel-Crafts acylation will not work. (Indeed, nitrobenzene is often used as a solvent for this reaction.)
2. The aromatic ring should not be "super activated," that is, with amino ($-NH_2$) or phenolic ($-OH$) groups. One reason is that these groups, since they are basic toward aluminum chloride, tie up the catalyst. Another is that their benzene rings are simply too reactive.

If only formyl chloride existed, $H{-}\overset{O}{\overset{\|}{C}}{-}Cl$, the Friedel-Crafts acylation would also serve as a synthesis of aldehydes. The Gatterman-Koch reaction gets around this problem.

9.8 The Gatterman-Koch Reaction

If copper(II) chloride is present along with the other conditions for a Friedel-Crafts reaction, it will "fix" a mixture of carbon monoxide and hydrogen chloride to act as formyl chloride:

$$\underset{\underset{\downarrow\ CuCl_2}{CO + HCl}}{}\quad Ar{-}H + [H{-}\overset{\overset{\large O}{\|}}{C}{-}Cl] \xrightarrow[\text{Solvent: Nitrobenzene}]{AlCl_3} Ar{-}\overset{\overset{\large O}{\|}}{C}{-}H$$

Example:

$$CH_3{-}C_6H_4 + CO + HCl \xrightarrow[C_6H_5NO_2]{AlCl_3,\ CuCl_2} CH_3{-}C_6H_4{-}\overset{\overset{\large O}{\|}}{C}H$$

p-TOLUALDEHYDE (51%)

9.9 Hydrolysis of 1,1-Dihalides

Any synthesis of *gem*-dihalides is a potential synthesis of an aldehyde or ketone. *Gem*-dihalides can be hydrolyzed to *gem*-dialcohols, but these are unstable and spontaneously split out water to give the carbonyl compound. The method is used most often to make aromatic aldehydes:

$$Ar{-}CH_3 \xrightarrow[(X\,=\,Cl\text{ or }Br)]{X_2;\ UV} Ar{-}CH\begin{matrix}X\\X\end{matrix} \xrightarrow[\Delta]{H_2O} \left[Ar{-}CH\begin{matrix}O{-}H\\O{-}H\end{matrix}\right] \longrightarrow Ar{-}\overset{\overset{\large O}{\|}}{C}{-}H + H_2O$$

A BENZAL HALIDE

If Ar = $C_6H_5{-}$, 70% benzaldehyde

= $Cl{-}C_6H_4{-}$, 60% *p*-chlorobenzaldehyde

= $Br{-}C_6H_4{-}$, 69% *p*-bromobenzaldehyde

This method is one of the industrial syntheses of benzaldehyde (starting from toluene).

9.10 The Oxo Reaction

This is strictly an industrial synthesis of low-formula-weight aldehydes. (Usually the products are mixtures.) On paper it looks like the addition of formaldehyde to an alkene double bond. In practice it is brought about by heating a mixture of the alkene with carbon monoxide and hydrogen at high pressure in the presence of cobalt octacarbonyl. For example:

$$\underset{\text{PROPYLENE}}{CH_3CH{=}CH_2} + CO + H_2 \xrightarrow[\text{2200–3500 psi, }140^\circ]{[Co(CO_4)]_2} \underset{\text{BUTYRALDEHYDE}}{CH_3CH_2CH_2\overset{\overset{\large O}{\|}}{C}H} + \underset{\text{ISOBUTYRALDEHYDE}}{(CH_3)_2CH\overset{\overset{\large O}{\|}}{C}H}$$

By giving careful attention to the conditions as well as to the solvent used, the proportions of isomers in the products can be controlled. One important use of the butyraldehyde obtained this way is the manufacture of a chemical needed in fabricating plastics.

9.11 Laboratory versus Industrial Syntheses

A word should be added here concerning laboratory versus industrial syntheses of organic chemicals. The scientist in the laboratory is usually interested in reactions that give the highest yields of one particular compound in a method that as much as possible avoids extraordinary or costly equipment (such as high pressure apparatus) and a method that simplifies the problem of purifying a product. (Methods giving mixtures of isomers that are essentially impossible to separate are methods avoided whenever possible.) Such laboratory work sometimes leads to discovery of a substance of commercial value. Then product development, pilot plant work, market research, and many other studies take place leading to full-scale manufacture. These studies often reveal that it would be wise to invest heavily in finding the best possible catalyst, and industry can make such an investment. The studies may also show that a mixture of isomers (rather than one pure substance) is perfectly acceptable for the application desired. If so, methods can be used industrially that would be shunned in small scale laboratory work.

Chemical Properties of Aldehydes and Ketones

We are interested in the behavior of the aldehyde and the keto group toward oxidizing agents, reducing agents, and nucleophiles.

9.12 Oxidation

Here we see the main reason why aldehydes and ketones are given separate classes. Aldehydes are extremely easy to oxidize to carboxylic acids without losing any carbons from the chain. In fact, it is hard to store aldehydes because even air will slowly oxidize them. Ketones, in sharp contrast, are hard to oxidize. Open-chain ketones, when oxidized, suffer the breaking up of the carbon chain. Cyclic ketones undergo ring opening.

The ease with which the aldehyde group is oxidized makes it possible to detect its presence with reagents that leave other groups unaffected.

9.13 Tollens' Test. Silvering Mirrors

Tollens' reagent is a solution of the diammonia complex of the silver ion in dilute ammonium hydroxide. (Because the silver ion precipitates as silver oxide in the presence of hydroxide ion, it is complexed with ammonia molecules.) In general:

$$\underset{\text{ALDEHYDE}}{RCH{=}O} + 2Ag(NH_3)_2^+ + 3OH^- \longrightarrow \underset{\text{CARBOXYLATE ION}}{RCO_2^-} + 2Ag\downarrow + 2H_2O + 4NH_3$$

Specific example:

$$\underset{\text{ACETALDEHYDE}}{CH_3CH{=}O} + 2Ag(NH_3)_2^+ + 3OH^- \longrightarrow \underset{\text{ACETATE ION}}{CH_3CO_2^-} + \underset{\text{SILVER}}{2Ag\downarrow} + 2H_2O + 4NH_3$$

Any good test for a functional group must provide easily observed, positive evidence that reaction occurs with that functional group. The positive evidence in the Tollens' test is the appearance of metallic silver in a previously clear, colorless solution. If the inner wall of the test vessel is clean and grease free, the silver comes out as a beautiful mirror, a reaction that serves as the basis for silvering mirrors. If the glass surface is not clean, the silver separates as a gray, finely divided, powdery precipitate. Tollens' reagent does not keep well and is prepared just before use. (If the reagent dries up, the residue could explode.)

9.14 Benedict's Test and Fehling's Test

Benedict's solution and Fehling's solution are both alkaline and contain the copper(II) ion, Cu^{2+}, stabilized by a complexing agent. This agent is the citrate ion in Benedict's solution and the tartrate ion in Fehling's solution. Without these complexing agents, copper(II) ions would be precipitated as copper(II) hydroxide by the alkali present. Benedict's solution is more frequently used because it stores well. Fehling's solution must be prepared just before use.

With either solution the Cu^{2+} ion is the oxidizing agent. In the presence of certain easily oxidized groups it is reduced to the copper(I) state, but copper(I) ions cannot be solubilized by either citrate or tartrate ions and a precipitate of copper(I) oxide, Cu_2O, forms. What we see in a positive test, then, is a change from the brilliant blue color of the test solution to the bright orange-red color of precipitated copper(I) oxide. The principal systems that give this result are α-hydroxy aldehydes, α-keto aldehydes, and α-hydroxy ketones. The first two, α-hydroxy aldehydes and α-hydroxy ketones, are

$$\underset{\alpha\text{-HYDROXY ALDEHYDE}}{\mathrm{R\underset{\underset{OH}{|}}{C}H\overset{\overset{O}{\|}}{C}H}} \qquad \underset{\alpha\text{-KETO ALDEHYDE}}{\mathrm{R{-}\overset{\overset{O}{\|}}{C}{-}\overset{\overset{O}{\|}}{C}{-}H}} \qquad \underset{\alpha\text{-HYDROXY KETONE}}{\mathrm{R\underset{\underset{OH}{|}}{C}H\overset{\overset{O}{\|}}{C}R}}$$

common to the sugar family. (Glucose is an α-hydroxy aldehyde; fructose is an α-hydroxy ketone.) Benedict's test, or Fehling's test, is a common method for detecting the presence of glucose in urine. Normally, urine does not contain glucose, but in certain conditions—for example, diabetes—it does. A positive test for glucose varies from a bright green color (0.25% glucose), to yellow-orange (1% glucose), to brick red (over 2% glucose).

Some ordinary aliphatic aldehydes react with these reagents, but the changes are complex. A dark, gummy precipitate may form, which apparently is not copper(I) oxide, while the color of the test solution remains blue or becomes any hue from yellow to green to brown. Aromatic aldehydes, in general, do not react.[4]

9.15 Haloform Reaction. Iodoform Test

Methyl ketones react with the halogens in the presence of sodium hydroxide and form trihalomethanes (e.g., chloroform, bromoform, or iodoform) plus the salt of that carboxylic acid resulting from loss of the methyl carbon:

[4] R. Daniels, C. C. Rush, and L. Bauer, *Journal of Chemical Education,* Vol. 37, 1960, page 205.

$$R{-}\overset{O}{\overset{\|}{C}}{-}CH_3 + 3X_2 + 4OH^- \longrightarrow R{-}\overset{O}{\overset{\|}{C}}{-}O^- + \underset{\text{HALOFORM}}{CHX_3} + 3H_2O + 3X^-$$

(X = Cl, Br, or I)

When iodine is the halogen, the product is iodoform, a yellow solid with a characteristic odor and color. Its formation in the iodoform test is the basis for the detection of methyl ketones. In a positive test the deep brown color of molecular iodine disappears, and the yellow iodoform is evident.

The test is also positive for methylcarbinols, as structurally defined below. The systems giving the test are, therefore:

$$\underset{\substack{\text{METHYLCARBINOLS} \\ \text{(OR ETHANOL, IF R = H)}}}{(H)R{-}\overset{OH}{\overset{|}{C}H}{-}CH_3} \qquad \underset{\substack{\text{METHYL KETONES} \\ \text{(OR ACETALDEHYDE, IF R = H)}}}{(H)R{-}\overset{O}{\overset{\|}{C}}{-}CH_3}$$

Methylcarbinols are believed to be oxidized by the reagent to their corresponding methyl ketones, which are then further attacked. The reaction with methyl ketones apparently involves halogenation of the methyl group followed by cleavage of the molecule:

$$R{-}\overset{O}{\overset{\|}{C}}{-}CH_3 \xrightarrow{I_2,\ NaOH,\ H_2O} R{-}\overset{O}{\overset{\|}{C}}{-}CI_3$$

$$R{-}\overset{O}{\overset{\|}{C}}{-}CI_3 + OH^- \longrightarrow (R{-}\overset{O}{\overset{\|}{C}}{-}O{-}H) + :CI_3^-$$

$$\downarrow \qquad\qquad \downarrow$$

$$\underset{\text{CARBOXYLATE ION}}{R{-}\overset{O}{\overset{\|}{C}}{-}O^-} + \underset{\text{IODOFORM}}{HCI_3}$$

Exercise 9.7 Which compounds will give positive iodoform tests?

(a) 1-methylcyclohexanol (cyclohexane ring bearing OH and CH_3 on the same carbon)

(b) $C_6H_5\overset{O}{\overset{\|}{C}}CH_3$

(c) $C_6H_5CH_2OH$

(d) cyclohexyl$-\overset{OH}{\overset{|}{C}}HCH_3$

9.16 Strong Oxidizing Agents

Aldehydes are of course readily oxidized to carboxylic acids by permanganate ion, dichromate ion, and other oxidizing agents. Ketones also eventually yield to such reagents, but except in isolated instances the reaction is not particularly useful. Ketones tend to fragment somewhat randomly on both sides of the carbonyl carbon:

$$R{-}\overset{O}{\overset{\|}{C}}{-}R' \xrightarrow{\text{Strong oxidizing agent, heat}} RCO_2H + R'CO_2H \quad \text{(as well as acids of lower formula weights)}$$

9.17 Reduction

Several methods are available for reducing aldehydes and ketones to their corresponding primary and secondary alcohols.

Catalytic Hydrogenation

In general:

$$\underset{\text{ALDEHYDE}}{R{-}\overset{\overset{\displaystyle O}{\|}}{C}{-}H} + H_2 \xrightarrow[\text{Pressure, heat}]{\text{Ni}} \underset{1°\ \text{ALCOHOL}}{R{-}CH_2{-}OH}$$

$$\underset{\text{KETONE}}{R{-}\overset{\overset{\displaystyle O}{\|}}{C}{-}R'} + H_2 \xrightarrow[\text{Pressure, heat}]{\text{Ni}} \underset{2°\ \text{ALCOHOL}}{R{-}\overset{\overset{\displaystyle OH}{|}}{C}H{-}R'}$$

Specific examples:

$$\underset{n\text{-BUTYRALDEHYDE}}{CH_3CH_2CH_2CH{=}O} + H_2 \longrightarrow \underset{n\text{-BUTYL ALCOHOL (85\%)}}{CH_3CH_2CH_2CH_2OH}$$

$$\underset{\text{ACETONE}}{CH_3\overset{\overset{\displaystyle O}{\|}}{C}CH_3} + H_2 \longrightarrow \underset{\text{ISOPROPYL ALCOHOL (100\%)}}{CH_3\overset{\overset{\displaystyle OH}{|}}{C}HCH_3}$$

Metal Hydrides

Metal hydrides such as lithium aluminum hydride ($LiAlH_4$) and sodium borohydride ($NaBH_4$) are excellent reducing agents for aldehydes and ketones.

$$\underset{\text{HEPTANAL}}{CH_3(CH_2)_5\overset{\overset{\displaystyle O}{\|}}{C}H} \xrightarrow[\text{2. } H_2O]{\text{1. } LiAlH_4\text{, ether}} \underset{\text{1-HEPTANOL (86\%)}}{CH_3(CH_2)_5CH_2OH}$$

$$\underset{\text{CYCLOPENTANONE}}{C_5H_8{=}O} \xrightarrow[\text{2. } H^+]{\text{1. } NaBH_4} \underset{\text{CYCLOPENTANOL (90\%)}}{C_5H_9{-}OH}$$

In both cases, the metal hydride transfers hydride ion, H:⁻, to the carbonyl carbon, and the anion of the alcohol is produced. To simplify let us use M—H as our metal hydride:

$$\underset{\text{ALDEHYDE OR KETONE}}{{-}\overset{\overset{\displaystyle :O:^{\delta-}}{\|}}{C^{\delta+}}{-}} \quad H{:}M \longrightarrow M^+ + \underset{\text{ANION OF ALCOHOL}}{{-}\overset{\overset{\displaystyle :\ddot{O}:^-}{|}}{\underset{\underset{\displaystyle H}{|}}{C}}{-}} \xrightarrow{H^+} \underset{\text{ALCOHOL}}{{-}\overset{\overset{\displaystyle OH}{|}}{\underset{\underset{\displaystyle H}{|}}{C}}{-}}$$

Lithium aluminum hydride is a more powerful reducing agent than sodium borohydride. (It is also far more dangerous to handle; it will ignite spontaneously in air, for example.) Unlike lithium aluminum hydride, sodium borohydride will not reduce such other common, unsaturated groups as the following:

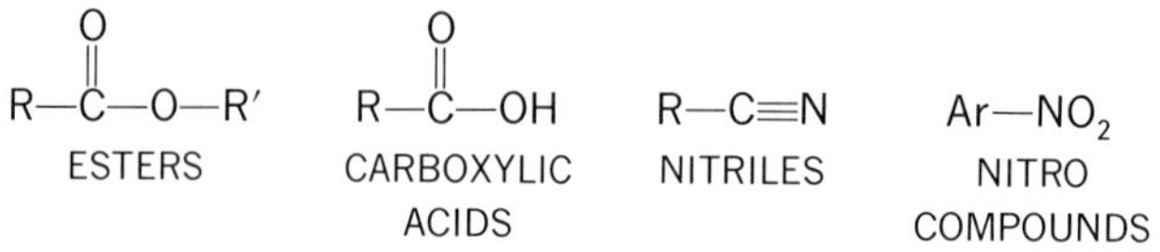

$$\underset{\text{ESTERS}}{R{-}\overset{O}{\overset{\|}{C}}{-}O{-}R'} \qquad \underset{\text{CARBOXYLIC ACIDS}}{R{-}\overset{O}{\overset{\|}{C}}{-}OH} \qquad \underset{\text{NITRILES}}{R{-}C{\equiv}N} \qquad \underset{\text{NITRO COMPOUNDS}}{Ar{-}NO_2}$$

Addition of Water and Alcohols[5]

9.18 Hydrates

Most aldehydes react in aqueous solutions with water to establish an equilibrium mixture with the hydrated form:

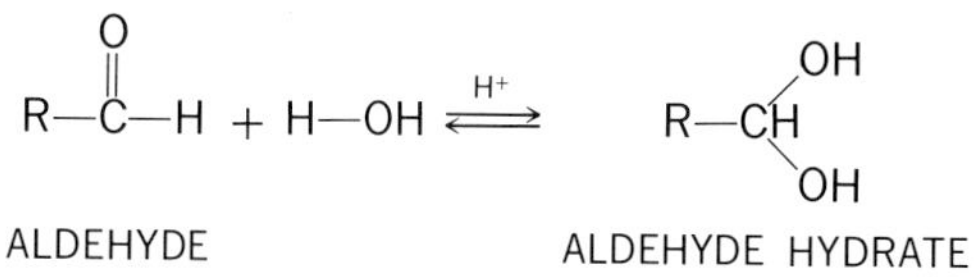

$$\underset{\text{ALDEHYDE}}{R{-}\overset{O}{\overset{\|}{C}}{-}H} + H{-}OH \overset{H^+}{\rightleftharpoons} \underset{\text{ALDEHYDE HYDRATE}}{R{-}CH(OH)_2}$$

Formaldehyde is especially prone to form its hydrate. Water probably adds as follows:

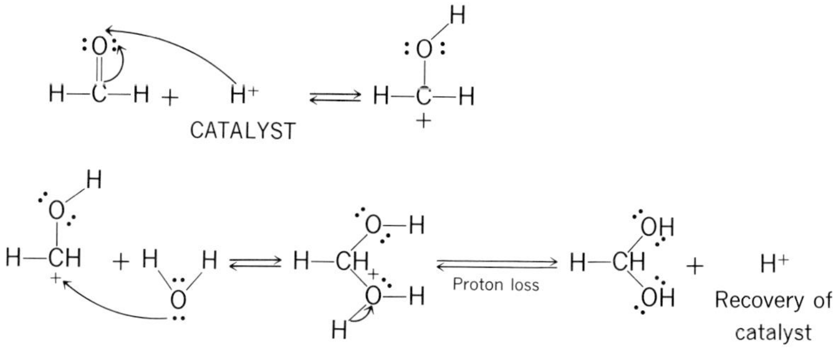

$$H{-}\overset{:O:}{\overset{\|}{C}}{-}H + \underset{\text{CATALYST}}{H^+} \rightleftharpoons H{-}\overset{H{-}\ddot{O}:}{\underset{+}{C}}{-}H$$

$$H{-}\overset{+}{C}H(\ddot{O}H) + H{-}\ddot{O}{-}H \rightleftharpoons H{-}CH(\ddot{O}H)(\overset{+}{O}H_2) \xrightleftharpoons{\text{Proton loss}} H{-}CH(\ddot{O}H)_2 + \underset{\text{Recovery of catalyst}}{H^+}$$

9.19 Hemiacetals and Acetals

In a similar manner, solutions of most aldehydes in alcohols consist of equilibrium mixtures:

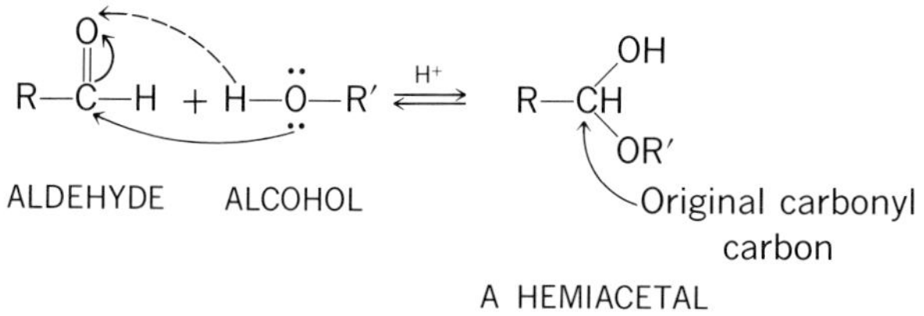

$$\underset{\text{ALDEHYDE}}{R{-}\overset{O}{\overset{\|}{C}}{-}H} + \underset{\text{ALCOHOL}}{H{-}\ddot{O}{-}R'} \overset{H^+}{\rightleftharpoons} \underset{\text{A HEMIACETAL}}{R{-}CH(OH)(OR')}$$

(Original carbonyl carbon)

From the alcohol the hydrogen goes to the carbonyl oxygen; the alcohol oxygen goes to the carbonyl carbon. Most hemiacetals cannot be isolated. Efforts to do so cause them to convert back to the original aldehyde and alcohol. In fact, the

[5]Sections 9.18–9.19 are needed for our study of carbohydrates, Chapter 14. If Chapter 14 will not be included in your course, these sections may be omitted; or they could be delayed until Chapter 14 is reached.

hemiacetal group is often called a *potential aldehyde group*. The hemiacetals that can be isolated are, however, very important, for they occur among carbohydrate molecules.

What we have said about aldehydes applies to some extent with ketones. The main difference is that ketones are less prone to add water or alcohols. In ketones there is one more bulky group at the carbonyl carbon, a group that discourages the addition of water or alcohols in two ways. It gets in the way, sterically; and its inductive effect reduces the partial positive charge on the carbonyl carbon. We do not even consider hydrates of ketones. However we may speak of *hemiketals,* which are just like hemiacetals except they come from ketones. We shall encounter one that is stable when we study the structure of fructose, one of the important sugars, in Section 14.16.

Ordinary alcohols interact and change to ordinary ethers only under rather extreme conditions (cf. Section 6.33). As we shall see next, however, the —OH group in a hemiacetal or hemiketal is easily converted into a second ether linkage.

Hemiacetals or hemiketals can be converted into acetals or ketals by the action of additional alcohol in the presence of an acid catalyst such as dry hydrogen chloride. We illustrate with acetals:

$$\underset{\text{HEMIACETAL}}{\text{R—CH(O—H)(O—R}')} + \text{H—O—R}' \xrightarrow[\text{(via resonance-stabilized carbonium ion)}]{\text{H}^+} \underset{\text{ACETAL}}{\text{R—CH(O—R}')\text{(O—R}')} + H_2O$$

To drive the equilibrium to the right we could use an excess of HOR′ and or remove H_2O as it forms.

The key to this behavior, at least under certain conditions, is the relative stability of the carbonium ion that forms in acid. We see this for hemiacetals as follows:

$$\underset{\text{HEMIACETAL}}{\text{R—CH(OH)(OR}')} \xrightarrow[\text{(H}_2\text{O leaves)}]{\text{H}^+} H_2O + \left[\text{R—}\overset{+}{\text{C}}\text{H—}\ddot{\text{O}}\text{—R}' \longleftrightarrow \text{R—CH}=\overset{+}{\text{O}}\text{—R}\right] \equiv \underset{\text{Hybrid}}{\text{R—}\overset{\delta+}{\text{C}}\text{H}\text{---}\overset{\delta+}{\text{O}}\text{—R}'}$$

CARBONIUM ION

It is a resonance-stabilized hybrid of these two contributors. In the second of the two, all atoms have octets.

Then,

$$\text{H—}\ddot{\text{O}}\text{—R}' + \text{R—}\overset{\delta+}{\text{C}}\text{H}\text{---}\overset{\delta+}{\text{O}}\text{—R}' \longrightarrow \text{R—CH(}\overset{+}{\text{O}}\text{HR}')\text{—}\ddot{\text{O}}\text{—R}' \xrightarrow{-\text{H}^+} \underset{\text{ACETAL}}{\text{R—CH(O—R}')\text{—}\ddot{\text{O}}\text{—R}'}$$

This carbonium ion is much more stable than ordinary carbonium ions where delocalization by resonance cannot occur.

Acetals and ketals are stable in neutral and alkaline media. They can be isolated and stored, in great contrast to the hemiacetals and hemiketals. The one reaction of importance to our future studies is their ready *acid-catalyzed* hydrolysis, which is the reverse of their formation. First, the hemiacetal stage is reached, then the original alcohol and aldehyde (or ketone) form.

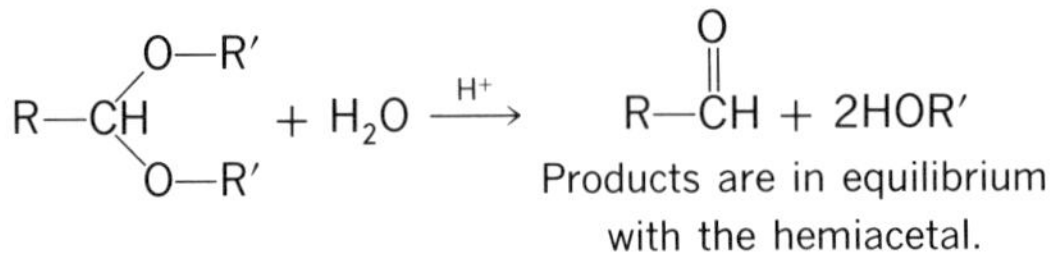

$$R{-}CH(O{-}R')_2 + H_2O \xrightarrow{H^+} R{-}\overset{\overset{O}{\|}}{C}H + 2HOR'$$

Products are in equilibrium with the hemiacetal.

Specific example:

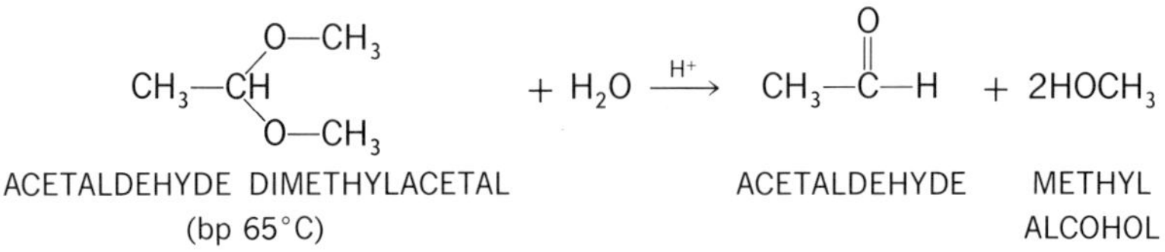

$$CH_3{-}CH(O{-}CH_3)_2 + H_2O \xrightarrow{H^+} CH_3{-}\overset{\overset{O}{\|}}{C}{-}H + 2HOCH_3$$

ACETALDEHYDE DIMETHYLACETAL (bp 65°C) — ACETALDEHYDE — METHYL ALCOHOL

Exercise 9.8 Write the structure of the hemiacetal that will exist in equilibrium with each of the following pairs of compounds.

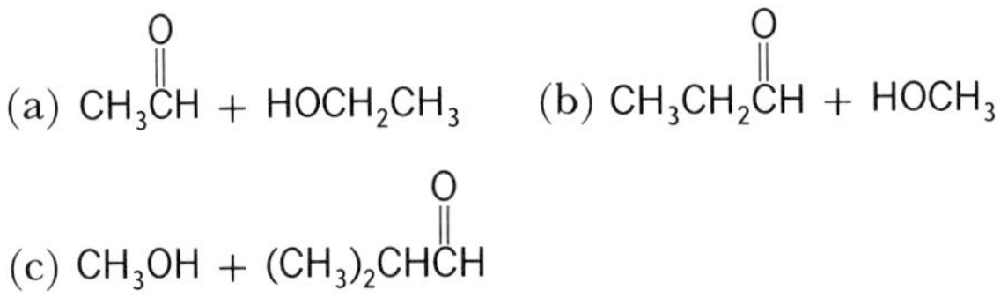

(a) $CH_3\overset{\overset{O}{\|}}{C}H + HOCH_2CH_3$ (b) $CH_3CH_2\overset{\overset{O}{\|}}{C}H + HOCH_3$

(c) $CH_3OH + (CH_3)_2CH\overset{\overset{O}{\|}}{C}H$

Exercise 9.9 Write the structure of the acetal that will form if the aldehyde in each of the parts of exercise 9.8 combines with two molecules of the alcohol that is shown with it.

Exercise 9.10 Write the structure of the original aldehyde and the original alcohol that will form if the following acetals are hydrolyzed. One example is not an acetal. Which one is it?

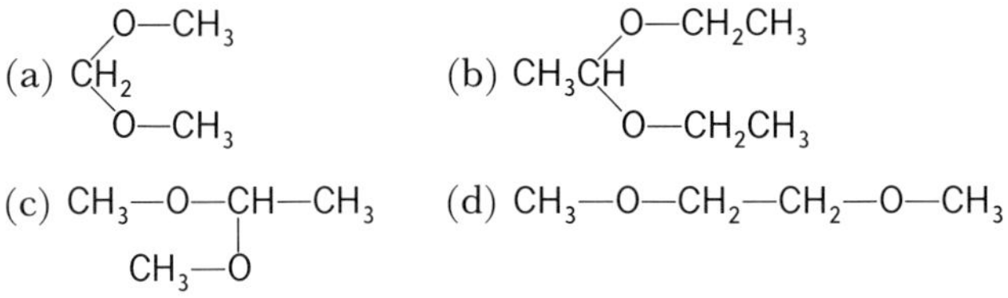

(a) $CH_2(O{-}CH_3)_2$ (b) $CH_3CH(O{-}CH_2CH_3)_2$

(c) $CH_3{-}O{-}CH(O{-}CH_3){-}CH_3$ (d) $CH_3{-}O{-}CH_2{-}CH_2{-}O{-}CH_3$

9.20 Condensation Reactions Between Aldehydes or Ketones and Derivatives of Ammonia

Aldehydes and ketones undergo a variety of reactions of the following general type:

$$\rangle C{=}O + H_2\ddot{N}{-}G \longrightarrow \rangle C{=}\ddot{N}{-}G + H_2O$$

The group, —*G*, attached to nitrogen may vary as illustrated in the examples that follow. The reaction is usually catalyzed by acids, but the exact pH that works best varies with the nature of the group. Shown below is a probable mechanism.

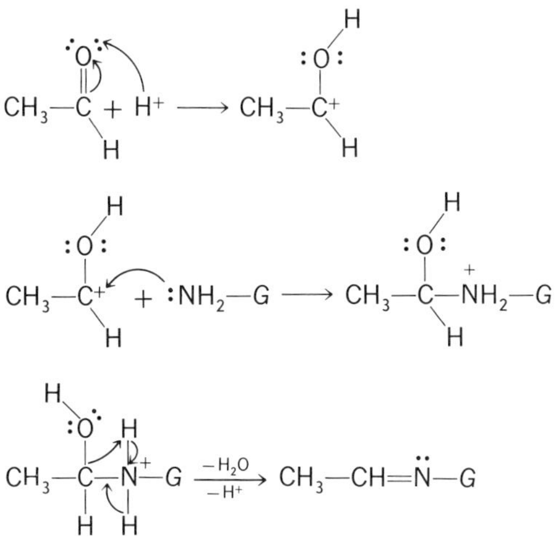

$$CH_3{-}\overset{\ddot{O}:}{\overset{\|}{C}}{-}H + H^+ \longrightarrow CH_3{-}\overset{:\ddot{O}{-}H}{\overset{|}{C^+}}{-}H$$

This reaction makes the carbonyl carbon more attractive to an electron-rich species.

$$CH_3{-}\overset{:\ddot{O}{-}H}{\overset{|}{C^+}}{-}H + :NH_2{-}G \longrightarrow CH_3{-}\overset{:\ddot{O}{-}H}{\overset{|}{C}}(H){-}\overset{+}{N}H_2{-}G$$

The nucleophilic amino group attacks.

$$CH_3{-}\overset{H{-}\ddot{O}:}{\overset{|}{C}}(H){-}\overset{+}{N}(H)(H){-}G \xrightarrow[-H^+]{-H_2O} CH_3{-}CH{=}\ddot{N}{-}G$$

Both water and the proton are lost.

Examples. All reactions are general for both aldehydes and ketones.

1. Formation of **substituted imines:** G = alkyl.

$$\underset{\text{AN ALDEHYDE}}{R{-}CH{=}O} + \underset{\text{A 1° AMINE}}{NH_2{-}R'} \longrightarrow \underset{\substack{\text{IMINES}\\ \text{(Only rarely}\\ \text{stable)}}}{R{-}CH{=}N{-}R} + H_2O$$

2. Formation of **phenylhydrazones:** $G = {-}NHC_6H_5$.

$$R{-}\underset{R'}{\underset{|}{C}}{=}O + \underset{\text{PHENYLHYDRAZINE}}{NH_2NHC_6H_5} \longrightarrow \underset{\text{A PHENYLHYDRAZONE}}{R{-}\underset{R'}{\underset{|}{C}}{=}NNHC_6H_5} + H_2O$$

3. Formation of **2,4-dinitrophenylhydrazones:** $G = {-}NH{-}C_6H_3(NO_2)_2$ (2,4-dinitrophenyl)

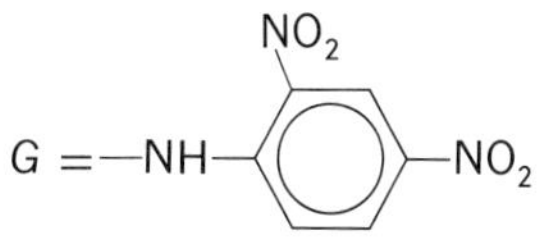

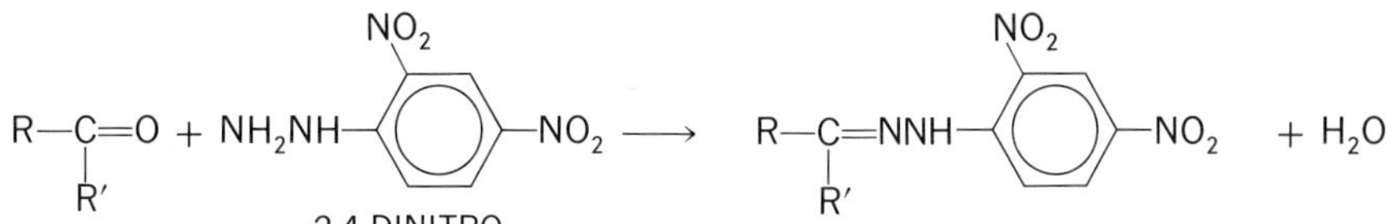

$$R{-}\underset{R'}{\underset{|}{C}}{=}O + \underset{\text{2,4-DINITROPHENYLHYDRAZINE}}{NH_2NH{-}C_6H_3(NO_2)_2} \longrightarrow \underset{\text{A 2,4-DINITROPHENYLHYDRAZONE}}{R{-}\underset{R'}{\underset{|}{C}}{=}NNH{-}C_6H_3(NO_2)_2} + H_2O$$

4. Formation of **semicarbazones:** $G = {-}NH{-}\overset{O}{\overset{\|}{C}}{-}NH_2$.

$$R{-}CH{=}O + \underset{\text{SEMICARBAZIDE}}{NH_2NH\overset{O}{\overset{\|}{C}}NH_2} \longrightarrow \underset{\text{A SEMICARBAZONE}}{RCH{=}NNH\overset{O}{\overset{\|}{C}}NH_2} + H_2O$$

5. Formation of **oximes:** $G = -OH$.

$$\mathrm{RCH{=}O} + \underset{\text{HYDROXYLAMINE}}{\mathrm{NH_2OH}} \longrightarrow \underset{\text{AN OXIME}}{\mathrm{RCH{=}NOH}} + \mathrm{H_2O}$$

9.21 The Use of Solid Derivatives to Identify Aldehydes and Ketones

When a chemist is faced with the task of identifying an aldehyde or a ketone, the problem is complicated by the fact that a great number of these compounds are liquids. Experimentally, solids are usually easier to purify than liquids. Therefore the carbonyl compound is often converted into a solid derivative whose melting point identifies the original material. The oximes, semicarbazones, phenylhydrazones, and especially the 2,4-dinitrophenylhydrazones of nearly all aldehydes and ketones are crystalline solids with well-defined melting points—their most serviceable property. These carbonyl derivatives may be used for identification purposes.

Extensive tables of known carbonyl compounds are available, with each family in a separate table and listed in their order of increasing boiling points or melting points. Table 9.5 is a portion of the table of liquid aldehydes. The boiling point alone is generally not enough evidence to prove an identification. Therefore, the "unknown" may be converted into a solid derivative. Often two derivatives are made, and if their melting points "check," the identification is proved.

A chemist has many other approaches. He will check for coincidence between his unknown and data for a known compound on many points: boiling point, density, melting point, various kinds of spectra, refractive index, solid derivatives, and any number of chemical properties. Of particular and ever-increasing importance are four spectra: infrared (IR), nuclear magnetic resonance (NMR), ultravi-

Table 9.5
Derivatives of Some Aldehydes[a]

Aldehyde	Bp (°C)	Melting Points of Derivatives (°C)		
		Oxime	Semicarbazone	2,4-Dinitrophenylhydrazone
Phenylacetaldehyde	194	103	156	121
Salicylaldehyde	196	57	231	252 (decomposes)
m-Tolualdehyde	199	60	213	211
o-Tolualdehyde	200	49	208	195
p-Tolualdehyde	204	79	221	239
Citronellal	206	Liquid	82	77

[a] Data from R. L. Shriner, R. C. Fuson, and D. Y. Curtin, *The Systematic Identification of Organic Compounds,* fifth edition, John Wiley and Sons, New York, 1964, page 320.

olet (UV), and mass spectra. (These are described in an elementary way in Chapter 17, put where we can take advantage of our introduction to all the major families. However, there is no strong reason to prevent your browsing in that chapter now.)

Reactions at the α-Position

9.22 General Considerations. Enols and Enolate Anions

The carbonyl group is a proton acceptor. There are unshared electrons on the oxygen. Moreover, because of oxygen's electronegativity, the carbonyl oxygen has a partial negative charge, as we saw in Section 9.1. On these grounds we *must expect* the following behavior:

$$\overset{\delta-}{:\!O\!:}=\overset{\delta+}{C} + H\!-\!A \rightleftharpoons \left[\overset{+}{O}(H)=C \longleftrightarrow O(H)-\overset{+}{C} \right] \equiv \overset{\delta+}{O}(H)\cdots C^{\delta+}$$

(Weak $\delta+$) (An acid) Protonated carbonyl group (Resonance stabilized) (Larger $\delta+$)

Let us now give this protonated carbonyl group an alpha hydrogen and see where it might lead. A carbon bearing a rather large partial positive charge is close by the alpha hydrogen. That charge makes that carbon electron withdrawing, and that weakens the bond holding the alpha hydrogen.

$$H^+ + -\underset{H}{\overset{|}{C}}-\overset{O}{\overset{\|}{C}}- \rightleftharpoons \left[-\underset{H}{\overset{|}{C}}-\overset{\overset{+}{O}-H}{\overset{\|}{C}}- \longleftrightarrow -\underset{H}{\overset{|}{C}}-\overset{O-H}{\overset{|}{\overset{+}{C}}}- \right] \rightleftharpoons \;\; >C=C<^{OH} + H^+$$

KETO FORM AN ENOL (-ENE + -OL)

The equilibrium actually overwhelmingly favors the keto form, but some enol is present, and some reactions of aldehydes and ketones go through the enol form.

An *enol* is a compound whose molecules contain an "ol" function joined to an "ene" function. Acetone, for example, contains a trace of enol:

$$CH_3-\overset{O}{\overset{\|}{C}}-CH_3 \rightleftharpoons CH_2=\overset{OH}{\overset{|}{C}}-CH_3$$

ACETONE ENOL OF ACETONE

If we generate an enol indirectly we cannot isolate it; it changes to the keto form. Thus, if we add water to the triple bond of propyne we get the enol of acetone which immediately changes to acetone.

$$CH_3-C\equiv C-H + H_2O \xrightarrow[H_2SO_4]{Hg^{2+}} \left(CH_3-\overset{O-H}{\overset{|}{C}}=CH_2 \right) \longrightarrow CH_3-\overset{O}{\overset{\|}{C}}-CH_3$$

PROPYNE ENOL ACETONE

An enol and its keto form (or aldehydo form) are isomers. Because they have a special relation—one changes more or less rapidly into the other—they are given a special type-name. Enols and their more stable forms are called **tautomers.** The phenomenon itself—the existence of these kinds of isomers—is called **tautomerism.**

Acids catalyze the formation of the equilibrium which contains some of the enol form of aldehydes and ketones. Alkaline substances force the formation of a similar equilibrium, one producing not the enol itself but, instead, its salt:

$$-\overset{|}{\underset{\underset{H}{|}}{C}}-\overset{\overset{:O:}{\|}}{C}- \rightleftharpoons H_2O \text{ (or HOR)} + \left[-\overset{|}{\underset{\cdot\cdot}{C}}-\overset{\overset{:O:}{\|}}{C}- \longleftrightarrow \;\rangle C=C\langle^{\ddot{O}:^{-}}\right] \equiv \;\rangle \overset{}{\underset{\delta-}{C}}\cdots C\langle^{O^{\delta-}}$$

$+\ ^{-}OH$ (or ^{-}OR)
Strong bases

ENOLATE ANION
(resonance-stabilized anion)

Only if there is a hydrogen on the *alpha* position will this work. Only from that system can an anion form that, like an allyl system, can acquire a measure of stability via resonance. The anion is formally the salt of an enol, and its general name is *enolate anion.*

Now let us see what these enols or their anions can do. One reaction is with halogen (Cl_2, Br_2 and I_2). We can have halogenation, without free radicals, at an alpha C—H bond in carbonyl compounds.

9.23 Halogenation of Aldehydes and Ketones

We shall illustrate with the bromination of acetone.

Acid catalysis:

$$CH_3-\overset{\overset{O}{\|}}{C}-CH_3 \;\xrightleftharpoons{\text{Acid catalysis}}\; CH_3-\overset{\overset{OH}{|}}{C}=CH_2 \;\xrightarrow[Br^-]{Br_2}\; \left(CH_3-\overset{\overset{O-H}{|}}{\underset{+}{C}}-\underset{\underset{Br}{|}}{CH_2}\right) \xrightarrow{-H^+} CH_3-\overset{\overset{O}{\|}}{C}-CH_2-Br$$

ACETONE ... ENOL FORM

Base catalysis:

$$CH_3-\overset{\overset{O}{\|}}{C}-CH_3 \;\xrightleftharpoons{\text{Base catalysis}}\; CH_3-\overset{\overset{O^-}{|}}{C}=CH_2 \;\xrightarrow[Br^-]{Br_2}\; \left(CH_3-\overset{\overset{:\ddot{O}:^-}{|}}{\underset{+}{C}}-\underset{\underset{Br}{|}}{CH_2}\right) \longrightarrow CH_3-\overset{\overset{O}{\|}}{C}-CH_2Br$$

ACETONE ... ENOLATE ANION

Further bromination is possible if the conditions are right. (Cf. the iodoform test, Section 9.15.)

9.24 The Aldol Condensation

This reaction, generally carried out under basic conditions, involves enolate anions as nucleophiles. But first let us see what happens.

When heated in the presence of 10% sodium hydroxide, acetaldehyde reacts to form β-hydroxybutyraldehyde, a reaction called the *aldol condensation.*

$$2CH_3CH=O \xrightarrow[\text{Heat}]{\text{NaOH(aq)}} CH_3\overset{\overset{OH}{|}}{C}HCH_2CH=O$$

ACETALDEHYDE → β-HYDROXYBUTYRALDEHYDE (ALDOL)

In general:

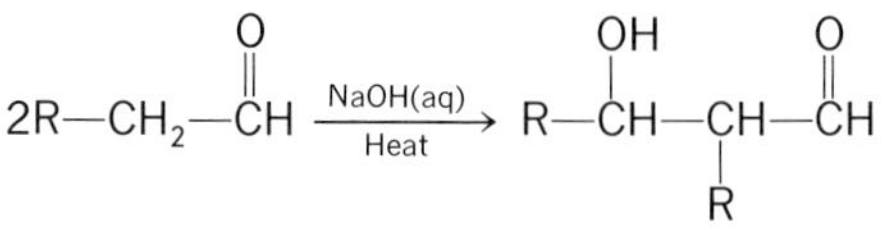

$$2R{-}CH_2{-}\overset{O}{\overset{\|}{C}}H \xrightarrow[\text{Heat}]{\text{NaOH(aq)}} R{-}\overset{OH}{\overset{|}{C}}H{-}\underset{R}{\underset{|}{C}}H{-}\overset{O}{\overset{\|}{C}}H$$

This reaction is illustrative of a large number that have the following feature in common:

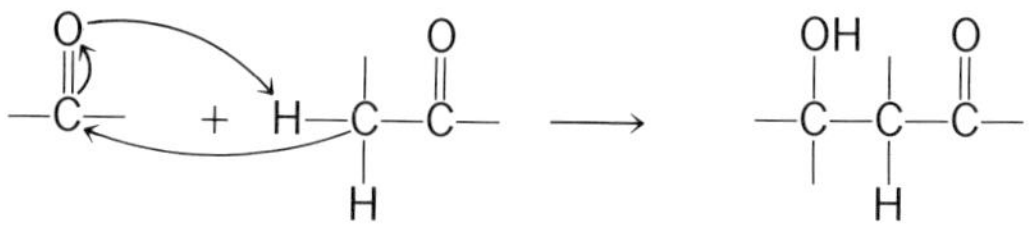

$$-\overset{O}{\overset{\|}{C}}- \;+\; H{-}\underset{H}{\underset{|}{\overset{|}{C}}}{-}\overset{O}{\overset{\|}{C}}- \longrightarrow -\underset{|}{\overset{OH}{\overset{|}{C}}}{-}\underset{H}{\underset{|}{\overset{|}{C}}}{-}\overset{O}{\overset{\|}{C}}-$$

CARBONYL COMPOUND + CARBONYL COMPOUND WITH TWO α-HYDROGENS → A β-HYDROXY CARBONYL COMPOUND

In words, one aldehyde adds across the carbonyl group of another. Most other examples of this type of reaction continue beyond the β-hydroxy carbonyl stage. Dehydration occurs and an unsaturated carbonyl compound forms:

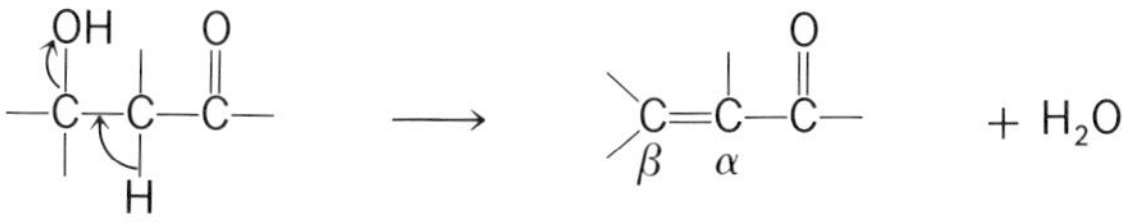

$$-\underset{|}{\overset{OH}{\overset{|}{C}}}{-}\underset{H}{\underset{|}{\overset{|}{C}}}{-}\overset{O}{\overset{\|}{C}}- \longrightarrow \;>\underset{\beta}{C}{=}\underset{\alpha}{\overset{|}{C}}{-}\overset{O}{\overset{\|}{C}}- \quad + H_2O$$

A β-HYDROXY CARBONYL COMPOUND → AN α,β-UNSATURATED CARBONYL COMPOUND

Before examining specific examples, let us first see how the reaction occurs.

9.25 General Mechanism for Aldol-Type Condensations

Step 1. An acid-base equilibrium between the base and a molecule of carbonyl compound is formed. The base takes a proton from the α-position and only from the α-position.

In general:

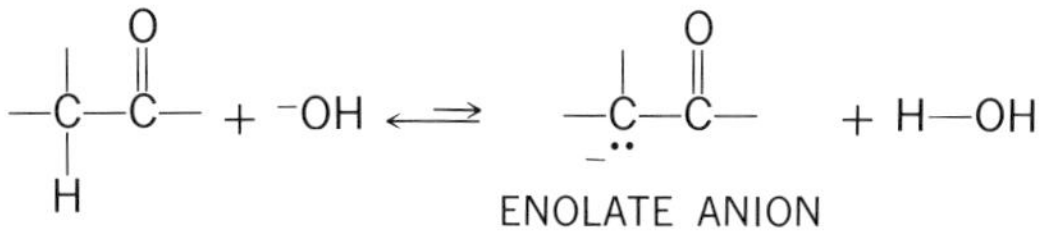

$$-\underset{H}{\underset{|}{\overset{|}{C}}}{-}\overset{O}{\overset{\|}{C}}- + {}^{-}OH \rightleftharpoons -\underset{-\,\cdot\cdot}{\overset{|}{C}}{-}\overset{O}{\overset{\|}{C}}- \quad + H{-}OH$$

ENOLATE ANION

Example:

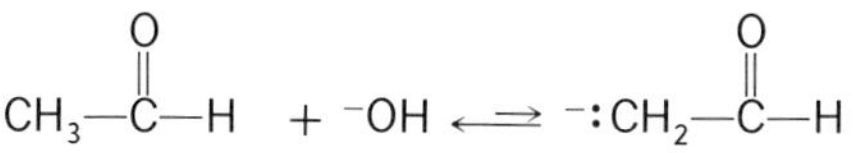

$$\underset{\text{ACETALDEHYDE}}{CH_3{-}\overset{O}{\overset{\|}{C}}{-}H} + {}^{-}OH \rightleftharpoons {}^{-}{:}CH_2{-}\overset{O}{\overset{\|}{C}}{-}H$$

Step 2. The enolate anion attacks the carbonyl carbon of another carbonyl-containing molecule.

In general:

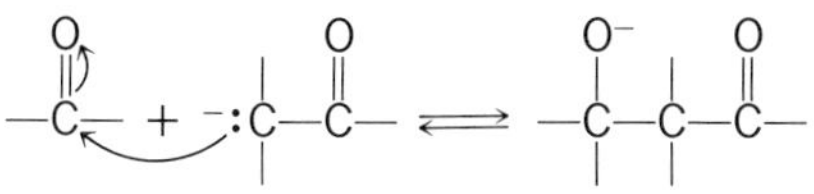

$$-\overset{O}{\overset{\|}{C}}- + {}^{-}{:}\underset{|}{\overset{|}{C}}{-}\overset{O}{\overset{\|}{C}}- \rightleftharpoons -\underset{|}{\overset{O^-}{\overset{|}{C}}}{-}\underset{|}{\overset{|}{C}}{-}\overset{O}{\overset{\|}{C}}-$$

A NEW ANION

$$\text{Example: } CH_3-\overset{\overset{O}{\|}}{C}-H + {}^{-}\!:CH_2-\overset{\overset{O}{\|}}{C}-H \rightleftharpoons \underset{\text{AN ALKOXIDE ION}}{CH_3\overset{\overset{O^-}{|}}{C}H-CH_2-\overset{\overset{O}{\|}}{C}-H}$$

Step 3. An acid-base equilibration between the alkoxide ion formed in step 2 and a proton donor (such as water).

$$\text{In general: } -\overset{\overset{O^-}{|}}{\underset{|}{C}}-\underset{|}{C}-\overset{\overset{O}{\|}}{C}- + H-OH \rightleftharpoons -\overset{\overset{OH}{|}}{\underset{|}{C}}-\underset{|}{C}-\overset{\overset{O}{\|}}{C}- + \underset{\text{(RECOVERED CATALYST)}}{{}^{-}OH}$$

$$\text{Example: } CH_3-\overset{\overset{O^-}{|}}{C}H-CH_2-\overset{\overset{O}{\|}}{C}-H + H-OH \rightleftharpoons \underset{\substack{\beta\text{-HYDROXYBUTYRALDEHYDE} \\ \text{(ALDOL)}}}{CH_3-\overset{\overset{OH}{|}}{C}H-CH_2-\overset{\overset{O}{\|}}{C}-H} + {}^{-}OH$$

Aldol is a specific name for β-hydroxybutyraldehyde and a general name for β-hydroxy aldehydes. Aldols are easily dehydrated. In some cases dehydration cannot be prevented; it occurs spontaneously. Aldols from simple, aliphatic aldehydes generally need an acid catalyst for their dehydration.

$$\text{In general: } -\overset{\overset{OH}{|}}{\underset{|}{C}}-\overset{|}{\underset{\underset{H}{|}}{C}}-\overset{\overset{O}{\|}}{C}- \longrightarrow \underset{\substack{\text{AN } \alpha,\beta\text{-UNSATURATED} \\ \text{CARBONYL COMPOUND}}}{-\overset{|}{C}=\overset{|}{C}-\overset{\overset{O}{\|}}{C}-} + H_2O$$

$$\text{Example: } CH_3-\overset{\overset{OH}{|}}{C}H-CH_2-\overset{\overset{O}{\|}}{C}-H \xrightarrow[\Delta]{H^+} CH_3-CH=CH-\overset{\overset{O}{\|}}{C}-H + H_2O$$

9.26 Crossed Aldol Condensations

If we heat a mixture of two different aldehydes in the presence of aqueous sodium hydroxide, four products form in a "crossed" aldol condensation. "Crossing over" inevitably occurs in the random interactions. For example, with a mixture of acetaldehyde and propionaldehyde in the presence of a base, two anions are possible:

$$\underset{\text{ACETALDEHYDE}}{CH_3-\overset{\overset{O}{\|}}{C}-H} + {}^{-}OH \rightleftharpoons {}^{-}\!:CH_2-\overset{\overset{O}{\|}}{C}-H$$

$$\underset{\text{PROPIONALDEHYDE}}{CH_3-CH_2-\overset{\overset{O}{\|}}{C}-H} + {}^{-}OH \rightleftharpoons {}^{-}\!:\overset{\overset{CH_3}{|}}{C}H-\overset{\overset{O}{\|}}{C}-H$$

Note that the event occurs at the α-carbon.

These two anions form in the presence of molecules of both of the unchanged aldehydes. A carbonyl carbon is a carbonyl carbon, and the anions will attack any accidentally near them:

$$CH_3-\overset{O}{\overset{\|}{C}}-H \xleftarrow{(a)} \bar{:}CH_2-\overset{O}{\overset{\|}{C}}-H \xrightarrow[\text{2 and 3}]{\text{(a) steps}} CH_3\overset{OH}{\overset{|}{C}}H-CH_2\overset{O}{\overset{\|}{C}}-H$$

β-HYDROXYBUTYRALDEHYDE

or

$$CH_3CH_2\overset{O}{\overset{\|}{C}}-H \xleftarrow{(b)} \bar{:}CH_2-\overset{O}{\overset{\|}{C}}-H \xrightarrow{(b)} CH_3CH_2\overset{OH}{\overset{|}{C}}H-CH_2\overset{O}{\overset{\|}{C}}-H$$

β-HYDROXYVALERALDEHYDE

$$CH_3-\overset{O}{\overset{\|}{C}}-H \xleftarrow{(c)} \bar{:}\overset{CH_3}{\overset{|}{C}}H-\overset{O}{\overset{\|}{C}}-H \xrightarrow{(c)} CH_3\overset{OH}{\overset{|}{C}}H-\overset{CH_3}{\overset{|}{C}}H-\overset{O}{\overset{\|}{C}}-H$$

β-HYDROXY-α-METHYLBUTYRALDEHYDE

or

$$CH_3CH_2-\overset{O}{\overset{\|}{C}}-H \xleftarrow{(d)} \bar{:}\overset{CH_3}{\overset{|}{C}}H-\overset{O}{\overset{\|}{C}}-H \xrightarrow{(d)} CH_3CH_2\overset{OH}{\overset{|}{C}}H-\overset{CH_3}{\overset{|}{C}}H-\overset{O}{\overset{\|}{C}}-H$$

β-HYDROXY-α-METHYLVALERALDEHYDE

Thus, whenever *both* aldehydes have α-hydrogens, the crossed aldol gives a mixture of four aldols. They are difficult to separate, however, and the method is seldom useful in synthesis. The problem is somewhat simplified if only one of the aldehydes has an α-hydrogen. Then only one anion can form in the first step. By carefully regulating the experimental conditions, the statistical chances of this anion attacking the carbonyl of the *other* compound can be made great enough to render such a crossed aldol condensation useful. We do this by adding the one with an α-hydrogen *to* an alkaline mixture of the one without an α-hydrogen.

Examples of Crossed Aldol Condensations

In these examples we assume that the fourth step—loss of water—takes place, and usually it does. The net effect in each is the following:

$$>C{=}O + H_2\overset{R}{\overset{|}{C}}-\overset{O}{\overset{\|}{C}}-H \xrightarrow{\text{Base}} >C{=}\overset{R}{\overset{|}{C}}-\overset{O}{\overset{\|}{C}}-H$$

CARBONYL WITH NO α-HYDROGENS; CARBONYL WITH *TWO* α-HYDROGENS; α,β-UNSATURATED CARBONYL COMPOUND

$$C_6H_5-\overset{H}{\overset{|}{C}}{=}O + CH_3-\overset{O}{\overset{\|}{C}}-H \xrightarrow[20^\circ C]{OH^-} C_6H_5-CH{=}CH-\overset{O}{\overset{\|}{C}}-H$$

BENZALDEHYDE; ACETALDEHYDE; CINNAMALDEHYDE

$$C_6H_5-\overset{H}{\overset{|}{C}}{=}O + CH_3-\overset{O}{\overset{\|}{C}}-CH_3 \xrightarrow[100^\circ C]{OH^-} C_6H_5-CH{=}CH-\overset{O}{\overset{\|}{C}}-CH_3$$

ACETONE; BENZALACETONE

$$C_6H_5-\overset{H}{\overset{|}{C}}{=}O + CH_3-\overset{O}{\overset{\|}{C}}-C_6H_5 \xrightarrow[20^\circ C]{OH^-} C_6H_5-CH{=}CH-\overset{O}{\overset{\|}{C}}-C_6H_5$$

ACETOPHENONE; BENZALACETOPHENONE

$$C_6H_5{-}\overset{\displaystyle H}{\overset{|}{C}}{=}O + \underset{\displaystyle C_6H_5}{\underset{|}{CH_2}}{-}\overset{\displaystyle O}{\overset{\|}{C}}{-}H \xrightarrow[20^\circ C]{OH^-} C_6H_5{-}CH{=}\underset{\displaystyle C_6H_5}{\underset{|}{C}}{-}\overset{\displaystyle O}{\overset{\|}{C}}{-}H$$

PHENYLACETALDEHYDE — α-PHENYLCINNAMALDEHYDE

Exercise 9.11 Write the structure of the product of the aldol condensation involving each of the following aldehydes: (a) propionaldehyde, (b) butyraldehyde, and (c) phenylacetaldehyde ($C_6H_5CH_2CH{=}O$).

Exercise 9.12 Write the structure of the product of the crossed aldol condensation between the following pairs of compounds: (a) benzaldehyde and propionaldehyde, (b) benzaldehyde and butyraldehyde, and (c) benzaldehyde and $CH_3CH_2\overset{O}{\overset{\|}{C}}C_6H_5$ (propiophenone).

9.27 Important Aldehydes and Ketones

Formaldehyde

At room temperature formaldehyde is a gas with a very irritating and distinctive odor. In one industrial method it is prepared by passing a mixture of methyl alcohol and air over a silver catalyst at a temperature of over 600°C. The newly formed formaldehyde together with unchanged methanol is absorbed in water until the solution has a concentration of about 37% formaldehyde. In this form it is marketed as Formalin. Little free formaldehyde exists in Formalin. The hydrated form $CH_2(OH)_2$ (cf. Section 9.18), and a polymeric form, $HO(CH_2O)_nH$, with n having an average value of 3, are the major organic constituents.

Formaldehyde, as its Formalin solution, is used as a disinfectant and as a preservative for biological samples and specimens. The largest amounts of Formalin, however, are consumed in the manufacture of various resins. Bakelite, a phenol-formaldehyde plastic developed in a commercially useful way by L. H. Baekeland (see Figure 9.1), and melamine resins, made from melamine and formaldehyde, are the most familiar. Exceptional resistance to heat and water makes melamine resin a favored material for plastic dinnerware (see Figure 9.2). Formaldehyde is also used in the manufacture of high explosives such as cyclonite and pentaerythritol tetranitrate (PETN).

Acetaldehyde

Acetaldehyde is manufactured by a variety of processes: by the hydration of acetylene,

$$H{-}C{\equiv}C{-}H + H{-}OH \xrightarrow[Hg^{2+}]{H^+} \left[H{-}\overset{\displaystyle H}{\overset{|}{C}}{=}\overset{\displaystyle OH}{\overset{|}{C}}{-}H\right] \longrightarrow CH_3{-}\overset{\displaystyle O}{\overset{\|}{C}}{-}H$$

ACETYLENE — AN ENOL — ACETALDEHYDE

by the air oxidation of ethanol,

$$CH_3{-}CH_2{-}OH \xrightarrow[\text{Ag, heat}]{O_2} CH_3{-}\overset{\overset{\displaystyle O}{\|}}{C}{-}H$$

and by the controlled air oxidation of propane-butane fractions of natural gas. It is an important raw material for the manufacture of acetic acid, acetic anhydride, and ethyl acetate.

A tetramer of acetaldehyde, commercially known as *metaldehyde,* is sold as a pesticide to control garden slugs and snails.

Acetone

One of the most important organic solvents, acetone not only dissolves a wide variety of organic substances but is also miscible with water in all proportions.

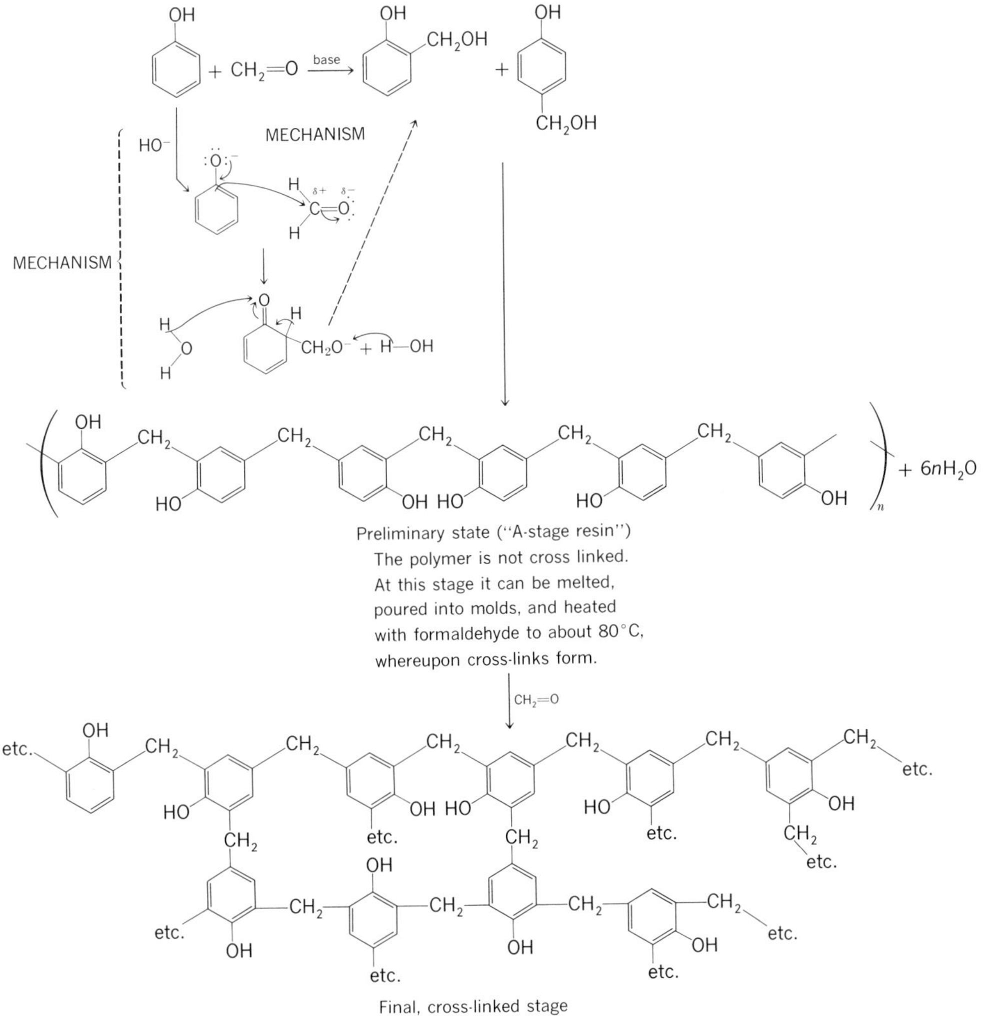

base
MECHANISM
MECHANISM
Preliminary state ("A-stage resin")
The polymer is not cross linked.
At this stage it can be melted, poured into molds, and heated with formaldehyde to about 80°C, whereupon cross-links form.
Final, cross-linked stage

Figure 9.1
Bakelite, a phenol-formaldehyde resin. Only the principal structural features are indicated here.

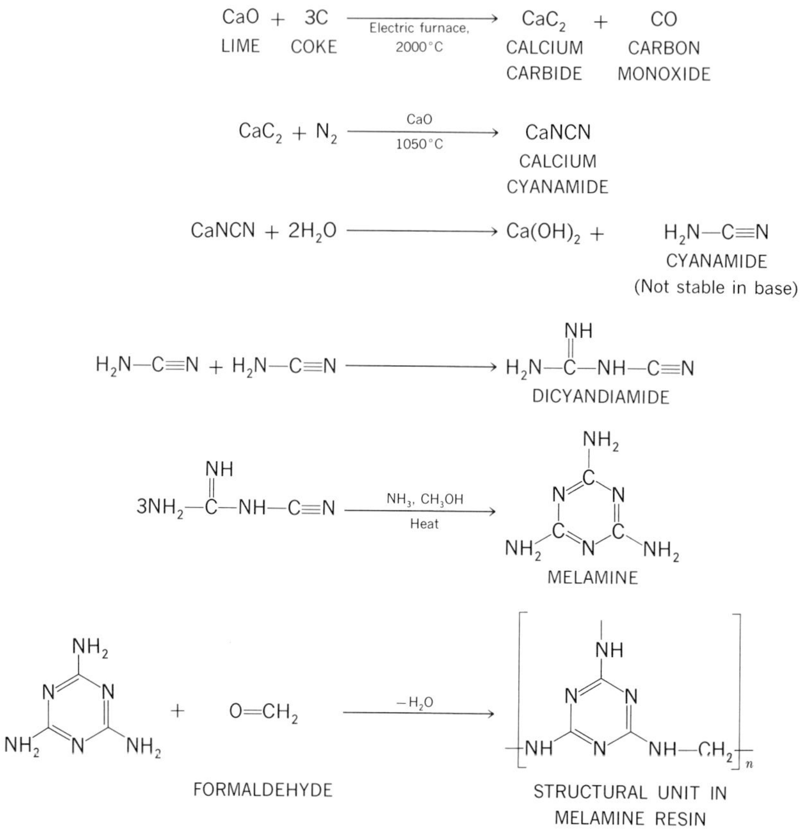

$$\underset{\text{LIME}}{CaO} + \underset{\text{COKE}}{3C} \xrightarrow[\text{2000°C}]{\text{Electric furnace,}} \underset{\text{CALCIUM CARBIDE}}{CaC_2} + \underset{\text{CARBON MONOXIDE}}{CO}$$

$$CaC_2 + N_2 \xrightarrow[\text{1050°C}]{\text{CaO}} \underset{\text{CALCIUM CYANAMIDE}}{CaNCN}$$

$$CaNCN + 2H_2O \longrightarrow Ca(OH)_2 + \underset{\text{CYANAMIDE (Not stable in base)}}{H_2N—C{\equiv}N}$$

$$H_2N—C{\equiv}N + H_2N—C{\equiv}N \longrightarrow \underset{\text{DICYANDIAMIDE}}{H_2N—C({=}NH)—NH—C{\equiv}N}$$

$$3NH_2—C({=}NH)—NH—C{\equiv}N \xrightarrow[\text{Heat}]{NH_3,\ CH_3OH} \text{MELAMINE}$$

$$\text{Melamine} + \underset{\text{FORMALDEHYDE}}{O{=}CH_2} \xrightarrow{-H_2O} \text{STRUCTURAL UNIT IN MELAMINE RESIN}$$

Figure 9.2
Melamine resin. Its manufacture illustrates one of several industrial processes that start with abundant, relatively inexpensive natural resources: lime from limestone, coke from coal and wood, nitrogen from air, energy from hydroelectric sources, and other materials.

Industrially, acetone is used as a solvent in processing cellulose acetate (for fiber), cellulose nitrate (an explosive), and acetylene (which would present great hazards shipped in the undiluted form). Acetone is made in many ways, including the air oxidation of isopropyl alcohol at elevated temperatures.

Glucose and Fructose

Glucose in one form is a pentahydroxy aldehyde, fructose a pentahydroxy ketone. Of obvious importance in nutrition and metabolism, they will be discussed more fully in Chapter 14.

Brief Summary: Chemical and Physical Properties

1. Aldehydes and ketones have molecules less polar than those of alcohols. Being able to accept hydrogen bonds from water, low-formula-weight aldehydes and ketones are soluble in water.

2. Aldehydes may be synthesized by the following:
   (a) oxidation of 1° alcohols using:
      1. A strong oxidizing agent (e.g., CrO_3) provided the aldehyde is sufficiently more volatile than the alcohol to be separated as it forms.
      2. Hot copper as a catalyst for removing H_2.
   (b) Gatterman-Koch reaction, for aromatic aldehydes from aromatic hydrocarbons (reagents: CO, HCl and $CuCl_2/AlCl_3$).
   (c) Oxo reaction, an industrial synthesis from alkenes (reagents: CO, H_2, catalyst).
   (d) Hydrolysis of 1,1-dihalocompounds, usually of the benzal halide type: $ArCHX_2$ (reagents: water and base).
   (e) Aldol condensation and crossed aldol condensations whereby we make bigger aldehydes from smaller ones.
3. Ketones may be synthesized by the following:
   (a) Oxidation of 2° alcohols. (For various ways, see 2a above.)
   (b) Friedel-Crafts acylation of aromatic systems (reagents: acyl halide and $AlCl_3$).
   (c) Hydrolysis of gem dihalides ($—CX_2—$) other than 1,1-dihalides (which hydrolyze to aldehydes).
   (d) Aldol-type condensations (and crossed aldols) to give (usually) α,β-unsaturated ketones (or aldehydes) depending on the choice of starting materials.

4. We may summarize chemical properties of aldehydes and ketones as follows. (The best way to drill yourself on these is to write specific equations illustrating each one.)

	Chemical Behavior	
Reagent System	Aldehydes	Ketones
Oxidation	Easily oxidized to carboxylic acids (cf. also Tollens', Benedict's, and Fehling's tests)	Much more stable than aldehydes; when oxidized, undergo fragmentation
Reduction	Change to 1° alcohols	Change to 2° alcohols
Water	No major reaction; unstable hydrates appear	No major reaction; essentially no hydrates form
Alcohols	Unstable hemiacetals, then (if H^+ is present) acetals	Unstable hemiketals then (if H^+ is present) ketals
Aqueous alkali	Aldol or crossed aldol condensations	Aldol or crossed aldol condensations
Halogen (in presence of H^+ or OH^-)	α-Haloaldehydes form	α-Haloketones form

Reagent System $:NH_2—G$ G =	Amine Derivatives: $>C=N—G$ Name of Derivative	Name of $NH_2—G$
—H or —R	Imines	Ammonia or amines
—OH	Oximes	Hydroxylamine
$—NH\overset{O}{\overset{\parallel}{C}}NH_2$	Semicarbazones	Semicarbazide
—NH—Ar	Phenylhydrazones (or related compounds)	Phenylhydrazine

5. Simple hemiacetals (or hemiketals) exist only in a mixture of the parent alcohol and aldehyde (or ketone).

6. Acetals and ketals are stable in water that is neutral or alkaline. They are hydrolyzed by action of dilute acids reverting back to the original alcohols and carbonyl compounds.

7. Aldehydes and ketones exist in the presence of traces of their tautomers, called enols.

Problems and Exercises

1. Write the structures for each of the following.
   - (a) 2,3-dimethylhexanal
   - (b) acetic anhydride
   - (c) chloroacetic acid
   - (d) ethyl propionate
   - (e) di-*n*-butyl ketone
   - (f) sodium isobutyrate
   - (g) α-methylbutyryl chloride
   - (h) diisopropyl ketone
   - (i) α,β-dibromopropionaldehyde
   - (j) *p*-bromobenzaldehyde
   - (k) *p*-chloroacetophenone
   - (l) acetone semicarbazone
2. If the common name for $CH_3CH_2CH_2CH_2CO_2H$ is valeric acid, write common and IUPAC names for the following.

   (a) $CH_3CH_2CH_2CH_2\overset{O}{\overset{\parallel}{C}}H$ (b) $CH_3O\overset{O}{\overset{\parallel}{C}}CH_2CH_2CH_2CH_3$

   (c) $Cl\overset{O}{\overset{\parallel}{C}}CH_2CH_2CH_2CH_3$ (d) $CH_3CH_2CH_2CH_2\overset{O}{\overset{\parallel}{C}}NH_2$

   (e) $CH_3CH_2CH_2CH_2\overset{O}{\overset{\parallel}{C}}O^-Na^+$ (f) $CH_3CH_2CH_2CH_2\overset{O}{\overset{\parallel}{C}}C_6H_5$

3. If the structure of benzoic acid is $C_6H_5—\overset{O}{\overset{\parallel}{C}}—OH$ (benzene ring drawn), write the structures of:

   (a) benzoyl chloride (b) benzaldehyde (c) phenyl benzoate
   (d) ethyl *p*-nitrobenzoate (e) benzamide (f) benzoic anhydride

4. Write common names for each of the following.

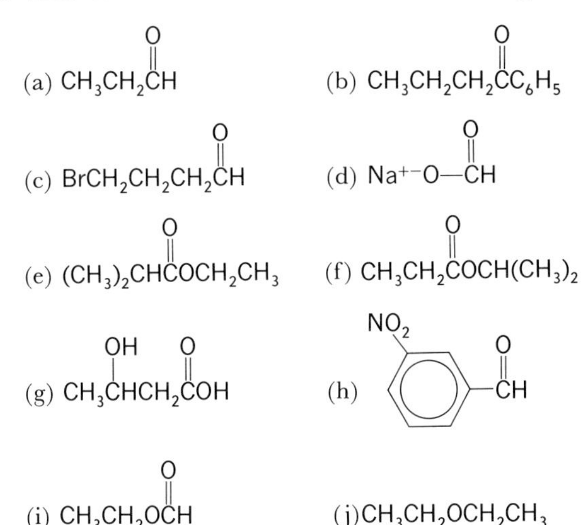

5. Write the structure(s) of the principal organic product(s) that might reasonably be expected to form if *acetaldehyde* were subjected to each of the following reagents and conditions.

(a) H_2, Ni, pressure, heat
(b) excess CH_3OH, dry HCl
(c) CH_3OH (as a solvent)
(d) NH_2OH
(e) 10% NaOH, heat
(f) $NH_2NH\overset{O}{\overset{\|}{C}}NH_2$
(g) Tollens' reagent
(h) CrO_3, H^+
(i) excess I_2, NaOH
(j) $NH_2NHC_6H_5$

6. Repeat question 3 using *acetone* instead of acetaldehyde.

7. Write the structures of the principal organic products that would form in each of the following reactions.

(a) $(CH_3)_2CHOH \xrightarrow[H^+]{CrO_3}$

(b) $C_6H_6 + CH_3\overset{O}{\overset{\|}{C}}Cl \xrightarrow{AlCl_3}$

(c) $CH_3\overset{O}{\overset{\|}{C}}H + H_2 \xrightarrow{\text{Catalyst}}$

(d) $CH_3CH_2CH_2CH_2OH \xrightarrow[\text{Heat}]{Cu}$

(e) 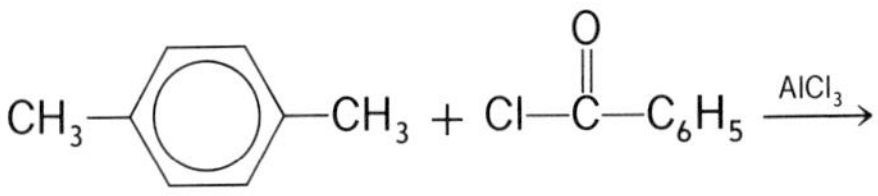 $CH_3-C_6H_4-CH_3 + Cl-\overset{O}{\overset{\|}{C}}-C_6H_5 \xrightarrow{AlCl_3}$ (para-xylene, drawn as a benzene ring)

(f) $C_6H_5-\underset{OH}{\underset{|}{C}H}-C_6H_5 \xrightarrow{CrO_3}$

(g) $CH_3-\overset{O}{\overset{\|}{C}}-C_6H_5 \xrightarrow[\text{NaOH, heat}]{I_2}$

(h) $C_6H_5-\overset{O}{\overset{\|}{C}}-H + NH_2OH \xrightarrow{H^+}$

(i) $C_6H_5-\overset{O}{\overset{\|}{C}}-H + CH_3-\overset{O}{\overset{\|}{C}}-H \xrightarrow{OH^-}$

(j) $CH_3\overset{O}{\overset{\|}{C}}CH_3 + NH_2NH\overset{O}{\overset{\|}{C}}NH_2 \xrightarrow{H^+}$

8. Which of the following would be expected to give a positive Benedict or Fehling test?

(a) $CH_3\underset{OH}{\underset{|}{C}}H\overset{O}{\overset{\|}{C}}H$ (b) $CH_3\overset{O}{\overset{\|}{C}}CH_2\overset{O}{\overset{\|}{C}}CH_3$ (c) $CH_3\overset{O}{\overset{\|}{C}}-\overset{O}{\overset{\|}{C}}H$ (d) $CH_3\underset{OH}{\underset{|}{C}}HCH_2CH_2\overset{O}{\overset{\|}{C}}H$

(e) $CH_3\underset{OH}{\underset{|}{C}}HCH_2CH_2\overset{O}{\overset{\|}{C}}CH_3$ (f) $CH_3CH_2\underset{OH}{\underset{|}{C}}H\overset{O}{\overset{\|}{C}}CH_3$

9. Shown below is the structure of a hypothetical molecule.

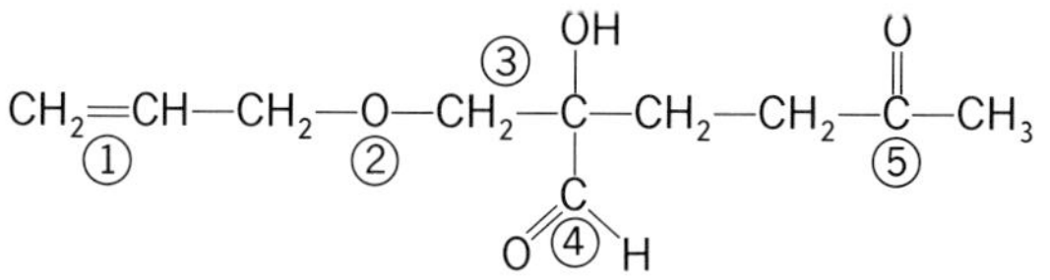

$$\underset{①}{CH_2{=}CH}-CH_2-\underset{②}{O}-CH_2-\overset{OH}{\overset{|}{\underset{\underset{O{=}\overset{④}{C}-H}{|}}{\overset{③}{C}}}}-CH_2-CH_2-\underset{⑤}{\overset{O}{\overset{\|}{C}}}-CH_3$$

Predict the chemical reactions it would probably undergo if it were subject to the reagents and conditions listed. To simplify the writing of the equations illustrating these reactions, isolate the portion of the molecule that would be involved, replace the other noninvolved portions by symbols such as *G*, *G′*, *G″*, etc., and then write the equation (see example). You are expected to make reasonable predictions based only on the reactions we have studied thus far. If no reaction is to be predicted, write "None."

Example. Reagent: excess H_2, Ni, heat, pressure. We approach this problem in the following steps.

(a) First identify *by name* the functional groups in the molecule. (By doing this you take advantage of the way you mentally store information about the reactions of functional groups. Thus the *name* for $-CH{=}O$ is aldehyde or aldehyde group. In your mental "file" on this group there should be such statements as "Aldehydes can be reduced to 1° alcohols" or "Aldehydes can be easily oxidized to acids," etc.)

In the structure given, you should recognize ① alkene, ② ether, ③ 3° alcohol, ④ aldehyde, ③ and ④ α-hydroxy aldehyde, ⑤ ketone.

(b) To continue the example, ask yourself how each group would respond to the reagent given.

1. Alkenes will add hydrogen catalytically under heat and pressure.
2. Ethers do not react with hydrogen.
3. There is no reaction of 3° alcohols with hydrogen.
4. Aldehydes can be reduced to 1° alcohols.
5. Ketones can be reduced to 2° alcohols.

(c) Third, write the equations. For example,

$$CH_2{=}CH{-}G + H_2 \xrightarrow[\text{Heat, pressure}]{\text{Ni}} CH_3{-}CH_2{-}G$$

where *G* in this reaction is

$$-CH_2OCH_2\overset{\overset{\displaystyle CH{=}O}{|}}{\underset{\underset{\displaystyle OH}{|}}{C}}CH_2CH_2\overset{\overset{\displaystyle O}{\|}}{C}CH_3$$

$$G'{-}\overset{\overset{\displaystyle O}{\|}}{C}H + H_2 \xrightarrow[\text{Heat, pressure}]{\text{Ni}} G'{-}CH_2OH \quad \text{(What is } G' \text{ in this reaction?)}$$

$$G''{-}\overset{\overset{\displaystyle O}{\|}}{C}CH_3 + H_2 \xrightarrow[\text{Heat, pressure}]{\text{Ni}} G''{-}\overset{\overset{\displaystyle OH}{|}}{C}HCH_3$$

The reagents and conditions for this exercise are:

1. Br_2 in CCl_4 solution, room temperature.
2. Excess CH_3OH in the presence of dry HCl.
3. Tollens' reagent.
4. Benedict's reagent (do not attempt to write an equation).
5. Phenylhydrazine.
6. Dilute sodium hydroxide at room temperature.

10. Repeat the directions given in exercise 6 for the following hypothetical structure and the reagents given.

$$CH_3{-}O{-}\overset{\overset{\displaystyle O{-}CH_3}{|}}{C}H{-}CH_2{-}O{-}CH_2CH_2\overset{\overset{\displaystyle O}{\|}}{C}H$$

(a) Tollens' reagent (which is a basic solution).
(b) Excess water, H^+ catalyst.
(c) MnO_4^-, OH^-.

chapter ten

Organic Acids and Their Derivatives

The two major families of organic acids are the carboxylic acids and the sulfonic acids. Both can be converted to corresponding salts, acid chlorides, esters and amides, and these can be changed back to the parent acids. See Table 10.1.

10.1 Carboxylic Acids

Some typical carboxylic acids are shown in Table 10.2. They are commonly called *fatty acids* because many can be obtained by the hydrolysis of animal fats (or vegetable oils).

Formic acid is a sharp-smelling, irritating liquid responsible for the sting of certain ants and the nettle.

Acetic acid is what gives tartness to vinegar, where its concentration is 4–5%. Because pure acetic acid (mp 16.6°C or 63°F) congeals to an icelike solid in a cool room, it is often called *glacial acetic acid.* Over 1.2 million tons of acetic acid were manufactured in 1973 in the United States. It is used principally to make cellulose acetates, polyvinyl acetates, and simple acetate esters used as solvents or flavoring agents. One method for making it begins with calcium carbide, which is hydrolyzed to acetylene (Section 3.35). Water is added to acetylene to produce acetaldehyde, which is air-oxidized to acetic acid.

$$\underset{\text{CALCIUM CARBIDE}}{CaC_2} \xrightarrow{H_2O} \underset{\text{ACETYLENE}}{HC{\equiv}CH} \xrightarrow[(Hg^{2+},\ H^+)]{H_2O} \underset{\text{AN ENOL}}{[CH_2{=}CHOH]} \longrightarrow \underset{\text{ACETALDEHYDE}}{CH_3\overset{\overset{O}{\|}}{C}H} \xrightarrow{\text{air}} \underset{\text{ACETIC ACID}}{CH_3\overset{\overset{O}{\|}}{C}OH}$$

Butyric acid causes the odor of rancid butter. What is "strong" about valeric acid (Latin *valerum,* to be strong) is its odor. The same is true of caproic, caprylic and capric acids, named from the Latin, *caper,* meaning "goat."

Table 10.1
The Two Major Organic Acids and Their Derivatives

	Carboxylic Acid		Sulfonic Acid
	$-\overset{O}{\overset{\|\|}{C}}-O-H$		$-\overset{O}{\overset{\uparrow}{\underset{\underset{O}{\downarrow}}{S}}}-OH$
Carboxylate ion	$-\overset{O}{\overset{\|\|}{C}}-O^-$	Sulfonate ion	$-\overset{O}{\overset{\uparrow}{\underset{\underset{O}{\downarrow}}{S}}}-O^-$
Acyl chloride	$-\overset{O}{\overset{\|\|}{C}}-Cl$	Sulfonyl chloride	$-\overset{O}{\overset{\uparrow}{\underset{\underset{O}{\downarrow}}{S}}}-Cl$
Ester	$-\overset{O}{\overset{\|\|}{C}}-O-\overset{\|}{\underset{\|}{C}}-$	Sulfonate ester	$-\overset{O}{\overset{\uparrow}{\underset{\underset{O}{\downarrow}}{S}}}-O-\overset{\|}{\underset{\|}{C}}-$
Amide	$-\overset{O}{\overset{\|\|}{C}}-N\langle$	Sulfonamide	$-\overset{O}{\overset{\uparrow}{\underset{\underset{O}{\downarrow}}{S}}}-N\langle$

$$HO-\overset{O}{\overset{\|\|}{C}}-\overset{O}{\overset{\|\|}{C}}-OH \qquad HO-\overset{CH_2CO_2H}{\overset{|}{\underset{\underset{CH_2CO_2H}{|}}{C}}}-CO_2H \qquad CH_3\underset{\underset{OH}{|}}{C}HCO_2H$$

OXALIC ACID CITRIC ACID LACTIC ACID

Oxalic acid gives the tart flavor to rhubarb. Because it is both an acid and a reducing agent, aqueous solutions of oxalic acid will remove stains caused by inks based on iron compounds.

The tart taste of citrus fruits is caused by citric acid. Its salt, sodium citrate, is sometimes added to drawn, whole blood to keep it from clotting. So important is citric acid in metabolism that one major biochemical sequence is called the "citric acid cycle."

Lactic acid gives the tart taste to sour milk. During periods of strenuous exercise, glucose units in the body are changed to lactic acid, and this acid is believed to be responsible for the soreness of muscles.

10.2 Syntheses of Carboxylic Acids

Oxidation of 1° Alcohols

$RCH_2OH \xrightarrow{(O)} RCO_2H$ (O) = Any strong oxidizing agent (e.g., CrO_3, MnO_4^-) (Discussed in Sec. 6.29)

Oxidation of Aldehydes

$$\mathrm{RCHO} \xrightarrow{(O)} \mathrm{RCO_2H} \qquad (O) = \text{Any moderate or strong oxidizing agent}$$

We studied this in Section 9.12.

These two methods leave both the length and the basic arrangement of the carbon skeleton unchanged. The next two methods increase the number of carbons by one.

The Grignard Synthesis

In general:

$$\mathrm{R{-}X + Mg} \xrightarrow[\text{Step 1}]{\text{Ether}} \mathrm{R{-}Mg{-}X} \xrightarrow[\text{Step 2}]{CO_2} \mathrm{R{-}\overset{\overset{\large O}{\|}}{C}{-}O^{-}Mg^{2+}X^{-}} \xrightarrow[\text{Step 3}]{H^+} \mathrm{R{-}\overset{\overset{\large O}{\|}}{C}{-}O{-}H}$$

This reaction can be presented in another way:

$$\mathrm{R{-}X} \xrightarrow[\text{3. H}^+ \text{ (e.g., HCl)}]{\text{1. Mg, ether} \atop \text{2. } CO_2} \mathrm{R{-}CO_2H}$$

The three steps are carried out successively in the same reaction vessel. Specific examples are:

$$\underset{\text{ETHYL BROMIDE}}{\mathrm{CH_3CH_2Br}} \xrightarrow[\text{3. H}^+]{\text{1. Mg, ether} \atop \text{2. } CO_2 \text{ at } -20°\text{C}} \underset{\text{PROPIONIC ACID (72\%)}}{\mathrm{CH_3CH_2CO_2H}}$$

$$\underset{t\text{-BUTYL CHLORIDE}}{\mathrm{(CH_3)_3CCl}} \xrightarrow[\text{3. H}^+]{\text{1. Mg, ether} \atop \text{2. } CO_2 \text{ at } 0°\text{C}} \underset{\text{2,2-DIMETHYLPROPANOIC ACID (70\%)}}{\mathrm{(CH_3)_3CCO_2H}}$$

$$\underset{\text{BROMOBENZENE}}{\mathrm{C_6H_5Br}} \xrightarrow[\text{3. H}^+]{\text{1. Mg, ether} \atop \text{2. } CO_2 \text{ (cold)}} \underset{\text{BENZOIC ACID (90\%)}}{\mathrm{C_6H_5CO_2H}}$$

Alkyl halides are usually obtained from corresponding alcohols (Section 6.23); aryl halides are made either by direct halogenation of a benzene ring or in a Sandmeyer reaction (Section 8.15).

The carbonation of a Grignard reagent is analogous to its addition to any carbonyl group:

$$\underset{\text{APPROXIMATION OF THE GRIGNARD REAGENT}}{\mathrm{R{:}^{-}Mg^{2+}X^{-}}} + \mathrm{O^{\delta-}{=}C^{\delta+}{=}O^{\delta-}} \longrightarrow \underset{\text{MIXED SALT OF THE ACID}}{\mathrm{R{-}\overset{\overset{\large O}{\|}}{C}{-}O^{-}Mg^{2+}X^{-}}} \xrightarrow{H^+} \mathrm{R{-}\overset{\overset{\large O}{\|}}{C}{-}O{-}H}$$

Table 10.2

Carboxylic Acids

	n	Structure	Name	Origin of Name	Mp (°C)	Bp (°C)	Solubility (in g/100 g water at 20°C)	K_a (at 25°C)
Straight-chain saturated acids $C_nH_{2n}O_2$	1	HCO_2H	Formic acid (methanoic)	L. *formica*, ant	8	101	∞	1.77×10^{-4} (20°C)
	2	CH_3CO_2H	Acetic acid (ethanoic)	L. *acetum*, vinegar	17	118	∞	1.76×10^{-5}
	3	$CH_3CH_2CO_2H$	Propionic acid (propanoic)	Gr. *proto*, first *pion*, fat	−21	141	∞	1.34×10^{-5}
	4	$CH_3CH_2CH_2CO_2H$	Butyric acid (butanoic)	L. *butyrum*, butter	−6	164	∞	1.54×10^{-5}
	5	$CH_3(CH_2)_3CO_2H$	Valeric acid (pentanoic)	L. *valere*, to be strong (valerian root)	−35	186	4.97	1.52×10^{-5}
	6	$CH_3(CH_2)_4CO_2H$	Caproic acid (hexanoic)	L. *caper*, goat	−3	205	1.08	1.31×10^{-5}
	7	$CH_3(CH_2)_5CO_2H$	Enanthic acid (heptanoic)	Gr. *oenanthe*, vine blossom	−9	223	0.26	1.28×10^{-5}
	8	$CH_3(CH_2)_6CO_2H$	Caprylic acid (octanoic)	L. *caper*, goat	16	238	0.07	1.28×10^{-5}
	9	$CH_3(CH_2)_7CO_2H$	Pelargonic acid (nonanoic)	pelargonium (geranium family)	15	254	0.03	1.09×10^{-5}
	10	$CH_3(CH_2)_8CO_2H$	Capric acid (decanoic)	L. *caper*, goat	32	270	0.015	1.43×10^{-5}
	12	$CH_3(CH_2)_{10}CO_2H$	Lauric acid (dodecanoic)	Laurel	44	—	0.006	—
	14	$CH_3(CH_2)_{12}CO_2H$	Myristic acid (tetradecanoic)	Myristica (nutmeg)	54	—	0.002	—
	16	$CH_3(CH_2)_{14}CO_2H$	Palmitic acid (hexadecanoic)	Palm oil	63	—	0.0007	—
	18	$CH_3(CH_2)_{16}CO_2H$	Stearic acid (octadecanoic)	Gr. *stear*, tallow	70	—	0.0003	—
Miscellaneous carboxylic acids		$C_6H_5CO_2H$	Benzoic acid	Gum benzoin	122	249	0.34 (25°C)	6.46×10^{-5}
		$C_6H_5CH{=}CHCO_2H$	Cinnamic acid (*trans*)	Cinnamon	42	—	0.04	3.65×10^{-5}
		$CH_2{=}CHCO_2H$	Acrylic acid	L. *acer*, sharp	13	141	soluble	5.6×10^{-5}
		(benzene ring) —CO_2H, OH	Salicylic acid	L. *salix*, willow	159	211	0.22 (25°C)	1.1×10^{-3} (19°C)

Hydrolysis of Nitriles

In general:

$$\underset{\text{A NITRILE}}{R{-}C{\equiv}N} \xrightarrow[\text{Heat}]{H_2O,\ H^+} R{-}\overset{O}{\overset{\|}{C}}{-}OH\ (+NH_4^+)$$

The reaction is also promoted by alkali, in which case the initial product is $R{-}CO_2^-$.

Specific examples:

$$\underset{\substack{\text{VALERONITRILE}\\ \text{(PENTANENITRILE)}}}{CH_3CH_2CH_2CH_2C{\equiv}N} + NaOH(aq) \xrightarrow[\text{followed by } H^+]{H_2O} \underset{\substack{\text{VALERIC ACID (81\%)}\\ \text{(PENTANOIC ACID)}}}{CH_3CH_2CH_2CH_2\overset{O}{\overset{\|}{C}}OH} + NH_3$$

$$\underset{\substack{\text{PHENYLACETONITRILE}\\ \text{(PHENYLETHANENITRILE)}}}{C_6H_5CH_2C{\equiv}N} + H_2O \xrightarrow{H^+} \underset{\substack{\text{PHENYLACETIC ACID (78\%)}\\ \text{(PHENYLETHANOIC ACID)}}}{C_6H_5CH_2\overset{O}{\overset{\|}{C}}OH} + NH_4Cl$$

Nitriles are made by the action of cyanide ion on an alkyl halide (generally 1° and sometimes 2° halides; 3° halides are changed to alkenes). See Section 7.8. Nitriles of aromatic acids are made from aromatic amines which are diazotized and treated with copper(I) cyanide. See Section 8.15.

Naming Nitriles

They are given common names after the acid to which they can be hydrolyzed. For example:

CH_3CN	CH_3CH_2CN	$(CH_3)_2CHCN$	C_6H_5CN
ACETONITRILE	PROPIONITRILE	ISOBUTYRONITRILE	BENZONITRILE
(ETHANENITRILE)	(PROPANENITRILE)	2-METHYLPROPANENITRILE	

For the IUPAC name, change the name of the alkane having exactly the same number of carbons as the nitrile by simply adding the suffix "-nitrile." In complicated cases, the —CN group is treated as a substituent with the name "cyano."

Exercise 10.1 Outline the synthesis that converts the given starting material into the acid specified. Some syntheses involve more than one step.

(a) Ethanol into acetic acid.
(b) Ethanol into propionic acid.
(c) *t*-Butyl alcohol into 2,2-dimethylpropanoic acid.
(d) Aniline into benzoic acid.
(e) Isopropyl chloride into isobutyric acid.

Hydrolysis of Other Acid Derivatives

These are discussed in separate sections and are listed here solely for reference.

Salts of acids (10.7) $R{-}CO_2^-Na^+ \xrightarrow{H^+} (-Na^+)$

Acid chlorides (10.16) $R{-}C(=O){-}Cl \xrightarrow{H_2O} (-HCl)$

Acid anhydrides (10.16) $R{-}C(=O){-}O{-}C(=O){-}R \xrightarrow{H_2O} R{-}C(=O){-}OH$

Esters (10.18) $R{-}C(=O){-}O{-}R' \xrightarrow{H_2O} (-R'OH)$

Amides (10.20) $R{-}C(=O){-}NH_2 \xrightarrow{H_2O} (-NH_3)$

(All of these give $R{-}C(=O){-}OH$.)

10.3 Physical Properties of Carboxylic Acids

The carboxyl group confers considerable polarity to molecules possessing it. A carboxylic acid has a higher boiling point than an alcohol of the same formula weight, as these data show:

CH_3CH_2OH (f. wt. 46), bp 78°C $CH_3CH_2CH_2OH$ (f. wt. 60), bp 97°C

HCO_2H (f. wt. 46), bp 101°C CH_3CO_2H (f. wt. 60), bp 118°C

The reason is that pairs of carboxylic acid molecules can be held together by two hydrogen bonds:

$$R{-}C(=O^{\delta-}\cdots H^{\delta+}{-}O){-}O{-}H^{\delta+}\cdots O^{\delta-}{=}C{-}R$$

Because a carboxylic acid has these dimeric forms (Greek: *di-*, two; *meros,* part), its effective formula weight is higher than the structural formula implies. Its boiling point is thus higher than we would otherwise expect. Other hydrogen-bonded polymers are also present.

The first four members of the homologous series of carboxylic acids are soluble in water. The carboxyl group can both donate and accept hydrogen bonds to and from water molecules. As the hydrocarbon chain in the acid lengthens, the solubility of the acid in water falls off sharply.

10.4 Acidity of Carboxylic Acids

In an aqueous medium molecules of carboxylic acids interact with water molecules. In general:

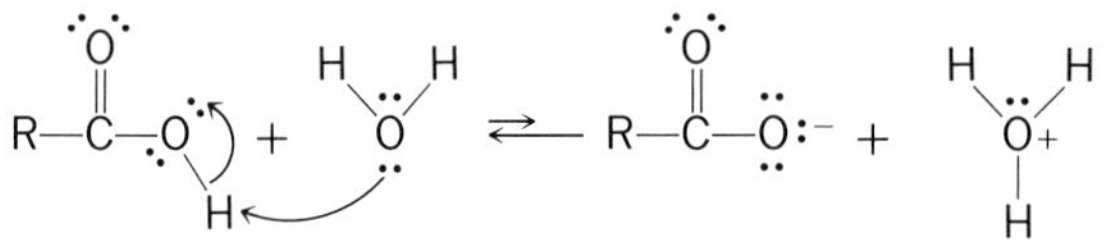

$$R-C(=O)-O-H + H_2O \rightleftharpoons R-C(=O)-O^- + H_3O^+$$

Weaker acid | Weaker base | Stronger base | Stronger acid

A 1-molar solution of acetic acid is ionized only to about 0.5% at room temperature. Evidently water molecules are not very strong proton acceptors in relation to carboxylic acids, which are relatively weak proton donors. Compared with alcohols, however, which also have the hydroxyl group,[1] carboxylic acids are stronger acids by several orders of magnitude.

Acidity Constants

Relative acidities of acids can be studied by comparing the acidity constants K_a for their reaction with water:

$$K_a = \frac{[RCO_2^-][H_3O^+]}{[RCO_2H]}$$

The higher the K_a, the stronger the acid. In Tables 10.2 and 10.3 several such values of K_a, most on the order of 10^{-5} (0.00001), are recorded. In marked contrast, 1° alcohols have K_a values on the order of 10^{-16}. To understand this great difference in acidity for two classes possessing hydroxyl groups, we must look to structural differences.

10.5 Structure and Acidity

The ionization of an alcohol molecule produces an alkoxide ion. A carboxylate ion is produced from an acid.

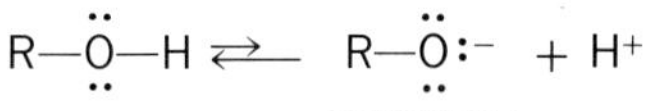

$$R-O-H \rightleftharpoons R-O^- + H^+$$

ALKOXIDE ION

$$R-C(=O)-O-H \rightleftharpoons R-C(=O)-O^- + H^+$$

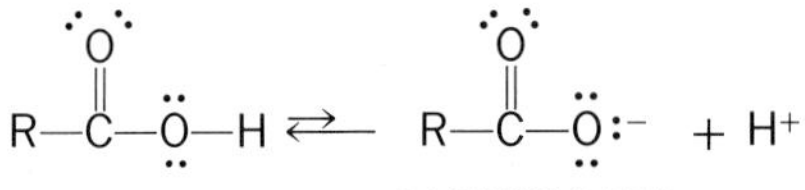

CARBOXYLATE ION

The classical structure for an alkoxide ion adequately represents it. We cannot draw any other reasonable resonance structures for it.

[1]Here, as always, hydroxyl *group* is not to be confused with hydroxide *ion*.

Table 10.3
Structure and Acidity of Substituted Carboxylic Acids

	Structure	K_a (25°C)
α-Haloacetic acids	$F—CH_2CO_2H$	219×10^{-5}
	$Cl—CH_2CO_2H$	155×10^{-5}
	$Br—CH_2CO_2H$	138×10^{-5}
	$I—CH_2CO_2H$	75×10^{-5}
	$H—CH_2CO_2H$	1.8×10^{-5}
Chloroacetic acids	$H—CH_2CO_2H$	1.8×10^{-5}
	$Cl—CH_2CO_2H$	155×10^{-5}
	Cl_2CHCO_2H	$5{,}100 \times 10^{-5}$
	Cl_3CCO_2H	$90{,}000 \times 10^{-5}$
Monochlorobutyric acids acids	α-chloro $CH_3CH_2CH(Cl)CO_2H$	139×10^{-5}
	β-chloro $CH_3CH(Cl)CH_2CO_2H$	8.9×10^{-5}
	γ-chloro $CH_2(Cl)CH_2CH_2CO_2H$	3.8×10^{-5}
	no chloro $CH_3CH_2CH_2CO_2H$	1.54×10^{-5}

For the carboxylate ion, on the other hand, two reasonable structures—differing only in locations of electrons—can be drawn, **1** and **2**, and the ion is therefore stabilized by resonance.

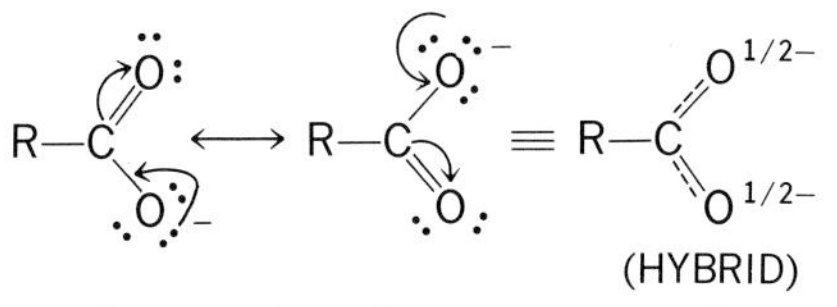

$$R—C(=O)O^- \longleftrightarrow R—C(O^-)=O \equiv R—C(O^{1/2-})(O^{1/2-})$$

1   **2**   (HYBRID) **3**

The net effect is that, compared with the molecule from which it forms, the carboxylate ion is more stable than an alkoxide ion. This means that the energy change for the ionization of an acid is more favorable than that for ionization of an alcohol. Therefore the carboxylic acid is a stronger acid than the alcohol.

10.6 The Inductive Effect and Acidity

The K_a values for α-haloacetic acids given in Table 10.3 reveal that as the halogen substituent changes from —I to —Br to —Cl to —F, the acidity *increases*. This order is the same as the order of the relative electronegativities of the halogens. We have learned that one way to stabilize a charge is to disperse it. Electron-withdrawing substituents such as halogens have this kind of inductive effect. Hence, their presence on the α-carbon of acetic acid makes the acid stronger.

$$X \leftarrow C \genfrac{}{}{0pt}{}{\overset{\ \ }{\diagup\!\!\diagup} O^{1/2-}}{\diagdown\!\!\diagdown O^{1/2-}} \qquad X \rightarrow C \genfrac{}{}{0pt}{}{\diagup\!\!\diagup O^{1/2-}}{\diagdown\!\!\diagdown O^{1/2-}}$$

If X is a group that withdraws electrons, the anion is stabilized further; hence the acid is a stronger acid.

If X tends to release electrons the anion is destabilized; negative charge is being forced into a region already negatively charged, making the acid weaker.

As more and more electron-withdrawing substituents are placed on the α-carbon of acetic acid, its acidity rises sharply. The K_a values for mono-, di-, and trichloroacetic acids (Table 10.3) reveal this. Trichloroacetic acid is almost as strong as a mineral acid and is sometimes used in its place as an acid catalyst.

As the substituent exerting an inductive effect becomes farther and farther removed from the negatively charged end of the carboxylate ion, its inductive influence on that group drops off markedly. The K_a values for the three monochlorobutyric acids in Table 10.3 illustrate this.

10.7 Salts of Carboxylic Acids

The K_a value of an acid is a measure of the ability of the carboxylic acid group to donate a proton to a water molecule. In this role the neutral water molecule acts as a base, a proton acceptor, and a very weak one at that.

In contrast, the hydroxide ion is a far stronger base. Aqueous sodium hydroxide, for example, can quantitatively convert carboxylic acids into their salts. In general:

$$\underset{\text{Stronger acid}}{R\text{—}\overset{O}{\overset{\|}{C}}\text{—O—H}} + \underset{\text{Stronger base}}{Na^+OH^-} \rightleftharpoons \underset{\text{Weaker base}}{R\text{—}\overset{O}{\overset{\|}{C}}\text{—}O^-Na^+} + \underset{\text{Weaker acid}}{H\text{—}OH}$$

Specific examples:

$$\underset{\text{ACETIC ACID}}{CH_3\text{—}\overset{O}{\overset{\|}{C}}\text{—O—H}} + \underset{\text{SODIUM HYDROXIDE}}{Na^+OH^-} \rightleftharpoons \underset{\text{SODIUM ACETATE}}{CH_3\text{—}\overset{O}{\overset{\|}{C}}\text{—}O^-Na^+} + \underset{\text{WATER}}{H\text{—}OH}$$

$$\underset{\substack{\text{STEARIC ACID}\\ \text{(Insoluble in water)}}}{CH_3(CH_2)_{16}\overset{O}{\overset{\|}{C}}\text{—O—H}} + Na^+OH^- \rightleftharpoons \underset{\substack{\text{SODIUM STEARATE}\\ \text{(Soluble in water, a soap)}}}{CH_3(CH_2)_{16}\overset{O}{\overset{\|}{C}}\text{—}O^-Na^+} + H\text{—}OH$$

$$\underset{\substack{\text{BENZOIC ACID} \\ \text{(Insoluble in water)}}}{C_6H_5\overset{\overset{\displaystyle O}{\|}}{C}—O—H} + Na^+OH^- \rightleftharpoons \underset{\substack{\text{SODIUM BENZOATE} \\ \text{(Soluble in water)}}}{C_6H_5\overset{\overset{\displaystyle O}{\|}}{C}—O^-Na^+} + H—OH$$

The salts that form in these reactions are isolated by evaporating the water. All are crystalline solids with relatively high melting points; many decompose before they melt. They are fully ionic substances with the properties to be expected of such materials: solubility in water, insolubility in nonpolar solvents, and relatively high melting points. Table 10.4 lists a few representative examples.

Since they form from substances that are not good proton donors, carboxylate ions (RCO_2^-) must be good proton acceptors. In other words, they are quite good bases, especially toward a good proton donor such as hydronium ion:

$$\underset{\substack{\text{CARBOXYLATE} \\ \text{ION} \\ \text{(Stronger} \\ \text{base)}}}{R—\overset{\overset{\displaystyle O}{\|}}{C}—O^-} + \underset{\substack{\text{HYDRONIUM} \\ \text{ION} \\ \text{(Stronger} \\ \text{acid)}}}{H—\overset{+}{O}H_2} \rightleftharpoons \underset{\substack{\text{CARBOXYLIC} \\ \text{ACID} \\ \text{(Weaker} \\ \text{acid)}}}{R—\overset{\overset{\displaystyle O}{\|}}{C}—O—H} + \underset{\substack{\text{WATER} \\ \\ \text{(Weaker} \\ \text{base)}}}{H—OH}$$

For example,

$$\underset{\substack{\text{SODIUM BENZOATE} \\ \text{(Soluble in water)}}}{C_6H_5CO_2^-Na^+} + \underset{\substack{\text{HYDROCHLORIC} \\ \text{ACID}}}{HCl} \xrightarrow{\text{Water}} \underset{\substack{\text{BENZOIC ACID} \\ \text{(Insoluble in water)}}}{C_6H_5CO_2H} + \underset{\text{SODIUM CHLORIDE}}{Na^+Cl^-}$$

Table 10.4
Salts of Carboxylic Acids

Name	Structure	Mp (°C)	Solubility	
			Water	Ether
Sodium formate	$HCO_2^-Na^+$	253	Soluble	Insoluble
Sodium acetate	$CH_3CO_2^-Na^+$	323	Soluble	Insoluble
Sodium propionate	$CH_3CH_2CO_2^-Na^+$	—	Soluble	Insoluble
Sodium benzoate	$C_6H_5CO_2^-Na^+$	—	66 g/100 cc	Insoluble
Sodium salicylate	$o\text{-}HOC_6H_4CO_2^-Na^+$ (benzene ring with $CO_2^-Na^+$ and OH)	—	111 g/100 cc	Insoluble

10.8 Reactions of Carboxylic Acids

The changes of the carboxyl group of principal concern in our study are of the type:

$$R\text{—}\overset{\displaystyle O}{\overset{\|}{C}}\text{—}OH \xrightarrow[\text{Reagent}]{} R\text{—}\overset{\displaystyle O}{\overset{\|}{C}}\text{—}G$$

where G is —Cl, —OR′, —NH_2, —NHR′, or —NHR'_2. By action of some reagent the —OH group may be replaced by one of these groups. We shall study these transformations next.

10.9 Synthesis of Acid Chlorides

Action of any one of three reagents—thionyl chloride ($SOCl_2$), phosphorus trichloride (PCl_3), or phosphorus pentachloride (PCl_5)—on a carboxylic acid converts it to its acid chloride. In general:

$$R\text{—}\overset{\displaystyle O}{\overset{\|}{C}}\text{—}OH \xrightarrow[PCl_5]{SOCl_2 \text{ or } PCl_3 \text{ or}} R\text{—}\overset{\displaystyle O}{\overset{\|}{C}}\text{—}Cl$$

Specific examples:

$$CH_3\overset{\displaystyle O}{\overset{\|}{C}}\text{—}OH \xrightarrow{PCl_3} CH_3\overset{\displaystyle O}{\overset{\|}{C}}\text{—}Cl \quad (+\ H_3PO_3)$$

ACETIC ACID → ACETYL CHLORIDE (67%)

$$C_6H_5\overset{\displaystyle O}{\overset{\|}{C}}\text{—}OH \xrightarrow{SOCl_2} C_6H_5\overset{\displaystyle O}{\overset{\|}{C}}\text{—}Cl \quad (+\ SO_2 + HCl)$$

BENZOIC ACID → BENZOYL CHLORIDE (91%)

$$CH_3(CH_2)_6\overset{\displaystyle O}{\overset{\|}{C}}\text{—}OH \xrightarrow{PCl_5} CH_3(CH_2)_6\overset{\displaystyle O}{\overset{\|}{C}}\text{—}Cl \quad (+\ POCl_3 + HCl)$$

CAPRYLIC ACID → CAPRYLYL CHLORIDE (82%)

10.10 Synthesis of Acid Anhydrides

The most important aliphatic anhydride is acetic anhydride, and one of the ways it may be prepared is by the reaction of acetic acid with acetyl chloride:

$$CH_3\overset{\displaystyle O}{\overset{\|}{C}}\text{—}OH + Cl\text{—}\overset{\displaystyle O}{\overset{\|}{C}}CH_3 \xrightarrow{\text{Pyridine}} CH_3\overset{\displaystyle O}{\overset{\|}{C}}\text{—}O\text{—}\overset{\displaystyle O}{\overset{\|}{C}}CH_3 + \text{Pyridine}\cdot HCl$$

ACETIC ANHYDRIDE

This method is quite general.

10.11 Synthesis of Esters

Esters may be made in a variety of ways. Direct esterification is presented here. In general:

$$R-\overset{O}{\overset{\|}{C}}-OH + H-O-R' \overset{H^+}{\rightleftharpoons} R-\overset{O}{\overset{\|}{C}}-O-R' + H-OH$$

Specific examples:

$$\underset{\text{ACETIC ACID}}{CH_3\overset{O}{\overset{\|}{C}}OH} + \underset{\text{METHYL ALCOHOL}}{HOCH_3} \underset{\text{Heat}}{\overset{H^+}{\rightleftharpoons}} \underset{\text{METHYL ACETATE}}{CH_3\overset{O}{\overset{\|}{C}}OCH_3} + H_2O$$

$$\underset{\text{PROPIONIC ACID}}{CH_3CH_2\overset{O}{\overset{\|}{C}}OH} + \underset{\text{ETHYL ALCOHOL}}{HOCH_2CH_3} \underset{\text{Heat}}{\overset{H^+}{\rightleftharpoons}} \underset{\text{ETHYL PROPIONATE}}{CH_3CH_2\overset{O}{\overset{\|}{C}}OCH_2CH_3} + H_2O$$

$$\underset{\text{BENZOIC ACID}}{C_6H_5\overset{O}{\overset{\|}{C}}OH} + \underset{\text{METHYL ALCOHOL}}{CH_3OH} \underset{\text{Heat}}{\overset{H^+}{\rightleftharpoons}} \underset{\text{METHYL BENZOATE}}{C_6H_5\overset{O}{\overset{\|}{C}}OCH_3} + H_2O$$

When a straight-chain acid and a 1° alcohol are mixed in a 1:1 mole ratio and heated, about one-third mole percent of acid and alcohol is still present at the eventual equilibrium. About two-thirds mole percent changes to ester and water. The equilibrium may be shifted in favor of the ester by using a large excess either of alcohol (usually the cheaper starting material) or of acid or by removing the water (or ester) as it forms.

10.12 Mechanism of Direct Esterification

If methanol enriched in ^{18}O is allowed to react with benzoic acid, the ^{18}O appears in the ester, methyl benzoate, instead of in the water:

$$C_6H_5C(=O){\not-}O-H + H{\not-}{}^{18}O-CH_3 \xrightarrow{H^+} C_6H_5C(=O)-{}^{18}O-CH_3 + H-OH$$

This means that in direct esterification it is the C—OH bond of the acid and the H—O bond in the alcohol that break.

A mineral acid is normally used to catalyze the reaction. It protonates the carboxyl group, which increases the size of the positive charge on the carbonyl carbon. That makes this carbon more attractive to nucleophilic attack by an alcohol molecule.

$$R-\overset{\delta+}{C}(=\overset{\delta-}{\ddot{O}}:)-\ddot{O}H + \underset{\text{CATALYST}}{H^+} \underset{}{\overset{\text{Protonation}}{\rightleftharpoons}} \left[R-\overset{+}{C}(-\ddot{O}-H)-\ddot{O}-H\right]$$

Partial plus, mildly reactive toward a nucleophile (i.e., an electron-rich reagent)

Full plus, much more reactive toward a nucleophile

After this protonation of the acid, a succession of reactions takes place, all involving equilibria.

$$R{-}C(=\ddot{O})OH + H^+ \underset{\text{Protonation}}{\rightleftharpoons} \left[R{-}\overset{+}{C}(\ddot{O}H)_2\right] + H{-}\ddot{O}{-}R' \underset{\substack{\text{Attack by}\\ \text{nucleophile,}\\ R\ddot{O}H}}{\rightleftharpoons}$$

$$\left[R{-}C(:\ddot{O}H)_2{-}\overset{+}{O}(H){-}R'\right] \underset{\substack{\text{Proton}\\ \text{shift}}}{\rightleftharpoons} \left[R{-}C(\ddot{O}H)(\ddot{O}{-}R'){-}\overset{+}{O}H_2\right] \xrightarrow{\text{Ejection of } H_2O \text{ and } H^+} R{-}C(=\ddot{O})\ddot{O}{-}R' + H_2\ddot{O} + H^+$$

An ester can be converted to its original acid and alcohol by action of excess water in the presence of a strong acid catalyst. The mechanism for the hydrolysis of an ester is the exact reverse of esterification.

Exercise 10.2 Write the structure and the common name of the ester that could be prepared from each of the following pairs of compounds.

(a) Acetic acid and ethyl alcohol.
(b) Benzoic acid and isopropyl alcohol.
(c) Isobutyric acid and propyl alcohol.
(d) Propionic acid and methyl alcohol.

10.13 Synthesis of Amides

Under appropriate conditions ammonia, 1° amines, or 2° amines react with a carboxylic acid and water splits out.

$$\underset{\text{CARBOXYLIC ACID}}{R{-}\overset{O}{\overset{\|}{C}}{-}OH} + \underset{NH_3,\ NH_2R',\ \text{or } NHR'_2}{H{-}N\langle} \underset{\Delta}{\rightleftharpoons} \underset{\text{AMIDE}}{R{-}\overset{O}{\overset{\|}{C}}{-}N\langle} + H{-}OH$$

Example:

$$\underset{\text{ACETIC ACID}}{CH_3CO_2H} + \underset{\text{AMMONIA}}{NH_3} \rightleftharpoons \underset{\text{AMMONIUM ACETATE}}{CH_3CO_2^-NH_4^+} \xrightarrow[\text{acetic acid, } 110°C]{\text{Excess}} \underset{\substack{\text{ACETAMIDE}\\ (90\%)}}{CH_3\overset{O}{\overset{\|}{C}}NH_2} + H_2O$$

This method of preparing amides is not often used in the laboratory. Usually the acid is first converted to its acid chloride, which is then allowed to react with ammonia (or 1° or 2° amines). The enhanced reactivity of the acid chloride makes up for the inconvenience of using two steps instead of one.

Example:

$$CH_3CO_2H \xrightarrow{SOCl_2} CH_3COCl \xrightarrow{NH_3} CH_3\overset{O}{\overset{\|}{C}}NH_2\ (+\ NH_4Cl)$$

Exercise 10.3 Write the structure and common name of the amide that would be expected to form (if any) from the following pairs of compounds.

(a) Acetic acid and methylamine.
(b) Butyric acid and diethylamine.
(c) Benzoic acid and aniline.
(d) Propionic acid and trimethylamine.

10.14 Summary

The principal general reactions of carboxylic acids are the following:

$$R{-}\overset{O}{\overset{\|}{C}}{-}OH \xrightleftharpoons{H_2O} R{-}\overset{O}{\overset{\|}{C}}{-}O^- + H_3O^+$$ Small percentage ionization in water (weak acids)

$$R{-}\overset{O}{\overset{\|}{C}}{-}OH \xrightarrow{M^+OH^-} R{-}\overset{O}{\overset{\|}{C}}{-}O^-M^+ + H_2O$$ Formation of metallic salts (M = a metal) This is simply acid-base neutralization.

$$R{-}\overset{O}{\overset{\|}{C}}{-}OH \xrightarrow[\text{or } PCl_5]{SOCl_2,\text{ or } PCl_3,} R{-}\overset{O}{\overset{\|}{C}}{-}Cl$$ Formation of acid chloride

$$R{-}\overset{O}{\overset{\|}{C}}{-}OH \xrightleftharpoons[H^+]{H{-}O{-}R'} R{-}\overset{O}{\overset{\|}{C}}{-}O{-}R' + H_2O$$ Esterification

$$R{-}\overset{O}{\overset{\|}{C}}{-}OH \xrightleftharpoons{H{-}N\langle\quad\text{Heat}} R{-}\overset{O}{\overset{\|}{C}}{-}N\langle + H_2O$$ Formation of amide

10.15 Derivatives of Carboxylic Acids. General Principles

Much of the chemistry of acid derivatives consists of converting one into another or into the parent acid. Even though we must deal with several types of compounds, their reactions are so similar that it would be best to note these general patterns first. The common theme running throughout the chemistry of these substances is the following kind of nucleophilic substitution.

$$R{-}C\begin{matrix}\nearrow O\\ \searrow G\end{matrix} + :Z \longrightarrow R{-}C\begin{matrix}\nearrow O\\ \searrow Z\end{matrix} + :G$$

where G = —OH (as in carboxylic acids)
—Cl (as in acid chlorides)

—O—CO—R (as in an acid anhydride)
—O—R′ (as in an ester)
—NH_2 (or —NHR or —NR_2, as in amides)
:*Z* = a nucleophile that may be a neutral species or negatively charged (e.g., OH^- or H_2O, ^-OR or HOR, NH_3, etc.)

These are all substitution reactions at the *unsaturated* carbon of the carbonyl group. They usually go much more readily than analogous substitutions at a saturated carbon (as in S_N2 events of alkyl halides or alcohols.) One reason is the fact that the carbonyl carbon goes from an essentially trivalent state to a temporarily tetravalent state as the intermediate forms:

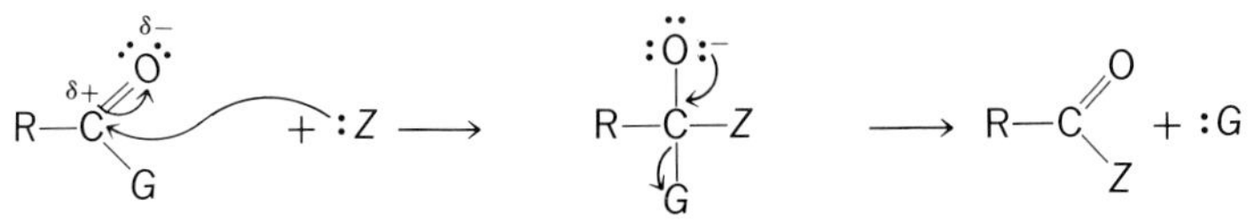

R—C(=O)—G + :Z ⟶ R—C(—O:⁻)(—Z)—G ⟶ R—C(=O)—Z + :G

Carbonyl carbon is trivalent in the sense that it is holding only three groups.

The former carbonyl carbon is tetravalent.

It is restored to its trivalent state.

In contrast, S_N2 substitution at a saturated carbon requires a shift from a tetravalent carbon to one that is temporarily pentavalent (holding five groups):

G—R + :Z ⟶ [G···R···Z] ⟶ G: + R—Z

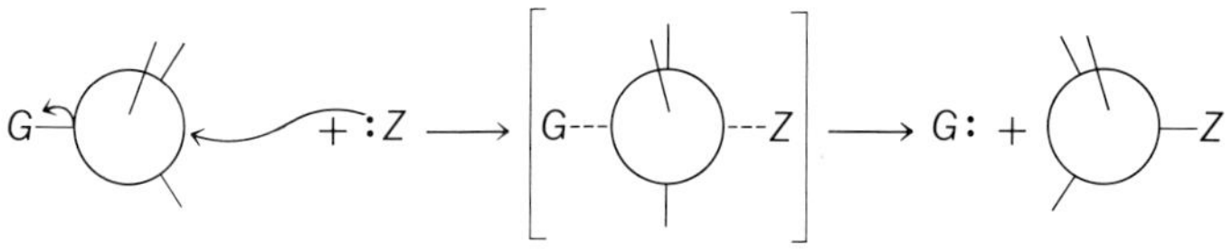

G—C + :Z ⟶ [G---C---Z] ⟶ G: + C—Z

The carbon holding —G is tetravalent.

In the transition state the carbon is holding five groups.

The severe crowding of groups gives this transition state a higher internal energy than would be involved when a nucleophile attacks a carbonyl group.

There is another factor making acid derivatives R—C(=O)—*G* more reactive than their counterparts in saturated systems, R—*G*. In acid derivatives, *two* groups exert electron withdrawal from the carbonyl carbon. In R—*G*, only the group, *G*, acts this way. Hence the relative size of the partial positive charge on a carbonyl carbon is usually greater than that on the carbon holding *G* in R—*G*.

If the reactions of acid derivatives are catalyzed by mineral acids, and many are, the carbonyl carbon is made even more positively charged:

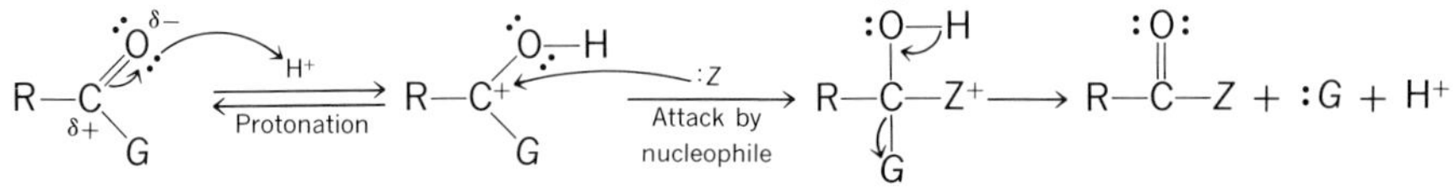

R—C(=O)—G ⇌ (H⁺, Protonation) R—C⁺(—O—H)—G ⟶ (:Z, Attack by nucleophile) R—C(—OH)(—Z⁺)—G ⟶ R—C(=O)—Z + :G + H⁺

Nucleophilic substitution at a carbonyl carbon therefore has two advantages over S_N2 events to alkyl halides. The carbonyl carbon is more wide open to attack, and it is more attractive to the attacking species—the nucleophile.

Within the series of acid derivatives, relative reactivities toward a common nucleophile (e.g., water) are generally in the following order:

Acid chlorides $\gtrsim$ acid anhydrides > esters > amides
MOST REACTIVE — LEAST REACTIVE

——— Decreasing reactivity ———→

10.16 Acid Chlorides and Anhydrides

Some examples of these substances are given in Table 10.5. Since the reactions of these two types of derivatives are so similar, they will be considered together. The principal reactions are outlined in the following equations.

Hydrolysis.
Conversion to Original Acid

$$R-\overset{O}{\overset{\|}{C}}-Cl + H_2O \longrightarrow R-\overset{O}{\overset{\|}{C}}-OH + H-Cl$$

ACID CHLORIDE

$$R-\overset{O}{\overset{\|}{C}}-O-\overset{O}{\overset{\|}{C}}-R + H_2O \longrightarrow R-\overset{O}{\overset{\|}{C}}-OH + HO-\overset{O}{\overset{\|}{C}}-R$$

ANHYDRIDE

Table 10.5
Acid Chlorides and Anhydrides

Name	Structure	Bp (°C)
Acetyl chloride	CH_3COCl	51
Propionyl chloride	CH_3CH_2COCl	80
Benzoyl chloride	C_6H_5COCl	197
Acetic anhydride	$CH_3\overset{O}{\overset{\|}{C}}-O-\overset{O}{\overset{\|}{C}}CH_3$	136
Benzoic anhydride	$C_6H_5\overset{O}{\overset{\|}{C}}-O-\overset{O}{\overset{\|}{C}}C_6H_5$	42 (mp)
Phthalic anhydride	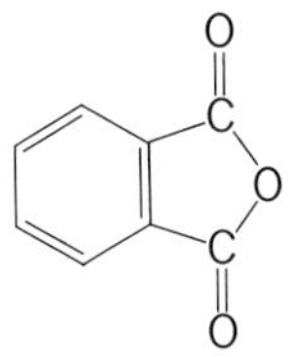 (benzene ring fused to a five-membered ring: C(=O)–O–C(=O))	131 (mp)

Alcoholysis. Ester Formation

$$R{-}\overset{O}{\overset{\|}{C}}{-}Cl + H{-}O{-}R' \longrightarrow \underset{\text{ESTER}}{R{-}\overset{O}{\overset{\|}{C}}{-}O{-}R'} + H{-}Cl$$

$$R{-}\overset{O}{\overset{\|}{C}}{-}O{-}\overset{O}{\overset{\|}{C}}{-}R + H{-}O{-}R' \longrightarrow \underset{\text{ESTER}}{R{-}\overset{O}{\overset{\|}{C}}{-}O{-}R'} + H{-}O{-}\overset{O}{\overset{\|}{C}}{-}R$$

Ammonolysis. Amide Formation

$$R{-}\overset{O}{\overset{\|}{C}}{-}Cl + \underset{\text{(excess)}}{H{-}N\begin{matrix}H(R)\\H(R')\end{matrix}} \longrightarrow \underset{\text{AMIDE}}{R{-}\overset{O}{\overset{\|}{C}}{-}N\begin{matrix}H(R)\\H(R')\end{matrix}} + HCl$$ (as its salt with the excess ammonia or amine)

$$R{-}\overset{O}{\overset{\|}{C}}{-}O{-}\overset{O}{\overset{\|}{C}}{-}R + \underset{\text{(excess)}}{H{-}N\begin{matrix}H(R)\\H(R')\end{matrix}} \longrightarrow \underset{\text{AMIDE}}{R{-}\overset{O}{\overset{\|}{C}}{-}N\begin{matrix}H(R)\\H(R')\end{matrix}} + HO\overset{O}{\overset{\|}{C}}R$$ (as a salt; see previous reaction)

Exercise 10.4 Complete the following equations by writing the structures of the products that would be expected to form in each. If no reaction is to be expected, write "No reaction."

(a) $CH_3\overset{O}{\overset{\|}{C}}{-}Cl + H_2O \longrightarrow$

(b) $C_6H_5\overset{O}{\overset{\|}{C}}{-}Cl + CH_3OH \longrightarrow$

(c) $CH_3\overset{O}{\overset{\|}{C}}{-}O{-}\overset{O}{\overset{\|}{C}}CH_3 + CH_3CH_2OH \longrightarrow$

(d) $CH_3CH_2\overset{O}{\overset{\|}{C}}Cl + NH_3$ (excess) $\longrightarrow$

(e) $C_6H_5CH_2\overset{O}{\overset{\|}{C}}{-}Cl + (CH_3)_2NH$ (excess) $\longrightarrow$

(f) $CH_3\overset{O}{\overset{\|}{C}}{-}Cl + (CH_3)_3N \longrightarrow$

Exercise 10.5 Write a mechanism for the reaction of water with an acid chloride, for the reaction of an alcohol with an anhydride, and for the reaction of ammonia with an acid chloride.

10.17 Anhydrides of Phosphoric Acid

The reactions of acid chlorides and anhydrides are quite exothermic, and those shown occur readily. The counterpart of anhydride chemistry of particular importance in biochemistry involves derivatives of phosphoric acid.

If a molecule of water splits out between two molecules of phosphoric acid, H_3PO_4, a substance known as pyrophosphoric acid, $H_4P_2O_7$, results.

$$HO-P(=O)(OH)-O-H + H-O-P(=O)(OH)-OH \xrightarrow{215^\circ C} HO-P(=O)(OH)-O-P(=O)(OH)-OH + H_2O$$

PHOSPHORIC ACID — PYROPHOSPHORIC ACID (mp 61 °C)

This is an acid in which all four hydrogens are replaceable and, in addition, pyrophosphoric acid is also very much like an anhydride.

$$-C(=O)-O-C(=O)- \qquad -P(=O)(-)-O-P(=O)(-)-$$

SKELETON OF A CARBOXYLIC ACID ANHYDRIDE — SKELETON OF A PHOSPHORIC ACID ANHYDRIDE

The system occurs widely in living cells in the form of compounds of the following general type. They are shown in the acid forms, but usually they are present as singly or doubly ionized particles.

$$RO-P(=O)(OH)-O-P(=O)(OH)-OH$$

MONOESTER OF PYROPHOSPHORIC ACID (A diphosphate)[2]

For example: $(CH_3)_2C{=}CHCH_2O-P(=O)(OH)-O-P(=O)(OH)-OH$

ISOPENTENYL PYROPHOSPHATE

Pyrophosphate esters are simultaneously esters, acids, and anhydrides. As anhydrides they participate in typical reactions of these compounds, for example, hydrolysis:

$$RO-P(=O)(OH)-O-P(=O)(OH)-OH + H_2O \longrightarrow RO-P(=O)(OH)-OH + HO-P(=O)(OH)-OH$$

DIPHOSPHATE ESTER — MONOPHOSPHATE ESTER — PHOSPHORIC ACID

[2]Organic chemists and biochemists frequently refer to these compounds as diphosphates. Note carefully that in this and similar contexts the term does *not* mean two separate PO_4^{3-} groups.

Triphosphate derivatives are also known and are essentially double anhydrides, plus being esters as well as acids.

$$RO-\overset{\overset{O}{\|}}{\underset{\underset{OH}{|}}{P}}-O-\overset{\overset{O}{\|}}{\underset{\underset{OH}{|}}{P}}-O-\overset{\overset{O}{\|}}{\underset{\underset{OH}{|}}{P}}-OH$$

A TRIPHOSPHATE ESTER

10.18 Esters of Carboxylic Acids

Physical Properties

Since molecules of an ester cannot be hydrogen bond donors, they behave as less polar compounds than carboxylic acids with respect to solubilities and boiling points. The ethyl esters of the fatty acids shown in Table 10.6 all boil at lower

Table 10.6
Esters of Carboxylic Acids

	Name	Structure	Mp (°C)	Bp (°C)	Solubility (in g/100 g water at 20°C)
Ethyl esters of straight chain carboxylic acids, $RCO_2C_2H_5$	Ethyl formate	$HCO_2C_2H_5$	−79	54	Soluble
	Ethyl acetate	$CH_3CO_2C_2H_5$	−82	77	7.39 (25°C)
	Ethyl propionate	$CH_3CH_2CO_2C_2H_5$	−73	99	1.75
	Ethyl butyrate	$CH_3(CH_2)_2CO_2C_2H_5$	−93	120	0.51
	Ethyl valerate	$CH_3(CH_2)_3CO_2C_2H_5$	−91	145	0.22
	Ethyl caproate	$CH_3(CH_2)_4CO_2C_2H_5$	−68	168	0.063
	Ethyl enanthate	$CH_3(CH_2)_5CO_2C_2H_5$	−66	189	0.030
	Ethyl caprylate	$CH_3(CH_2)_6CO_2C_2H_5$	−43	208	0.007
	Ethyl pelargonate	$CH_3(CH_2)_7CO_2C_2H_5$	−45	222	0.003
	Ethyl caproate	$CH_3(CH_2)_8CO_2C_2H_5$	−20	245	0.0015
Esters of acetic acid, CH_3CO_2R	Methyl acetate	$CH_3CO_2CH_3$	−99	57	24.4
	Ethyl acetate	$CH_3CO_2CH_2CH_3$	−82	77	7.39 (25°C)
	n-Propyl acetate	$CH_3CO_2CH_2CH_2CH_3$	−93	102	1.89
	n-Butyl acetate	$CH_3CO_2CH_2CH_2CH_2CH_3$	−78	125	1.0 (22°C)
Miscellaneous esters	Methyl acrylate	$CH_2{=}CHCO_2CH_3$		80	5.2
	Methyl benzoate	$C_6H_5CO_2CH_3$	−12	199	Insoluble
	Methyl salicylate	$C_6H_4(OH)CO_2CH_3$ (ring with $-CO_2CH_3$ and ortho OH)	−9	223	Insoluble
	Acetylsalicylic acid	$C_6H_4(CO_2H)OCOCH_3$ (ring with CO_2H and ortho $O\overset{O}{\overset{\|}{C}}CH_3$)	135		
	"Waxes"	$CH_3(CH_2)_nCO_2(CH_2)_nCH_3$	*n* = 23–33: carnauba wax; = 25–27: beeswax; = 14–15: spermaceti		

Table 10.7
Fragrances or Flavors Associated with Some Esters

Name	Structure	Source or Flavor
Ethyl formate	$HCO_2CH_2CH_5$	Rum
Isobutyl formate	$HCO_2CH_2CH(CH_3)_2$	Raspberries
n-Pentyl acetate (*n*-amyl acetate)	$CH_3CO_2CH_2CH_2CH_2CH_2CH_3$	Bananas
Isopentyl acetate (isoamyl acetate)	$CH_3CO_2CH_2CH_2CH(CH_3)_2$	Pears
n-Octyl acetate	$CH_3CO_2(CH_2)_7CH_3$	Oranges
Ethyl butyrate	$CH_3CH_2CH_2CO_2CH_2CH_5$	Pineapples
n-Pentyl butyrate	$CH_3CH_2CH_2CO_2(CH_2)_4CH_3$	Apricots
Methyl salicylate	$C_6H_4(-CO_2CH_3)(OH)$	Oil of wintergreen

temperatures and have lower solubilities in water than the parent acids. Some lower-formula-weight esters have very fragrant odors (Table 10.7), in sharp contrast with those of the parent acids.

Chemical Properties

Esters undergo reactions of hydrolysis, alcoholysis, and ammonolysis, all of which involve breaking the carbonyl-to-oxygen bond.

Hydrolysis

In general:

$$R-\overset{O}{\overset{\|}{C}}-OR' + H-\ddot{\underset{\cdot\cdot}{O}}H \text{ (excess)} \xrightarrow{H^+} R-\overset{O}{\overset{\|}{C}}-O-H + H-O-R'$$

Specific examples:

$$\underset{\text{ETHYL ACETATE}}{CH_3\overset{O}{\overset{\|}{C}}-OCH_2CH_3} + H_2O \xrightarrow{H^+} \underset{\text{ACETIC ACID}}{CH_3\overset{O}{\overset{\|}{C}}OH} + \underset{\text{ETHYL ALCOHOL}}{CH_3CH_2OH}$$

$$\underset{\text{METHYL BENZOATE}}{C_6H_5\overset{O}{\overset{\|}{C}}-OCH_3} + H_2O \xrightarrow{H^+} \underset{\text{BENZOIC ACID}}{C_6H_5\overset{O}{\overset{\|}{C}}OH} + \underset{\text{METHYL ALCOHOL}}{CH_3OH}$$

Alcoholysis (Transesterification, or Ester Interchange)

In general:

$$R{-}\overset{\overset{\large O}{\|}}{C}{-}O{-}R' + \underset{\text{(Large excess)}}{R''OH} \xrightarrow{H^+} \underset{\text{NEW ESTER}}{R{-}\overset{\overset{\large O}{\|}}{C}{-}O{-}R''} + H{-}O{-}R'$$

Specific example:

$$\underset{\text{METHYL ACRYLATE}}{CH_2{=}CH\overset{\overset{\large O}{\|}}{C}OCH_3} + \underset{\substack{\text{ETHYL ALCOHOL}\\ \text{(Large excess)}}}{HOCH_2CH_3} \xrightarrow{H^+} \underset{\text{ETHYL ACRYLATE (99\%)}}{CH_2{=}CH\overset{\overset{\large O}{\|}}{C}OCH_2CH_3} + CH_3OH$$

This method is chosen only when circumstances make the more usual synthesis of an ester impractical. Acrylic acid, for example, polymerizes quite readily. Methyl acrylate (see the example just given) is not quite as susceptible to polymerization, and other esters of acrylic acid (e.g., ethyl acrylate in the example) are easily made by alcoholysis of methyl acrylate. This general type of reaction is usually called transesterification to indicate transferral of an ester group from one alcohol portion to another.

Ammonolysis

In general:

$$R{-}\overset{\overset{\large O}{\|}}{C}{-}O{-}R' + NH_3 \longrightarrow R{-}\overset{\overset{\large O}{\|}}{C}{-}NH_2 + H{-}O{-}R'$$

Specific examples:

$$\underset{\text{ETHYL NICOTINATE}}{C_5H_4N{-}\overset{\overset{\large O}{\|}}{C}{-}OCH_2CH_3} + NH_3 \xrightarrow[H_2O]{} \underset{\text{NICOTINAMIDE (78\%)}}{C_5H_4N{-}\overset{\overset{\large O}{\|}}{C}{-}NH_2} + HOCH_2CH_3$$

$$\underset{\text{ETHYL MALONATE}}{CH_2(\overset{\overset{\large O}{\|}}{C}{-}OCH_2CH_3)_2} + 2NH_3 \longrightarrow \underset{\text{MALONDIAMIDE (99\%)}}{CH_2(\overset{\overset{\large O}{\|}}{C}{-}NH_2)_2} + 2HOCH_2CH_3$$

Amides are normally prepared by the action of ammonia (or an amine) on an acid chloride or anhydride. Some of these derivatives, however, are not easily made. Ammonolysis of an ester is therefore a useful alternative.

Saponification

This important reaction of esters is a slight variation of ester hydrolysis. Ester hydrolysis occurs in the presence of an acid catalyst (or an enzyme); saponification occurs in the presence of aqueous sodium or potassium hydroxide. Essentially the same products form, except that saponification produces not the free acid but its salt. In both hydrolysis and saponification the alcohol portion is liberated as the free alcohol. In general:

$$R-\overset{O}{\overset{\|}{C}}-O-R' + NaOH \xrightarrow[\text{Heat}]{H_2O} R-\overset{O}{\overset{\|}{C}}-O^-Na^+ + HOR'$$

Specific examples (compare these with the examples of ester hydrolysis):

$$\underset{\text{ETHYL ACETATE}}{CH_3\overset{O}{\overset{\|}{C}}OCH_2CH_3} + NaOH \xrightarrow[\text{Heat}]{H_2O} \underset{\text{SODIUM ACETATE}}{CH_3\overset{O}{\overset{\|}{C}}O^-Na^+} + \underset{\text{ETHYL ALCOHOL}}{CH_3CH_2OH}$$

$$\underset{\text{METHYL BENZOATE}}{C_6H_5\overset{O}{\overset{\|}{C}}OCH_3} + NaOH \xrightarrow[\text{Heat}]{H_2O} \underset{\text{SODIUM BENZOATE}}{C_6H_5\overset{O}{\overset{\|}{C}}O^-Na^+} + \underset{\text{METHYL ALCOHOL}}{CH_3OH}$$

The term *saponification* comes from the Latin *sapo* and *onis,* "soap" and "-fy"—that is, to make soap. Ordinary soap is a mixture of sodium salts of long-chain carboxylic acids.

Exercise 10.6 Write the structures of the products that would be expected to occur in each of the following situations. If no reaction is to be expected, write "No reaction."

(a) $CH_3CH_2\overset{O}{\overset{\|}{C}}OCH_3 + \underset{\text{(Excess)}}{H_2O} \xrightarrow[\text{Heat}]{H^+}$

(b) $CH_3CH_2O\overset{O}{\overset{\|}{C}}CH_3 + NaOH \xrightarrow[\text{Heat}]{H_2O}$

(c) $C_6H_5\overset{O}{\overset{\|}{C}}OCH(CH_3)_2 + \underset{\text{(Large excess)}}{NH_3} \xrightarrow[\text{Heat}]{H_2O}$

(d) $CH_2{=}\overset{CH_3}{\overset{|}{C}}-\underset{O}{\underset{\|}{C}}OCH_3 + \underset{\text{(Excess)}}{CH_3CH_2CH_2OH} \xrightarrow[\text{Heat}]{H^+}$

(e)
$$\begin{array}{l} CH_3(CH_2)_{14}\overset{O}{\overset{\|}{C}}-O-CH_2 \\ \qquad\qquad\qquad\quad\;\; | \\ CH_3(CH_2)_{12}\overset{O}{\overset{\|}{C}}-O-CH \\ \qquad\qquad\qquad\quad\;\; | \\ CH_3(CH_2)_{16}\overset{O}{\overset{\|}{C}}-O-CH_2 \end{array} + \underset{\text{(Large excess)}}{CH_3OH} \xrightarrow[\text{Heat}]{H^+}$$

(f) $CH_3\overset{CH_3}{\overset{|}{C}}HCH_2\overset{O}{\overset{\|}{C}}-O-CH_2CH_3 + H_2O \xrightarrow[\text{Heat}]{H^+}$

(g) $CH_3CH_2CH_2OC(=O)—CH(CH_3)CH_3 + H_2O \xrightarrow[\text{Heat}]{H^+}$

(h) $CH_3C(=O)—O—CH_2—CH_2—O—C(=O)CH_3 + NaOH$ (Excess) $\xrightarrow[\text{Heat}]{H_2O}$

10.19 Important Individual Esters

The Salicylates

Salicyclic acid can function either as an acid or as a phenol, for it possesses both groups. Salicyclic acid, its esters, and its salts, taken internally, have both an analgesic effect (depressing sensitivity to pain) and an antipyretic action (reducing fever). As analgesics they raise the threshold of pain by depressing pain centers in the thalamus region of the brain. As antipyretics they increase perspiration as well as circulation of the blood in capillaries near the surface of the skin. Both mechanisms have cooling effects. Salicyclic acid itself irritates the moist membranes lining the mouth, gullet, and the stomach because it is too acidic. The sodium or calcium salts, however, are less irritating. In the early history of aspirin the salts had to be administered in aqueous solutions that had a very disagreeable sweetish taste. The search for a more palatable form led to the discovery of acetyl salicyclic acid, or aspirin, first sold by the German company Bayer in 1899 (which still owns the name Aspirin as a trademark in Germany). Aspirin is the most widely used drug in the world. During the early 1970s the annual United States production exceeded 33 million pounds.

$C_6H_4(CO_2H)(OH)$ — SALICYLIC ACID

$C_6H_4(CO_2^-Na^+)(OH)$ — SODIUM SALICYLATE

$C_6H_4(CO_2H)(OC(=O)CH_3)$ — ACETYLSALICYLIC ACID ("ASPIRIN")

$C_6H_4(C(=O)OCH_3)(OH)$ — METHYL SALICYLATE ("OIL OF WINTERGREEN")

$C_6H_4(C(=O)OC_6H_5)(OH)$ — PHENYL SALICYLATE ("SALOL")

Methyl salicylate is used in liniments, for it is readily absorbed through the skin. Phenyl salicylate is sometimes a constituent of ointments that protect the skin against ultraviolet rays.

Dacron

Dacron is the trade name for the fiber made from polyethylene terephthalate. This polymer may also be made into exceptionally clear, durable films, and in this form it is known as Mylar. Other trade names for polyethylene terephthalate are Terylene and Cronar. It is made by an ester interchange (transesterification) between ethylene glycol and dimethyl terephthalate.

$$CH_3O\overset{O}{\overset{\|}{C}}-C_6H_4-\overset{O}{\overset{\|}{C}}OCH_3 + HOCH_2CH_2OH + CH_3O\overset{O}{\overset{\|}{C}}-C_6H_4-\overset{O}{\overset{\|}{C}}OCH_3 + HOCH_2CH_2OH$$

DIMETHYL TEREPHTHALATE   ETHYLENE GLYCOL

$$\downarrow -CH_3OH$$

$$\text{etc.}-O\overset{O}{\overset{\|}{C}}-C_6H_4-\overset{O}{\overset{\|}{C}}\left(OCH_2CH_2O\overset{O}{\overset{\|}{C}}-C_6H_4-\overset{O}{\overset{\|}{C}}\right)_n OCH_2CH_2O-\text{etc.}$$

REPEATING UNIT IN DACRON

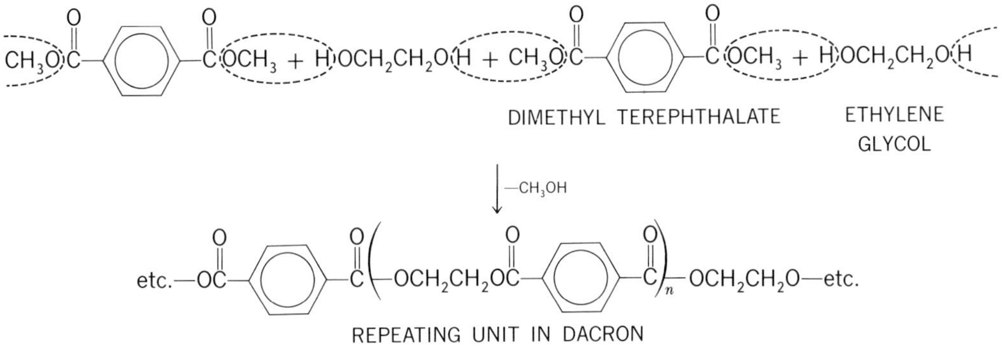

Glyptal Resins

Ethylene glycol has but two hydroxyl groups and can form only a linear polymer with terephthalic acid. Glycerol, however, has three hydroxyl groups, the third group making the formation of cross-linked polymers possible. The glyptal resin, prized as a tough, durable surface coating in automobile and other finishes, forms when phthalic anhydride and glycerol interact in a mole ratio of 3:2 (Figure 10.1).

Acrylates

Esters of acrylic acid and methacrylic acid can be made to polymerize in a fashion exactly analogous to vinyl polymerization. The products are well-known plastics.

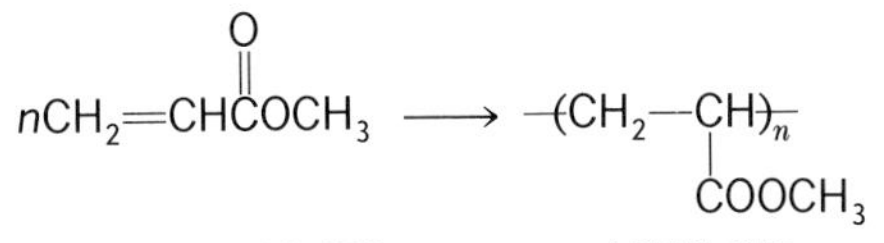

$$nCH_2{=}CH\overset{O}{\overset{\|}{C}}OCH_3 \longrightarrow -\!\!\left(CH_2-\underset{COOCH_3}{\underset{|}{C}H}\right)_n\!\!-$$

METHYL ACRYLATE   ACRYLOID

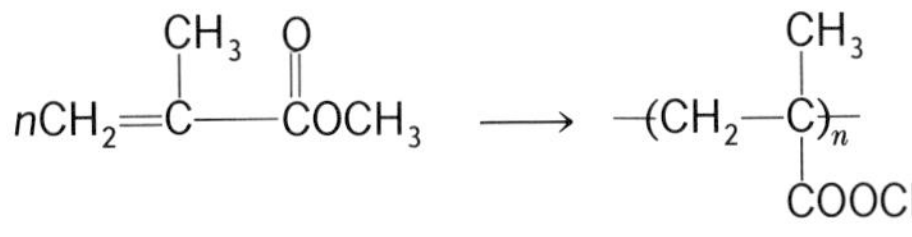

$$nCH_2{=}\overset{CH_3}{\overset{|}{C}}-\overset{O}{\overset{\|}{C}}OCH_3 \longrightarrow -\!\!\left(CH_2-\underset{COOCH_3}{\underset{|}{\overset{CH_3}{\overset{|}{C}}}}\right)_n\!\!-$$

METHYL METHACRYLATE   PLEXIGLAS, LUCITE

Esters of Inorganic Oxygen-Acids

The one or more —OH groups in the molecule of an inorganic oxygen acid can, like those in the molecule of a carboxylic acid, be replaced by —OR. For this reason several types of esters of inorganic oxygen acids are known, many of which are important and useful materials. Some are formed simply by the reaction of an alcohol with the inorganic acid.

High-molecular-weight alcohols react with sulfuric acid to yield alkyl hydrogen

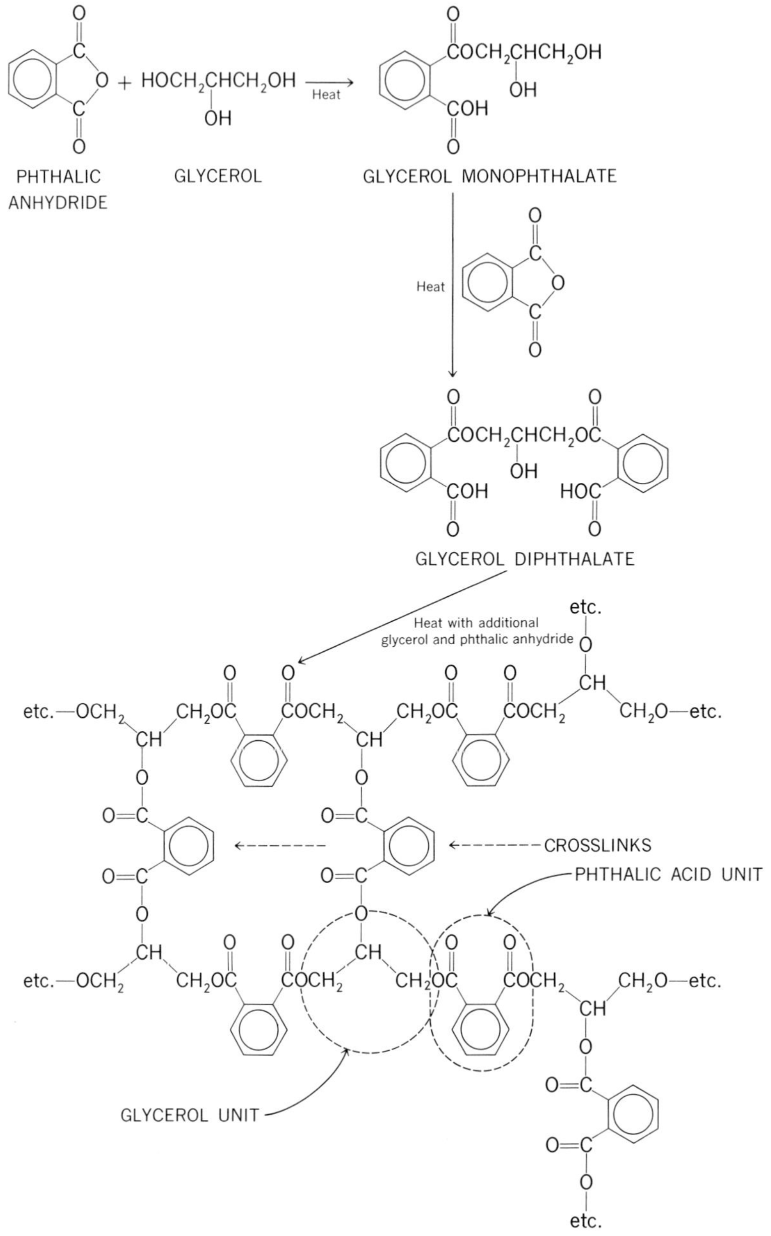

PHTHALIC ANHYDRIDE + GLYCEROL $\xrightarrow{\text{Heat}}$ GLYCEROL MONOPHTHALATE

GLYCEROL MONOPHTHALATE + phthalic anhydride $\xrightarrow{\text{Heat}}$ GLYCEROL DIPHTHALATE

GLYCEROL DIPHTHALATE → Heat with additional glycerol and phthalic anhydride

CROSSLINKS

PHTHALIC ACID UNIT

GLYCEROL UNIT

Figure 10.1
Glyptal resin, basic structural features. The cross-linking shown here is for illustrative purposes only. In the actual resin these links do not occur at every possible position.

sulfates, as illustrated by the synthesis of a common synthetic detergent, sodium lauryl sulfate:

$$CH_3(CH_2)_{10}CH_2OH + HO-\overset{\overset{\displaystyle O}{\uparrow}}{\underset{\underset{\displaystyle O}{\downarrow}}{S}}-OH \longrightarrow CH_3(CH_2)_{10}CH_2O-\overset{\overset{\displaystyle O}{\uparrow}}{\underset{\underset{\displaystyle O}{\downarrow}}{S}}-OH$$

LAURYL HYDROGEN SULFATE

$$\downarrow NaOH$$

$$CH_3(CH_2)_{10}CH_2O-\overset{\overset{\displaystyle O}{\uparrow}}{\underset{\underset{\displaystyle O}{\downarrow}}{S}}-O^-Na^+$$

SODIUM LAURYL SULFATE (DREFT®)

10.20 Amides

Physical Properties

With the exception of formamide, all the simple amides, those derived from ammonia, melt well above room temperature. Their molecules are apparently quite polar and can serve as both hydrogen bond acceptors and donors. Several are listed in Table 10.8.

Table 10.8
Amides of Carboxylic acids

Name	Structure	Mp (°C)	Bp (°C)	Solubility in Water
Formamide	$HCONH_2$	2.5	210 (decomposes)	∞
N-Methylformamide	$HCONHCH_3$	−5	131 (at 90 mm pressure)	Very soluble
N,N-Dimethylformamide	$HCON(CH_3)_2$	−61	153	∞
Acetamide	CH_3CONH_2	82	222	Very soluble
N-Methylacetamide	$CH_3CONHCH_3$	28	206	Very soluble
N,N-Dimethylacetamide	$CH_3CON(CH_3)_2$	−20	166	∞
Propionamide	$CH_3CH_2CONH_2$	79	213	Soluble
Butyramide	$CH_3(CH_2)_2CONH_2$	115	216	Soluble
Valeramide	$CH_3(CH_2)_3CONH_2$	106	Sublimes	Soluble
Caproamide	$CH_3(CH_2)_4CONH_2$	100	Sublimes	Slightly soluble
Benzamide	$C_6H_5-CONH_2$	133	290	Slightly soluble

Chemical Properties. Hydrolysis

$$R{-}\overset{\overset{\displaystyle O}{\|}}{C}{-}NH_2 + H_2O \xrightarrow[\text{(slowly by heat alone)}]{\text{Promoted by acids or bases}} R{-}\overset{\overset{\displaystyle O}{\|}}{C}{-}OH + NH_3$$

$$R{-}\overset{\overset{\displaystyle O}{\|}}{C}{-}NHR' + H_2O \xrightarrow{H^+} R{-}\overset{\overset{\displaystyle O}{\|}}{C}{-}OH + R'\overset{+}{N}H_3$$

$$R{-}\overset{\overset{\displaystyle O}{\|}}{C}{-}NR'_2 + H_2O \xrightarrow{OH^-} R{-}\overset{\overset{\displaystyle O}{\|}}{C}{-}O^- + R'_2NH$$

It is the carbonyl-to-nitrogen bond that breaks. The reaction usually requires prolonged refluxing of the amide with either acid or base, for they are quite unreactive. *In vivo,* however, enzymes are available for catalyzing the reaction. The digestion of proteins is nothing more than hydrolysis of the amide linkages of proteins.

Exercise 10.7 Write the structures of the products of the hydrolysis of each of the following.

(a) $CH_3\overset{\overset{\displaystyle O}{\|}}{C}NH_2$ (b) $CH_3NH\overset{\overset{\displaystyle O}{\|}}{C}CH_2CH_3$ (c) $C_6H_5NH\overset{\overset{\displaystyle O}{\|}}{C}CH_3$

(d) $CH_3CH_2\overset{\overset{\displaystyle O}{\|}}{C}N(CH_3)_2$ (e) $NH_2CH_2\overset{\overset{\displaystyle O}{\|}}{C}NH\underset{\underset{\displaystyle CH_3}{|}}{C}H\overset{\overset{\displaystyle O}{\|}}{C}OH$

10.21 Important Individual Amides

Nicotinamide (Niacinamide)

(3-pyridyl)$-\overset{\overset{\displaystyle O}{\|}}{C}-NH_2$

This amide is one of the B vitamins, and it is involved in two important coenzymes needed for biological oxidations: NAD and NADP (i.e., nicotinamide adenine dinucleotide and nicotinamide adenine dinucleotide phosphate).

Nylon

The word "nylon" is the family name for a group of polyamides capable of being drawn into strong fibers. Perhaps the most famous member of the family is nylon 66, so named because each of the two monomers has six carbons per molecule.

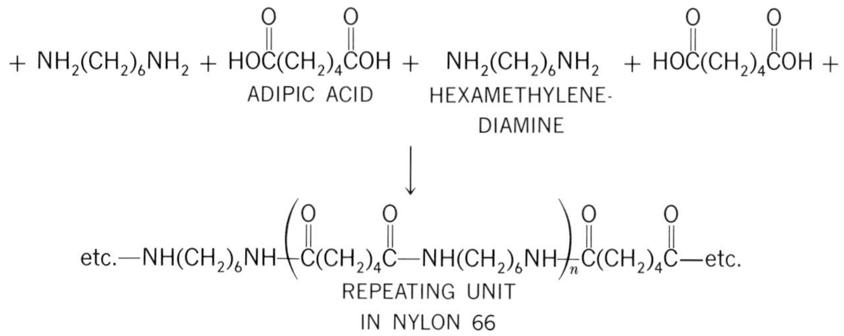

$$+ NH_2(CH_2)_6NH_2 + \underset{\text{ADIPIC ACID}}{HO\overset{O}{\overset{\|}{C}}(CH_2)_4\overset{O}{\overset{\|}{C}}OH} + \underset{\text{HEXAMETHYLENE-DIAMINE}}{NH_2(CH_2)_6NH_2} + HO\overset{O}{\overset{\|}{C}}(CH_2)_4\overset{O}{\overset{\|}{C}}OH +$$

$$\downarrow$$

$$\text{etc.}—NH(CH_2)_6NH\underbrace{\left(\overset{O}{\overset{\|}{C}}(CH_2)_4\overset{O}{\overset{\|}{C}}—NH(CH_2)_6NH\right)_n}_{\text{REPEATING UNIT IN NYLON 66}}\overset{O}{\overset{\|}{C}}(CH_2)_4\overset{O}{\overset{\|}{C}}—\text{etc.}$$

To be useful as a fiber, each molecule in a batch of nylon 66 should contain from 50 to 90 of each of the monomer units. Shorter molecules make weak or brittle fibers.

The unusual resistance of nylon 66 to breakage by stretching, mechanical abrasion, moisture, action of mild acids and alkalis, light, dry-cleaning solvents, mildew, bacteria, and rotting conditions make it one of the most desirable of all fibers. It is more resistant to burning than wool, rayon, cotton, or silk and is almost as immune to insect attack as Fiberglas. No known insects metabolize nylon molecules, although some will cut their way through a nylon fabric if they are entrapped by it!

10.22 Sulfonic Acids

Sulfonic acids must not be confused with esters of sulfuric acid (e.g., alkyl hydrogen sulfates). Sulfinic acids are also known but are rare. The system does occur among intermediates of the metabolism of organic sulfur compounds. We shall very briefly

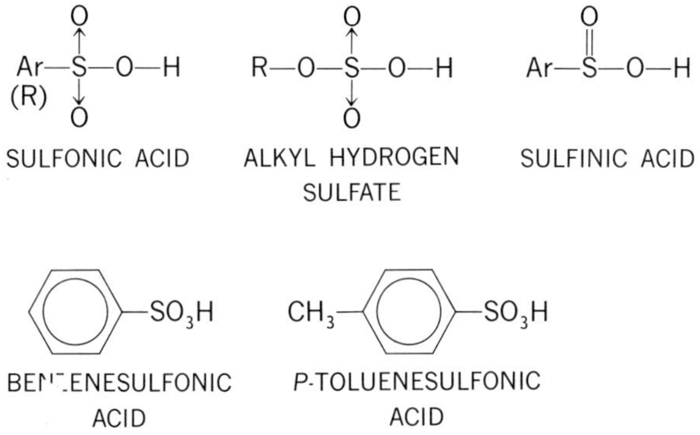

$$\underset{\text{SULFONIC ACID}}{\underset{(R)}{Ar}—\overset{O}{\overset{\uparrow}{\underset{\underset{O}{\downarrow}}{S}}}—O—H} \qquad \underset{\text{ALKYL HYDROGEN SULFATE}}{R—O—\overset{O}{\overset{\uparrow}{\underset{\underset{O}{\downarrow}}{S}}}—O—H} \qquad \underset{\text{SULFINIC ACID}}{Ar—\overset{O}{\overset{\|}{S}}—O—H}$$

$$\underset{\text{BENZENESULFONIC ACID}}{C_6H_5—SO_3H} \qquad \underset{\text{P-TOLUENESULFONIC ACID}}{CH_3—C_6H_4—SO_3H}$$

review some of the chemistry of the aromatic sulfonic acids and their derivatives here. The parent acids generally are made by the sulfonation of aromatic compounds (Section 4.3).

The sulfonic acids are the strongest of the organic acids, being roughly similar to phosphoric acid.

Acid	K_a (25°)
H_3PO_4	0.14
$C_6H_5SO_3H$	0.2
$C_6H_5CO_2H$	0.000014

The sulfonic acid group, being very polar and capable of ionizing, is an excellent helper in making substances soluble in water. The corresponding sodium salts, like nearly all sodium salts are generally water soluble.

10.23 Derivatives of Sulfonic Acids

The esters and the amides are nearly always made from the sulfonyl chlorides. The latter may be made in the same way acyl chlorides are made—by the action of $SOCl_2$ or PCl_5 on the parent acid.

$$ArSO_3H \xrightarrow{SOCl_2} ArSO_2Cl$$

Or they may be made directly by action of chlorosulfonic acid on the aromatic compound.

$$\underset{\text{}}{C_6H_6} + \underset{\text{CHLOROSULFONIC ACID}}{2HOSO_2Cl} \longrightarrow \underset{\text{BENZENESULFONYL CHLORIDE}}{C_6H_5SO_2Cl} + H_2SO_4 + HCl$$

Moreover, just as acyl chlorides may be changed to esters by reacting with alcohols or to amides by reacting with ammonia (or amines), so too with the sulfonyl chlorides.

$$ArSO_2Cl \xrightarrow{ROH} Ar{-}\overset{\overset{O}{\uparrow}}{\underset{\underset{O}{\downarrow}}{S}}{-}O{-}R + HCl$$

ALKYL SULFONATES

$$ArSO_2Cl \xrightarrow[\text{or } RNH_2 \text{ or } R_2NH]{NH_3} Ar{-}\overset{\overset{O}{\uparrow}}{\underset{\underset{O}{\downarrow}}{S}}{-}\overset{\overset{H(R)}{|}}{N}{-}H(R) \quad (+\ HCl \longrightarrow \text{neutralized by ammonia or amine})$$

SULFONAMIDES

Both the sulfonate esters and amides can be hydrolyzed, but the reactions are much slower than for the hydrolysis of ester or amides of carboxylic acids.

10.24 Acidity of Sulfonamides

The sulfonyl group is a powerful electron-withdrawing group, much more so than the carbonyl group. As a result sulfonamides still having at least one hydrogen on nitrogen are weak acids. They will form salts with sodium or potassium hydroxide.

$$Ar{-}\overset{O\uparrow}{\underset{\downarrow O}{S}}{-}\overset{H}{\overset{|}{N}}{-}H + NaOH_{aq} \rightleftharpoons Ar{-}\overset{O\uparrow}{\underset{\downarrow O}{S}}{-}\overset{H}{\overset{|}{N}}{}^{-}Na^{+} + H_2O$$

Among trivalent nitrogen compounds we have the following order or relative basicity:

$$R{-}\ddot{N}H_2 \qquad R{-}\overset{O}{\overset{\|}{C}}{-}\ddot{N}H_2 \qquad Ar{-}\overset{O\uparrow}{\underset{\downarrow O}{S}}{-}\ddot{N}H_2$$

BASIC NEUTRAL ACIDIC

← Increasing basicity

The order follows exactly the order of electron availability on nitrogen.

10.25 Sulfa Drugs

Wonder drugs of World War II and several years after, the sulfa drugs have more and more given way to other antibiotics such as penicillin and aureomycin. They are still used, however; in 1972, 5.5 million pounds of sulfa drugs were manufactured in the United States. We shall study their general synthesis because it illustrates some principles of organic chemistry.

The synthesis begins with aniline. Unfortunately its molecules have rings so reactive that direct high-yield sulfonation is difficult. To solve this problem aniline is first changed to an amide, acetanilide. Because the acetyl group withdraws electron density from the nitrogen, electron density from nitrogen to the ring is not released as much as in the case of the parent aniline. The ring is now only moderately activated. We say that the ring has been protected and that the acetyl group is a *protecting group*. (It also protects the $-NH_2$ group, too.) We remove the acetyl group later.

$$C_6H_5{-}NH_2 + (CH_3\overset{O}{\overset{\|}{C}}{-})_2O \longrightarrow C_6H_5{-}NH\overset{O}{\overset{\|}{C}}CH_3 \xrightarrow{2ClSO_3H} Cl{-}\overset{O\uparrow}{\underset{\downarrow O}{S}}{-}C_6H_4{-}NH\overset{O}{\overset{\|}{C}}CH_3$$

ANILINE ACETIC ANHYDRIDE ACETANILIDE

$$Cl{-}SO_2{-}C_6H_4{-}NHCOCH_3 \xrightarrow{NH_3} NH_2{-}\overset{O\uparrow}{\underset{\downarrow O}{S}}{-}C_6H_4{-}NH\overset{O}{\overset{\|}{C}}CH_3 \xrightarrow[HCl,\Delta]{H_2O} NH_2{-}\overset{O\uparrow}{\underset{\downarrow O}{S}}{-}C_6H_4{-}NH_2$$

SULFANILAMIDE

$$Cl{-}SO_2{-}C_6H_4{-}NHCOCH_3 \xrightarrow{RNH_2} R{-}NH{-}\overset{O\uparrow}{\underset{\downarrow O}{S}}{-}C_6H_4{-}NH\overset{O}{\overset{\|}{C}}CH_3 \xrightarrow[HCl,\Delta]{H_2O} R{-}NH{-}\overset{O\uparrow}{\underset{\downarrow O}{S}}{-}C_6H_4{-}NH_2$$

(route to all other sulfa drugs)

SUBSTITUTED SULFANILAMIDES

The last step works—selective hydrolysis of the acetamide and not the sulfonamide end—because sulfonamides are much harder to hydrolyze than amides of carboxylic acids.

Summary of Important Chemical and Physical Properties

1. Hydrogen bonding between molecules of carboxylic acids is largely responsible for their having boiling points higher than one would predict on the basis of formula weights.

2. Carboxylic acids may be made by:
   (a) oxidation of 1° alcohols
   (b) oxidation of aldehydes
   (c) carbonation of a Grignard reagent
   (d) hydrolysis of a nitrile
   (e) hydrolysis of other acid derivatives (ester, amides, acid chlorides and anhydrides)

3. The carboxyl group is a solubility "switch" for substances whose molecules have it.
   (a) If the aqueous medium is made alkaline, the carboxyl group changes to the anionic form, $—CO_2^-$, and the substance is more soluble in water.
   (b) If the medium is made acidic, RCO_2^- becomes RCO_2H again and the substance is less soluble in water.

4. Electron-withdrawing groups (e.g., halogen) near a carboxyl group make the acid more acidic.

5. The carboxyl group can be changed:
   (a) to an acid chloride (e.g., by $SOCl_2$ or PCl_5)
   (b) to an ester (action of an alcohol in the presence of an acid catalyst)
   (c) to an anhydride (by action of an acid chloride in pyridine) (The important anhydrides are those of acetic and benzoic acid as well as those of dicarboxylic acids—see Sections 12.4 and 12.5. These are generally available commercially.)
   (d) to an amide (by heating ammonium salts of the acid or using 1° or 2° amines in the place of ammonia)
   (e) to salts (by simple neutralization with alkali)

6. The important reactions of acid derivatives—the acid chlorides, anhydrides, esters, and amides—are with water, alcohols, and ammonia (or 1° or 2° amines).
   (a) Water converts them back to the parent acids.
   (b) Alcohols convert them to esters (transesterification, in the case of an alcohol acting on an ester).
   (c) Ammonia (or 1° or 2° amines) changes them to amides.

7. The acid chlorides and anhydrides are the most reactive to these nucleophiles. Acid chlorides are much more reactive than corresponding 1° alkyl chlorides; esters are more reactive than ethers; amides are more reactive than amines.

8. Amides of carboxylic acids are neutral compounds.

9. Like any other esters, those of inorganic acids can also be hydrolyzed.

10. Sulfonic acids are the strongest of the organic acids.

11. Derivatives of sulfonic acids—the esters and amides—cannot be made directly from the acids. Instead, they must be made from the sulfonyl chloride by action of alcohols or amines, respectively.

12. Sulfonamides, unlike amides of carboxylic acids, are slightly acidic and can be converted into salts by sodium or potassium hydroxide.

13. Sulfonamides are much more difficult to hydrolyze than amides of carboxylic acids under the same conditions.

Problems and Exercises

1. Write the structure of each of the following compounds.
   (a) acetamide (b) methyl propionate
   (c) acetyl chloride (d) acetic anhydride
   (e) *N*-methylformamide (f) isopropyl *n*-butyrate
   (g) glycerol triacetate

2. Write common names for each of the following compounds.

(a) $CH_3CH_2CH_2CH_2O\overset{O}{\overset{\|}{C}}CH_2CH_3$ (b) $CH_3CH_2CH_2\overset{O}{\overset{\|}{C}}NH_2$

(c) $CH_3\overset{O}{\overset{\|}{C}}CH_2\overset{O}{\overset{\|}{C}}OH$ (d) $CH_3CH_2\overset{O}{\overset{\|}{C}}{-}Cl$

(e) $C_6H_5\overset{O}{\overset{\|}{C}}OCH_2CH_3$ (f) $CH_3CH_2OCH_2CH_3$

(g) $CH_3CH_2CH_2CH_2NH_2$ (h) $CH_3CH_2CH_2CH_2CH_2\overset{O}{\overset{\|}{C}}OH$

(i) $(CH_3)_2CHO\overset{O}{\overset{\|}{C}}H$ (j) 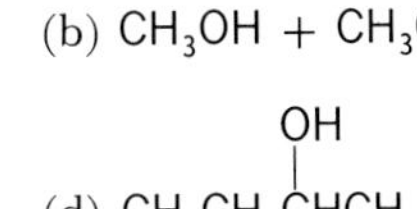 $Br{-}C_6H_4{-}SO_3H$ (para-substituted benzene ring)

3. Complete the following equations by writing the structure(s) of the principal organic product(s) that are expected to form. If no reaction is expected, write "No reaction."

(a) $CH_3\overset{O}{\overset{\|}{C}}H \xrightarrow{K_2Cr_2O_7}$ (b) $CH_3OH + CH_3\overset{O}{\overset{\|}{C}}OH \xrightarrow[\text{Heat}]{H^+}$

(c) $CH_3CH_2CH_3 + H_2SO_4 \longrightarrow$ (d) $CH_3CH_2\overset{OH}{\overset{|}{C}}HCH_3 \xrightarrow[\text{Heat}]{KMnO_4}$

(e) $CH_3CH_2\overset{O}{\overset{\|}{C}}NH_2 + H_2O \xrightarrow[\text{Heat}]{H^+}$ (f) $CH_3\overset{OH}{\overset{|}{C}}HCH_3 \xrightarrow[\text{Heat}]{H_2SO_4}$

(g) $CH_3O\overset{O}{\overset{\|}{C}}CH(CH_3)_2 + H_2O \xrightarrow[\text{Heat}]{H^+}$ (h) $CH_3CH_2CH{=}CH_2 + HCl \longrightarrow$

(i) $CH_3CH(OCH_3)_2 + H_2O \xrightarrow[\text{Heat}]{H^+}$ (j) $CH_3\overset{O}{\overset{\|}{C}}Cl + CH_3OH \longrightarrow$

(k) $CH_3CH_2OCH_2CH_3 + H_2O \xrightarrow{^-OH}$

(l) $CH_3CH_2\overset{O}{\overset{\|}{C}}H + 2CH_3OH \xrightarrow{HCl}$

(m) $CH_3CH{=}CHCH_3 + H_2 \xrightarrow[\text{Heat, pressure}]{Ni}$

(n) $CH_3\overset{OH}{\underset{CH_3}{C}}CH_2CH_3 \xrightarrow{K_2Cr_2O_7}$

(o) $(CH_3)_2CH\overset{O}{\overset{\|}{C}}OH + CH_3OH \xrightarrow{H^+}$

(p) $CH_3CH_2Br + NH_3 \longrightarrow$

(q) $CH_3CH_2OH \xrightarrow{KMnO_4}$

(r) $CH_3\underset{Br}{C}HCH_3 + KOH \xrightarrow{\text{Alcohol}}$

(s) $C_6H_6 + Br_2 \xrightarrow{Fe}$

(t) $CH_3CH_2NH\overset{O}{\overset{\|}{C}}CH_3 + H_2O \xrightarrow[\text{Heat}]{}$

(u) $CH_3\overset{O}{\overset{\|}{C}}CH_3 \xrightarrow{KMnO_4}$

(v) $NH_2CH_2CH_2CH_3 + HCl(aq) \longrightarrow$

(w) $CH_3CH_2CH_2CH_3 + H_2O \xrightarrow{H^+}$

(x) $CH_3C_6H_5 + HNO_3 \xrightarrow[\text{Heat}]{H_2SO_4}$

(y) $CH_3OCH_2\overset{O}{\overset{\|}{C}}CH_3 + H_2O \xrightarrow[\text{Heat}]{H^+}$

(z) $NH_2CH_2\overset{O}{\overset{\|}{C}}NHCH_2\overset{O}{\overset{\|}{C}}OH + H_2O \xrightarrow[\text{Heat}]{}$

(aa) $CH_3\overset{O}{\overset{\|}{C}}OH + NaOH(aq) \longrightarrow$

(bb) $CH_3CH_2\overset{CH_3}{\underset{H}{\overset{+}{N}}}{-}H\ Cl^- + NaOH(aq) \longrightarrow$

(cc) $CH_3OCH_2\underset{CH_3O}{C}HCH_3 + H_2O \xrightarrow[\text{Heat}]{}$

(dd) $CH_3NH_2 + CH_3\overset{O}{\overset{\|}{C}}OH \xrightarrow[\text{Heat}]{}$

(ee) $CH_3C{\equiv}CCH_3 + \underset{\text{(excess)}}{H_2} \xrightarrow{Ni}$

(ff) $C_6H_5O\overset{O}{\overset{\|}{C}}CH_3 + H_2O \xrightarrow{H^+}$

(gg) $CH_3\overset{O}{\overset{\|}{C}}OCH_2CH_2O\overset{O}{\overset{\|}{C}}CH_3 + H_2O \xrightarrow[\text{Heat}]{H^+}$

(hh) $CH_3CH_2Br + Mg \xrightarrow[\text{Ether}]{}$

(ii) Product of (hh) + $CH_3\overset{O}{\overset{\|}{C}}CH_3 \xrightarrow[\text{then } H^+]{}$

(jj) $CH_3CH{=}CHO\overset{O}{\overset{\|}{C}}CH_3 + H_2O \xrightarrow[\text{Heat}]{H^+}$

4. Assume that you have available any alcohol of four carbons or fewer plus any needed inorganic reagents. Outline steps for preparing each of the following compounds from such starting materials.

(a) 2-butene	(b) 1-bromobutane	(c) 1-pentanol
(d) ethylamine	(e) acetone	(f) acetic acid
(g) methyl propionate	(h) *n*-butyramide	(i) *n*-octane

5. Arrange the following compounds in order of increasing boiling point. (You should be able to do this without consulting tables.) Explain your answer.

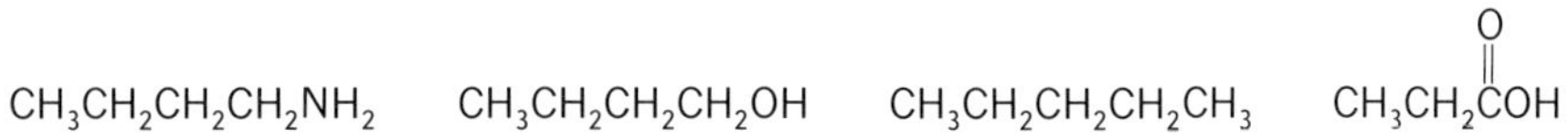

$CH_3CH_2CH_2CH_2NH_2$ (a) $CH_3CH_2CH_2CH_2OH$ (b) $CH_3CH_2CH_2CH_2CH_3$ (c) $CH_3CH_2\overset{O}{\overset{\|}{C}}OH$ (d)

6. Arrange the following compounds in order of increasing solubility in water, and outline the reasons for the order you select.

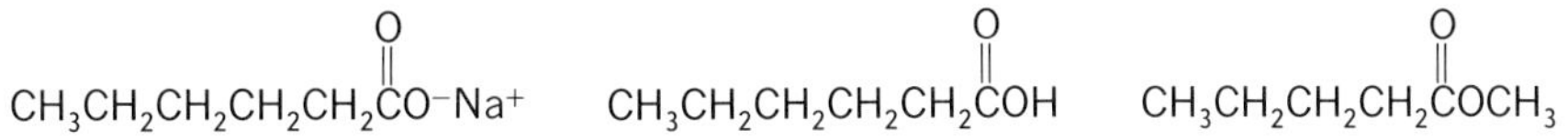

$CH_3CH_2CH_2CH_2CH_2\overset{O}{\overset{\|}{C}}O^-Na^+$ (a) $CH_3CH_2CH_2CH_2CH_2\overset{O}{\overset{\|}{C}}OH$ (b) $CH_3CH_2CH_2CH_2\overset{O}{\overset{\|}{C}}OCH_3$ (c)

$CH_3CH_2CH_2CH_2CH_2OCH_3$ (d) $CH_3CH_2CH_2CH_2CH_2CH_3$ (e)

7. Arrange the following compounds in order of increasing acidity, and outline the reasons for the order you select.

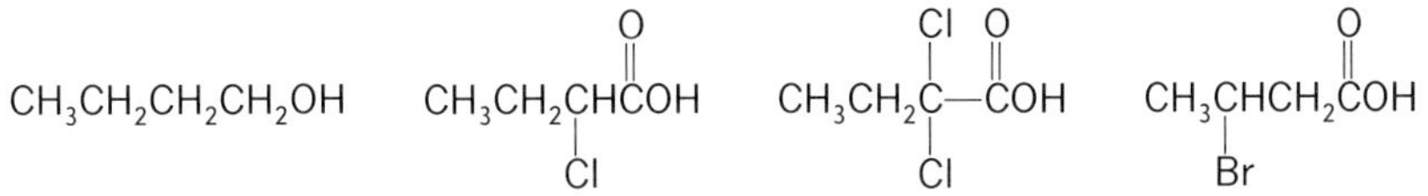

$CH_3CH_2CH_2CH_2OH$ (a) $CH_3CH_2\underset{Cl}{\underset{|}{C}}H\overset{O}{\overset{\|}{C}}OH$ (b) $CH_3CH_2\overset{Cl}{\overset{|}{\underset{Cl}{\underset{|}{C}}}}—\overset{O}{\overset{\|}{C}}OH$ (c) $CH_3\underset{Br}{\underset{|}{C}}HCH_2\overset{O}{\overset{\|}{C}}OH$ (d)

8. Arrange the following compounds in their order of increasing basicity, and outline the reasons for the order you select.

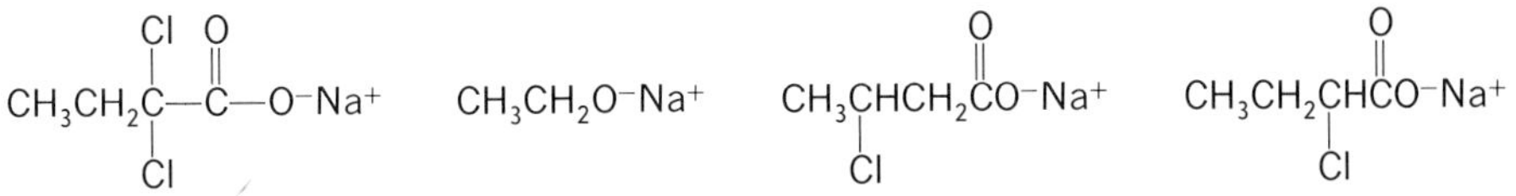

$CH_3CH_2\overset{Cl}{\overset{|}{\underset{Cl}{\underset{|}{C}}}}—\overset{O}{\overset{\|}{C}}—O^-Na^+$ (a) $CH_3CH_2O^-Na^+$ (b) $CH_3\underset{Cl}{\underset{|}{C}}HCH_2\overset{O}{\overset{\|}{C}}O^-Na^+$ (c) $CH_3CH_2\underset{Cl}{\underset{|}{C}}H\overset{O}{\overset{\|}{C}}O^-Na^+$ (d)

9. Write detailed, step-by-step mechanisms for the following reactions: (a) acid-catalyzed hydrolysis of ethyl propionate and (b) aldol condensation of *n*-butyraldehyde.

10. Explain how *N,N*-dimethylacetamide melts at a temperature so much lower than that of acetamide. (See Table 10.8)

chapter eleven
Lipids

11.1 What Lipids Are

Lipids are defined in a rather peculiar way—in terms of an experimental operation, not in terms of molecular structure. The operation is that of extracting with a nonpolar solvent. When plant or animal material is crushed and ground with nonpolar solvents such as benzene, chloroform, or carbon tetrachloride (fat solvents), the portion that dissolves in the solvent is classified as lipid. Carbohydrates, proteins, and inorganic substances are insoluble. Depending on the plant or animal origin, the extracted material may include such a wide variety of compounds that it is difficult to make a precise structural definition. Included in the group could be neutral fats containing only carbon, hydrogen, and oxygen; phosphorus-containing compounds called phospholipids, which usually also contain nitrogen; aliphatic alcohols; waxes; steroids; terpenes; and "derived lipids," those substances resulting from partial or complete hydrolysis of some of the foregoing.

Simple Lipids

11.2 The Triglycerides

The most abundant group of lipids are the simple lipids or triglycerides. Included in this group are common substances such as lard, tallow and butterfat (the animal fats), and olive oil, cottonseed oil, corn oil, peanut oil, linseed oil, coconut oil, and soybean oil (the vegetable oils). Their molecules consist of esters of the trihydroxy alcohol, glycerol, with various long-chain fatty acids. The three acyl units in a typical triglyceride molecule are not identical. They usually are from three different fatty acids.

Fats and oils of whatever source are mixtures of different glyceride molecules. The differences are in specific fatty acids incorporated, not in the general structural features. Thus we cannot write the structure of cottonseed oil, for example. We are limited to describing what a typical molecule is like. One such molecule is shown in structure **1.** In a particular fat or oil, certain fatty acids tend to predominate, certain others are either absent or are present in trace amounts, and virtually all the molecules are triglycerides. Data for several fats and oils are in Table 11.1.

11.3 Fatty Acids

Fatty acids obtained from the lipids of most plants and animals tend to share the following characteristics.

1. They are usually monocarboxylic acids, $R—CO_2H$.
2. The R— group is usually an unbranched chain.
3. The number of carbon atoms is almost always even.
4. The R— group may be saturated, or it may have one or more double bonds, which are *cis*.

The most abundant saturated fatty acids are palmitic acid, $CH_3(CH_2)_{14}CO_2H$, and stearic acid, $CH_3(CH_2)_{16}CO_2H$, having 16 and 18 carbons, respectively. Others are included in Table 10.2, the acids above acetic with an even number of carbons, but they are present in only small amounts.

The most frequently occurring unsaturated fatty acids are listed in Table 11.2. The 18-carbon skeleton of stearic acid is duplicated in oleic, linoleic, and linolenic acids. Oleic acid is the most abundant and most widely distributed fatty acid in nature.

The properties of the fatty acids are those to be expected of compounds having a carboxyl group, double bonds (in some), and long hydrocarbon chains. They are insoluble in water and soluble in nonpolar solvents. They can form salts, be

Table 11.1
Composition of the Fatty Acids Obtained by Hydrolysis of Common Neutral Fats and Oils

				Average Composition of Fatty Acids (%)					
	Fat or Oil	Iodine Number	Saponification Value	Myristic Acid	Palmitic Acid	Stearic Acid	Oleic Acid	Linoleic Acid	Others
Animal fats	Butter	25–40	215–235	8–15	25–29	9–12	18–33	2–4	a
	Lard	45–70	193–200	1–2	25–30	12–18	48–60	6–12	b
	Beef tallow	30–45	190–200	2–5	24–34	15–30	35–45	1–3	b
Vegetable oils	Olive	75–95	185–200	0–1	5–15	1–4	67–84	8–12	
	Peanut	85–100	186–195	—	7–12	2–6	30–60	20–38	
	Corn	115–130	188–194	1–2	7–11	3–4	25–35	50–60	
	Cottonseed	100–117	191–196	1–2	18–25	1–2	17–38	45–55	
	Soybean	125–140	190–194	1–2	6–10	2–4	20–30	50–58	c
	Linseed	175–205	190–196	—	4–7	2–4	14–30	14–25	d
Marine oils	Whale	110–150	188–194	5–10	10–20	2–5	33–40		e
	Fish	120–180	185–195	6–8	10–25	1–3			e

[a] Three to four percent butyric acid, 1 to 2% caprylic acid, 2 to 3% capric acid, 2 to 5% lauric acid.
[b] One percent linolenic acid.
[c] Five to ten percent linolenic acid.
[d] Forty-five to sixty percent linolenic acid.
[e] Large amounts of other highly unsaturated fatty acids.

Table 11.2
Common Unsaturated Fatty Acids

Name	Number of Double Bonds	Total Number of Carbons	Structure	Mp (°C)
Palmitoleic acid	1	16	$CH_3(CH_2)_5CH{=}CH(CH_2)_7CO_2H$	32
Oleic acid	1	18	$CH_3(CH_2)_7CH{=}CH(CH_2)_7CO_2H$	4
Linoleic acid	2	18	$CH_3(CH_2)_4CH{=}CHCH_2CH{=}CH(CH_2)_7CO_2H$	−5
Linolenic acid	3	18	$CH_3CH_2CH{=}CHCH_2CH{=}CHCH_2CH{=}CH(CH_2)_7CO_2H$	−11
Arachidonic acid	4	20	$CH_3(CH_2)_4CH{=}CHCH_2CH{=}CHCH_2CH{=}CHCH_2CH{=}CH(CH_2)_3CO_2H$	−50

esterified, and be reduced to the corresponding long-chain alcohols. Where present, alkene linkages react with bromine and take up hydrogen in the presence of a catalyst.

11.4 Iodine Number

The degree of unsaturation in a lipid is measured by its iodine number. The iodine number is the number of grams of iodine that would add to the double bonds present in 100 g of the lipid if iodine itself could add to alkenes. The reagent used is I—Br, iodine bromide. The data are calculated as if iodine had been used. Saturated fatty acids, having no alkene linkages, have zero iodine numbers. Oleic acid has an iodine number of 90, linoleic acid 181, and linolenic acid 274. Animal fats have low iodine numbers while vegetable oils (the polyunsaturated oils of countless advertisements) higher values, as the data of Table 11.1 show. The iodine number of the mixed triglyceride of structure **1** is calculated to be 89, in the range of the iodine number of olive oil or peanut oil.

Chemical Properties of Triglycerides

11.5 Hydrolysis in the Presence of Enzymes

Enzymes in the digestive tracts of human beings and animals act as efficient catalysts for the hydrolysis of the ester links of triglycerides. In general:

$$\underset{\text{TRIGLYCERIDE}}{\begin{matrix} RC(=O)\text{—}OCH_2 \\ R'C(=O)\text{—}OCH \\ R''C(=O)\text{—}OCH_2 \end{matrix}} + 3H_2O \xrightarrow{\text{Enzyme}} \underset{\text{FATTY ACIDS}}{\begin{matrix} RC(=O)\text{—}OH \\ + \\ R'C(=O)\text{—}OH \\ + \\ R''C(=O)\text{—}OH \end{matrix}} + \underset{\text{GLYCEROL}}{\begin{matrix} HOCH_2 \\ HOCH \\ HOCH_2 \end{matrix}}$$

Example:

$$CH_3(CH_2)_7CH{=}CH(CH_2)_7\overset{O}{\overset{\|}{C}}{-}OCH_2$$
$$CH_3(CH_2)_{14}\overset{O}{\overset{\|}{C}}{-}OCH + 3H_2O \xrightarrow{\text{Enzyme}}$$
$$CH_3(CH_2)_4CH{=}CHCH_2CH{=}CH(CH_2)_7\overset{O}{\overset{\|}{C}}{-}OCH_2$$
1

$$CH_3(CH_2)_7CH{=}CH(CH_2)_7\overset{O}{\overset{\|}{C}}OH$$
OLEIC ACID
+
$$\text{GLYCEROL} + CH_3(CH_2)_{14}\overset{O}{\overset{\|}{C}}OH$$
PALMITIC ACID
+
$$CH_3(CH_2)_4CH{=}CHCH_2CH{=}CH(CH_2)_7\overset{O}{\overset{\|}{C}}OH$$
LINOLEIC ACID

The enzyme-catalyzed hydrolysis of triglycerides is the only reaction they undergo during digestion.

11.6 Saponification

When the ester links in triglycerides are saponified by the action of a strong base (e.g., NaOH, KOH), glycerol plus the salts of the fatty acids are produced. These salts are soaps, and how they exert detergent action will be described later in this chapter. In general:

$$R\overset{O}{\overset{\|}{C}}{-}OCH_2$$
$$R'\overset{O}{\overset{\|}{C}}{-}OCH + 3NaOH \xrightarrow[\text{Heat}]{}$$
$$R''\overset{O}{\overset{\|}{C}}{-}OCH_2$$

$$R\overset{O}{\overset{\|}{C}}O^-Na^+ + R'\overset{O}{\overset{\|}{C}}O^-Na^+ + R''\overset{O}{\overset{\|}{C}}O^-Na^+ \quad + \quad HOCH_2{-}HOCH{-}HOCH_2$$
MIXTURE OF SALTS — GLYCEROL

Example:

$$CH_3(CH_2)_7CH{=}CH(CH_2)_7\overset{O}{\overset{\|}{C}}{-}OCH_2$$
$$CH_3(CH_2)_{14}\overset{O}{\overset{\|}{C}}{-}OCH + 3NaOH \xrightarrow[\text{Heat}]{}$$
$$CH_3(CH_2)_4CH{=}CHCH_2CH{=}CH(CH_2)_7\overset{O}{\overset{\|}{C}}{-}OCH_2$$
1

$$CH_3(CH_2)_7CH{=}CH(CH_2)_7\overset{O}{\overset{\|}{C}}O^-Na^+ \qquad HOCH$$
SODIUM OLEATE
+
$$CH_3(CH_2)_{14}\overset{O}{\overset{\|}{C}}O^-Na^+ + HOCH$$
SODIUM PALMITATE
+
$$CH_3(CH_2)_4CH{=}CHCH_2CH{=}CH(CH_2)_7\overset{O}{\overset{\|}{C}}O^-Na^+ \qquad HOCH_2$$
SODIUM LINOLEATE — GLYCEROL

11.7 Saponification Value of a Lipid

Alkali (NaOH or KOH) is consumed by the saponification of a fat or oil. How much is used for one gram of fat or oil, for example, can be easily measured by the techniques of quantitative analysis. From these facts come one common quality control measurement for a lipid, that of its saponification value (or saponification number). The saponification value of a lipid is the number of milligrams of potassium hydroxide needed to saponify one gram.

What is the meaning and usefulness of this determination? For one thing, if the lipid is adulterated with something that cannot be saponified, the measured saponification number for that sample will be unusually low. Otherwise, for unadulterated lipids, the higher the saponification value the more the short-chain fatty acids are present (in esterified form, of course). (We leave the proof of that as an end-of-chapter exercise.)

11.8 Hydrogenation

If some of the double bonds in vegetable oils were hydrogenated, the oils would become like animal fats. They would be solids at room temperature, for example. Complete hydrogenation to an iodine value of zero is not desirable, for the product would be as brittle and unpalatable as tallow. Manufacturers of commercial hydrogenated vegetable oils such as Crisco, Fluffo, Mixo, and Spry limit the degree of hydrogenation. Some double bonds are left. The product has a lower iodine number (e.g., about 50 to 60), a higher degree of saturation, and a melting point that makes it a creamy solid at room temperature, similar to lard or butterfat. The oils of soybean and cottonseed, more abundant than fats, are common raw materials for hydrogenated products. If just one molecule of hydrogen were added to the molecule of the mixed triglyceride of structure **1**, the iodine number would drop from 89 to 59, from the olive or peanut-oil range to that of lard. The peanut oil in popular brands of peanut butter has been partially hydrogenated.

Oleomargarine is made from hydrogenated, carefully selected and highly refined oils and fats. One goal is to produce a final product that will melt readily on the tongue, a feature that makes butter so desirable.

11.9 Rancidity

When fats and oils are left exposed to warm, moist air for any length of time, changes produce disagreeable flavors and odors. That is, the material becomes *rancid.* Two kinds of reactions are chiefly responsible: hydrolysis of ester links and oxidation of double bonds.

The hydrolysis of butterfat would produce a variety of relatively volatile and odorous fatty acids, as the data in the first footnote of Table 11.1 indicate. Water for such hydrolysis is, of course, present in butter, and airborne bacteria furnish enzymes. If the reaction producing rancidity is oxidation, attack by atmospheric oxygen occurs at unsaturated side chains in mixed triglycerides. Eventually, short-chain and volatile carboxylic acids and aldehydes are formed. Both types of substances have extremely disagreeable odors and flavors.

11.10 Hardening of Oils. "Drying"

Oxygen attacks unsaturated glycerides not only at their double bonds but also at carbons attached to them, at allylic positions. Linolenic acid has four such activated sites, marked by asterisks in the following structure:

$$CH_3CH_2\overset{*}{C}H{=}CHCH_2\overset{*}{C}H{=}CHCH_2\overset{*}{C}H{=}CHCH_2(\overset{*}{C}H_2)_6CO_2H$$

LINOLENIC ACID (asterisks mark allylic sites)

Oxygen attacks these positions to form, first, hydroperoxides (—OOH):

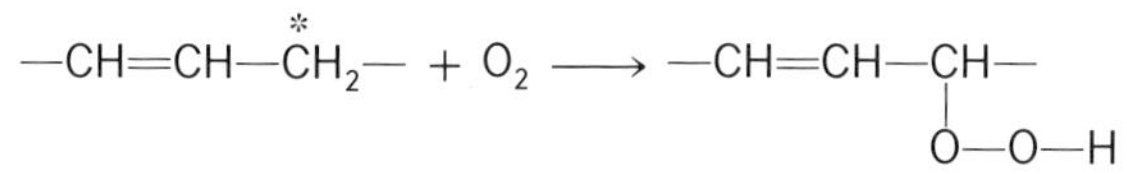

$$-CH{=}CH-\overset{*}{C}H_2- + O_2 \longrightarrow -CH{=}CH-\underset{\displaystyle O-O-H}{\underset{|}{C}H}-$$

Portion of a side chain in an unsaturated triglyceride — Hydroperoxy group introduced at an allylic site

Once these hydroperoxy groups are randomly introduced, they begin to interact with unchanged allylic sites to form peroxide bridges, building a vast interlacing network as illustrated in Figure 11.1. As these chemical changes occur, important modifications take place in the material. If it has been spread on a surface, a film that is dry, tough, and durable forms, constituting a painted surface. The most highly unsaturated vegetable oils work best, because they have the most allylic sites. Linseed oil, which is quite rich in linolenic acid, is the most widely used drying oil. Fresh (raw) linseed oil still dries too slowly, however, and is made faster drying by heating it with driers, such as lead, manganese, or cobalt salts of certain organic acids. Pigments and dyes are added to give the desired color to the paint. Nearly a billion pounds of drying oils, mostly linseed oil, are used each year in the United States.

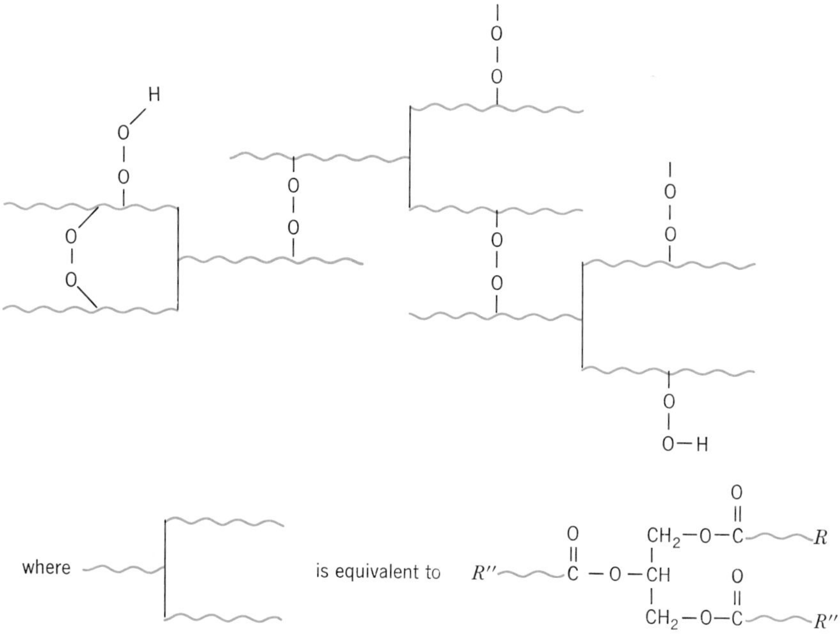

Figure 11.1
When unsaturated glycerides harden, as in the drying of a paint, peroxide cross-links ultimately develop.

11.11 How Detergents Work

Triglycerides and other types of "greases," such as hydrocarbon oils, are the binding agents that hold dirt to surfaces. The problem of cleansing reduces itself to finding a way to loosen or to dissolve this "glue." If it can no longer stick to the surface, neither can the dirt particles. Water alone is a poor cleansing agent, for its molecules are so polar that they stick to each other through hydrogen bonds rather than penetrate into a nonpolar region such as a surface film of grease.

Molecules of a typical soap and a typical synthetic detergent are shown in Figure 11.2. Each has a long, nonpolar hydrocarbon "tail" and a very polar, water-soluble "head." The tail should be easily soluble in nonpolar materials (triglycerides and greases), and the head should be quite soluble in water.

When a detergent, either a soap or a *syndet,* a synthetic detergent, is added to water and the solution makes contact with lipid or greasy material, the action shown in Figure 11.3 takes place. The grease is emulsified and therefore loosened.

One of the advantages of the syndets is that they are not precipitated by the heavier metal ions found in hard water, Ca^{2+}, MG^{2+}, and Fe^{2+} or Fe^{3+} or both. Salts between any of these ions and the fatty acids are insoluble. If ordinary soap is added to hard water, the familiar scum that forms is a mixture of these salts. For example:

$$2RCO_2^- + Ca^{2+} \longrightarrow \underset{\text{"SOAP SCUM"}}{(RCO_2)_2Ca\downarrow}$$

11.12 Biodegradability

Although ordinary soaps have their disadvantages in hard water, they cause virtually no problem when waste water containing them is discharged into sewage disposal systems and eventually into the ground. They are *biodegradable,* that is, microorganisms metabolize them. The search for economical detergents that would not precipitate in hard water led to a variety of types as several companies worked

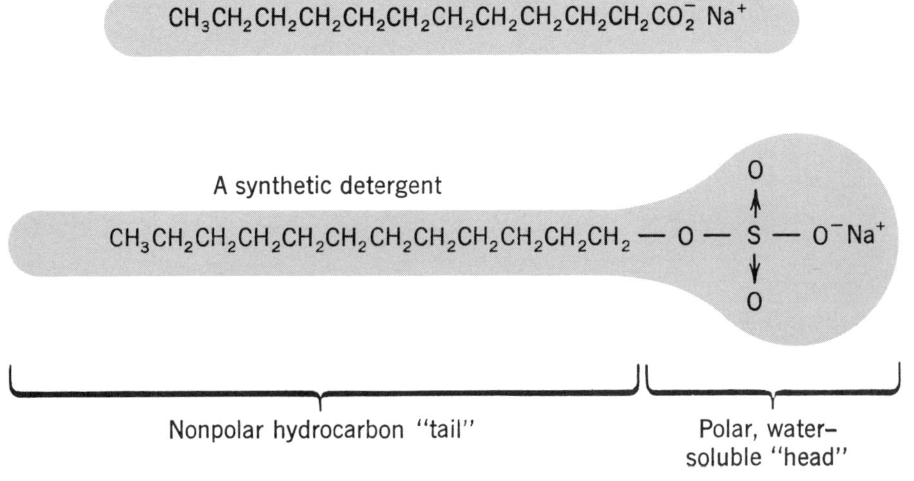

Figure 11.2
Two organic salts that have detergent properties.

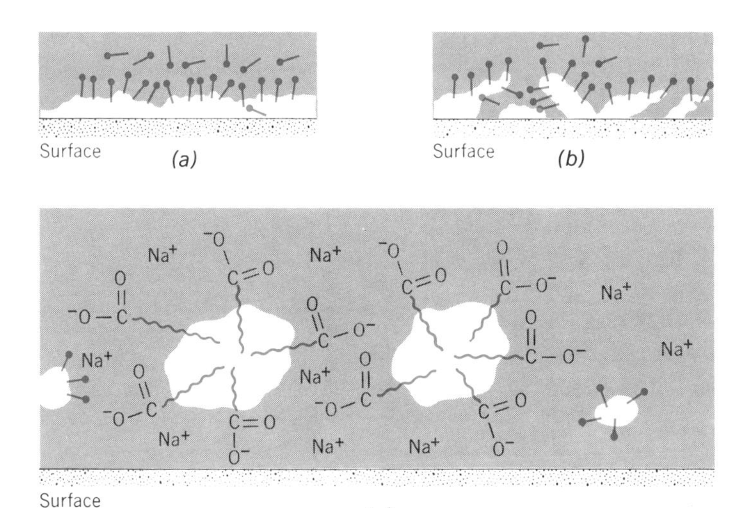

Figure 11.3
How a detergent works. (*a*) The nonpolar "heads" of detergent molecules become embedded in the grease layer, also relatively nonpolar. (*b*) The polar (or ionic) "heads" of the detergent particles are in the water and they tend to urge the grease layer away from the surface. (*c*) Eventually grease globules form, each pincushioned by negatively charged groups. The globules therefore repel each other; they cannot coalesce. This means they are emulsified and are now easily washed away.

on unique products that they could patent. A few such types that were placed on the market are shown in Table 11.3. Whether these syndets were biodegradable was not considered a serious question until the sewage disposal systems of major metropolitan areas slowly became glutted with the foam and suds of undegraded detergents. Homeowners with private wells reported that their tap water foamed.

The enzyme systems of microorganisms are not equipped to degrade the highly

Table 11.3
Biodegradability of Representative Types of Synthetic Detergents

Types of Synthetic Detergents	Examples to Illustrate Structural Features	Biodegradable?
Sodium alkyl sulfate	$CH_3(CH_2)_{10}CH_2OSO_2O^-Na^+$	Yes
Sodium alkylbenzene sulfonate, tetrapropylene-based	$CH_3CH(CH_3)CH_2CH(CH_3)CH_2CH(CH_3)CH_2CH(CH_3)$–$C_6H_4$–$S(\rightarrow O)_2$–$O^-Na^+$	No
Sodium alkylbenzene sulfonate, *n*-paraffin or α-olefin-based	$CH_3(CH_2)_nCH(CH_3)$–C_6H_4–$S(\rightarrow O)_2$–O^-Na^+ ($n = 7 - 11$)	Largely so
Sodium alkane sulfonate	$C_nH_{2n+1}SO_3^-Na^+$ (chain is largely straight, n is 15 to 18)	Almost completely

branched side chains of the tetrapropylene-based alkylbenzene sulfonates (Table 11.3). However, by 1966 the major American syndet makers had switched to *n*-paraffin-based alkylbenzene sulfonates or similar materials. Although these are not completely degradable in groundwater, microorganisms manage to reduce their concentrations to acceptable levels.

Complex Lipids

11.13 Glycerol-Based Phospholipids

In the simple triglycerides just discussed, all three hydroxyl groups of glycerol are esterified with various fatty acids. In the molecules of the glycerol-based phospholipids, one of these hydroxyl groups is esterified with phosphoric acid that is further esterified with a particular amino alcohol. The nature of the groups attached at the remaining two —OHs of glycerol determines the subclass.

11.14 Phosphatides

In the phosphatides these two —OHs have become esterified with fatty acids. They are derivatives of phosphatidic acid. Specific phosphatides are formed by esterifying the phosphate unit with choline, ethanolamine, or serine. See Figure 11.4 for some of the types of phosphatides.

The phosphatides are frequently associated with membranes. They apparently are essential to the structure of cell walls and to the membranes enclosing cell nuclei, cytoplasmic organelles (e.g., mitochondria), and the endoplasmic reticulum. They also are metabolically more stable than ordinary triglycerides. Thus in animals that are starving the reserves of triglycerides (fats) become depleted as they are withdrawn for energy, but a certain amount of phosphatide remains, as it must since it is important to holding cells together.

11.15 Plasmalogens

The plasmalogens, another type of glycerol-based phospholipid, are widely distributed in animal tissue, particularly in the myelin sheaths of nerves and in heart and skeletal muscle and in the brain.

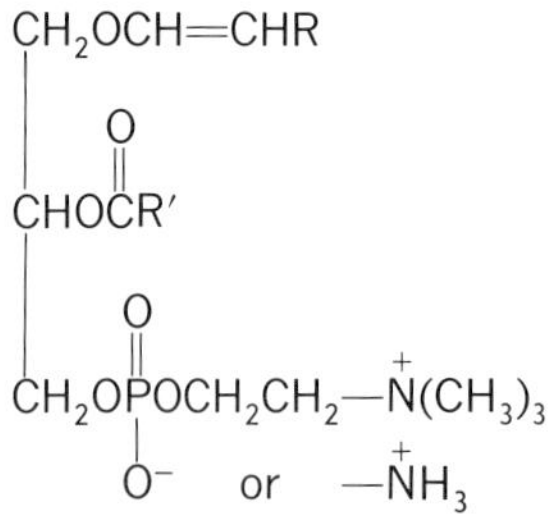

$$
\begin{array}{l}
CH_2OCH{=}CHR \\
|\quad\; O \\
|\quad\; \| \\
CHOCR' \\
|\quad\;\; O \\
|\quad\;\; \| \qquad\qquad\qquad\; + \\
CH_2OPOCH_2CH_2{-}N(CH_3)_3 \\
\qquad | \qquad\qquad\qquad\;\; + \\
\qquad O^- \quad \text{or} \quad {-}NH_3
\end{array}
$$

PLASMALOGENS

11.16 Sphingosine-Based Lipids. Sphingolipids

Brain and nerve tissue as well as the lungs and spleen contain a lipid based not on glycerol but on an unsaturated amino alcohol, sphingosine. Figure 11.5 explains

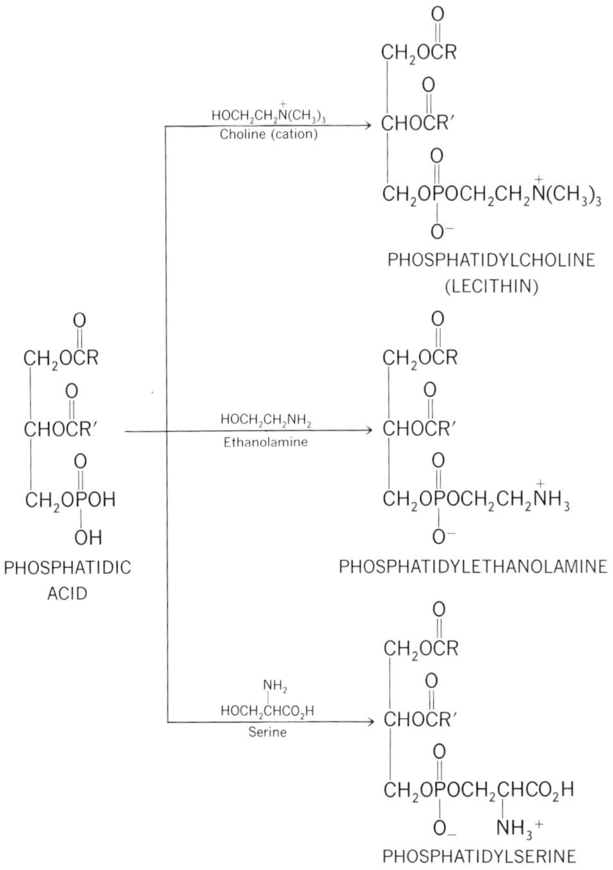

$$\begin{array}{l} \overset{O}{\overset{\|}{CH_2OCR}} \\ | \quad \overset{O}{\overset{\|}{}} \\ CHOCR' \\ | \quad \overset{O}{\overset{\|}{}} \\ CH_2OPOH \\ \qquad | \\ \qquad OH \end{array}$$

PHOSPHATIDIC ACID

$\xrightarrow[\text{Choline (cation)}]{HOCH_2CH_2\overset{+}{N}(CH_3)_3}$ CH_2OCOR, $CHOCOR'$, $CH_2OP(O)(O^-)OCH_2CH_2\overset{+}{N}(CH_3)_3$

PHOSPHATIDYLCHOLINE (LECITHIN)

$\xrightarrow[\text{Ethanolamine}]{HOCH_2CH_2NH_2}$ CH_2OCOR, $CHOCOR'$, $CH_2OP(O)(O^-)OCH_2CH_2\overset{+}{N}H_3$

PHOSPHATIDYLETHANOLAMINE

$\xrightarrow[\text{Serine}]{HOCH_2CH(NH_2)CO_2H}$ CH_2OCOR, $CHOCOR'$, $CH_2OP(O)(O^-)OCH_2CH(NH_3^+)CO_2H$

PHOSPHATIDYLSERINE

Figure 11.4
The principal phosphatides are derivatives of phosphatidic acid.

the structures of two types of sphingolipids, the sphingomyelins and the cerebrosides. The cerebrosides are interesting lipids in that they contain a sugar unit, usually galactose. Their presence in brain and nerve tissue indicates their importance, but their exact function is not yet fully understood.

11.17 Nonsaponifiable Lipids

When you extract plant or animal material with a nonpolar solvent you extract any and all substances that are nonpolar, too. You get not only the triglycerides and some complex lipids, you also obtain (depending on the plant or animal material) substances that cannot be saponified. Two important groups are the terpenes and the steroids.

11.18 Terpenes

The pleasant odors and flavors associated with many plants reveal the presence of volatile compounds of 10 or 15 carbons called *terpenes*. Terpenes are isolated

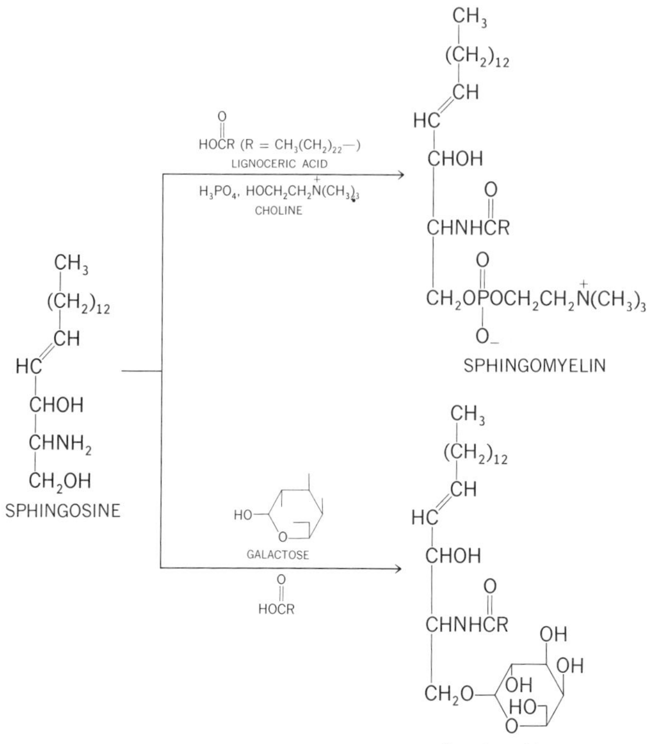

$CH_3(CH_2)_{12}CH{=}CH{-}CHOH{-}CHNH_2{-}CH_2OH$

SPHINGOSINE

$\xrightarrow{HOCR\ (R = CH_3(CH_2)_{22}-)\ \text{LIGNOCERIC ACID};\ H_3PO_4,\ HOCH_2CH_2\overset{+}{N}(CH_3)_3\ \text{CHOLINE}}$

$CH_3(CH_2)_{12}CH{=}CH{-}CHOH{-}CH(NHCOR){-}CH_2OP(O)(O^-)OCH_2CH_2\overset{+}{N}(CH_3)_3$

SPHINGOMYELIN

$\xrightarrow{\text{GALACTOSE};\ HOCR}$

$CH_3(CH_2)_{12}CH{=}CH{-}CHOH{-}CH(NHCOR){-}CH_2O{-}\text{galactose}$

Cerebrosides:

If $RCOH = CH_3(CH_2)_{22}CO_2H$ .... Kerasin
LIGNOCERIC ACID

$= CH_3(CH_2)_{21}CH(OH)CO_2H$ .... Phrenosin
CEREBRONIC ACID

$= CH_3(CH_2)_7CH{=}CH(CH_2)_{13}CO_2H$ ...... Nervon
NERVONIC ACID

$= CH_3(CH_2)_7CH{=}CH(CH_2)_{12}CH(OH)CO_2H$ .. Oxynervon
OXYNERVONIC ACID

Figure 11.5
The two common types of sphingolipids, sphingomyelin, and cerebrosides.

by extracting them with ether, or they may be obtained by steam distillation, a process in which steam is blown through a watery mixture of crushed portions of the plant (or the watery mixture may be boiled to generate the steam within the flask). To perfumers and food flavorers the materials isolated in this way are the *quinta essentia*, the essential parts of the plants, and for this reason they are called the essential oils.

11.19 Isoprene Rule

To the chemist one of the most interesting features of terpenes is the regular occurrence of isoprene units in their molecules. Two isoprene units give the

10-carbon terpenes, the structure to which the term "terpene" is properly restricted.

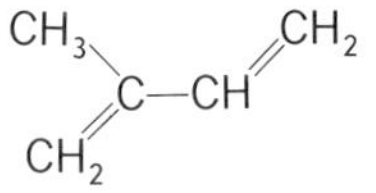

ISOPRENE (2-METHYL-1,3-BUTADIENE)

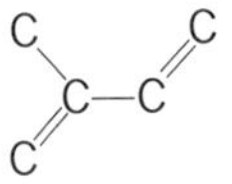

USEFUL CONDENSATIONS OF THE ISOPRENE STRUCTURE

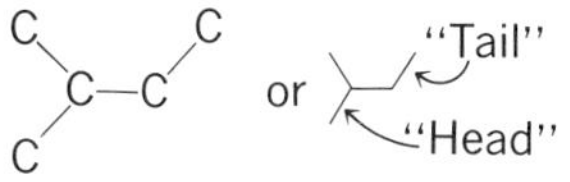

THE ISOPRENE "SKELETON"

The *sesquiterpenes* have three isoprene units making 15 carbons. *Diterpenes* are 20-carbon compounds composed of four isoprene units. The thirty-carbon substances (six isoprene units) are *triterpenes,* etc. *The terpene units usually occur in a "head-to-tail" arrangement,* as seen in the structure of myrcene, below. So often has this been observed that it is called the **isoprene rule.**

The structure of the terpene called *myrcene* (oil of bay)[1] illustrates the connection between isoprene skeletons and terpenes, with the dashed line or lines, here and elsewhere in this section, dividing the molecule into isoprene units linked "head-to-tail." Many terpenes such as *limonene* (oil of peppermint, oil of orange, oil of lemon) and *α-pinene* (oil of turpentine) are cyclic compounds. Important sesquiterpenes (15-carbon compounds) include *α-farnesene* (oil of citronella) and *zingiberene* (oil of ginger). The structure of rubber, Section 3.33, also illustrates the rule.

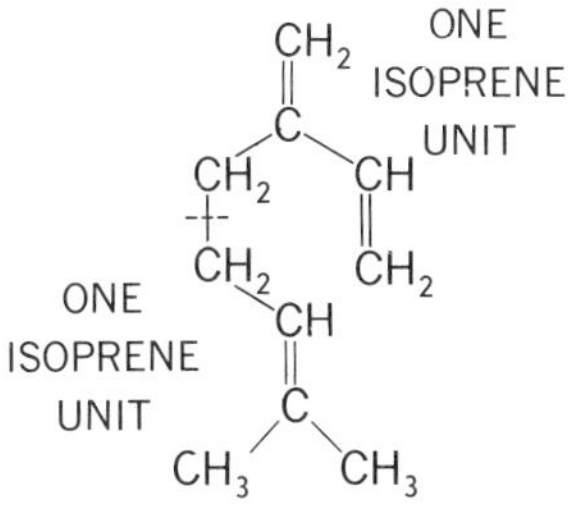

MYRCENE

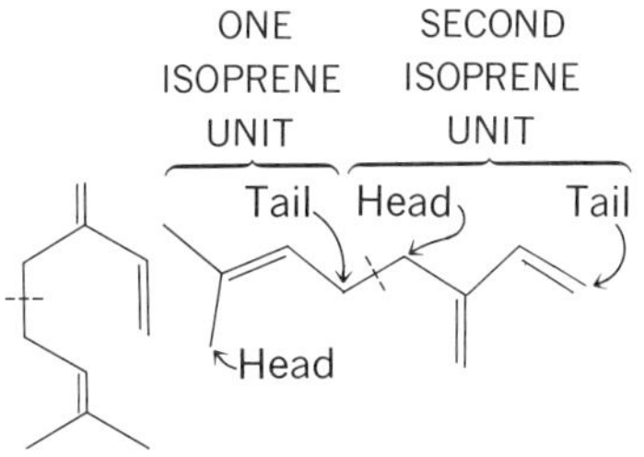

MYRCENE (Alternative symbols)

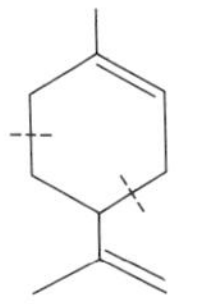

LIMONENE 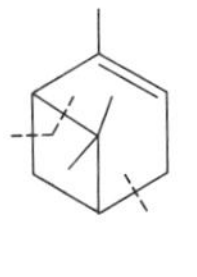 α-PINENE 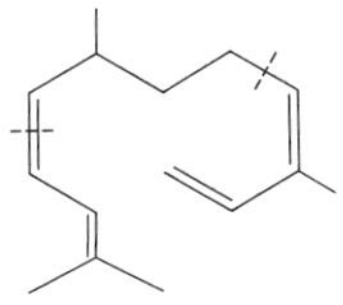 α-FARNESENE 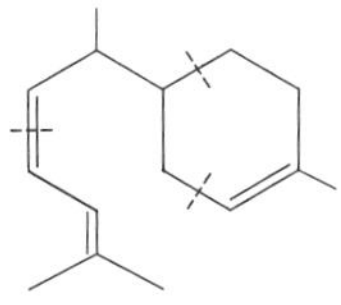 ZINGIBERENE

Squalene, a triterpene, is an unusual hydrocarbon found in shark-liver oil. In the multiple-step conversion of acetate units into cholesterol in the human body, squalene is one of the intermediates, as we shall see in Section 11.22. A violation of the isoprene rule is seen in squalene where, in the molecule's center, there is a joining of "tail-to-tail." Whenever the isoprene rule is violated it is usually in this way. Two fragments, each obeying this rule, are joined "tail-to-tail." Lycopene and β-carotene are other examples.

[1]"Oil of bay" is not a synonym for "myrcene," but oil of bay contains myrcene as one of its most characteristic components. In this section each terpene is named for the particular essential oil most clearly associated with it.

Tail Tail

$$\begin{array}{c}\mathrm{CH_3}\qquad\quad\mathrm{CH_3}\qquad\quad\mathrm{CH_3}\qquad\qquad\qquad\mathrm{CH_3}\qquad\quad\mathrm{CH_3}\qquad\quad\mathrm{CH_3}\\ \mathrm{CH_3C{=}CHCH_2CH_2C{=}CHCH_2CH_2C{=}CHCH_2CH_2CH{=}CCH_2CH_2CH{=}CCH_2CH_2CH{=}CCH_3}\end{array}$$

SQUALENE

The red-colored matter found in tomatoes is a tetraterpene, *lycopene.* Another tetraterpene, β-carotene, is largely responsible for the color of carrots, but it occurs in other plants as well where its color may be masked.

$$\begin{array}{c}\mathrm{CH_3}\qquad\quad\mathrm{CH_3}\qquad\qquad\qquad\mathrm{CH_3}\qquad\quad\mathrm{CH_3}\\ \mathrm{CH{=}CHC{=}CHCH{=}CHC{=}CHCH{=}CHCH{=}CCH{=}CHCH{=}CCH{=}CH}\end{array}$$

LYCOPENE

$$\begin{array}{c}\mathrm{CH_3}\qquad\quad\mathrm{CH_3}\qquad\qquad\qquad\mathrm{CH_3}\qquad\quad\mathrm{CH_3}\\ \mathrm{CH{=}CHC{=}CHCH{=}CHC{=}CHCH{=}CHCH{=}CCH{=}CHCH{=}CCH{=}CH}\end{array}$$

β-CAROTENE

The relation between the terpenoid compounds and isoprene is purely a formalism, for isoprene does not occur naturally and is not directly involved in the biosynthesis of any terpene. *Isopentenyl pyrophosphate* is the five-carbon intermediate through which isoprene units become linked together. (See Figure 11.8.)

11.20 Oxygen Derivatives of Terpenes

A great variety of alcohol, aldehyde, and ketone derivatives of the terpene hydrocarbons, isolated along with them, are essential contributors to the distinctive flavors and odors of many natural oils and perfumes. The structures of several are given in Figure 11.6. A good perfume, for example, attar of roses, is an extremely complex mixture of substances almost impossible to duplicate by synthetic means. Abietic acid is one of the principal constituents of rosin, a nonvolatile residue obtained when turpentine is extracted from pine. Its sodium salt is used in some laundry soaps.

11.21 Steroids

The steroids are generally characterized by a polycyclic carbon skeleton (Figure 11.7). The alcohol group, the keto group, and the double bond are common, but the ester linkage is seldom present.

The physiological effects of the steroids vary greatly from compound to compound, ranging from vitamin activity to the action of sex hormones. Some steroids are fat emulsifiers found in bile; others are important hormones; one stimulates the heart; another has been implicated in hardening of the arteries; still another ruptures red blood cells. The structures, names, and chief physiological properties of a few important steroids are given in Table 11.4.

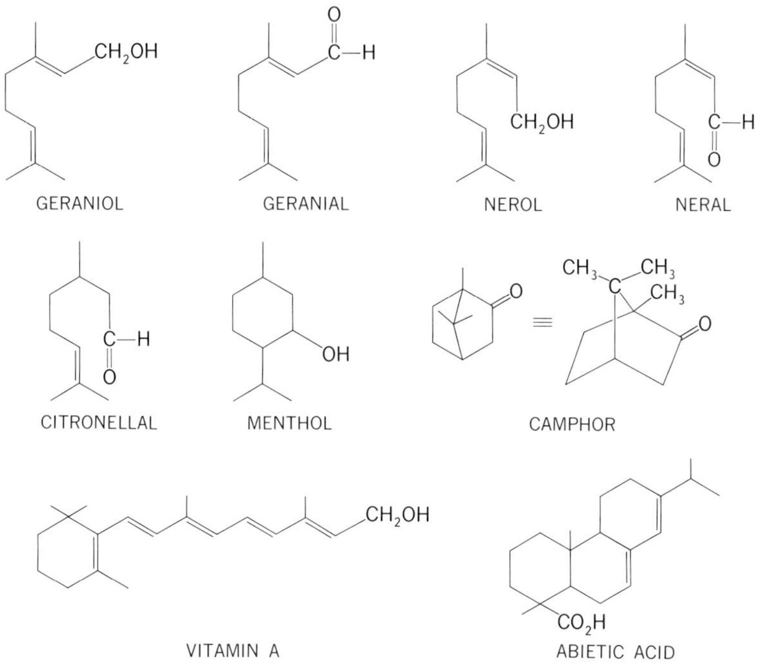
CH2OH
GERANIOL
O
C—H
GERANIAL
CH2OH
NEROL
C—H
O
NERAL
C—H
O
CITRONELLAL
OH
MENTHOL
O
CH3
CH3
C
CH3
O
CAMPHOR
CH2OH
VITAMIN A
CO2H
ABIETIC ACID

Figure 11.6
Some oxygen derivatives of terpenes.

11.22 Cholesterol

Cholesterol (Figure 11.7*c*) is a steroid alcohol or *sterol.* It is found in nearly all tissues of vertebrates, particularly in the brain and spinal cord, and is the main constituent of gallstones. In recent years scientists have found evidence associating cholesterol with circulatory problems such as hardening of the arteries.

The importance of cholesterol in human metabolism is seen in the fact that cholesterol is used to make several steroid hormones and bile acids. (Structures of some of these are given in Table 11.4.) These include progesterone, testosterone, estrone, and corticosterone.

How the body makes cholesterol has been established, and the main steps are shown in Figure 11.8. The conversion of the open-chain polyene, squalene, to the tetracyclic lanosterol is one of the most remarkable chemical events in living things,

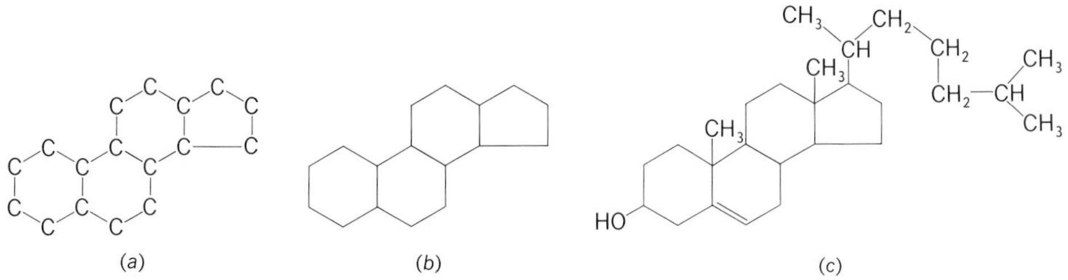
(a)
(b)
HO
CH3
CH3
CH3
CH
CH2
CH2
CH2—CH
CH3
CH3
(c)

Figure 11.7
(*a*) Carbon skeleton characteristic of steroids. (*b*) The most common way of representing the steroid "nucleus." (*c*) The steroid alcohol, cholesterol (Greek: *chole,* bile; *stereos,* solid; plus "-ol," alcohol).

a reaction of absorbing beauty. A series of simple, electron shifts and carbonium ion rearrangements of hydrogens and methyl groups closes four rings in one grand cascade.

Table 11.4
Important Individual Steroids

Vitamin D_2 precursor

Irradiation of this hormone, the commonest of all plant hormones, by ultraviolet light opens one of the rings to produce vitamin D_2.

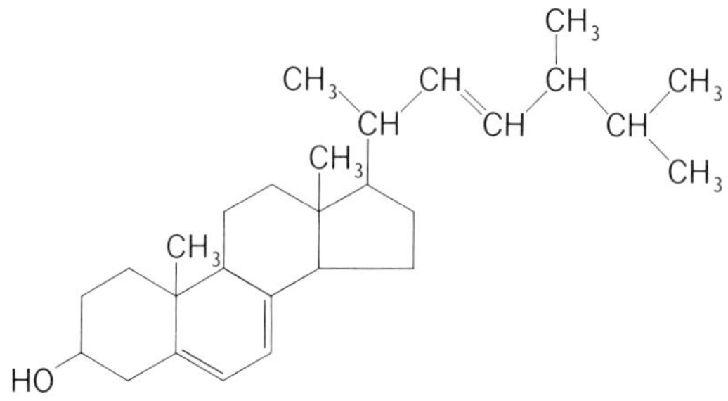

ERGOSTEROL

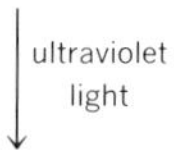

ultraviolet light

A deficiency of this antirachitic factor causes rickets, an infant and childhood disease characterized by faulty deposition of calcium phosphate and poor bone growth.

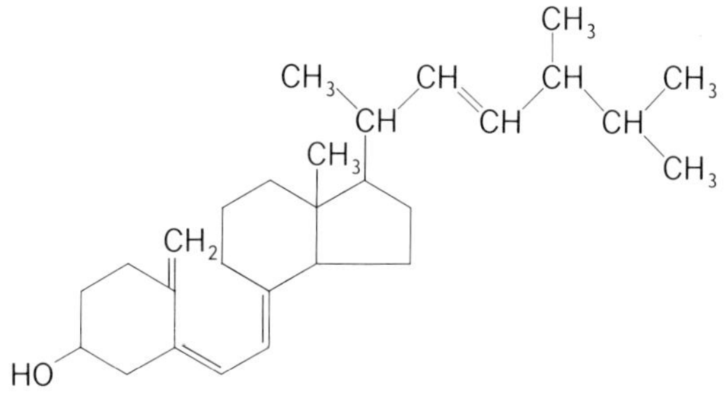

VITAMIN D_2

Bile acid

Cholic acid is found in bile in the form of its sodium salt. This and closely related salts are the bile salts that act as powerful emulsifiers of lipid material awaiting digestion in the upper intestinal tract. The sodium salt of cholic acid is soaplike because it has a very polar head and a large hydrocarbon tail.

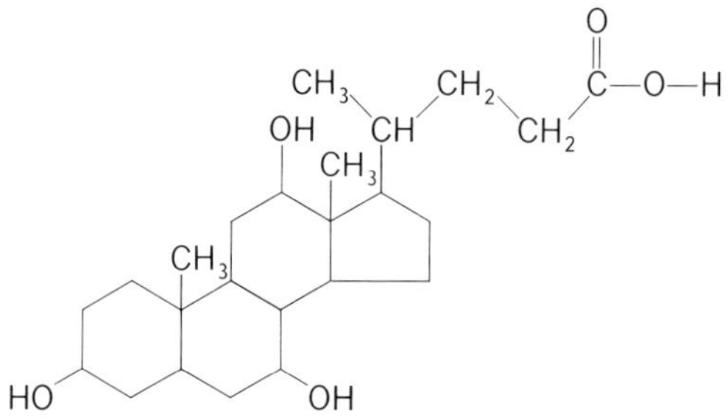

CHOLIC ACID

Table 11.4 (*Continued*)

Antiarthritic compound One of the 28 adrenal cortical hormones, cortisone is not only important in the control of carbohydrate metabolism but also effective in relieving the symptoms of rheumatoid arthritis.	CORTISONE
Cardiac aglycone Digitoxigenin is found in many poisonous plants, notably digitalis, as a complex glycoside. In small doses it stimulates the vagus mechanism and increases heart tone. In larger doses it is a potent poison.	DIGITOXIGENIN
Sex hormones This is one human estrogenic hormone.	ESTRONE
This human pregnancy hormone is secreted by the corpus luteum.	PROGESTERONE
This male sex hormone regulates the development of reproductive organs and secondary sex characteristics.	TESTOSTERONE

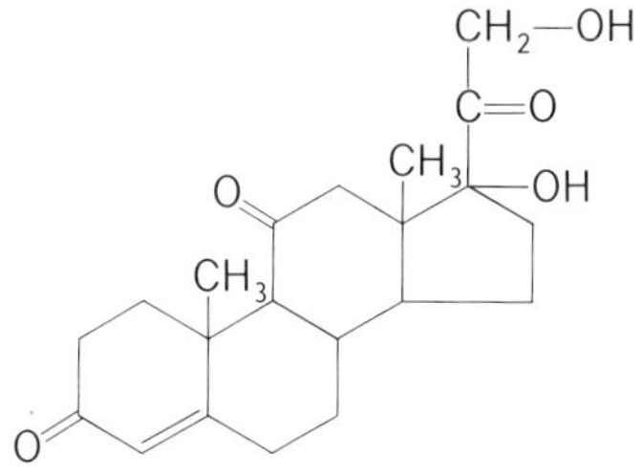

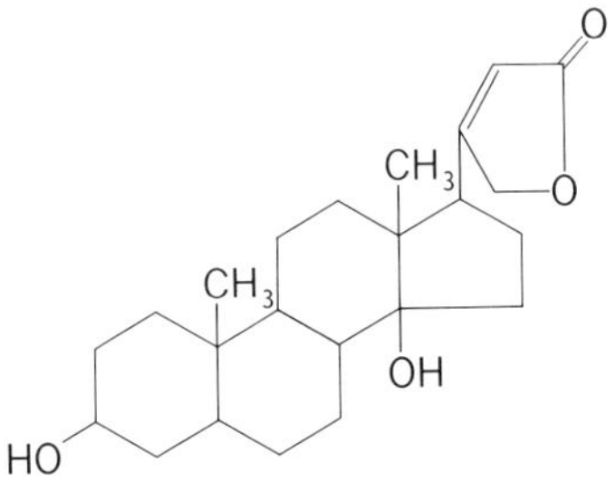

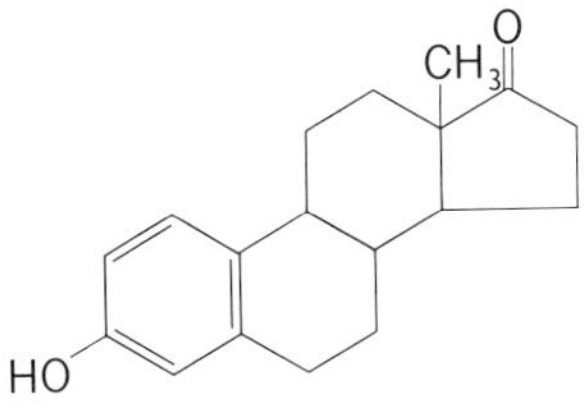

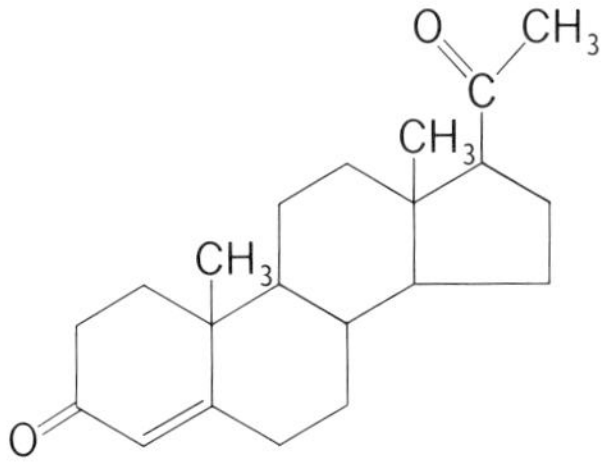

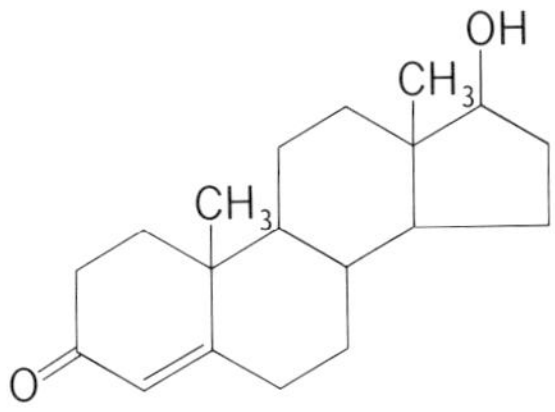

Table 11.4 (*Continued*)

Androsterone, a second male sex hormone, is less potent than testosterone	ANDROSTERONE
Synthetic hormones in fertility control Oral contraceptive pills contain one or two synthetic, hormonelike compounds. (Synthetic compounds have to be used because the real hormones, taken orally, are broken down by the body's enzymes.)	Synthetic estrogens If R = H, ETHYNYLESTRADIOL R = CH_3, MESTRANOL
Until about 1971 the most widely used pills contained an estrogen (0.05 mg or more) and a progestin (0.5 mg or more). The "minipill" appeared about then and it contains only a synthetic progestin (0.1 to 0.2 mg). Lacking the estrogen, its side effects appear to be reduced. (However, a small percentage of the progestin is changed to an estrogen in the body.)	Synthetic progestins NORETHYNODREL
The older, two-hormone pills acted primarily by inhibiting the release of the egg from the ovary. The minipill seems to act by hindering sperm migration into the oviducts.	If R = H, NORETHINDRONE R = $\overset{O}{\overset{\|\|}{C}}-CH_3$, NORETHINDRONE ACETATE ETHYNODIOL DIACETATE

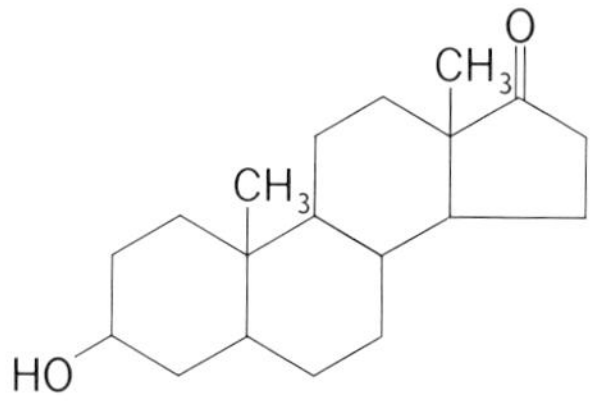

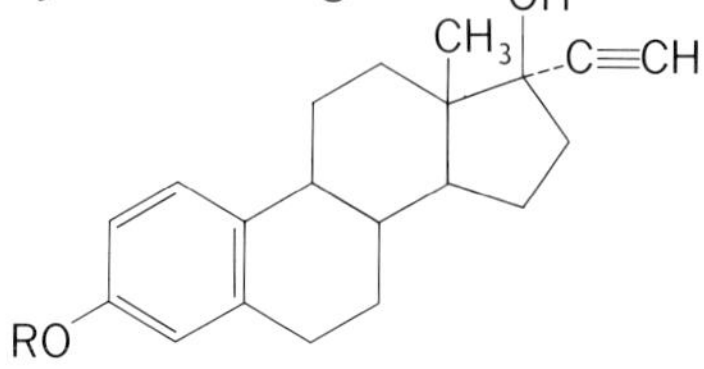

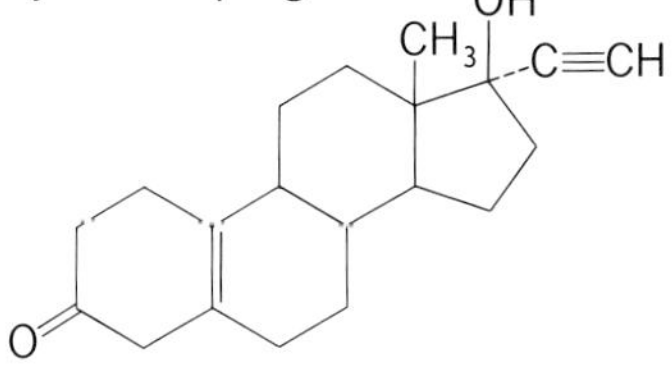

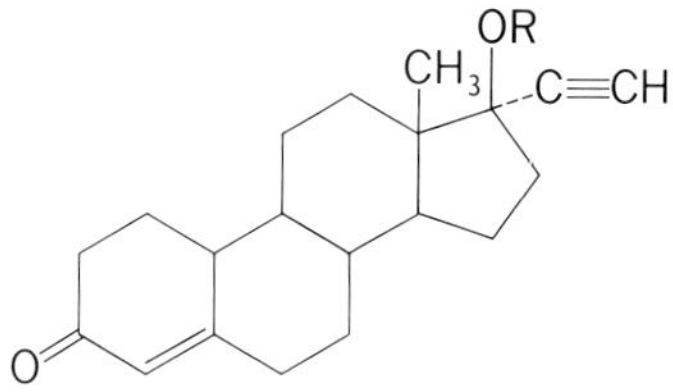

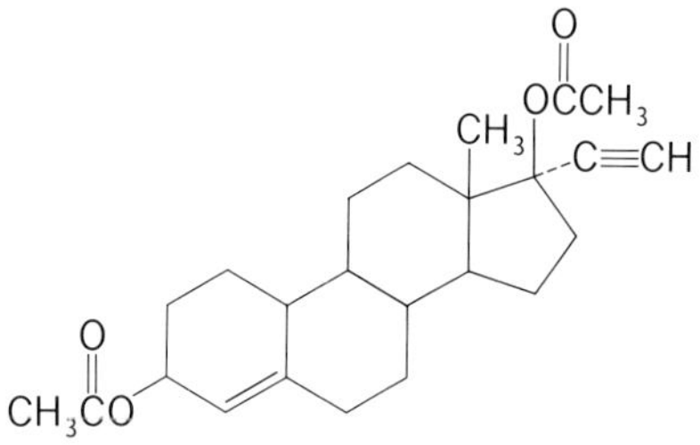

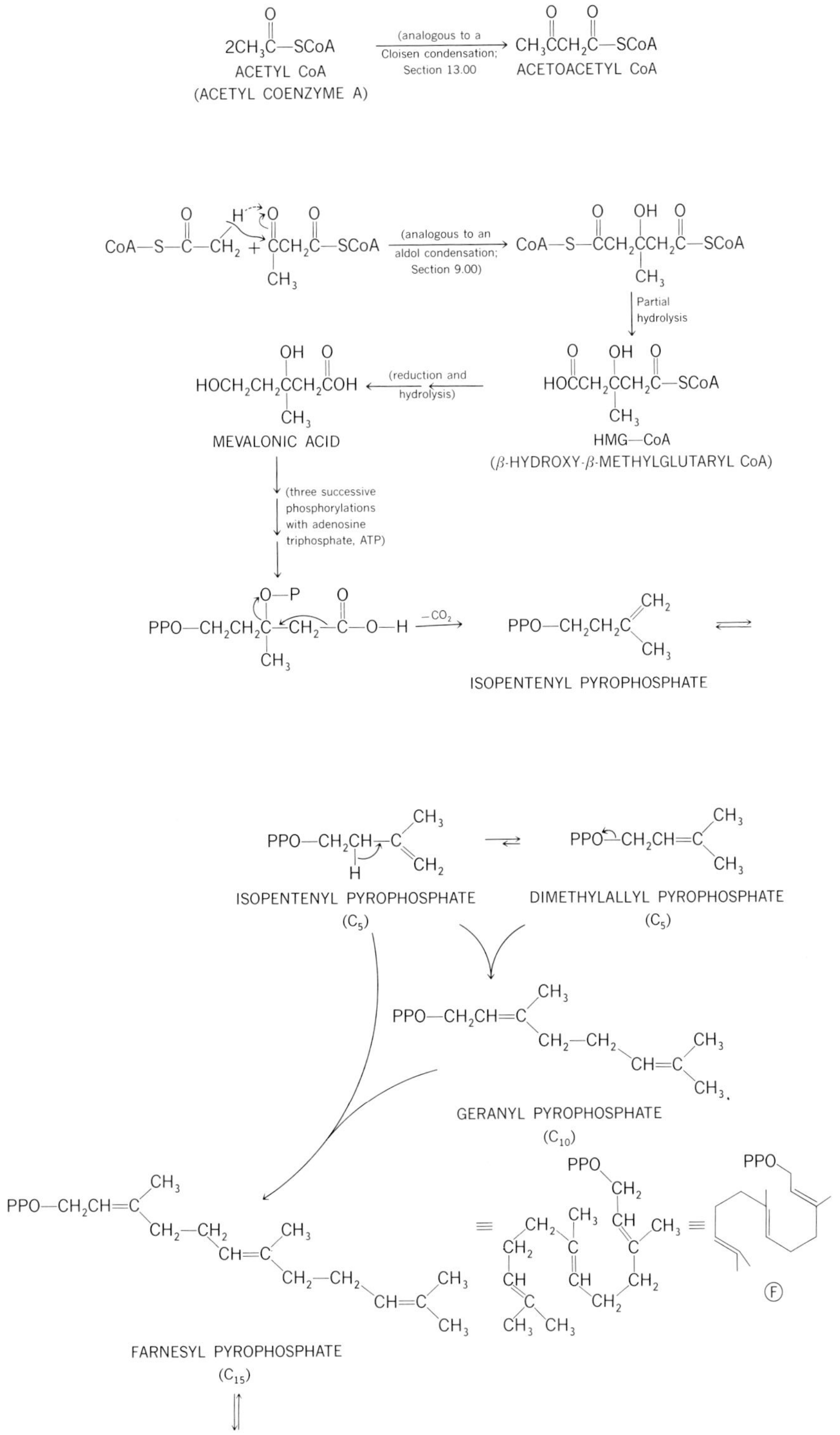

$$2CH_3\overset{O}{\overset{\|}{C}}\text{—SCoA} \xrightarrow[\text{Cloisen condensation; Section 13.00}]{\text{(analogous to a}} CH_3\overset{O}{\overset{\|}{C}}CH_2\overset{O}{\overset{\|}{C}}\text{—SCoA}$$

ACETYL CoA
(ACETYL COENZYME A)

ACETOACETYL CoA

$$\text{CoA—S—}\overset{O}{\overset{\|}{C}}\text{—}CH_2 + \overset{O}{\overset{\|}{C}}(CH_3)CH_2\overset{O}{\overset{\|}{C}}\text{—SCoA} \xrightarrow[\text{aldol condensation; Section 9.00)}]{\text{(analogous to an}} \text{CoA—S—}\overset{O}{\overset{\|}{C}}CH_2\overset{OH}{\overset{|}{C}}(CH_3)CH_2\overset{O}{\overset{\|}{C}}\text{—SCoA}$$

↓ Partial hydrolysis

$$HO\overset{O}{\overset{\|}{C}}CH_2\overset{OH}{\overset{|}{C}}(CH_3)CH_2\overset{O}{\overset{\|}{C}}\text{—SCoA}$$

HMG—CoA
(β-HYDROXY-β-METHYLGLUTARYL CoA)

$$\xrightarrow[\text{hydrolysis)}]{\text{(reduction and}} HOCH_2CH_2\overset{OH}{\overset{|}{C}}(CH_3)CH_2\overset{O}{\overset{\|}{C}}OH$$

MEVALONIC ACID

↓ (three successive phosphorylations with adenosine triphosphate, ATP)

$$\text{PPO—}CH_2CH_2\overset{O\text{—}P}{\overset{|}{C}}(CH_3)\text{—}CH_2\text{—}\overset{O}{\overset{\|}{C}}\text{—O—H} \xrightarrow{-CO_2} \text{PPO—}CH_2CH_2C(=CH_2)CH_3 \rightleftarrows$$

ISOPENTENYL PYROPHOSPHATE

$$\text{PPO—}CH_2CH(H)\text{—}C(CH_3)=CH_2 \rightleftarrows \text{PPO—}CH_2CH=C(CH_3)_2$$

ISOPENTENYL PYROPHOSPHATE
(C_5)

DIMETHYLALLYL PYROPHOSPHATE
(C_5)

$$\text{PPO—}CH_2CH=C(CH_3)\text{—}CH_2\text{—}CH_2\text{—}CH=C(CH_3)CH_3.$$

GERANYL PYROPHOSPHATE
(C_{10})

$$\text{PPO—}CH_2CH=C(CH_3)\text{—}CH_2\text{—}CH_2\text{—}CH=C(CH_3)\text{—}CH_2\text{—}CH_2\text{—}CH=C(CH_3)CH_3 \equiv \cdots \equiv$$

FARNESYL PYROPHOSPHATE
(C_{15})

Ⓕ

⇅

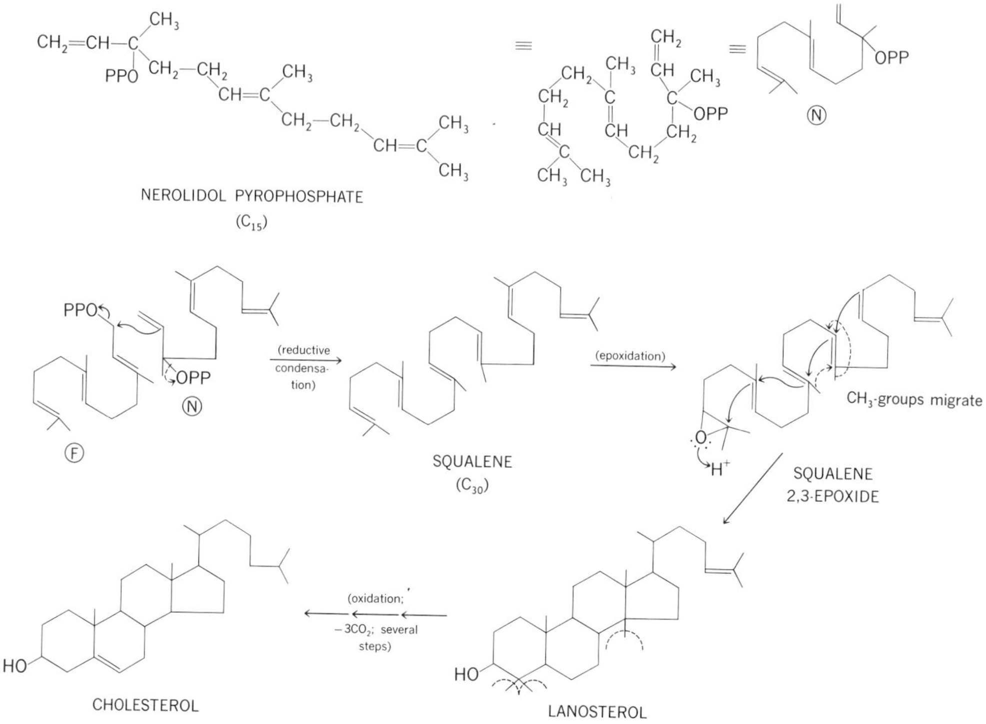

NEROLIDOL PYROPHOSPHATE (C_{15})

SQUALENE (C_{30})

SQUALENE 2,3-EPOXIDE

CHOLESTEROL

LANOSTEROL

Figure 11.8
Biosynthesis of cholesterol. It starts with two-carbon acetyl units attached by an esterlike link to coenzyme A. Three of these are joined to give the C-6 HMG-CoA, which is then reduced and hydrolyzed to yield mevalonic acid. What happens next—three successive phosphorylations—is analogous to cranking mevalonic acid up an energy hill. During the loss of CO_2, next, it is energetically easier to kick out "O-P" (a phosphate unit) than OH^-. With isopentenyl pyrophosphate we have reached the C-5 stage of the isoprene skeleton.

The rearrangement of a double bond gives an allylic system. Probably by means of some intermediate like an allylic carbonium ion, the two C-5 units join to give the C-10 geranyl compound. Another event like this leads to the C-15 farnesyl derivative, and this, by a small rearrangement gives some isomeric C-15 nerolidol derivative.

These two C-15 compounds join "end-to-end" to give the C-30 squalene. By oxidizing one double bond of squaline to give the squalene epoxide, the system is pushed to the energy edge of the ring-closing cascade. Methyl groups and hydride units must also rearrange during this extraordinary event. Lanosterol is formed and by a series of steps is changed to cholesterol.

Brief Summary: Important Chemical and Physical Properties

1. The fatty acids in lipids are long chain, even carbon-numbered carboxylic acids that may have one or more alkene groups.

2. Triglycerides ("glycerides," fats and oils) are esters of glycerol with fatty acids. Triglycerides can be:
   (a) hydrolyzed or digested to glycerol and fatty acids
   (b) saponified to glycerol and salts of fatty acids
   (c) hydrogenated to more saturated triglycerides

3. The degree of unsaturation of a lipid is measured by the iodine number, the number of grams of iodine that react with 100 grams of the lipid. The higher the iodine number the more unsaturated the lipid.

4. Some idea of the average lengths of fatty acid chains in triglycerides is given by the saponification value: the number of milligrams of potassium hydroxide needed to saponify one gram of lipid. The higher the saponification value, the shorter the average chain lengths.

5. The glycerides with high iodine numbers (e.g., linseed oil) are attacked by oxygen and harden. They are the basis for ordinary paints.

6. A detergent must have molecules (or ions) with long hydrocarbon chains and very polar or ionic "heads" to work.

7. If the hydrocarbon chain in detergent molecules (or ions) is full of alkyl branches, the detergent cannot be easily degraded by microorganisms in soil or ground water. The detergent is not biodegradable. (Ordinary soaps, salts of long chain fatty acids, are biodegradable.)

8. The phospholipids, especially important in the nervous system, have ordinary functional groups with ester links being the most common.

Phosphatides and plasmalogens can be hydrolyzed to glycerol, one or more fatty acids, phosphoric acid (or a salt of it), and an aminoalcohol or related compound.
Sphingolipids are based on sphingosine instead of glycerol.

9. The principal nonsaponifiable lipids are the steroids, which include several hormones, and the terpenes.

10. Terpenes are natural products whose molecules are based on isoprene units joined according to the isoprene rule (i.e., "head-to-tail"). (Large molecules sometimes have "tail-to-tail" junctures in the middle.)

Problems and Exercises

1. When hydrochloric acid is added to a solution of sodium stearate an organic compound precipitates. What is its structure and name?
2. Write the structure of a mixed glyceride between glycerol and each of the following three acids: palmitic acid, oleic acid, and linolenic acid.
   (a) Write an equation for the saponification of this triglyceride.
   (b) Write an equation for its digestion.
   (c) Write an equation for a reaction that would reduce the iodine number to zero.
   (d) Calculate the iodine number of this triglyceride. Is it likely to be a solid or a liquid at room temperature?

3. Given the structure of a C-10 compound how would you judge if the substance is or is not likely a terpene or derived from a terpene?
4. Which of the two structures is that of a terpene?

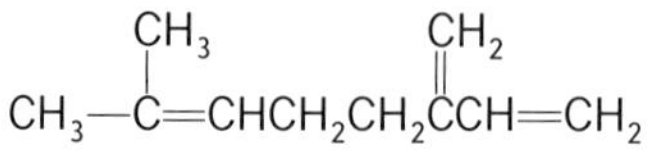

$$\underset{\text{(a)}}{CH_3{-}\overset{\overset{\displaystyle CH_3}{|}}{C}{=}CHCH_2CH_2\overset{\overset{\displaystyle CH_2}{\|}}{C}CH{=}CH_2} \qquad \underset{\text{(b)}}{CH_3CH{=}CH\overset{\overset{\displaystyle CH_3}{|}}{C}HCH_2\overset{\overset{\displaystyle CH_2}{\|}}{C}CH{=}CH_2}$$

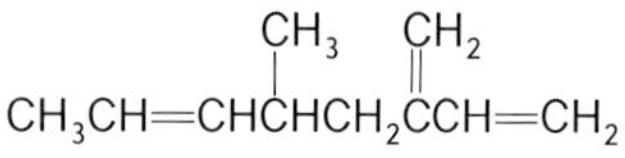

5. If linseed oil were hydrogenated, how would this affect its ability to act as a drying agent and why?
6. Plasmalogens can be hydrolyzed to a fatty acid, phosphoric acid, ethanolamine (or something similar), and to an unsaturated alcohol. (See general structure for plasmalogens in Section 11.15.) Actually the unsaturated alcohol is unstable and it spontaneously rearranges to what kind of compound? (Hint, see Section 9.22.)
7. Sodium acetate has essentially no detergent action compared to sodium laurate. Explain.
8. Prove the statement in Section 11.7, "the higher the saponification value the more the *short*-chain fatty acids are present" in a lipid sample.
9. Which would have the higher saponification value, glyceryl triacetate or glyceryl tripropionate?

chapter twelve
Difunctional Compounds

12.1 Symbols Without Surprises

Structural formulas are abstract symbols for molecules. Symbols are our servants. We invent them to be of help. We don't want them to mislead us. If they do, we usually discard them and make new ones. Some we keep. In spite of problems they are too useful to eliminate. Instead, we make mental adjustments; we learn how to make the necessary allowances. In this chapter we study some cases.

Ideally, we want a symbol for a functional group to stand for a fairly clear-cut, consistent set of chemical and physical properties. We want >C=C<, for example, to stand for "addition reactions," to name one relation. We'd rather not see that symbol in the structure of molecules of a compound if, in fact, the compound does not give addition reactions. Then we ran into C_6H_6, benzene. We decided (actually, of course, scientists worked this out long before us) to keep the old Kekulé structure with its three double bonds alternating with single bonds. Even though benzene does not give addition reactions—at least, not nearly as readily as do alkenes—there were too many uses for the Kekulé structure to throw it out entirely. We keep it when we want to apply the theory of resonance to aromatic compounds.

We learned that ethers, R—O—R, are extremely resistant to hydrolysis. When we saw the ether structure in acetals, however, we could not think in terms of "no hydrolysis." Acetals are easy to hydrolyze. We learned to make allowances for this special kind of "ether." (See Section 9.19.)

When we studied 1,3-butadiene, a conjugated diene, we had to learn to think in terms of 1,4-addition as well as 1,2-addition. We learned to make this kind of an allowance for conjugated systems.

When we studied alcohols we learned that we cannot expect its —OH groups to be acidic. But when this group occurs in phenols, it is acidic; when it is in carboxylic acids, it is more so; and when it is in sulfonic acids it's almost as acidic as sulfuric acid itself.

From these earlier encounters we should by now be used to the fact that virtually none of our symbols is so "perfect" that it leads to no surprises. Think back, however, on the examples just catalogued. In every case the molecules involved had at least two functional groups, and the two were near each other. There were three "double bonds," *conjugated* in benzene, two ether links *joined to the same carbon* in acetals, and two double bonds *in conjugation* in 1,3-dienes. The —OH group is *joined to unsaturated groups* in the benzene ring in phenols, a carbonyl group in carboxylic acids, and the sulfur-oxygen polar bond in sulfonic acids. That is the key; while most reactions will be "normal" we must *expect* some "unusual" behavior whenever two functional groups get near each other. We shall look at other important examples in this chapter. They have been selected largely for their relevance to biological chemistry and to organic synthesis.

12.2 Dicarboxylic Acids and Their Derivatives

Many important, naturally occurring carboxylic acids have two carboxyl groups per molecule. Several are listed in Table 12.1.

12.3 Some of the Expected Reactions of Dicarboxylic Acids

As the following equations illustrate we can expect dicarboxylic acids to form salts, esters, and acid chlorides.

$$HO\overset{O}{\overset{\|}{C}}CH_2CH_2\overset{O}{\overset{\|}{C}}OH + 2NaOH \longrightarrow \left(^{-}O\overset{O}{\overset{\|}{C}}CH_2CH_2\overset{O}{\overset{\|}{C}}O^{-}\right)2Na^{+} + 2H_2O$$

SUCCINIC ACID → DISODIUM SUCCINATE

$$HO\overset{O}{\overset{\|}{C}}(CH_2)_4\overset{O}{\overset{\|}{C}}OH + 2CH_3CH_2OH \xrightarrow[\Delta]{H^+} CH_3CH_2O\overset{O}{\overset{\|}{C}}(CH_2)_4\overset{O}{\overset{\|}{C}}OCH_2CH_3 + 2H_2O$$

ADIPIC ACID → DIETHYL ADIPATE

$$HO\overset{O}{\overset{\|}{C}}-C_6H_4-\overset{O}{\overset{\|}{C}}OH + 2SOCl_2 \longrightarrow Cl-\overset{O}{\overset{\|}{C}}-C_6H_4-\overset{O}{\overset{\|}{C}}-Cl + 2SO_2 + 2HCl$$

TEREPHTHALIC ACID → TEREPHTHALOYL CHLORIDE

The amides are known also, but they usually have to be made indirectly. For example, by ammonolysis of esters:

$$CH_3CH_2O\overset{O}{\overset{\|}{C}}CH_2\overset{O}{\overset{\|}{C}}OCH_2CH_3 + NH_3 \longrightarrow NH_2\overset{O}{\overset{\|}{C}}CH_2\overset{O}{\overset{\|}{C}}NH_2 + 2CH_3CH_2OH$$

DIETHYL MALONATE + (Excess) → MALONDIAMIDE

12.4 Effect of Heat on Dicarboxylic Acids

The course of the reaction depends entirely on the closeness of the two carboxyl groups to each other. The behavior of β-carboxylic acids—that is, malonic and substituted malonic acids—is particularly important.

Table 12.1
Dicarboxylic Acids

	n	Structure	Name	Origin of Name	Mp (°C)	K_a (25°C) First Proton	K_a (25°C) Second Proton
Straight-chain saturated dicarboxylic acids, $C_nH_{2n-2}O_4$	2	HO_2CCO_2H	Oxalic acid (ethanedioic)	L.*oxalis*, sorrel	190 (decomposes)	$5{,}900 \times 10^{-5}$	6.4×10^{-5}
	3	$HO_2CCH_2CO_2H$	Malonic acid (propanedioic)	L.*malum*, apple	136	149×10^{-5}	0.20×10^{-5}
	4	$HO_2C(CH_2)_2CO_2H$	Succinic acid (butanedioic)	L.*succinum*, amber	182	6.9×10^{-5}	0.25×10^{-5}
	5	$HO_2C(CH_2)_3CO_2H$	Glutaric acid (pentanedioic)	L.*gluten*, glue	98	4.6×10^{-5}	0.39×10^{-5}
	6	$HO_2C(CH_2)_4CO_2H$	Adipic acid (hexanedioic)	L.*adipis*, fat	153	3.7×10^{-5}	0.24×10^{-5}
Miscellaneous dicarboxylic acids		$HCCO_2H$ ‖ $HCCO_2H$	Maleic acid	L.*malum*, apple	131°	$1{,}420 \times 10^{-5}$	$-.08 \times 10^{-5}$
		HO_2CCH ‖ $HCCO_2H$	Fumaric acid	Fumitory (a garden annual)	Sublimes above 200	93×10^{-5}	3.6×10^{-5}
		$C_6H_4(CO_2H)_2$ (ortho)	Phthalic acid	naphthalene	231	130×10^{-5}	0.39×10^{-5}
		$C_6H_4(CO_2H)_2$ (meta)	Isophthalic acid		345	29×10^{-5}	0.25×10^{-5}
		$C_6H_4(CO_2H)_2$ (para)	Terephthalic acid		Sublimes	31×10^{-5}	1.5×10^{-5} (16°C)

1. Oxalic acid:

$$HO_2C{-}CO_2H \xrightarrow{150°C} HCO_2H + CO_2$$

OXALIC ACID → FORMIC ACID + CARBON DIOXIDE

2. Malonic acid (or any β-dicarboxylic acid):

$$HO_2\overset{\beta}{C}\overset{\alpha}{C}H_2CO_2H \xrightarrow{140°C} CH_3\overset{O}{\overset{\|}{C}}OH + CO_2$$

ACETIC ACID

$$\text{(H)R(H)R}'C(COOH)_2 \xrightarrow{\text{Heat}} \text{(H)R(H)R}'CH{-}\overset{O}{\overset{\|}{C}}OH + CO_2$$

SUBSTITUTED MALONIC ACID → SUBSTITUTED ACETIC ACID

3. Succinic acid (or any γ-dicarboxylic acid):

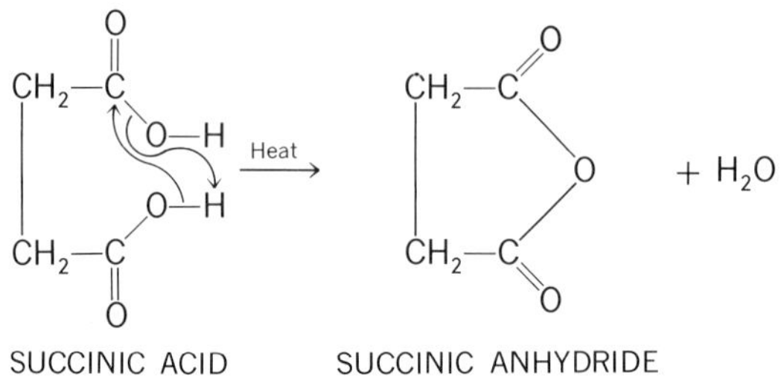

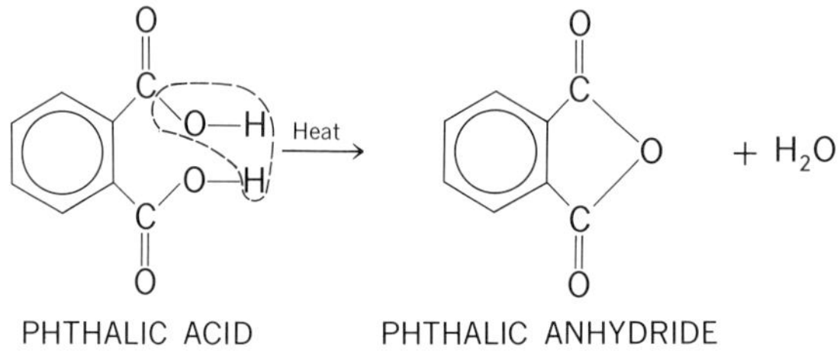

$$\underset{\text{SUCCINIC ACID}}{\begin{matrix} CH_2-C(=O)-O-H \\ | \\ CH_2-C(=O)-O-H \end{matrix}} \xrightarrow{\text{Heat}} \underset{\text{SUCCINIC ANHYDRIDE}}{\begin{matrix} CH_2-C(=O) \\ | \quad\quad \diagdown O \\ CH_2-C(=O) \diagup \end{matrix}} + H_2O$$

$$\text{PHTHALIC ACID} \xrightarrow{\text{Heat}} \text{PHTHALIC ANHYDRIDE} + H_2O$$

We shall not go farther than this, to the longer-chain dicarboxylic acids. (Glutaric acid can be changed to a cyclic anhydride, too, one with a six-membered ring. Beyond that it becomes a bit more complicated.)

12.5 Reactions of Cyclic Anhydrides

These are true anhydrides (Section 10.16). Alcohols change them to esters; ammonia and amines change them to amides; water hydrolyzes them. The difference between cyclic and open-chain anhydrides is that the other product in all these reactions—something with a free carboxyl group—is still tied to the molecule after cyclic anhydrides have been opened. The examples illustrate this.

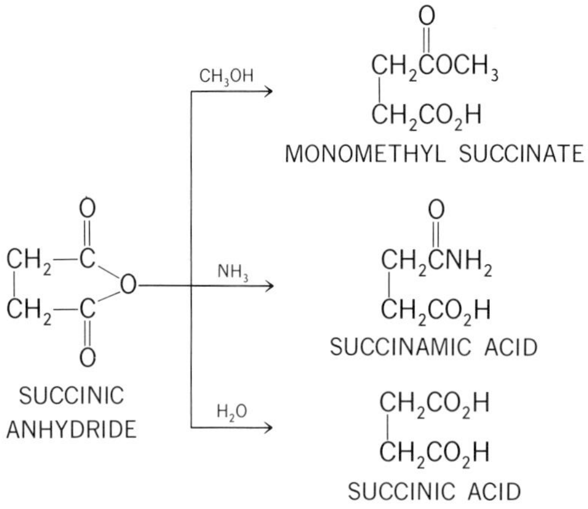

$$\text{SUCCINIC ANHYDRIDE} \xrightarrow{CH_3OH} \underset{\text{MONOMETHYL SUCCINATE}}{\begin{matrix} CH_2COCH_3 \ (C{=}O) \\ | \\ CH_2CO_2H \end{matrix}}$$

$$\text{SUCCINIC ANHYDRIDE} \xrightarrow{NH_3} \underset{\text{SUCCINAMIC ACID}}{\begin{matrix} CH_2CNH_2 \ (C{=}O) \\ | \\ CH_2CO_2H \end{matrix}}$$

$$\text{SUCCINIC ANHYDRIDE} \xrightarrow{H_2O} \underset{\text{SUCCINIC ACID}}{\begin{matrix} CH_2CO_2H \\ | \\ CH_2CO_2H \end{matrix}}$$

12.6 Malonic Esters in Synthesis

Diethyl malonate usually is called simply "malonic ester." Its molecules have a methylene group (—CH_2—) flanked by carbonyl groups, that is, two unsaturated electron-withdrawing groups.

Any time a —CH_2— group or simply a —CH— group is flanked by two unsaturated, electron-withdrawing groups it will be acidic enough to give up a proton to a strong base. The flanking groups may be any two from the following list:

—CO_2R	(ester)	—CN	(cyano)
—COR	(ketone)	—NO_2	(nitro)

Examples are:

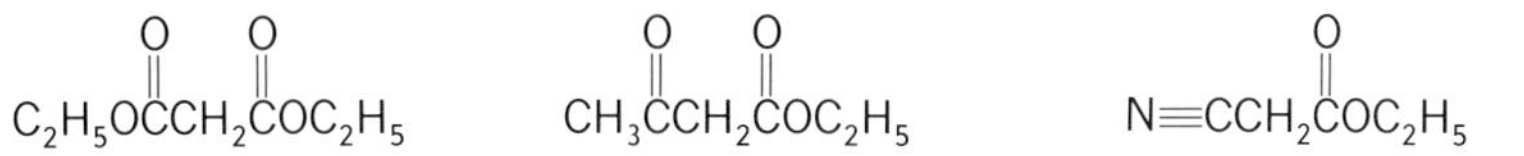

$$\underset{\text{DIETHYL MALONATE ("MALONIC ESTER")}}{C_2H_5O\overset{O}{\overset{\|}{C}}CH_2\overset{O}{\overset{\|}{C}}OC_2H_5} \qquad \underset{\text{ETHYL ACETOACETATE ("ACETOACETIC ESTER")}}{CH_3\overset{O}{\overset{\|}{C}}CH_2\overset{O}{\overset{\|}{C}}OC_2H_5} \qquad \underset{\text{ETHYL CYANOACETATE}}{N{\equiv}CCH_2\overset{O}{\overset{\|}{C}}OC_2H_5}$$

These are called "active" methylene compounds. The strong base most commonly used is sodium ethoxide in dry ethanol ("absolute" ethanol). For malonic ester the reaction is as follows:

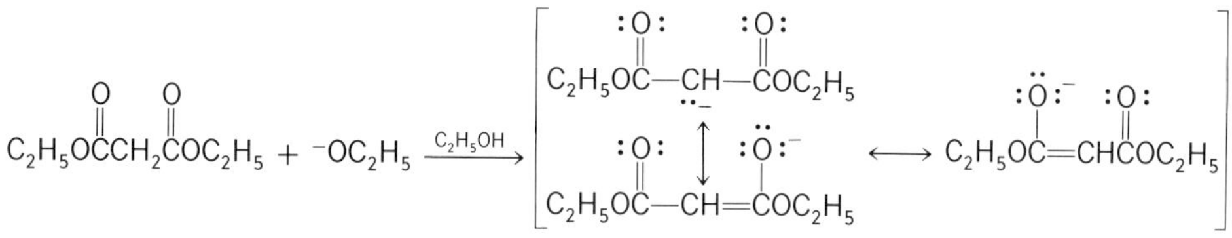

$$C_2H_5O\overset{O}{\overset{\|}{C}}CH_2\overset{O}{\overset{\|}{C}}OC_2H_5 + {}^-OC_2H_5 \xrightarrow{C_2H_5OH} \left[\begin{array}{c} C_2H_5O\overset{:O:}{\overset{\|}{C}}-\underset{\cdot\cdot}{\overset{}{C}}\!H^-\!-\overset{:O:}{\overset{\|}{C}}OC_2H_5 \\ \updownarrow \\ C_2H_5O\overset{:O:}{\overset{\|}{C}}-CH{=}\overset{:\ddot{O}:^-}{\overset{|}{C}}OC_2H_5 \end{array} \longleftrightarrow C_2H_5O\overset{:\ddot{O}:^-}{\overset{|}{C}}{=}CH\overset{:O:}{\overset{\|}{C}}OC_2H_5 \right]$$

The new anion is stabilized by resonance. The negative charge is delocalized to the two oxygens as shown. Only when the listed unsaturated, electron-withdrawing groups are *both* joined to the same —CH_2— (or —CH—) group is it possible for this much delocalization to take place. This much delocalization is apparently necessary to make it possible for even a strong base to pull H^+ quantitatively from an attachment to carbon.

What we have produced by the action of ethoxide ion is a new nucleophile, one that can attack 1° or 2° alkyl halides in a nucleophilic substitution reaction. For example:

$$\underset{\text{MALONIC ESTER}}{CH_2(CO_2C_2H_5)_2} + {}^-OC_2H_5 \longrightarrow \underset{\text{ANION}}{{}^-\!:CH(CO_2C_2H_5)_2} \xrightarrow{RX} \underset{\text{A MONOSUBSTITUTED MALONIC ESTER}}{R{-}CH(CO_2C_2H_5)_2} + X^-$$

What makes this a very useful reaction is that we can change the monosubstituted malonic ester to a monosubstituted acetic acid. We first hydrolyze the di-ester to a monosubstituted malonic acid. Next, we decarboxylate that (cf. Section 12.4). Thus:

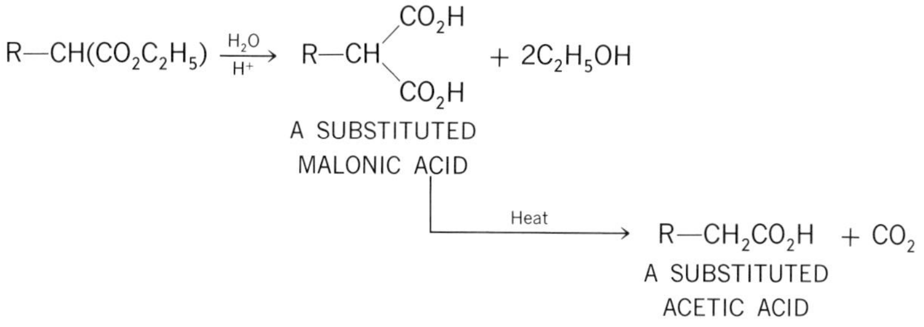

$$R{-}CH(CO_2C_2H_5) \xrightarrow[H^+]{H_2O} \underset{\text{A SUBSTITUTED MALONIC ACID}}{R{-}CH\begin{array}{l} CO_2H \\ CO_2H \end{array}} + 2C_2H_5OH$$

$$\xrightarrow{\text{Heat}} \underset{\text{A SUBSTITUTED ACETIC ACID}}{R{-}CH_2CO_2H} + CO_2$$

The overall change of malonic ester to a substituted acetic acid goes by the name of the **malonic ester synthesis.**

Exercise 12.1 Which organic halide would be used in the malonic ester synthesis of the following carboxylic acids?
(a) propionic acid (b) butyric acid
(c) isovaleric acid (d) 3-phenylpropanoic acid

The malonic-ester synthesis can be repeated, however, using monosubstituted malonic esters. The final products, then, will be disubstituted acetic acids. We shall not pursue this here but leave it instead to an exercise in class as time permits. No new principles are involved.

12.7 β-Keto Esters and Acids

A long-chain carboxylic acid molecule may also have a keto group anywhere down the chain. Some examples are:

$CH_3C(=O)CO_2H$	$CH_3C(=O)CH_2CO_2H$	$CH_3C(=O)CH_2CH_2CO_2H$
PYRUVIC ACID (AN α-KETOACID)	ACETOACETIC ACID (A β-KETOACID)	LEVULINIC ACID (A γ-KETOACID)

As we should expect, these compounds give reactions of ketones as well as reactions of carboxylic acids. We need not detail them here. Of all the types, only the β-ketoacids (and esters, etc.) are active methylene compounds. We shall limit our study to them.

12.8 Making β-Ketoesters. Claisen Condensation

Simple esters can be made to react with themselves. The reaction is very similar to the aldol condensation. In esters as in aldehydes and ketones, the carbonyl carbon is not the only reactive site. The α-position is also reactive. If a hydrogen is attached there, a strong base may take it; for example,

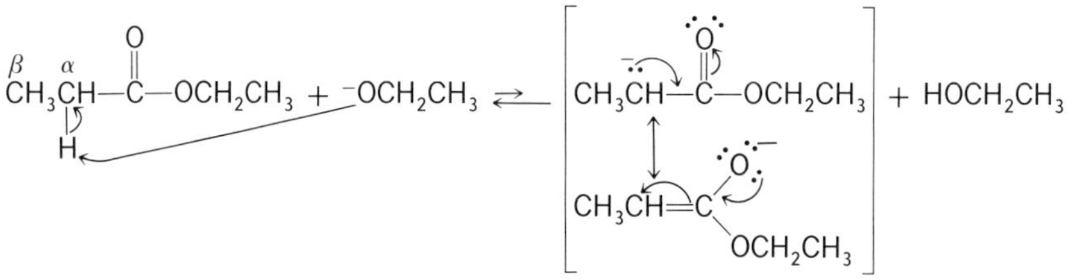

$$\underset{\text{ETHYL PROPIONATE}}{\overset{\beta\ \ \alpha}{CH_3CH(H)}-\overset{O}{\overset{\|}{C}}-OCH_2CH_3} + \underset{\text{ETHOXIDE ION}}{^-OCH_2CH_3} \rightleftharpoons \underset{\text{ANION OF ETHYL PROPIONATE}}{\left[CH_3\bar{C}H-\overset{O}{\overset{\|}{C}}-OCH_2CH_3 \longleftrightarrow CH_3CH=C(O^-)OCH_2CH_3\right]} + HOCH_2CH_3$$

Resonance stabilization of the anion helps to explain its formation.[1] Even so, it

[1]The reader may ask why the ethoxide ion does not attack the partially positively charged carbonyl carbon. Very likely it does, but no *new* product results. If such an attack causes ejection of the ethoxy group on the ester, the products are identical with the reactants. Such events undoubtedly occur but go nowhere.

is highly reactive. It can stabilize itself by abstracting a proton from a molecule of ethyl alcohol, regenerating the starting materials. (Hence the equilibrium arrows.) However, the anion of the ester is a new electron-rich species, a nucleophile. What other event might we *expect* to occur? If this anion encounters a molecule of unchanged ester, it could easily be attracted to the relatively electron-poor carbonyl carbon:

$$\underset{\text{UNCHANGED ESTER}}{CH_3CH_2\overset{\delta+}{\underset{}{C}}(=\overset{\delta-}{O})OCH_2CH_3} + \underset{\text{ANION OF THE ESTER}}{{}^{-}\!:\!CH(CH_3)\!-\!C(=O)\!-\!OCH_2CH_3} \rightleftharpoons \underset{\text{NEW ANION}}{\left[CH_3CH_2C(:\ddot{O}:^{-})(OCH_2CH_3)\!-\!CH(CH_3)\!-\!C(=O)OCH_2CH_3\right]}$$

The new anion might then stabilize itself by ejecting the ethoxide ion:

$$\left[CH_3CH_2C(:\ddot{O}:^{-})(OCH_2CH_3)\!-\!CH(CH_3)\!-\!C(=O)OCH_2CH_3\right] \rightleftharpoons \underset{\text{A }\beta\text{-KETO ESTER}}{CH_3CH_2C(=O)\!-\!CH(CH_3)\!-\!C(=O)OCH_2CH_3} + {}^{-}\!:\!\ddot{O}CH_2CH_3$$

The new carbon-carbon bond is shown in boldface. The product is a β-keto ester, an active methylene compound. Since the ethoxide ion, a very strong base, is still present, we must *expect* it to stabilize itself by taking a proton:

$$CH_3CH_2C(=O)\!-\!C(CH_3)(H)\!-\!C(=O)\!-\!OCH_2CH_3 \;\; {}^{-}OCH_2CH_3 \rightleftharpoons \left[\begin{array}{c} CH_3CH_2C(=O)\!-\!\ddot{C}^{-}(CH_3)\!-\!C(=\!:\!O\!:)OCH_2CH_3 \\ \updownarrow \\ CH_3CH_2C(=O)\!-\!C(CH_3)\!=\!C(:\ddot{O}:^{-})OCH_2CH_3 \\ \updownarrow \\ CH_3CH_2C(:\ddot{O}:^{-})\!=\!C(CH_3)\!-\!C(=O)OCH_2CH_3 \end{array}\right] + HOCH_2CH_3$$

The resonance contributors shown help to explain the stability of this new anion. If a relatively strong acid is now added, the electrically neutral β-keto ester will be liberated:

$$CH_3CH_2C(=O)\!-\!\ddot{C}^{-}(CH_3)\!-\!C(=O)OCH_2CH_3 + H^+ \longrightarrow CH_3CH_2C(=O)\!-\!CH(CH_3)\!-\!C(=O)OCH_2CH_3$$

This type of reaction wherein two esters, in the presence of a strong base, react to form a β-keto ester is known as the **Claisen ester condensation.**

Exercise 12.2 Which ethyl esters would be needed to make the following compounds by Claisen condensations?

(a) ethyl acetoacetate

(b) $C_6H_5CH_2C(=O)CH(C_6H_5)C(=O)OC_2H_5$

12.9 Properties of β-Keto Esters and β-Keto Acids. Hydrolysis. Decarboxylation

β-Keto esters are both ketones and esters, and they exhibit typical reactions of these functional groups. They can be hydrolyzed, for example, to β-keto acids:

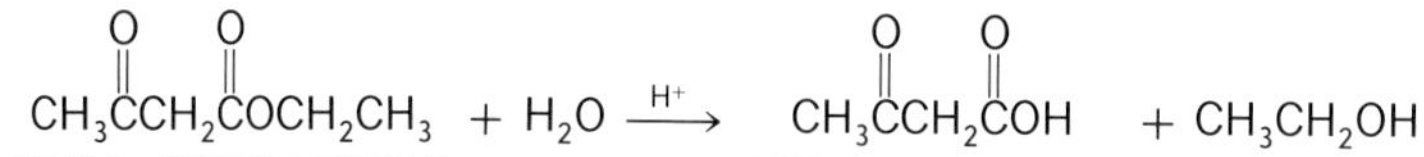

$$\underset{\text{ETHYL ACETOACETATE}}{CH_3\overset{O}{\overset{\|}{C}}CH_2\overset{O}{\overset{\|}{C}}OCH_2CH_3} + H_2O \xrightarrow{H^+} \underset{\text{ACETOACETIC ACID}}{CH_3\overset{O}{\overset{\|}{C}}CH_2\overset{O}{\overset{\|}{C}}OH} + CH_3CH_2OH$$

Such acids are subject to rather easy decarboxylation, in a manner analogous to that of β-dicarboxylic acids (malonic acid types).

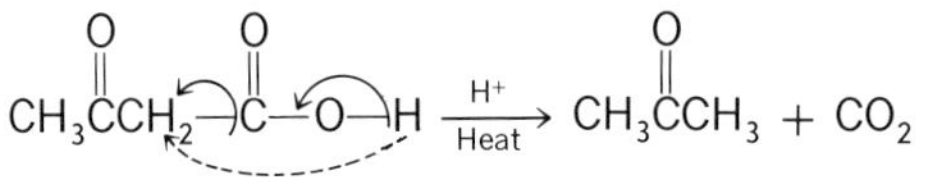

$$\underset{\text{ACETOACETIC ACID}}{CH_3\overset{O}{\overset{\|}{C}}CH_2{-}\overset{O}{\overset{\|}{C}}{-}O{-}H} \xrightarrow[\text{Heat}]{H^+} \underset{\text{ACETONE}}{CH_3\overset{O}{\overset{\|}{C}}CH_3} + CO_2$$

This specific reaction occurs in the body in connection with lipid metabolism.

12.10 Keto-Enol Tautomerism of β-Ketoesters

We first studied the phenomenon of keto-enol tautomerism in Section 9.22. We return to it here largely because β-ketoesters (and other active methylene compounds) are often substantially in the *enol* form.

All the simple carbonyl compounds we have studied contain only trace amounts of corresponding enol tautomers, normally in such minute quantities (about 10^{-4}% enol in acetone) that their presence may be ignored. When a molecule has two carbonyl groups *beta* to each other, however, the enol form can be stabilized by internal hydrogen bonding. Therefore the percent of enol form present in β-dicarbonyl compounds is substantially higher than that in other carbonyl compounds, as seen in the data for ethyl acetoacetate and acetylacetone.

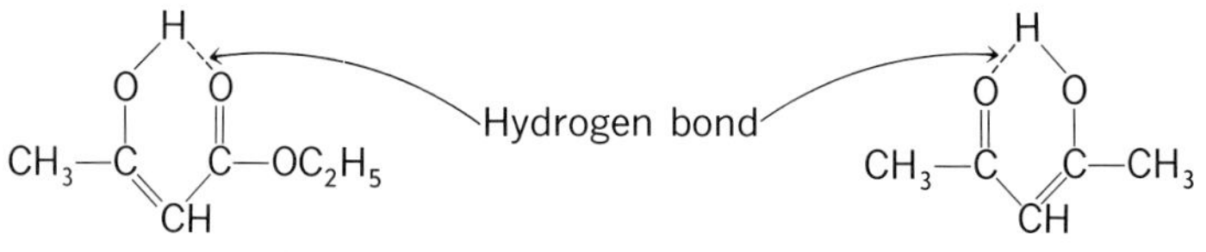

$$CH_3{-}\overset{O{-}H\cdots O}{C{=}CH{-}C}{-}OC_2H_5 \qquad \text{Hydrogen bond} \qquad CH_3{-}\overset{O\cdots H{-}O}{C{-}CH{=}C}{-}CH_3$$

ENOL OF ETHYL ACETOACETATE
(8% enol : 92% keto form)

ENOL OF ACETYLACETONE
(80% enol : 20% keto form)

12.11 The Acetoacetic Ester Synthesis

This series of reactions is exactly analogous to the malonic ester synthesis (Section 12.6). The end result is the synthesis of substituted acetones of the types:

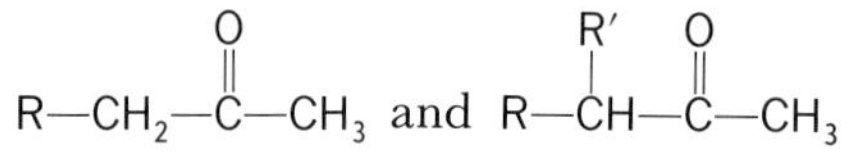

$$R{-}CH_2{-}\overset{O}{\overset{\|}{C}}{-}CH_3 \quad \text{and} \quad R{-}\overset{R'}{\overset{|}{C}H}{-}\overset{O}{\overset{\|}{C}}{-}CH_3$$

As in the malonic ester synthesis, one key intermediate is a carbanion obtained by action of strong base on an active methylene compound, this time—acetoacetic ester. We show here only the synthesis of a monosubstituted acetone.

$$CH_3-\overset{O}{\overset{\|}{C}}-CH_2-\overset{O}{\overset{\|}{C}}-OC_2H_5 + {}^{-}OC_2H_5 \xrightarrow{C_2H_5OH} CH_3-\overset{O}{\overset{\|}{C}}-\ddot{C}H^{-}-\overset{O}{\overset{\|}{C}}-OC_2H_5 \longleftrightarrow \text{(stabilized by resonance)}$$

ACETOACETIC ESTER (ETHYL ACETOACETATE)

$\downarrow$ R—X (R is 1° or 2°)
$\downarrow$ Acidify

$$CH_3\overset{O}{\overset{\|}{C}}-\underset{R}{\underset{|}{C}H}-\overset{O}{\overset{\|}{C}}-OC_2H_5 \xrightarrow[\text{then acidify}]{\text{dil NaOH}} CH_3\overset{O}{\overset{\|}{C}}-\underset{R}{\underset{|}{C}H}-\overset{O}{\overset{\|}{C}}-O-H + C_2H_5OH$$

$$\xrightarrow{\text{Heat}} CH_3\overset{O}{\overset{\|}{C}}CH_2R + CO_2$$

A MONOSUBSTITUTED ACETONE

Exercise 12.3 Which alkyl halides would be used to make each of the following compounds by the acetoacetic ester synthesis?

(a) $CH_3CH_2\overset{O}{\overset{\|}{C}}CH_3$

(b) $(CH_3)_2CHCH_2\overset{O}{\overset{\|}{C}}CH_3$

(c) $C_6H_5CH_2CH_2\overset{O}{\overset{\|}{C}}CH_3$

(d) cyclopentyl$-CH_2CH_2\overset{O}{\overset{\|}{C}}CH_3$

12.12 Hydroxy Acids and Their Derivatives

Hydroxy acids are both acids and alcohols. They generally give the reactions of these two groups.

The product of the dehydration of a hydroxy acid depends on the location of the —OH group relative to the CO_2H group. β-Hydroxy acids (or esters), like β-hydroxy aldehydes (and ketones), lose the elements of water very easily to form α,β-unsaturated acids (or esters) as the major products:

$$CH_3\underset{OH}{\underset{|}{C}H}-CH_2\overset{O}{\overset{\|}{C}}OH(R) \xrightarrow{H^+} CH_3CH{=}CH\overset{O}{\overset{\|}{C}}OH(R) + H_2O$$

α,β-UNSATURATED ACID (OR ESTER)

$$CH_3(CH_2)_{12}\overset{OH}{\overset{|}{C}H}CH_2\overset{O}{\overset{\|}{C}}OH \xrightarrow[\text{(occurs during metabolism of fatty acids)}]{\text{Enzyme}} CH_3(CH_2)_{12}CH{=}CH\overset{O}{\overset{\|}{C}}OH + H_2O$$

β-HYDROXYPALMITIC ACID → 2-HEXADECENOIC ACID

$$HO-\underset{CH_2COOH}{\underset{|}{\overset{CH_2COOH}{\overset{|}{C}}}}-COOH \xrightarrow[\text{(occurs early in the citric acid cycle)}]{\text{Enzyme}} \underset{CHCOOH}{\underset{\|}{\overset{CH_2COOH}{\overset{|}{C}}}}-COOH + H_2O$$

CITRIC ACID → ACONITIC ACID

12.13 Lactones

In γ-hydroxy acids and δ-hydroxy acids the two groups are positioned to favor the formation of a cyclic, or internal ester called a **lactone.** We know that five- and six-membered rings are stable systems, and have learned that alcohols and acids can be made to form esters. Apparently the hydroxyl group at the "tail" of the γ-hydroxy or δ-hydroxy acid molecule can more frequently find the carboxyl group at its "head" than a carboxyl group on a neighboring molecule.

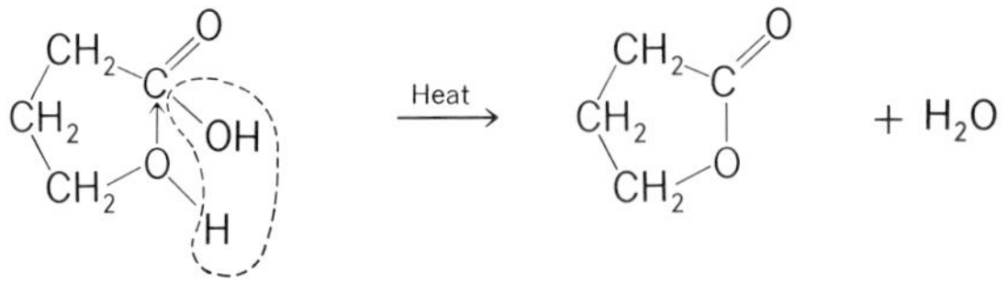

γ-HYDROXYBUTYRIC ACID $\xrightarrow{\text{Heat}}$ γ-BUTYROLACTONE $+ H_2O$

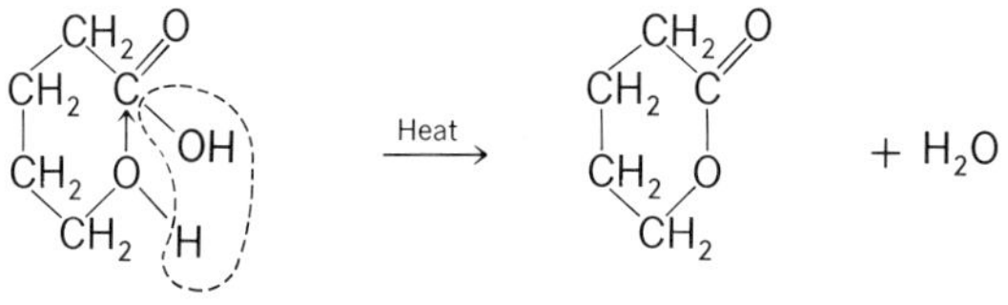

δ-HYDROXYVALERIC ACID $\xrightarrow{\text{Heat}}$ δ-VALEROLACTONE $+ H_2O$

These cyclic esters are true esters. They can be hydrolyzed back to the original hydroxy acids; they can be saponified to the salts of those acids.

12.14 α,β-Unsaturated Carbonyl Compounds

An α,β-unsaturated carbonyl compound has two double bonds in conjugation:

$$\overset{\beta}{>C}=\overset{\alpha}{C}<\!-C=O$$

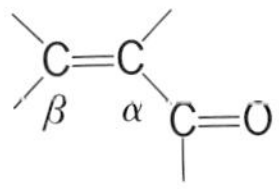

Many important substances having this system are given in Table 12.2.

Our study of 1,3-butadiene revealed an interesting property of a conjugated diene where *both* bonds are of the alkene type, namely, 1,4-addition. What happens when one of the two double bonds in conjugation is a carbonyl group? We shall learn here.

12.15 1,4-Addition

Acrylic acid adds hydrogen chloride in the following way:

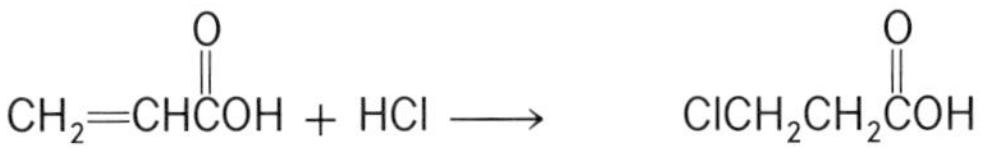

$$CH_2{=}CH\overset{O}{\overset{\|}{C}}OH + HCl \longrightarrow ClCH_2CH_2\overset{O}{\overset{\|}{C}}OH \quad \textit{not} \quad CH_3\underset{Cl}{\underset{|}{C}}H\overset{O}{\overset{\|}{C}}OH$$

ACRYLIC ACID — β-CHLOROPROPIONIC ACID — (Predicted if Markovnikov's Rule is applied)

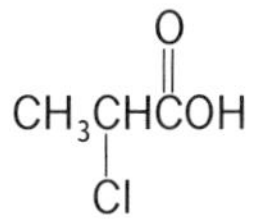

The reaction is in violation of Markovnikov's rule, but that rule was never meant to apply to anything other than simple alkenes anyway. To understand this reaction and similar reactions we have to go back to fundamentals.

Reactions, like spontaneous events in general, tend to go through paths of minimum energy. In the addition of hydrogen chloride to an alkene the more stable of two possible carbonium ions forms. Addition of hydrogen chloride to a conjugated system proceeds under the same constraint. The first step ①, below, is attack by a proton at the site richest in electrons, the oxygen of the carbonyl group, not the alkene group. The resonance-stabilized carbonium ion that forms has greater stability than any other that can form. This ion then picks up a chloride ion ② to produce an unstable enol that rearranges to the keto form ③. The first two steps are essentially identical to 1,4-addition, which we encountered earlier when studying butadiene (Section 3.29).

$$\overset{4}{CH_2}{=}\overset{3}{CH}{-}\overset{2}{C}(=\overset{1}{\ddot{O}}){-}OH + H{-}Cl$$

(path representing Markovnikov addition) H^+ ① H^+ (path through the more stable carbonium ion)

$$\left[CH_3{-}\overset{+}{C}H{-}C(=O){-}OH \;\nleftrightarrow\; CH_3{-}CH{=}C(\overset{+}{O}){-}OH \right]$$

Not stabilized by resonance; oxygen is more electronegative than carbon; the second form has one oxygen with a sextet of electrons and makes no contribution to the hybrid.

$$\left[CH_2{=}CH{-}\overset{+}{C}(\ddot{O}{-}H){-}OH \;\longleftrightarrow\; \overset{+}{C}H_2{-}CH{=}C(\ddot{O}{-}H){-}OH \right]$$

Resonance-stabilized intermediate carbonium ion

② Cl^-

$$\underset{\text{FINAL PRODUCT}}{Cl{-}CH_2{-}CH_2{-}C(=O){-}OH} \xleftarrow{③} \underset{\text{AN ENOL}}{Cl{-}CH_2{-}CH{=}C(\ddot{O}{-}H){-}OH}$$

The carbonyl group dominates the direction of the addition of reagents to the double bond of an α,β-unsaturated carbonyl compound. It can and does do so because conjugation connects it to the alkene group. Without the two being conjugated, they would react in isolation of each other. Conjugation, and the resonance made possible by it, gives molecules a pipeline through which functional groups can affect each others properties. That is the chief lesson we learn here.

12.16 Diels-Alder Reaction

This reaction makes six-membered rings from simpler compounds, usually from 1,3-dienes and α,β-unsaturated carbonyl compounds. For example:

Table 12.2
Some Important α,β-Unsaturated Carbonyl Compounds

		Structure	Use
Aldehydes	Acrolein	$CH_2{=}CH{-}CHO$	A toxic lachrymator Important industrial chemical
	Cinnamaldehyde	$C_6H_5CH{=}CHCHO$	In oil of cinnamon
Ketones	Methyl vinyl ketone	$CH_2{=}CH{-}CO{-}CH_3$	Used in synthesis of Vitamin A
Acids	Acrylic acid	$CH_2{=}CHCO_2H$	Raw material for its esters that are used to make acrylic fibers
	Sorbic acid	$CH_3CH{=}CHCH{=}CHCO_2H$	Preservative in food products; fungi inhibitor
	Cinnamic acid	$C_6H_5CH{=}CHCO_2H$ (*trans*)	Its esters are used in perfumes
Acid derivatives	Methyl acrylate	$CH_2{=}CHCO_2CH_3$	A raw material for acrylic fibers
	Methyl methacrylate	$CH_2{=}C(CH_3)CO_2CH_3$	Manufacture of Lucite® and Plexiglas®

A DIENE + A DIENOPHILE ⟶ A DIELS-ALDER ADDUCT

1,3-BUTADIENE + $CH_2{=}CH{-}CHO$ $\xrightarrow{\text{Heat}}$ 1,2,3,6-TETRAHYDROBENZALDEHYDE (100%)

A huge variety of reactants can be used, and the yields are usually very high. The systems that react with the diene are called dienophiles ("diene-loving"). Either diene or dienophile may be a cyclic compound. For example:

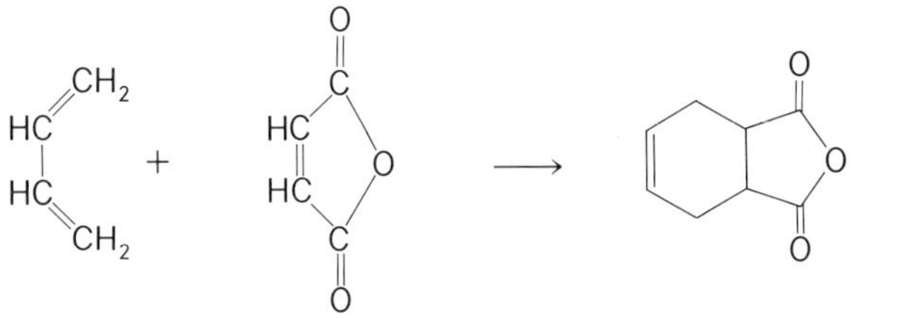

MALEIC ANHYDRIDE

TETRAHYDROPHTHALIC ANHYDRIDE

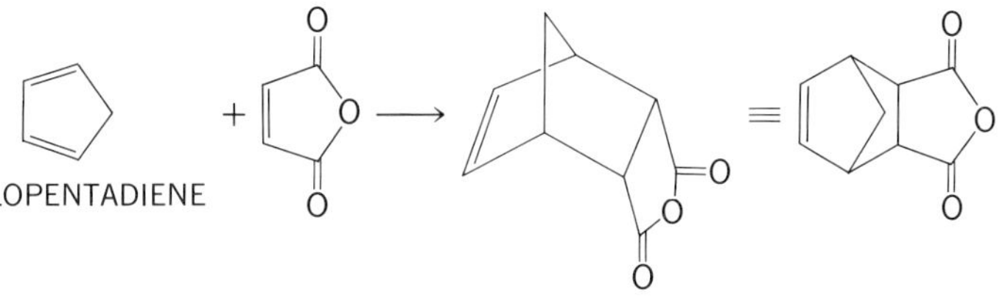

CYCLOPENTADIENE

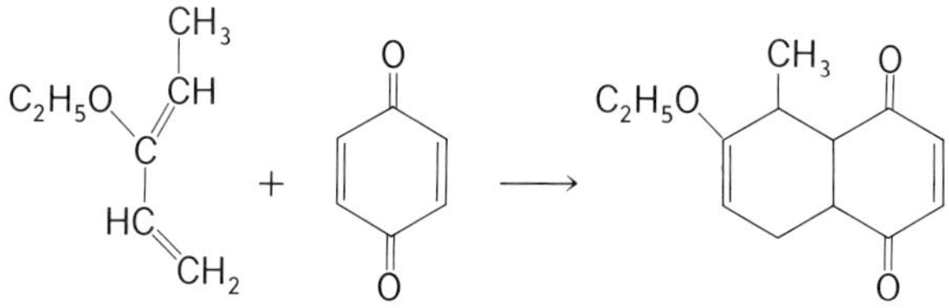

p-QUINONE

The last example was one of the steps in the total synthesis of cortisone by L. Sarett's group at Merck, Inc. This is mentioned because the Diels-Alder reaction has often been used in synthesizing important drugs and natural products. In fact, so valuable is the reaction that its discovery earned the 1950 Nobel prize in chemistry for its co-discoverers, Otto Diels and Kurt Alder.

The last example also introduced an example of a special type of α,β-unsaturated carbonyl compound, a quinone.

12.17 Quinones

Unsaturated, cyclic diketones of the two types shown are called quinones.

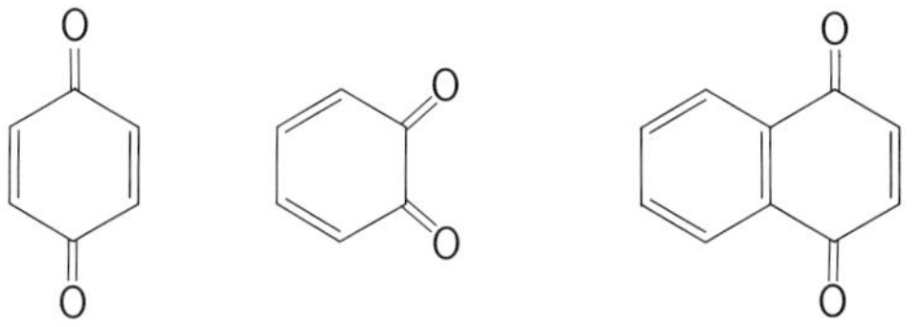

p-QUINONE *o*-QUINONE 1,4-NAPHTHOQUINONE

They are generally colored compounds. The *ortho*-quinones are less stable than their *para*- (or 1,4-) isomers. What is probably most important about these highly unsaturated systems is that they are just the gain of two electrons away from becoming fully aromatic. They are oxidizing agents "poised on the brink."

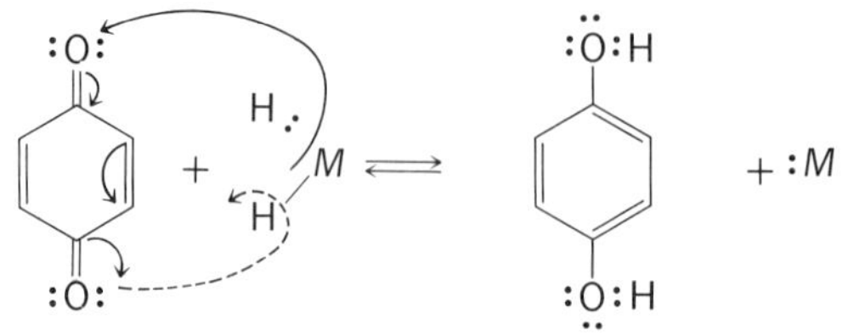

:O: + H:M ⇌ :O:H + :M

p-QUINONE HYDROQUINONE

MH_2 stands for any hydride donor (H:⁻), reducing agent. MH_2, for example, might be some metabolite in a cell, the reduced form. M: would be the oxidized form. Many inorganic reducing agents, for example, sulfite ion, SO_3^{2-}, also work. The reaction can be easily reversed if an oxidizing agent is made available. The ease of the reaction in either direction is related partly to the fact that both reactant and product are highly conjugated systems and therefore stabilized. Biological oxidation-reductions usually include an enzyme in which a quinone (coenzyme *Q*; ubiquinone) acts as the electron-transfer agent. Vitamin K is a quinone.

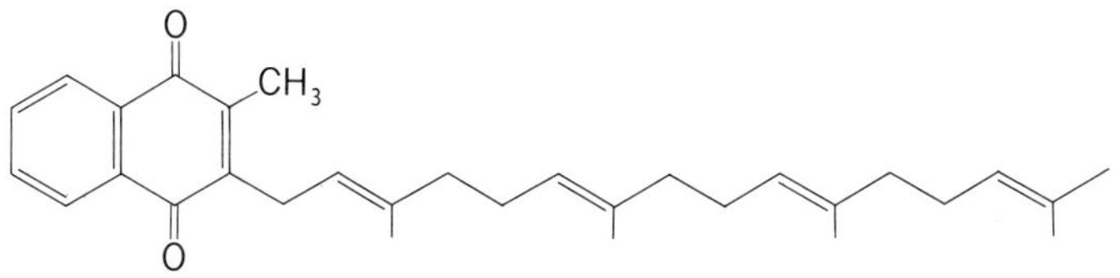

VITAMIN K_2
(PHYLLOQUINONE)

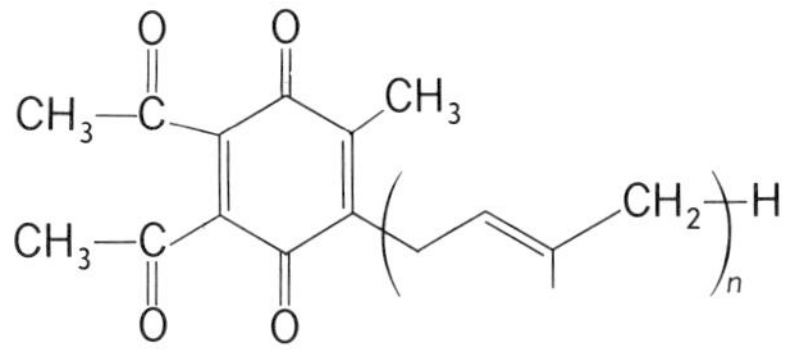

UBIQUINONES (n = 1, 2, . . . 10)
(COENZYME Q)

Brief Summary: Important Chemical and Physical Properties

1. Expect di- (and poly-) functional compounds to give most of the chemical reactions associated with their functional groups.

2. Two circumstances are especially effective in causing two functional groups on the same molecule to affect each others properties:

(a) Either the two groups are in conjugation in the starting materials (examples, 1,3-dienes or α,β-unsaturated carbonyl compounds) or the two groups become "connected" by resonance when an intermediate in a reaction forms (e.g., β-dicarbonyl compounds as in malonic or acetoacetic esters or any other active methylene compounds).

(b) The two groups are positioned just right to react with each other and form a cyclic compound (examples, γ- and δ- hydroxy acids or γ- or δ-dicarboxylic acids).

3. β-Dicarboxylic acids (malonic acids) and β-ketoacids readily decarboxylate.

4. Esters of β-dicarboxylic acids and of β-ketoacids—as well as of any other active methylene compounds—are acidic enough toward a base as strong as an alkoxide ion (e.g., $C_2H_5O^-$) to form enolate anions that can be alkylated by 1° or 2° alkyl halides.
   (a) By the malonic ester synthesis we make substituted acetic acids.
   (b) By the acetoacetic ester synthesis we make substituted acetones.

5. γ- and δ-dicarboxylic acids change to cyclic anhydrides when heated.

6. γ- and δ-hydroxyacids change to cyclic esters (lactones) when heated.

7. The Claisen condensation converts simple esters (with at least one α-hydrogen) to β-ketoesters.

8. Reagents that can add to an alkene group add to it in α,β-unsaturated carbonyl compounds in such a way that the negative part of the reagent always winds up on the β-carbon.

9. The Diels-Alder reaction is a way to make six-membered rings from 1,3-dienes and α,β-unsaturated carbonyl compounds (and other dienophiles).

10. Quinones and hydroquinones are easily interconverted by oxidation-reduction systems.

Problems and Exercises

1. Write structures for each compound.
   (a) diethyl succinate (b) α-methylmalonic acid
   (c) γ-butyrolactone (d) malonic ester
   (e) lactic acid (f) succinic anhydride
   (g) *p*-quinone (h) phthalic anhydride

2. Write common names for the following:

(a) $Na^{+}\,{}^{-}O\overset{O}{\overset{\|}{C}}CH_2CH_2\overset{O}{\overset{\|}{C}}O^{-}\,Na^{+}$

(b)

(c) $CH_2{=}CHCO_2H$

(d) $CH_3CH_2CH(CO_2H)_2$

(e)

(f) $HO\overset{O}{\overset{\|}{C}}(CH_2)_4\overset{O}{\overset{\|}{C}}OH$

(g) $CH_3\overset{O}{\overset{\|}{C}}CO_2H$

(h)

3. Write the steps in the mechanism of the Claisen condensation of $CH_3CH_2CO_2C_2H_5$ with ethoxide ion as the base.

4. If the Claisen condensation could be used to prepare any of the following compounds, write the structure of the simple ester needed. If this synthesis cannot be used, state so.

(a) $CH_3\overset{\overset{O}{\|}}{C}CH_2CO_2C_2H_5$ (c) $CH_3\overset{\overset{O}{\|}}{C}{-}\overset{\overset{O}{\|}}{C}OC_2H_5$

(b) $CH_3CH_2CH_2\overset{\overset{O}{\|}}{C}\underset{\underset{CH_2CH_3}{|}}{C}HCO_2C_2H_5$

5. Show by equations how to use the malonic ester synthesis to make each of the following. (Where this synthesis cannot be used, state so.)
(a) $CH_3CH_2CO_2H$ (b) $(CH_3)_2CHCH_2CO_2H$
(c) $C_6H_5CH_2CO_2H$ (d) $CH_2{=}CHCH_2CH_2CO_2H$
(e) $CH_2{=}CHCH_2CO_2H$

6. Show by equations how to use the acetoacetic ester synthesis to make each of the following. (Where this synthesis cannot be used, state so.)

(a) $CH_3\overset{\overset{O}{\|}}{C}CH_2CH(CH_3)_2$ (c) $CH_3\overset{\overset{O}{\|}}{C}CH_2C_6H_5$

(b) $CH_3\overset{\overset{O}{\|}}{C}CH_2CH_2CH{=}CH_2$ (d) $CH_3\overset{\overset{O}{\|}}{C}CH_2CH{=}CH_2$

7. Write the structures of the Diels-Alder adducts of each pair.

(a) $C_6H_5CH{=}CH{-}CH{=}CHC_6H_5$ + $CH_2{=}CH{-}CH{=}O$

(b) $CH_3CH{=}CH{-}CH{=}CHCH_3$ + (maleic anhydride ring: O, O, O)

8. Write the structure(s) of the organic product(s) that form in each case. If no reaction occurs, write "None."

(a) $CH_3\overset{\overset{O}{\|}}{C}OC_2H_5 \xrightarrow[2.\ H^+]{1.\ ^-OC_2H_5}$

(b) $CH_3\underset{\underset{OH}{|}}{C}HCH_2CH_2\overset{\overset{O}{\|}}{C}OH \xrightarrow[\Delta]{}$

(c) [succinic anhydride] $+ H_2O \xrightarrow{\Delta}$

(d) $HO\overset{O}{\overset{\|}{C}}CH_2CH_2CH_2\overset{O}{\overset{\|}{C}}OH + \text{excess } CH_3OH \xrightarrow[\Delta]{H^+}$

(e) $CH_3\overset{O}{\overset{\|}{C}}CH_2CH_2CO_2H + NH_2OH \xrightarrow[\Delta]{H^+}$

(f) $HOCH_2CH_2CH_2\overset{O}{\overset{\|}{C}}OCH_3 + H_2O \xrightarrow[\Delta]{H^+}$

(g) $HO\overset{O}{\overset{\|}{C}}\underset{CH_3}{\underset{|}{C}}HCH_2\underset{CH_3}{\underset{|}{C}}H\overset{O}{\overset{\|}{C}}OH \xrightarrow{\Delta}$

(h) [phthalic anhydride] $+ CH_3OH \xrightarrow{\Delta}$

(i) $CH_2{=}\overset{CH_3}{\overset{|}{C}}{-}\overset{O}{\overset{\|}{C}}{-}C_6H_5 + HCl_{(g)} \longrightarrow$

(j) $CH_2{=}CH{-}CH{=}CH_2 +$ [maleic anhydride] $\xrightarrow{\Delta}$

(k) [2-methyl-1,4-benzoquinone] $\xrightarrow{\text{Mild reducing agent}}$

9. What diene and dienophile would be needed to make the following?

[bicyclo[2.2.1]hept-2-ene bearing $-\overset{O}{\overset{\|}{C}}CH_3$]

chapter thirteen
Optical Isomerism

13.1 Types of Isomerism

Even though two (or more) compounds have the same general structure, they may be different in a very subtle way, a way even more subtle than *cis-trans* isomerism, which is merely a geometric difference. This new isomerism, optical isomerism, occurs widely in nature, and to interpret some of nature's workings we must understand it.

We have already encountered several ways in which substances have the same molecular formula and still are different. The more obvious examples were *chain isomers, position isomers,* and *functional isomers.*

Chain isomers: —C—C—C—C— (BUTANE) ; isobutane: three-carbon chain —C—C—C— with a —C— branch on the central carbon (ISOBUTANE)

Position isomers: $CH_3CH_2CH_2Cl$ (1-CHLOROPROPANE) ; CH_3CHCH_3 with Cl on the central carbon (2-CHLOROPROPANE)

Functional group isomers: CH_3CH_2OH (ETHYL ALCOHOL) ; $CH_3—O—CH_3$ (DIMETHYL ETHER)

Less obvious was stereoisomerism, and we have studied only one of the two types, *cis-trans* or geometrical.

CH_3 and CH_3 on the same side of C=C, H and H on the other: *cis*-2-BUTENE ; CH_3 and H on the upper side, H and CH_3 on the lower side: *trans*-2-BUTENE

The second kind of stereoisomerism has to do with the lack of symmetry in the molecule.

13.2 "One Substance, One Structure"

Chemists have long worked with the principle "one substance, one structure." If two samples of matter have identical physical and chemical properties, they must be identical at the level of their individual molecules. If two samples differ in even one way in fundamental properties, their individual molecules must also differ in some way.

Consider now a substance known as asparagine (ăs-păr′-ȧ-jĭn), a white solid with a bitter taste, first isolated in 1806 from the juice of asparagus. Its molecular

$$NH_2-\overset{\overset{\displaystyle O}{\|}}{C}-CH_2-\underset{\underset{\displaystyle NH_2}{|}}{CH}-\overset{\overset{\displaystyle O}{\|}}{C}-OH$$

1
ASPARAGINE

formula is $C_4H_8N_2O_3$, and its structure is now known to be **1**. Eighty years later, in 1886, a chemist isolated from sprouting vetch[1] a substance of the same molecular formula and structure, but it had a sweet taste. To have names for these two, we call the one from asparagus "L-asparagine" and the one from vetch sprouts "D-asparagine." (The small capitals, D and L, although arbitrary here, acquire meaning in Section 14.8.)

These two samples of asparagine are so similar they even have identical spectra. We must write the same structure, **1,** for each. Yet they have this dramatic difference in taste. One is sweet; one is bitter.

Taste is a chemical sense. The two samples of asparagine, therefore, do show one difference in a chemical property. Under the one-substance–one-structure doctrine, the molecules of D- and L-asparagine must be structurally different in some way.

13.3 Molecular Symmetry and Asymmetry

Two partial ball-and-stick models of asparagine molecules are shown in Figure 13.1*a*. Examine each. Assure yourself that each is structure **1**. Now let us simplify the structures. Let us go to part *b* of Figure 13.1. In each structure, the same four groups are attached to the central carbon. Yet the two structures are not identical!

For two structures to be identical it must be possible to imagine you can superimpose them, as illustrated in Figure 13.1*c*. *Superimposition is the fundamental criterion of identity for two structures.* If two structures are identical, each must have a rotational conformation (from twisting about single bonds) by which the two become superimposable. The two asparagine molecules are not identical, after all. Yet, they are related in a particularly interesting and significant way. They are

[1]Vetch is a member of a genus of herbs, some of which are useful as fodder.

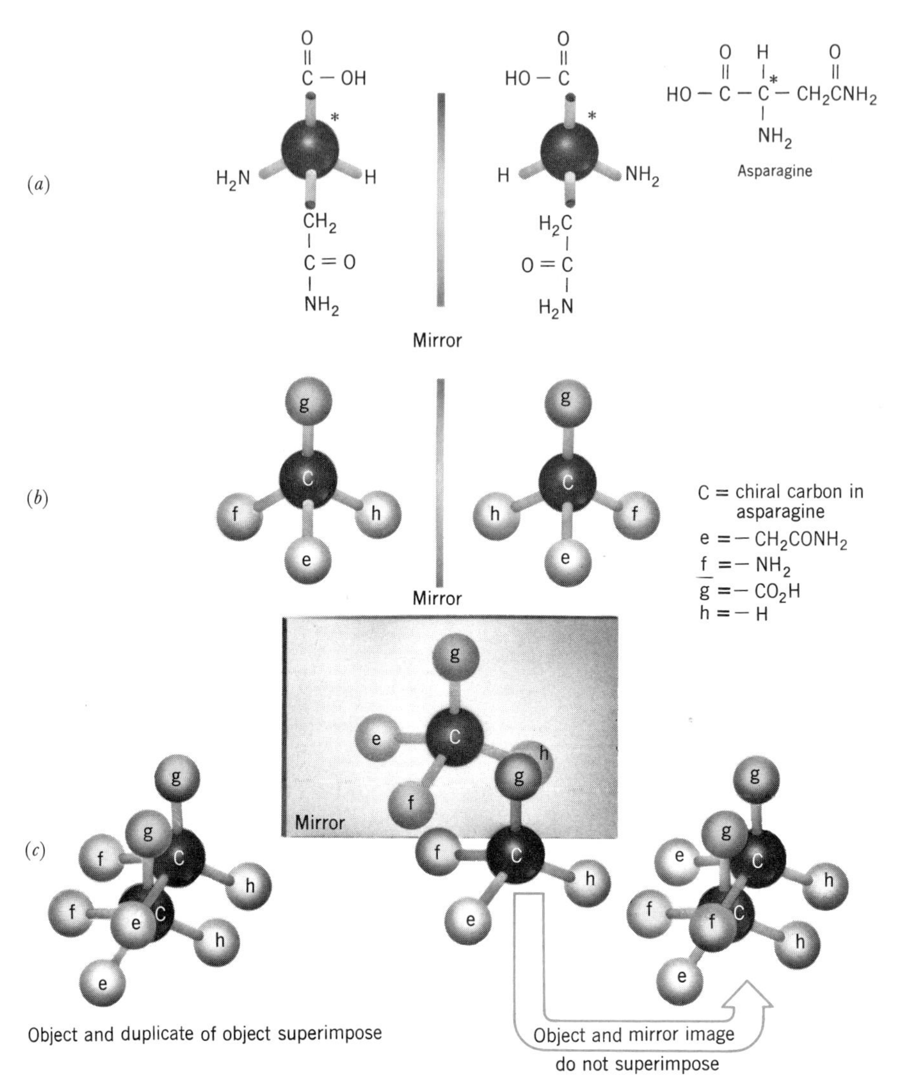

Figure 13.1
The enantiomers of asparagine are shown in part (*a*). When the groups attached to a chiral carbon do not themselves contribute further to chirality in the molecule, they may be more simply represented as in part (*b*). Checking superimposability requires the imaginary operation begun in part *c* but prevented by the impenetrability of wood.

related as an object to its mirror image. The fact that they cannot be superimposed raises a question. In exactly what way do they differ structurally?

The two asparagines differ only in the relative directions by which the four groups on the central carbon (Figure 13.1*b*) project into space. We say they have opposite *configurations*. We could call them configurational isomers. Instead, for historical reasons that will soon be apparent, we call the two *optical isomers*. Since there are several types of optical isomers we need a name for the type illustrated by the asparagines.

13.4 Enantiomers

Isomers that, like the two asparagines, are related as object and mirror image that cannot be superimposed are called *enantiomers,* from the Greek, *enantios,* "opposite," and *meros,* "part." To deal as swiftly as possible with one source of misunderstanding, *any* object will have a mirror image. Only when the object and mirror image cannot be superimposed are they said to be enantiomers. With this definition of enantiomers, optical isomers in general can be better defined. Optical isomers are members of a set of stereoisomers that includes at least two related as enantiomers.

Physical properties are identical for two enantiomers. This *must* be so. Within two enantiomers intranuclear distances and bond angles are identical, and polarities must therefore be identical. Melting points, boiling points, densities, solubilities, spectra, and, of course, molecular weights have to be the same for two enantiomers. In thinking about intranuclear distances and angles in enantiomers, compare your two hands, disregarding differences in fingerprints and wrinkles. The left hand and the right hand illustrate enantiomers. They, too, are related as object and mirror image that cannot be superimposed. They differ only in that holding the palms facing us, we "read" clockwise from thumb to little finger in the left hand and counterclockwise in the right hand.

13.5 Chirality and Chiral Molecules

Enantiomers have the structural property of opposite "handedness." The Greek word for "hand" is *cheir,* and from this we have the words *chiral* and *chirality.* We may call two enantiomers chiral or we may say they posses chirality.[2] In other words, they possess handedness. The opposite of chiral is *achiral;* it means symmetrical enough so that object and mirror images superimpose.

13.6 Enantiomers and Chiral Reagents

Chemical properties of enantiomers are also identical, provided that other reagents are symmetrical or achiral in the sense used here. If the reaction is with a chiral reagent, however, we observe striking differences in properties. A simple, widely used illustration involves a pair of gloves. Because two hands are related as enantiomers, so too are a pair of ordinary gloves:

$$\text{Right hand} + \text{right glove} \longrightarrow \text{gloved right hand}$$
$$\text{Right hand} + \text{left glove} \longrightarrow \text{no fit}$$

Taking the gloves as the chiral "reagents," we see that the ease and comfort with which each glove can "react" with each hand is certainly different. By contrast, if we use as the "reagent" system a pair of, for example, ski poles that are superimposable and achiral, no differences can be noted in the way these "react" with either hand.

[2]Teachers of organic chemistry will be keenly aware of the fact that very often objects having chirality have traditionally been called dissymmetric or (not as accurately) asymmetric. Both terms are widespread; both will be used for some time. Students will encounter them often in other chemical and biochemical literature. They should be learned, and with almost no danger of serious inaccuracy the terms chiral, asymetric, and dissymmetric may be considered to be synonyms. They are not, strictly speaking, but the distinctions are, for our purposes, unimportant.

13.7 Deciding Whether a Molecule is Symmetrical

Any structure that is identical with its mirror image has symmetry in some way. Three ways of having symmetry are recognized. However, for our purposes we need only one, the *plane of symmetry*.[3]

A plane of symmetry is an imaginary plane that divides an object into two halves that are mirror images of each other. Some simple planes of symmetry are shown in Figure 13.2. Some of the molecules have several planes, but we require only one. Just one guarantees that object and mirror image are superimposable and identical. On the other hand, if we cannot discover this (or any) symmetry in a structure, in at least one conformation of it, we are guaranteed that the object and mirror images are not superimposable and are optical isomers.

13.8 The Chiral Carbon—Another Shortcut

In nearly all examples of optical isomerism the molecules possess at least one carbon to which are attached four *different* atoms or groups. We saw this in asparagine (Figure 13.1). A carbon holding four different groups is defined as a chiral carbon. (In many references it will be called an *asymmetric* carbon.) A molecule with a chiral carbon will nearly always have no element of symmetry in any conformation.

The qualification "nearly always" that was used in these statements is an important one. Many substances with no chiral carbons have chiral molecules (Figure 13.3). In a few substances, the molecules contain chiral carbons but yet are symmetrical in some way (Figure 13.4). Some of these substances will be studied in more detail later. In general, if a molecule has *only one* chiral carbon, it will

[3]For the record, the other two elements of symmetry are a center of symmetry and a fourfold alternating (or mirror) axis of symmetry.

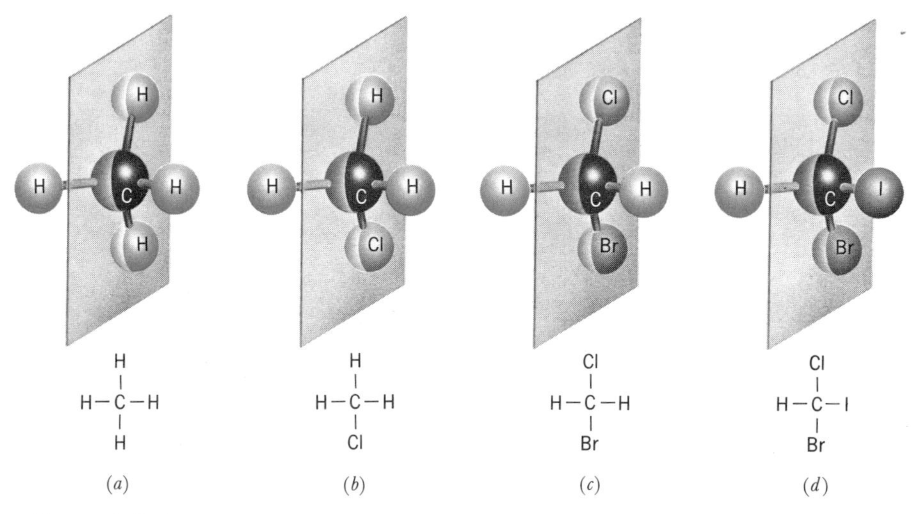

Figure 13.2
The plane of symmetry. (*a*) Methane, (*b*) methyl chloride, and (*c*) chlorobromomethane each possesses at least one plane of symmetry. This plane, serving as an imaginary mirror, splits each molecule into two halves, one being the mirror image of the other. No plane of symmetry exists for molecule (*d*), which is chiral. If a mirror image were made of it, the object and it would not be superimposable.

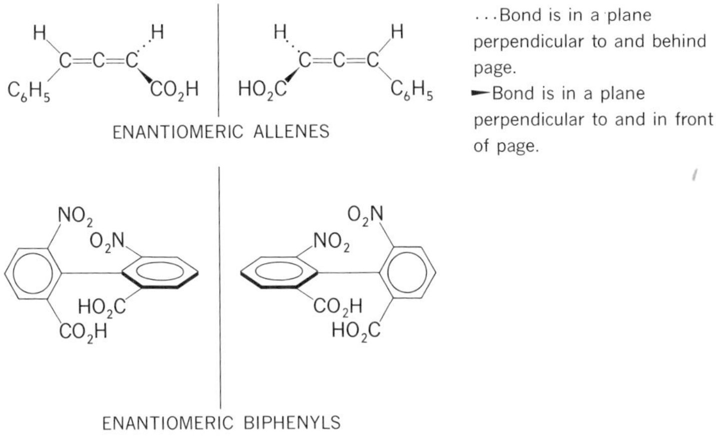

...Bond is in a plane perpendicular to and behind page.
—Bond is in a plane perpendicular to and in front of page.

ENANTIOMERIC ALLENES

ENANTIOMERIC BIPHENYLS

Figure 13.3
Chiral molecules having no chiral carbon atoms. In the allenes the rigid geometry about the ends of the double bonds makes chirality possible, provided that the end groups are different. In the biphenyls the bulkiness of the groups located at the ortho positions prevents free rotation about the single bond joining the two benzene rings. With the right choice of these bulky ortho groups, such biphenyl molecules can be chiral.

be chiral as a whole, and it will belong to a set of optical isomers. We need not qualify this statement with "nearly always."

If the molecules of a substance each have *n* different chiral carbons, the number of possible optical isomers is 2^n. (Two chiral carbons are "different" if the sets of four groups attached to each are different in at least one way.)

When it works, the concept of a chiral carbon is extremely useful. It must be remembered that the shortcut is fallible (cf. Figure 13.3 and 13.4), but with a little experience the exceptions are easily recognized. Moreover, they do not occur frequently. We turn next to the way we detect this kind of isomerism experimentally.

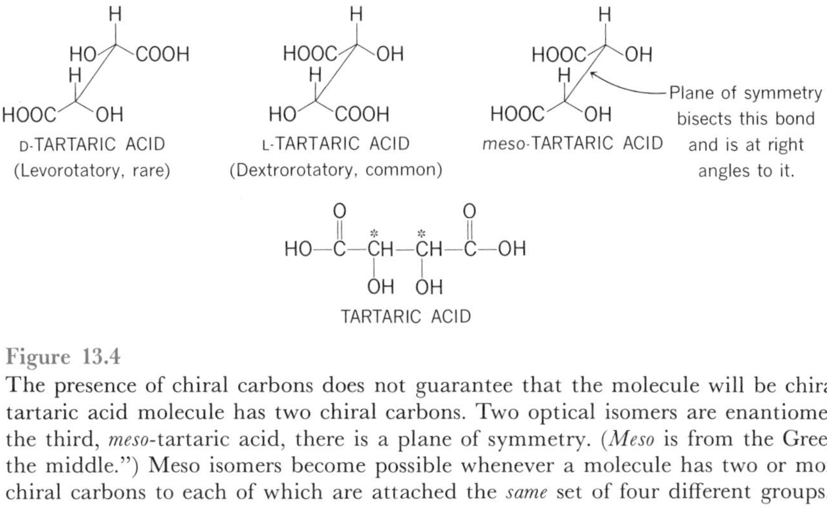

D-TARTARIC ACID (Levorotatory, rare)

L-TARTARIC ACID (Dextrorotatory, common)

meso-TARTARIC ACID

Plane of symmetry bisects this bond and is at right angles to it.

TARTARIC ACID

Figure 13.4
The presence of chiral carbons does not guarantee that the molecule will be chiral. The tartaric acid molecule has two chiral carbons. Two optical isomers are enantiomers. In the third, *meso*-tartaric acid, there is a plane of symmetry. (*Meso* is from the Greek, "in the middle.") Meso isomers become possible whenever a molecule has two or more chiral carbons to each of which are attached the *same* set of four different groups.

Exercise 13.1 Examine the given structures. Place an asterisk above each carbon to which four different groups are attached. If ball-and-stick sets or their equivalent are available to you, make molecular models of each structure and try to discover planes of symmetry. Make molecular models of the mirror images and check to see whether the objects and mirror images are superimposable.

```
(a) CH3CHCH3        (b) CH3CH—COOH      (c) HOOCCH2CH2CHCOOH
        |                  |                          |
        OH                 NH2                        NH2

    CH2—O—CH3              CH3
    |                      |
(d) CH—OH           (e) CH3CHCHCH3
    |                        |
    CH2—OH                   OH
```

Optical Activity

13.9 Polarized Light

Light is an electromagnetic radiation involving oscillations in the strengths of electric and magnetic fields set up by the light source. In ordinary light these oscillations or vibrations occur equally in all directions about the line that may be used to define the path of the light ray.

Certain materials affect ordinary light in a special manner. Polaroid film is an example. This material interacts with the oscillating electrical field of the incoming light in such a way that this field in the emerging light vibrates in one plane. The light is now *plane-polarized light* (Figure 13.5). If we look at some object through a Polaroid film and then place a second film in front of the first, we can rotate one film until the object can no longer be seen. If we now rotate one film by 90°, we'll see the object at maximum brightness again. It is as though the film acts as a lattice fence. It forces any light going through to vibrate in the plane

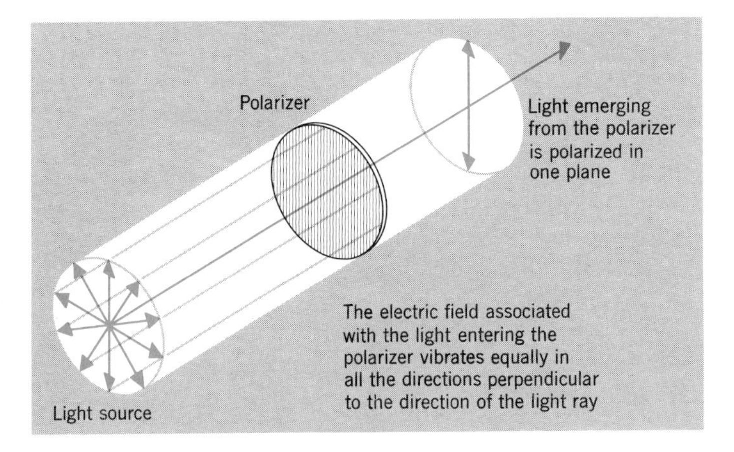

Figure 13.5
Light becomes polarized into vibrations in one plane when it passes through certain materials generally called polarizers. Polaroid film is an example.

of the long spaces between the slats. When this light reaches the second film or lattice fence, and the second slats are perpendicular to those of the first fence, the light cannot go any further. At intermediate angles fractional amounts of light can go through the second Polaroid film or fence.

13.10 Optical Rotation

If a solution of D-asparagine in water is placed in the path of plane-polarized light, the plane of polarization will be rotated. Any substance that will do this to plane-polarized light—rotate the plane—is said to be *optically active.*

13.11 The Polarimeter

The instrument that is used to detect and measure optical activity is called a polarimeter. See Figure 13.6. Its principal working parts consist of a **polarizer,** a tube for holding solutions in the light path, an **analyzer** (actually, just another polarizing device), and a scale for measuring degrees of rotation. As shown in Figure 13.6*a*, when polarizer and analyzer are "parallel," and the tube contains no optically active material, the polarized light goes through at maximum intensity to the observer. However, if a solution of an optically active material is placed in the light path, the plane-polarized light encounters molecules predominantly or wholly of just one chirality, and its plane of oscillation is rotated. The polarized light that leaves the solution is no longer "parallel" with the analyzer (cf. Figures 13.6*a* and *c*), and the intensity of the observed light is reduced. To restore the original intensity, the analyzer is rotated to the right or to the left until it is again parallel with the light *emerging from the tube.* As the operator looks toward the light source, he will, if he rotates the analyzer to the right, record the degrees as positive, and the optically active substance will be called **dextrorotatory.** If the rotation is counterclockwise, the degrees will be recorded as negative, and the substance will be called **levorotatory.** In the example of Figure 13.6*c*, the record will read $\alpha = -40°$, where α stands for the **optical rotation,** the observed number of degrees of rotation.

Since the optical rotation varies with both the temperature of the solution and the frequency of the light, the record actually notes both data, for example, $\alpha_{\mathrm{D}}^{20} = -40°$. The subscript D stands for the D-line of the spectrum of sodium, which is one particular frequency that electromagnetic radiation can have, and 20 means that the solution has a temperature of 20°C.

13.12 Specific Rotation

The optical rotation of a given solution depends on the population of chiral molecules that the light beam encounters. The longer the tube, the more will the plane of the polarized light be rotated. Or, if the tube length is held constant, the higher the concentration, the greater the observed rotation. To take these factors into account, we defined a new quantity called the specific rotation, which is the number of degrees rotation per unit concentration per unit of path length. The symbol for specific rotation is $[\alpha]_{\lambda}{}^{t°}$. By definition,

$$[\alpha]_{\lambda}{}^{t°} = \frac{100\alpha}{cl}$$

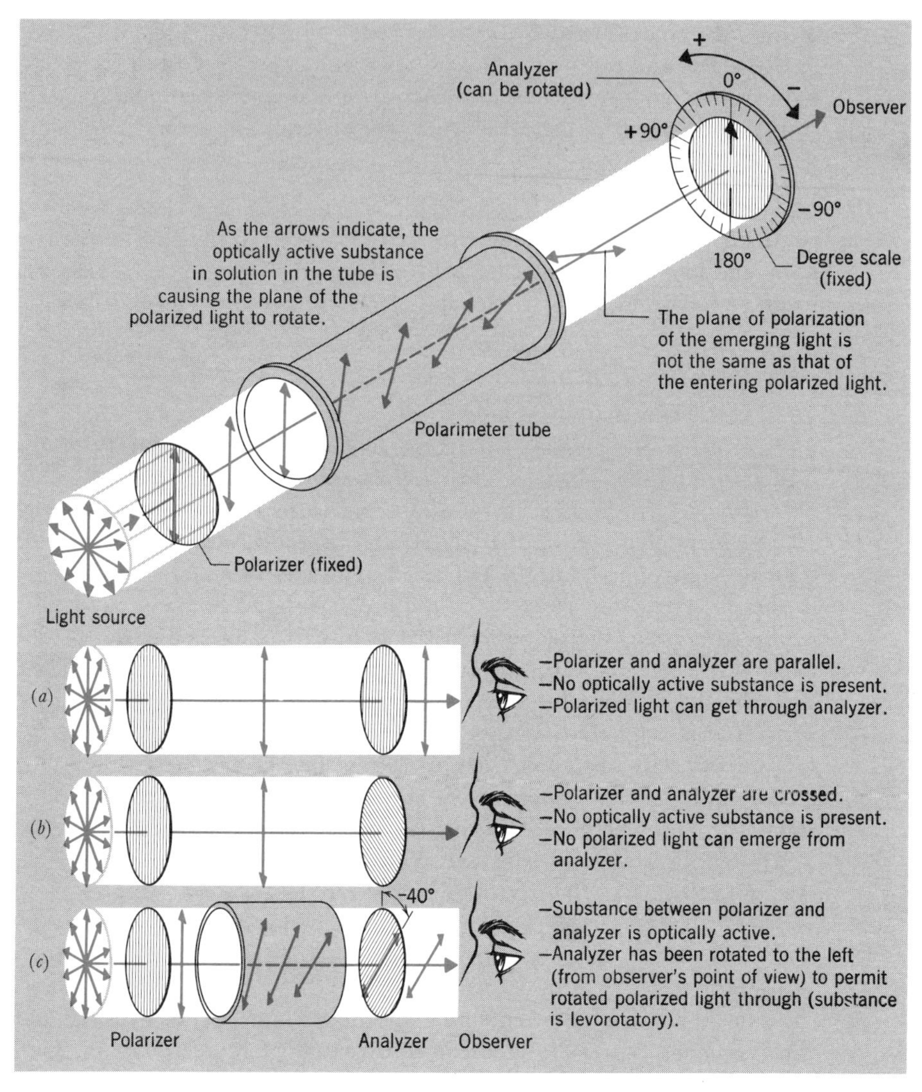

Figure 13.6
Principal working parts of a polarimeter and the measurement of optical rotation.

where c = concentration in grams per 100 cc of solution
l = length of the light path in the solution, measured in decimeters (1 decimeter = 10 cm)
α = observed rotation in degrees (plus or minus)
λ = wavelength of light
t° = temperature of the solution in degrees centigrade

The specific rotation of a compound is an important physical constant, comparable to its melting point or boiling point. It is one more physical characteristic that a chemist can use to identify a substance. Moreover, by measuring the actual rotation, α, of a solution of a substance of known specific rotation in a tube of fixed length, we can calculate the concentration of the solution. That is, with $[\alpha]$, α, and l known, the equation readily permits calculation for c.

Table 13.1 lists specific rotations for several materials, Table 13.2 physical constants for some sets of optical isomers. *Pairs of enantiomers differ physically only in the sign or direction in which each rotates the plane of plane-polarized light.* All other physical constants are the same, including even the number of degrees of rotation.

Other Kinds of Optical Isomers

13.13 Meso Compounds

In some sets of optical isomers, for example, the tartaric acids of Figure 13.4, one member is found to be optically inactive, even though it possesses chiral carbons. We are reminded that the fundamental criterion for chirality in a molecule is not the chiral carbon but rather the impossibility of superimposing object and mirror images. **The isomer that belongs to a set of optical isomers but is optically inactive itself is called the meso isomer.** The mirror image of *meso*-tartaric acid would be superimposable on the object. The acid also possesses within the molecule a plane of symmetry. Behind this plane is a reflection of what is in front. Hence, as one-half of the molecule tends to rotate polarized light to the left, for example, the other half cancels this with an equal rightward rotation. The net effect is optical inactivity.

Table 13.1
Specific Rotations of Various Substances

Physical State	Substance	Specific Rotation (sodium D light at 20°C in solvent specified)
Solutions[a]	Asparagine	+5.41 (water)
	Albumin (a protein)	−25 to −38 (water)
	Cholesterol	−31.61 (chloroform)
	Glucose	+52.5 (aged solution in water)
	Sucrose (table sugar)	+66.4 (water)
	3-Methyl-2-butanol ("active amyl alcohol")	+5.34 (ethyl alcohol)
	Quinine sulfate	−214 (water, at 17°C)
Pure liquids[a]	Turpentine	−37
	Cedar oil	−30 to −40
	Citron oil	+62 (at 15°C)
	Nicotine	−162
Pure solids[b]	Quartz	+21.7
	Cinnabar (HgS)	+32.5
	Sodium chlorate	+3.13

[a] Specific rotation in degrees per decimeter.
[b] Specific rotation in degrees per millimeter of path length in the crystal.

Table 13.2
Physical Constants of Optical Isomers

Set	Members of Set	Mp (°C)	$[\alpha]_D^{20}$ (degrees)	Miscellaneous
Mandelic acids	(+) Mandelic acid	132.8	+155.5	Solubility: 8.54 g/100 cc water (at 20°C)
	(−) Mandelic acid	132.8	−155.4	Solubility: 8.64 g/100 cc water (at 20°C)
Asparagines	(+) Asparagine	234.5 (decomposes)	+5.41	d_4^{15} = 1.534 g/cc
	(−) Asparagine	235 (decomposes)	−5.41	d_4^{15} = 1.54 g/cc
Tartaric acids	(+) Tartaric acid	170	+11.98	d_4^{15} = 1.760 g/cc
	(−) Tartaric acid	170	−11.98	d_4^{15} = 1.760 g/cc
	meso-Tartaric acid	140	0	d_4^{15} = 1.666 g/cc
	Racemic tartaric acid[a]	205	0	d_4^{15} = 1.687 g/cc

[a] This is a 50:50 "mixture" of the plus and minus forms of tartaric acid, which in this case happens to form what is known as a racemic compound.

13.14 Diastereomers

Geometrical isomers that are not related as object to mirror image are called diastereomers. Figure 13.7 illustrates examples, and we shall encounter them again in carbohydrate chemistry. For two molecules to qualify as diastereomers, they must have identical molecular formulas, they must have identical nucleus-to-nucleus sequences (without regard to geometry), but they may not be related as object to mirror image. Generally, they will have different chemical and physical properties.

13.15 Racemic Mixtures

A substance that is composed of a 50:50 mixture of enantiomers is called a racemic mixture. One enantiomer cancels the effects of the other on polarized light, and a racemic mixture is optically inactive, emphasizing the point that the *phenomenon* of optical activity relates to a measurement that we can make on a substance. The *explanation of the phenomenon*, in terms of chiral molecules, relates to our theory about it.

13.16 Configurational Changes in Chemical Reactions

A molecule of the compound *n*-butane has no chiral carbon, and it has a plane of symmetry. A molecule of bromine, Br_2, is also symmetrical. Yet if bromine can

$$CH_3CH_2CH_2CH_3 + Br \longrightarrow CH_3CH_2\underset{\displaystyle Br}{\underset{|}{C}}HCH_3 + HBr$$

n-BUTANE　　　　*sec*-BUTYL BROMIDE

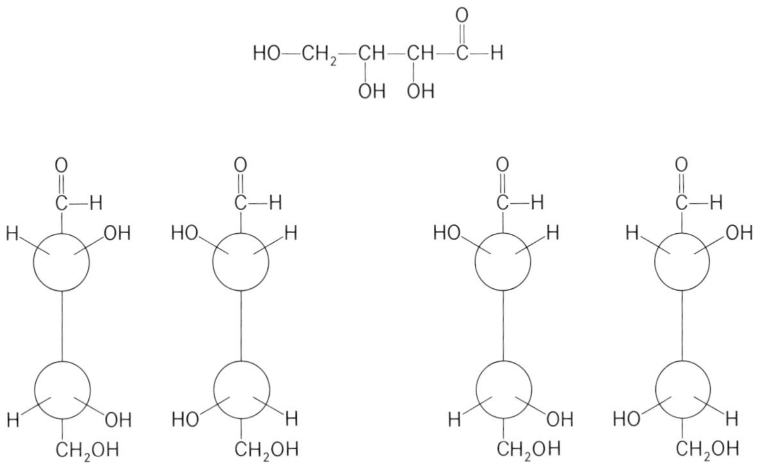

$HO-CH_2-CH(OH)-CH(OH)-C(=O)-H$

D-ERYTHROSE	L-ERYTHROSE	D-THREOSE	L-THREOSE
$[\alpha]_D^{20}$ − 14.8°	+14.8°	$[\alpha]_D^{20}$ + 19.6°	−19.6°

Figure 13.7
Optical isomers that illustrate pairs of enantiomers and diastereomers. D- and L-erythrose are related as enantiomers. Likewise, D- and L-threose are enantiomers. Any member of the first pair is a diastereomer of any member of the second pair. (Note the differences in specific rotations.) The meaning of "D" and "L" is discussed in Section 14.8.

be made to react with *n*-butane at the second carbon, this carbon becomes chiral in the product.

The question is whether the substance actually isolated as a product of this reaction and identified as *sec*-butyl bromide is optically active. The answer is *no*. We must distinguish between a *substance* that is optically active and one that has chiral molecules. Optical activity is something we can measure with a polarimeter, completely without regard for an interpretation of the observation. The reaction produces a chiral carbon but in such a way that the product is a 50:50 mixture of enantiomers, a racemic, or an optically inactive mixture. We can understand the reason for this outcome by examining Figure 13.8. In *n*-butane, at the site where bromine will eventually become affixed, are two equivalent hydrogens. In a substitution occurring to any one molecule there is an equal chance that either hydrogen will be the one replaced. In a collection approaching Avogadro's number in size, the statistics are such that production of a racemic mixture is a certainty.

It is generally true that optically active *substances* cannot be synthesized from optically inactive reagents. Chiral *molecules* can easily be made, but the final substance will be a racemic mixture.

13.17 Absolute Configuration

The last major topic involving optical isomerism is about the *absolute configuration* of enantiomers. How do we specify what specific "handedness" a particular enantiomer has? We say "right hand" or "left hand" when we refer to our two hands. What is done with enantiomers?

In biochemistry, the literature is filled with the use of one system, the designations D or L as the two families of chiral molecules. We shall study that system in the next chapter.

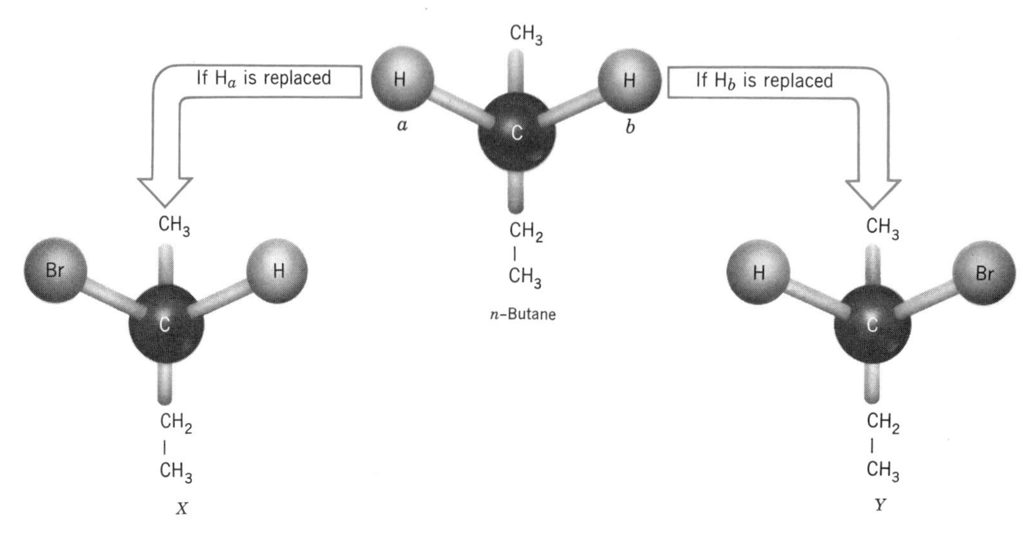

Figure 13.8
Illustrating a general fact: optically active substances cannot be prepared by the action of an optically inactive reagent on an optically inactive compound. In *n*-butane the two hydrogens that are candidates for replacement, if *sec*-butyl bromide is to form, have the same chance to be replaced. Therefore a 50 : 50 mixture of the two enantiomers of *sec*-butyl bromide, *X* and *Y*, is produced. Although the enantiomers would individually be optically active, the racemic mixture of them that actually forms is not.

Brief Summary

1. Stereoisomers are isomers whose molecules share the same molecular framework but differ in geometrical projections of groups in that framework.

2. The two broad categories of stereoisomers are
   (a) geometrical isomers (*cis-trans* relations)
   (b) optical isomers (chiral relations)

3. Enantiomers are chiral stereoisomers related as object and mirror image that cannot be superimposed.

4. Disastereomers are stereoisomers that are not enantiomers.

5. Meso compounds are achiral members of a set of optical isomers.

6. Racemic mixtures are 50:50 mixtures of enantiomers.

7. A chiral carbon is one to which four different groups are attached.

8. Optical activity is the ability to rotate the plane of plane-polarized light.

9. Enantiomers are identical in all properties except:
   (a) Their sign of rotation of plane-polarized light.
   (b) Their reactions with optically active substances.

10. Specific rotation, $[\alpha]$, is related to observed rotation, α, concentration (in g/100 cc) and path length (in dm) by the equation $[\alpha]_{\lambda}^{t} = \frac{100\,\alpha}{c \times l}$.

11. Substances that rotate the plane clockwise (when facing the light) arc dextrorotatory; the opposite is levorotatory.

Problems and Exercises

1. Write definitions for each of the following terms and when appropriate provide suitable illustrations.
   (a) plane of symmetry
   (b) optical isomers
   (c) enantiomers
   (d) superimposability
   (e) chiral carbon
   (f) optical activity
   (g) polarimeter
   (h) dextrorotatory
   (i) levorotatory
   (j) optical rotation (and symbol)
   (k) specific rotation
   (l) meso compound
   (m) diastereomers
   (n) racemic mixture

2. Examine the following structures and predict whether optical isomerism is a possibility.

   (a) 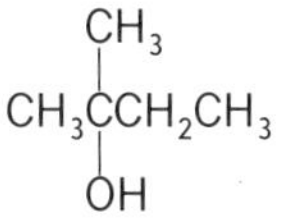 $CH_3C(CH_3)(OH)CH_2CH_3$
   (b) 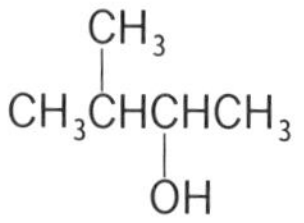 $CH_3CH(CH_3)CH(OH)CH_3$
   (c) 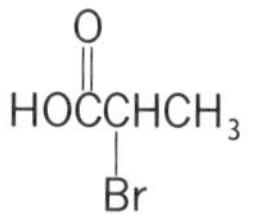 $HOC(=O)CH(Br)CH_3$
   (d) 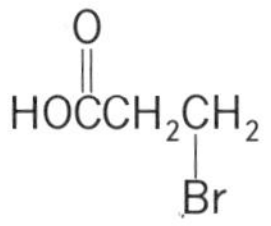 $HOC(=O)CH_2CH_2Br$
   (e) *cis*-1,2-cyclopentanediol
   (f) *trans*-1,2-cyclopentanediol

3. Explain why enantiomers *should* have identical physical properties (except for the sign of rotation of plane-polarized light).

4. Explain why diastereomers *should not* be expected to have identical physical properties (except coincidentally).

5. A solution of sucrose in water at 25°C in a tube 10-cm long gives an observed rotation of +2.0°. The specific rotation of sucrose in this solvent at this temperature is +66.4°. What is the concentration of the sucrose solution in grams per 100 cc?

chapter fourteen
Carbohydrates

14.1 Families of Carbohydrates

Carbohydrates are polyhydroxy aldehydes, polyhydroxy ketones, or substances that by simple hydrolysis yield these. Carbohydrates that cannot be hydrolyzed to simpler molecules are called **monosaccharides** or sometimes simple sugars. They are classified according to the following:

1. How many carbons there are in one molecule.

If the number of carbons is	the monosaccharide is a
3	triose
4	tetrose
5	pentose
6	hexose, etc.

2. Whether there is an aldehyde or a keto group present. If an aldehyde is present, the monosaccharide is an aldose. If a keto group is present, the monosaccharide is a ketose.

These terms may be combined. For example, a hexose that has an aldehyde group is called an aldohexose. A ketohexose possesses a keto group and a total of six carbons.

Carbohydrates that can be hydrolyzed to two monosaccharides are called **disaccharides.** Sucrose, maltose, and lactose are common examples. Starch and cellulose are common **polysaccharides;** their molecules yield many monosaccharide units when they are hydrolyzed.

Mono- and most disaccharides readily reduce Tollens', Benedict's and Fehling's reagents (Sections 9.13–9.14). Carbohydrates are therefore also classified as being either **reducing** or **non-reducing sugars.** Polysaccharides generally give negative tests with these reagents.

Monosaccharides

14.2 Glucose

This white, crystalline, dextrorotatory substance is perhaps the most abundant organic species on earth if we count all of its combined forms. Cellulose, a polysaccharide that yields nothing but glucose when it is hydrolyzed, is about 10% of all tree leaves of the world (dry weight basis), about 50% of the woody parts of plants, and nearly 100% of cotton. Starch, another polysaccharide, also yields nothing but glucose when it is hydrolyzed. Glucose and the ketohexose, fructose, are the major components of honey. Glucose is commonly found in plant juices. There are roughly 100 mg glucose per 100 cc of blood, which explains why glucose is often called "blood sugar" in clinical and medical work. Other nicknames for glucose are "corn sugar" and "dextrose" (after its dextrotatory property).

14.3 Evidence for the Structure of Glucose

Given a white powder with the molecular formula $C_6H_{12}O_6$ and several known physical properties, the chemist must still determine how the 6 carbons, 12 hydrogens, and 6 oxygens are put together. Are the carbons in a straight chain, a branched chain, or possibly a ring? What is the nature of the oxygens? Are they in hydroxyl groups, keto groups, aldehyde groups, or some other structural arrangement? The following reactions illustrate how information bearing on these questions was gathered.

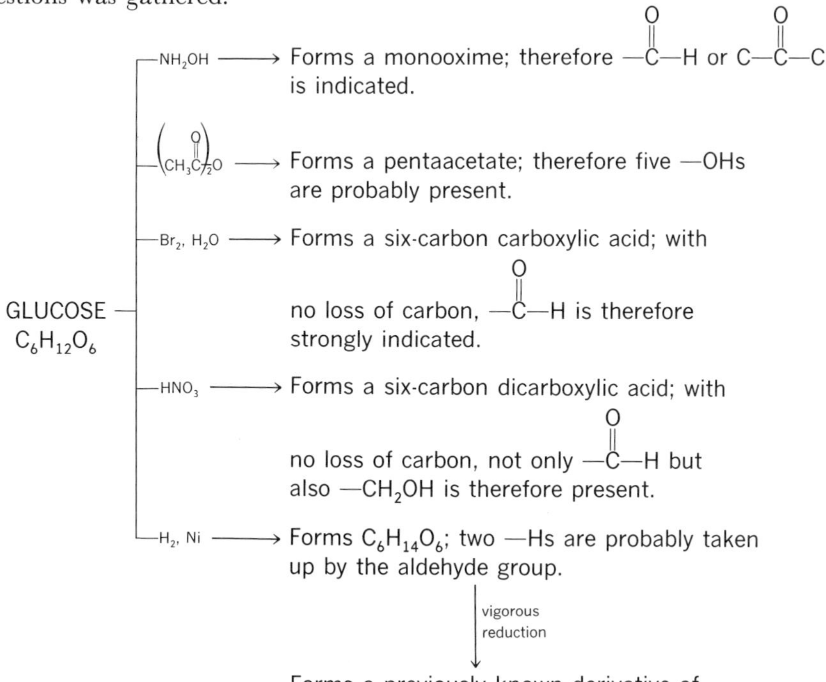

Of the six carbons in the straight chain, one is used in the aldehyde group, —CHO. The remaining five must accommodate the five hydroxyl groups indicated by the behavior toward acetic anhydride. Because chemists have rarely found a stable structure in which one carbon holds two hydroxyl groups, we distribute the five hydroxyl groups among the remaining five carbons. The general structure for glucose (without regard for optical isomers) must therefore be
(The optical isomers are discussed beginning with Section 14.6.)

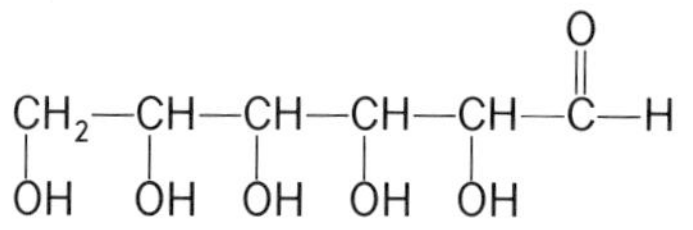

$$\underset{\text{OH}}{\text{CH}_2}-\underset{\text{OH}}{\text{CH}}-\underset{\text{OH}}{\text{CH}}-\underset{\text{OH}}{\text{CH}}-\underset{\text{OH}}{\text{CH}}-\overset{\text{O}}{\overset{\|}{\text{C}}}-\text{H}$$

14.4 Osazone Formation

On the basis of what we learned about the behavior of aldehydes and ketones toward phenylhydrazine (Section 9.20), we might expect aldoses and ketoses to form phenylhydrazones. In acid they do. We would not expect, however, that two more molecules of phenylhydrazine (if made available) will react further with each new phenylhydrazone molecule. Other products of the overall reaction are ammonia and aniline. This behavior is typical of α-hydroxy aldehydes and α-hydroxy ketones, and the double phenylhydrazones are called phenylosazones or *osazones*.

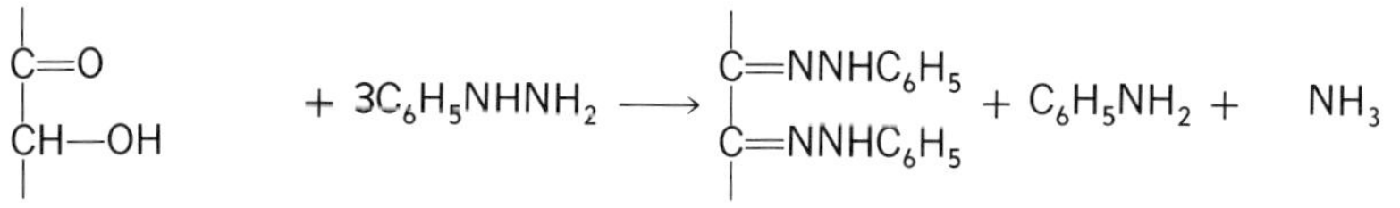

$$\begin{array}{c} | \\ \text{C}{=}\text{O} \\ | \\ \text{CH}{-}\text{OH} \\ | \end{array} + 3\text{C}_6\text{H}_5\text{NHNH}_2 \longrightarrow \begin{array}{c} | \\ \text{C}{=}\text{NNHC}_6\text{H}_5 \\ | \\ \text{C}{=}\text{NNHC}_6\text{H}_5 \\ | \end{array} + \text{C}_6\text{H}_5\text{NH}_2 + \text{NH}_3$$

α-HYDROXY ALDEHYDE OR KETONE — PHENYLHYDRAZINE — AN OSAZONE — ANILINE — AMMONIA

For example:

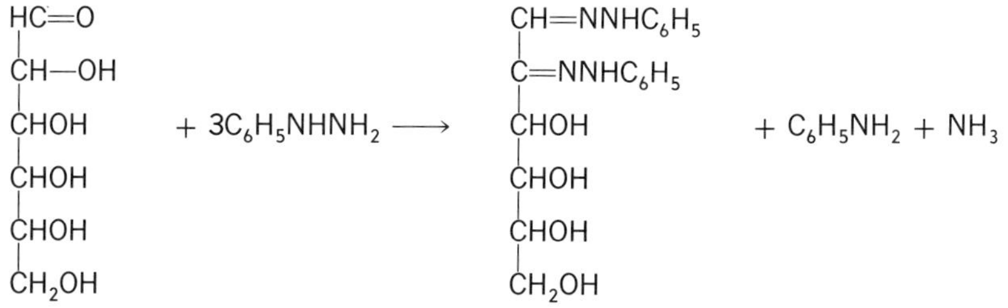

$$\begin{array}{l} \text{HC}{=}\text{O} \\ | \\ \text{CH}{-}\text{OH} \\ | \\ \text{CHOH} \\ | \\ \text{CHOH} \\ | \\ \text{CHOH} \\ | \\ \text{CH}_2\text{OH} \end{array} + 3\text{C}_6\text{H}_5\text{NHNH}_2 \longrightarrow \begin{array}{l} \text{CH}{=}\text{NNHC}_6\text{H}_5 \\ | \\ \text{C}{=}\text{NNHC}_6\text{H}_5 \\ | \\ \text{CHOH} \\ | \\ \text{CHOH} \\ | \\ \text{CHOH} \\ | \\ \text{CH}_2\text{OH} \end{array} + \text{C}_6\text{H}_5\text{NH}_2 + \text{NH}_3$$

AN ALDOHEXOSE — AN ALDOHEXOSAZONE

The phenylosazones are crystalline solids with characteristic crystalline shapes, decomposition points (rather than melting points), and specific optical rotations. Under carefully standardized conditions each aldose or ketose forms its osazone at its own characteristic rate. Preparing an osazone is therefore useful in identifying a sugar.

14.5 Thin-layer Chromatography

Thin-layer chromatography is another, even better technique for identifying sugars. See Figure 14.1. A solution of a mixture of sugars is allowed to move by

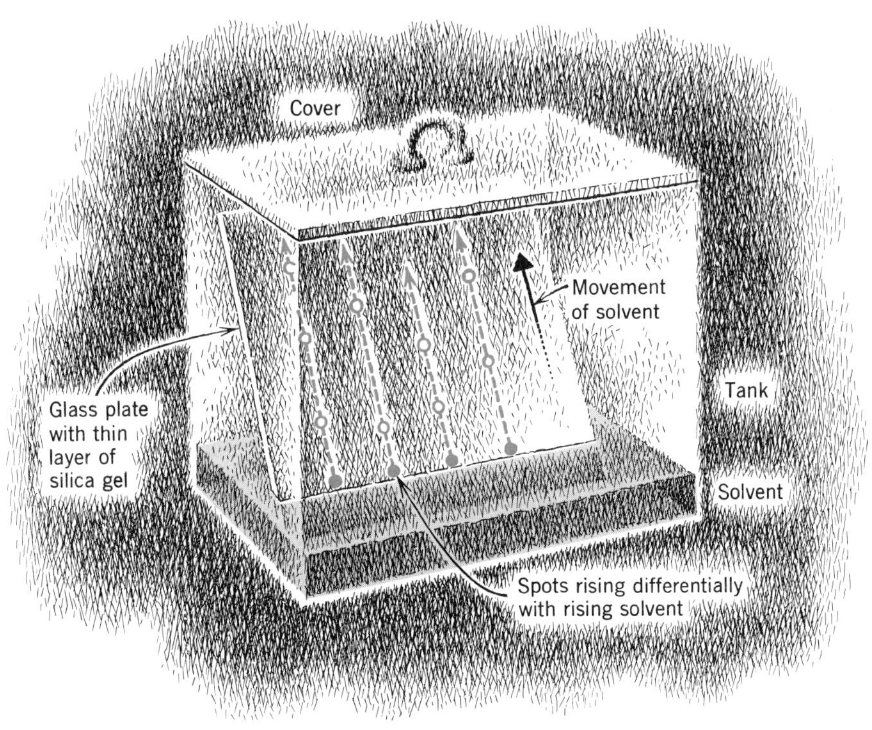

Figure 14.1
Thin-layer chromatography. The plate itself may be glass or an inert plastic. The thin layer of adsorbant may be any number of substances—silica gel, alumina, cellulose, to name three. The prepared plate is heated in an oven before use (and then cooled) to activate the adsorbant—from 100 to 250° depending on the desired degree of activation. The mixture or mixtures to be separated are dissolved in a solvent, and microliter portions are spotted on the starting line. When the spots have dried, the plate is put in an enclosed chamber having the developing solvent whose vapors saturate the chamber. As the solvent creeps up the thin layer the development of the spots takes place. (From E. E. Conn and P. K. Stumpf, *Outlines of Biochemistry,* third edition, 1973. John Wiley & Sons, Inc., used by permission.)

capillary action across a thin layer of silica, alumina, or some other adsorbing material that is deposited on glass or plastic. The sugars move less rapidly than the solvent, and because of differing abilities to be adsorbed by the layer, each sugar moves at a characteristic rate. In time the sugars are in separate regions of the thin layer, and special chemical sprays can then locate them. The relative rates of movement, easy to measure under standardized conditions, join the list of other physical constants that identify sugars. The members of many other classes of compounds are identified by thin-layer chromatography.

14.6 Optical Isomerism of the Aldohexoses

There are four chiral carbons in a molecule of glucose. Each carbon has its own set of four different groups. No set is identical with any other. In such circumstances the number of optical isomers is given simply by 2^n, where n = the number of different chiral carbons. Since $n = 4$ for glucose (or any aldohexose) there are 16

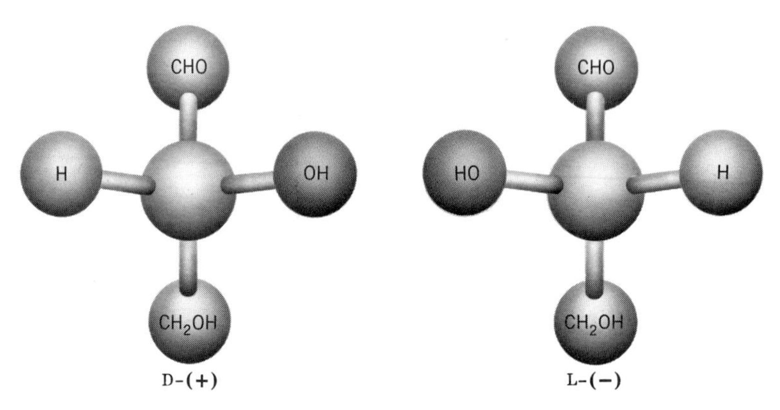

Figure 14.2
The absolute configurations of the enantiomers of glyceraldehyde. The dextrotatory enantiomer is labeled D-(+) and the levorotatory enantiomer, L-(−). These two are the "parents" or reference compounds for the two configurational families of carbohydrates, the D family and the L family.

optical isomers among all the aldohexoses. They occur as 8 pairs of enantiomers. Glucose is just one of them; galactose is another. But which? Since the types of chemical reactions given by glucose are given by all its optical isomers, these reactions provide no clue as to which is glucose. To pursue this further we must go back and take up a topic of central importance about optical isomerism in general, the question of absolute configuration.

14.7 Absolute Configuration

To simplify let us drop back down from the complexities of glucose to the simplest molecule that meets the definition of a monosaccharide and still has one chiral carbon. The simplest polyhydroxy aldehyde is glyceraldehyde; each molecule has one chiral carbon. It can therefore exist in two configurations related as object and a mirror image that cannot be superimposed (Figure 14.2). Until the advent of spectroscopic techniques, scientists could not determine which of two enantiomers was the dextrorotatory form and which the levorotatory. In other words, they did not know what the *absolute configuration* of the molecules in samples of each of the two glyceraldehydes was. They noted with considerable interest, however, that a sample of glyceraldehyde that was dextrotatory was oxidized to a glyceric acid that was levorotatory. The sample of glyceraldehyde that was levorotatory was oxidized to the dextrorotatory glyceric acid.

As we can see from the equation:

O ‖ C—H; H; OH; CH_2OH — (+)-GLYCERALDEHYDE

Oxidation (configuration does not change) →

O ‖ C—OH; H; OH; CH_2OH — (−)-GLYCERIC ACID

this oxidation does not change any of the bonds holding the four groups to the chiral carbon. Whatever the absolute configuration of the glyceraldehyde used as the reactant, the glyceric acid we get by oxidation must have the same.

Early in the history of organic chemistry an arbitrary decision about the absolute configuration of (+)-glyceraldehyde was made. It had a 50:50 chance of being correct. The molecules of the samples of glyceraldehyde that tested dextrorotatory were said to have the absolute configuration indicated in Figure 14.2. The guess turned out to be correct.

14.8 D and L Families

To deal with absolute configuration chemists long ago recognized the need to establish two reference families of compounds. It was agreed to place all substances with configurations related directly to (+)-glyceraldehyde in the D-*family*. All those related to (−)-glyceraldehyde were assigned to the L-*family*. The letters D and L are simply names for two families. The mirror image of any compound in the D-family will be in the L-family. These letters, pronounced simply "dee" and "ell," have nothing to do with specific rotation. They indicate something about absolute configuration. This system is not perfect, but it works very well in both carbohydrate and protein chemistry. To remedy defects of the D and L system another system for describing absolute configuration has been devised. In it the two configurational families are called R and S for reasons that you may discover for yourself in the Appendix. The newer biochemical literature as well as the literature of organic chemistry in general will be using the R and S system. It is completely unambiguous, and it depends in no way on having parent compounds.

14.9 Plane Projection Diagrams

To write perspective structures of molecules with four chiral centers we have a problem that is solved easily only by talented artists. While chemists insist that scientific work is one of mankind's high forms of artistry, few chemists claim to be able to draw well. It isn't too hard to draw a perspective of a molecule with one chiral carbon and give it a three-dimensional appearance, but when you have four—well, chemists decided long ago they had better devise another way. The plane projection diagram was one answer.

To make one of these diagrams requires only that you be able to draw reasonably straight lines at right angles to others. Besides that, you have to follow a few rules carefully. (These rules are phrased with their application to carbohydrate or protein chemistry especially in mind.)

1. The carbon skeleton that includes the chiral carbon or carbons is arranged *vertically*. Let that particular carbon that normally would be numbered 1 (for naming purposes) be at the top.
2. Represent each chiral carbon by a simple cross—that is, by an intersection of perpendicular lines.
3. The two halves of the horizontal line represent bonds to two groups that actually project forward, toward you, out of the plane from the chiral center.
4. The two halves—upper and lower—of the vertical line are the two bonds to two groups that are actually projecting backward, away from you, behind the plane of the paper from the chiral center.

5. Use two or more intersecting lines to represent two or more chiral centers, but always remember that at each intersection the vertical lines project backward, the horizontal, forward.

Figure 14.3 illustrates the application of these rules to the glyceraldehydes.

In Figure 14.4 we see optical isomers having two chiral carbons and how they appear in plane projection diagrams. Once we have plane projection diagrams we may easily draw their mirror images. There is one important rule governing the use of these diagrams. *We are not allowed to lift plane projection diagrams out of the plane of the paper.* We may only slide and rotate them within the plane when we want to check if an object is or is not superimpossible with its mirror image. (If you turn a plane projection diagram over, out of the plane, you actually take groups projecting in one direction and make them project into the opposite direction. That is why we have the rule given.)

14.10 Absolute Configuration of Natural Glucose

All of the naturally occurring aldoses belong to the D-family. By this we specifically mean something about the absolute configuration of one particular chiral carbon, the one farthest away from the carbonyl carbon. If the configuration at the chiral carbon farthest from the carbonyl group is like that of D-glyceraldehyde, the monosaccharide is in the D-famly; if that of L-glyceraldehyde, then

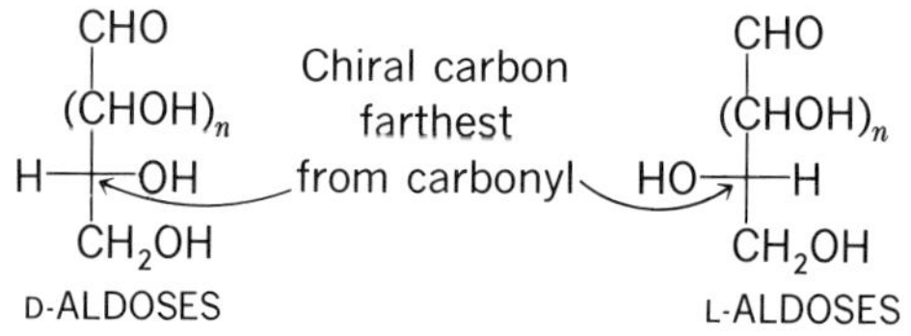

CHO
$(CHOH)_n$
H—C—OH
CH_2OH
D-ALDOSES

Chiral carbon farthest from carbonyl

CHO
$(CHOH)_n$
HO—C—H
CH_2OH
L-ALDOSES

If $n = 0$:	D-GLYCERALDEHYDE	If $n = 0$:	L-GLYCERALDEHYDE
$n = 1$:	D-ERYTHROSE OR D-THREOSE (FIG. 14.4)	$n = 1$:	L-ERYTHROSE OR L-THREOSE (FIG. 14.4)

it is in the L-family. Figure 14.5 shows the whole family of D-aldoses from D-glyceraldehyde through all the eight D-aldohexoses. The other eight optical isomers of the aldohexoses are the eight mirror images of those shown on the bottom row in Figure 14.5.

Exercise 14.1 (a) Write the plane projection structures of the phenylosazones of D-glucose and D-mannose. (b) How are these two structures related, as enantiomers, as diastereomers, or are they identical? (c) How are the phenylosazones of allose and altrose related? You should not have to write their structures to answer this. What you have learned from parts (a) and (b) should be enough. Diastereomers such as D-glucose and D-mannose that differ in their configurations only at just one of their chiral carbons are called *epimers.* We say that D-glucose and D-mannose are epimeric at C-2. Emil Fisher (Nobel prize, 1902) used osazones in sorting out the structures of aldoses. (d) What is the name of the C-2 epimer of D-galactose? (e) Of L-glucose?

The nutritionally important carbohydrates are all in the D-family. We cannot use the L-enantiomers; we lack enzymes (catalysts) with the proper "handedness"

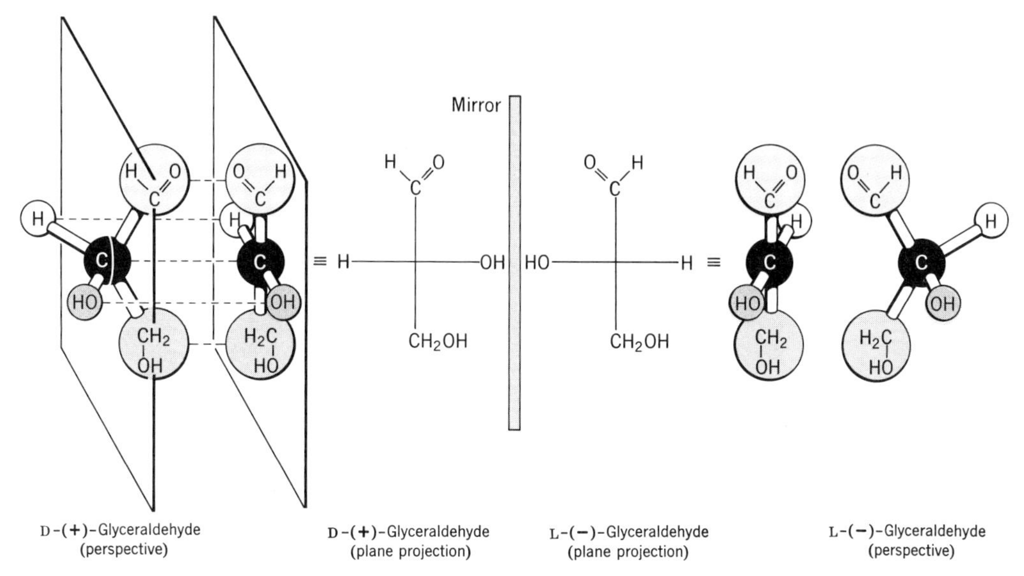

Figure 14.3
D- and L-glyceraldehyde. Relating their absolute configurations as seen in perspective drawings to their corresponding plane projection diagrams. Plane-projection diagrams of enantiomers will always relate as object to mirror image, as seen here. However, to check whether or not the two are superimposable you may *not* flap one or the other out of the plane. You are permitted only to slide and turn them around *in the plane.*

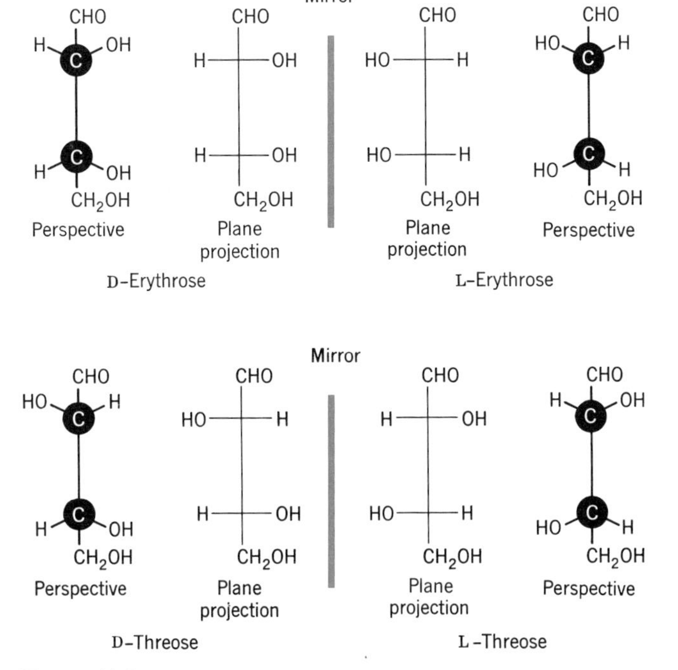

Figure 14.4
The enantiomers of erythrose and threose in perspective and plane-projection diagrams. These are the four aldotetroses. With two different chiral centers, the total number of optical isomers is $2^2 = 4$, occurring as two pairs of enantiomers.

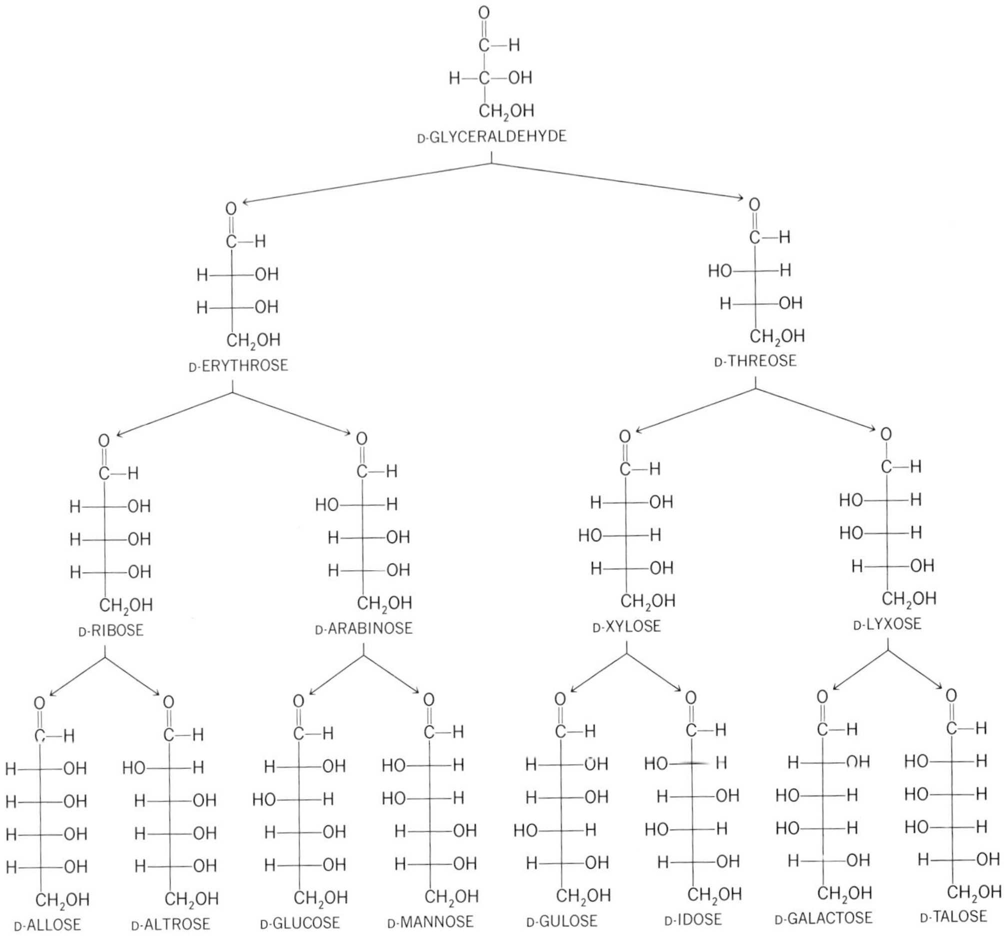

D-GLYCERALDEHYDE

D-ERYTHROSE D-THREOSE

D-RIBOSE D-ARABINOSE D-XYLOSE D-LYXOSE

D-ALLOSE D-ALTROSE D-GLUCOSE D-MANNOSE D-GULOSE D-IDOSE D-GALACTOSE D-TALOSE

Figure 14.5
The D-family of aldoses through the aldohexoses. In every structure the —OH group on the chiral carbon farthest from the carbonyl group projects to the right.

to act on the saccharides in the L-family. Therefore, for the rest of the book when we read "glucose" or any other sugar, we mean the D-family enantiomer. If we mean anything else we shall say so clearly.

Having said all this about the structure of glucose we must now turn to some facts that do not fit the picture. Glucose (and monosaccharides in general) have some properties that cannot be rationalized with the given structure. We shall look principally at two: a phenomenon called mutarotation and a reaction that produces acetal-like compounds in a peculiar way.

14.11 Mutarotation

When ordinary crystalline glucose is freshly dissolved in water, it shows a specific rotation $[\alpha]_D + 113°$. (Here the subscript D refers to the D-line of sodium vapor light.) While the solution remains at room temperature, this value gradually changes, stopping at $[\alpha]_D + 52°$. No deep-seated chemical change has happened because we can get back the original glucose from this aged solution. The change in rotatory power of a solution of an optically active compound is called *mutarotation.*

If we try to recover D-glucose from a solution that has fully mutarotated using a special method of recovery, we get a different form of D-glucose. If glucose is recovered from an aged solution by evaporating the solvent at a temperature above 98°C, this other form of glucose is obtained. Its specific rotation, in a freshly prepared solution, is $[\alpha]_D + 19°$, but this solution also mutarotates until a final value of $[\alpha]_D + 52°$ is obtained, the same final value cited for ordinary glucose. In fact, glucose with a specific rotation in a fresh solution of $[\alpha]_D + 113°$ can be separated from this second aged solution. The glucose that gives $[\alpha]_D + 113°$ in a fresh solution we call α-glucose. The other is called β-glucose. (More correctly: α-(D)-glucose and β-(D-glucose.) The structure of D-(+)-glucose that we have used thus far does not provide any obvious clue to explain this behavior.

The aldoses (from the aldotetroses and higher) as well as many other saccharides all exhibit mutarotation. Before dealing with an explanation of this phenomenon, let us study still another peculiar property of glucose which will yield to the same explanation.

14.12 Glucosides—Sugar Acetals

Another peculiarity of glucose is that conversion of its aldehyde group to an acetal requires only one mole of alcohol, instead of two, per mole of glucose (cf. Section 9.19). The product, called in the example methyl glucoside, is stable in a basic solution. It does not react positively to Tollens' test. In other words, it does not behave as though it were a potential aldehyde, that is, a hemiacetal. It behaves as though it were an acetal. The reaction shown can in fact be reversed. Like any acetal, methyl glucoside can be easily hydrolyzed. It does not mutarotate, however.

$$\underset{\text{GLUCOSE}}{C_6H_{12}O_6} + \underset{\text{METHYL ALCOHOL}}{HOCH_3} \xrightarrow{H^+} \underset{\text{METHYL GLUCOSIDE}}{C_6H_{11}O_5{-}OCH_3} + H_2O$$

Just as glucose can exist as α-glucose and β-glucose, methyl glucoside can be made in two forms: α-methyl glucoside and β-methyl glucoside. Let us now see how these properties make sense with a relatively small modification of our earlier structure of glucose. (The change will, at first, seem rather great, but nothing will be done to the basic configurations at each of the four chiral carbons.)

14.13 The Cyclic Structures of Glucose

Mutarotation and the peculiar way we get an acetal-like compound from glucose make sense if glucose is already actually a hemiacetal—a *cyclic* hemiacetal. See Figure 14.6. It would still be a "potential aldehyde" like all hemiacetals. It would therefore give all of the reactions we have studied for glucose.

When α-glucose mutarotates its cyclic molecules open up to the open form of

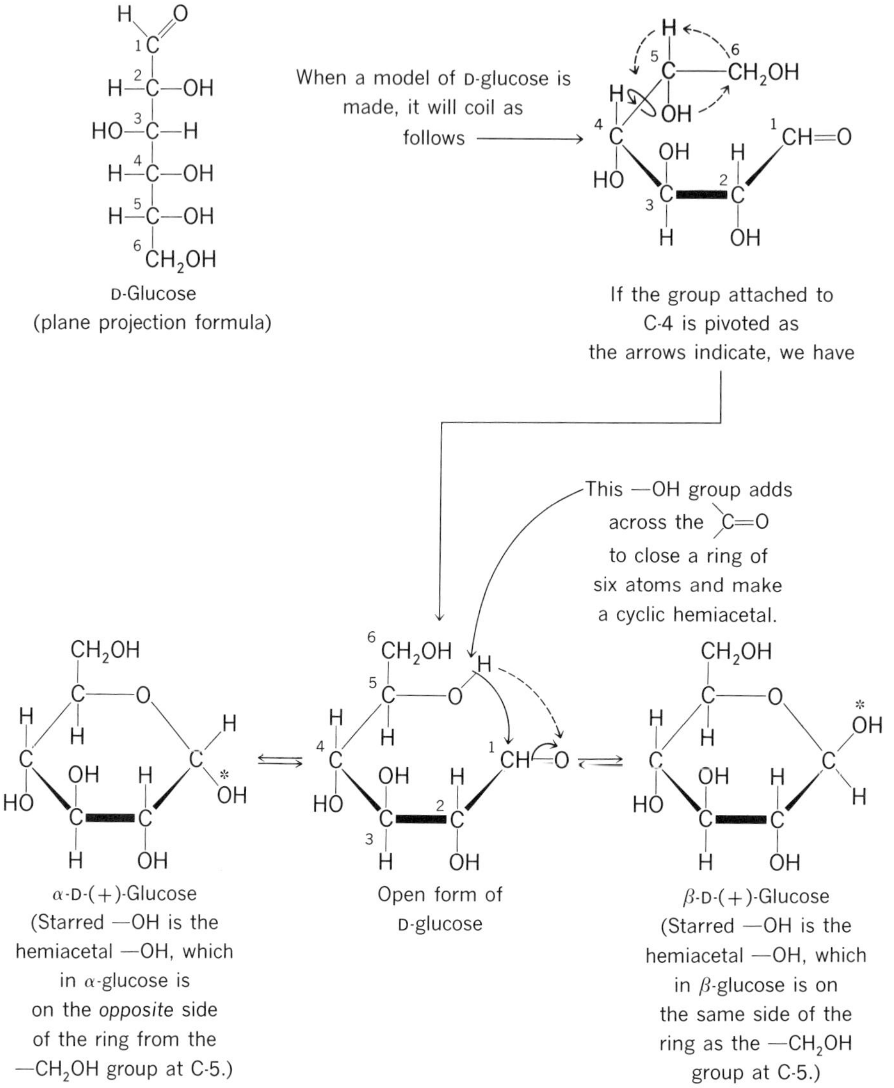

D-Glucose
(plane projection formula)

When a model of D-glucose is made, it will coil as follows

If the group attached to C-4 is pivoted as the arrows indicate, we have

This —OH group adds across the C=O to close a ring of six atoms and make a cyclic hemiacetal.

α-D-(+)-Glucose
(Starred —OH is the hemiacetal —OH, which in α-glucose is on the *opposite* side of the ring from the —CH_2OH group at C-5.)

Open form of D-glucose

β-D-(+)-Glucose
(Starred —OH is the hemiacetal —OH, which in β-glucose is on the same side of the ring as the —CH_2OH group at C-5.)

Figure 14.6
The two cyclic, hemiacetal forms of D-glucose and how they are related to the plane projection structure, top. (One final refinement in the model of glucose is in Figure 14.8.)

glucose. While opened, the aldehyde group can rotate about the C_1-C_2 bond. If ring closure occurs when the aldehyde group happens to be pointing one way (as in Figure 14.7) the α-form is made again. But in other open-form molecules the aldehyde group may happen to be pointing the other way at the moment of ring closure, and molecules of the β-form are made. Eventually an equilibrium mixture is established with all three kinds of molecules present. At equilibrium, of course, mutarotation stops; the specific rotation is stabilized at +52°. Whether you start with α- or β-glucose you get the identical mixture at equilibrium. It happens to consist of 36% α-form, 64% β-form, and only 0.02% of the open form.

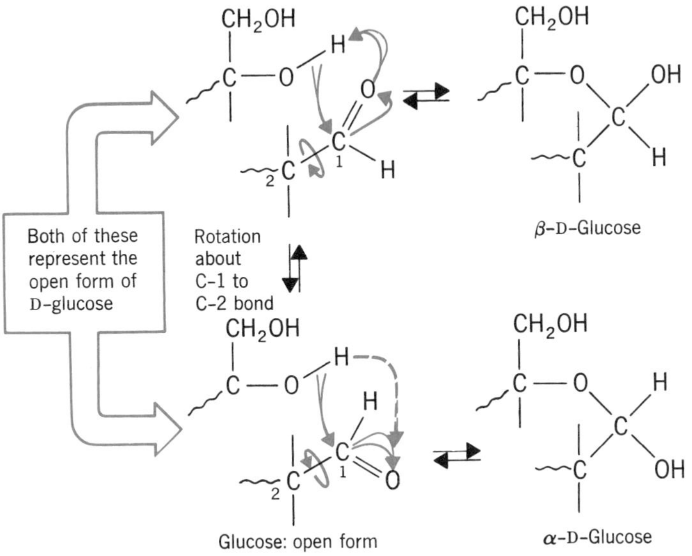

Figure 14.7
How the α- and the β-forms of D-glucose arise from a common intermediate, the open form. (Only key parts of glucose molecules are shown.)

Since glucose is nearly 100% in a cyclic, hemiacetal form, we need only one molecule more of an alcohol to get to the acetal stage. In sugar chemistry these cyclic acetals of glucose are called **glucosides.** (From other sugars we can prepare, for example, galactosides, mannosides, fructosides, etc., too. The general family name is **glycoside.**)

If we pick methyl alcohol as the additional alcohol for changing glucose to a glucoside, we have the same possibilities we had previously; we can get either diastereomer, an α- or a β-methyl glucoside.

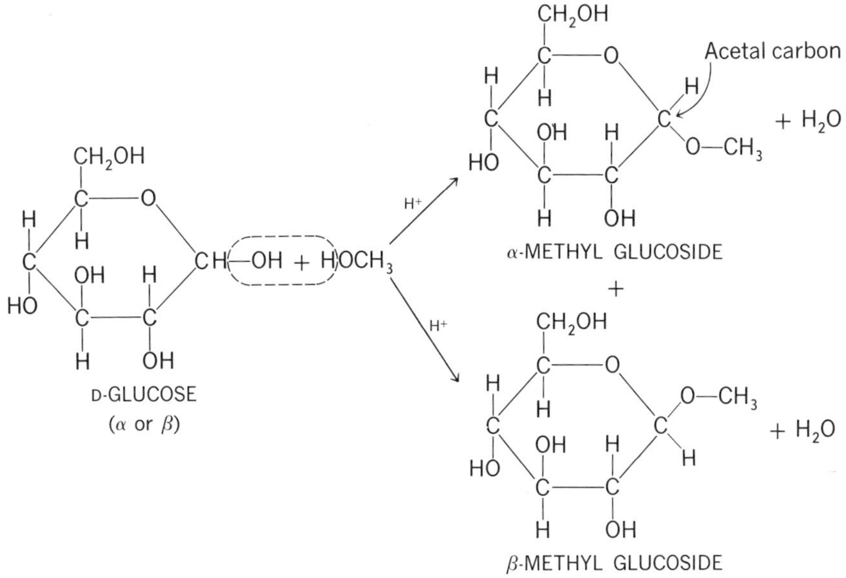

14.14 D-Glucose Rings in Chair Form

In Section 2.20 we learned that six-membered rings are not flat, that they exist in predominately chair conformations. The six-membered rings of the carbohydrates are just like that, too. Figure 14.8 gives the most accurate models for α- and β-D-glucose, and it further clarifies what is specifically meant by the "D" in cyclic D-glucose and what is meant by "α" or "β."

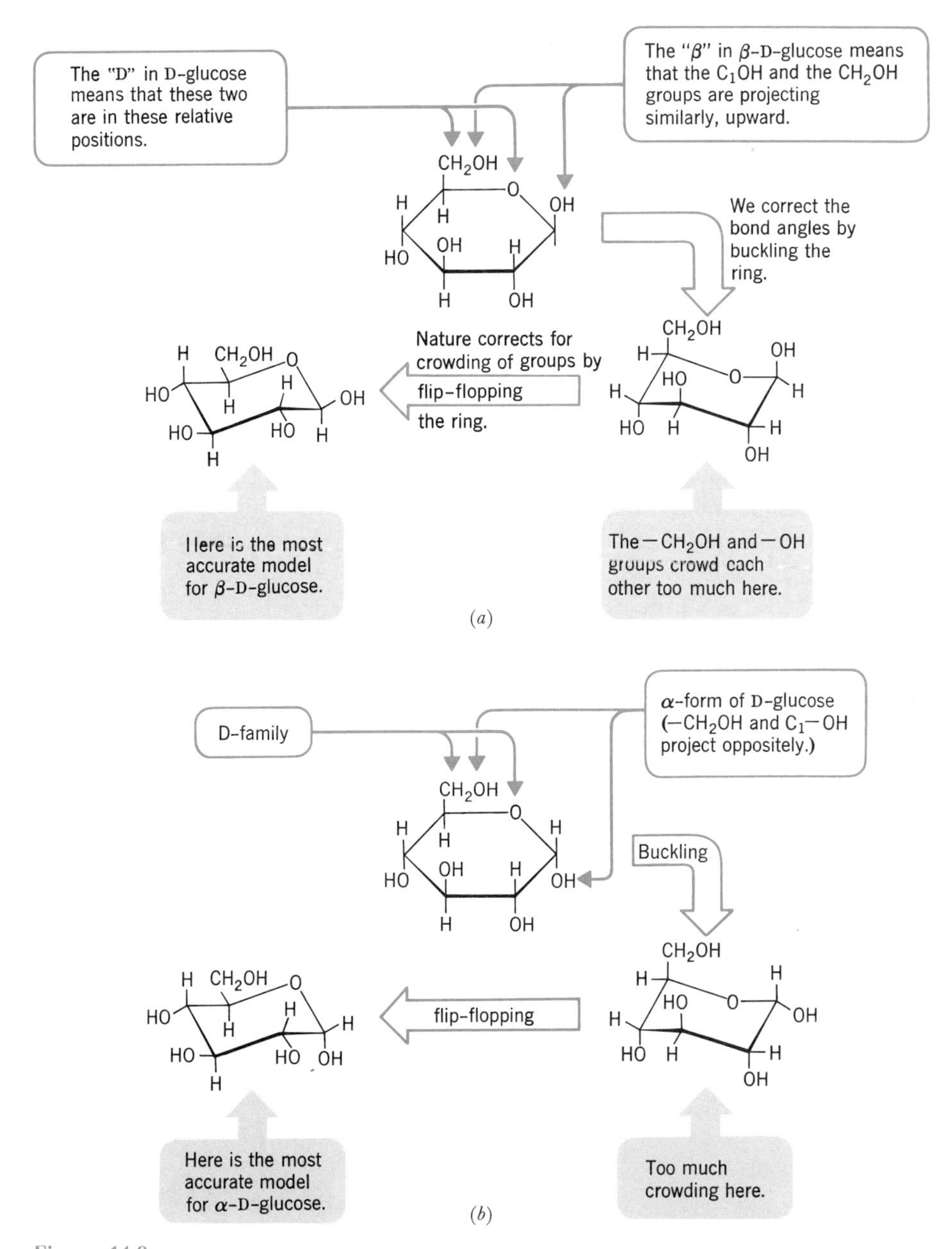

Figure 14.8
Models of β-D-glucose (*a*) and α-D-glucose (*b*) in their correct conformations (i. e., correct bond angles and minimum crowding) and their correct configurations (i. e., correct orientations at each chiral center).

1. First write a six-membered ring with an oxygen in the upper right-hand corner.
2. Next "anchor" the $-CH_2OH$ on the carbon to the left of the oxygen. (Let all the —Hs attached to ring carbons be "understood.")
3. Continue in a counterclockwise way around the ring, placing the —OHs first down, then up, then down.
4. Finally, at the last site on the trip, how the last —OH is positioned depends on whether the alpha or the beta form is to be written. The alpha is "down," the beta "up."

   If this detail is immaterial, or if the equilibrium mixture is intended, the structure may be written as.

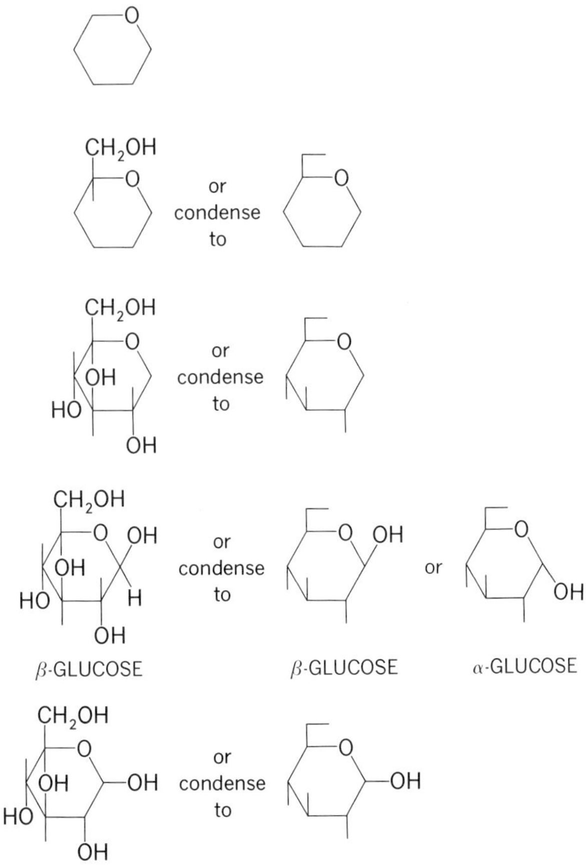

β-GLUCOSE β-GLUCOSE α-GLUCOSE

Figure 14.9
Condensing the structure of D-glucose in the cyclic forms.

Having looked at the truth, as best we know it, about the conformation and configuration of the D-glucoses we may now permit ourselves the ease of a simplification. We've done this often. When we write CH_3CH_3 for the structure of ethane we leave out everything about the geometry of the molecule or about its preferred rotational conformations. Of course, *after* we have studied those details, we really don't leave them out. We simply let them be understood because it is much easier to use simply CH_3CH_3 for most needs. What we need now is a simple representation of the two D-glucoses. Figure 14.9 shows how we construct them, and you should study that figure now and learn how to write that kind of structure of glucose.

14.15 Galactose

Most galactose occurs naturally in combined forms, especially in the disaccharide, lactose (milk sugar). Galactose, a diastereomer of glucose, differs from it only in the orientation of the C-4 hydroxyl. Like glucose, it is a reducing sugar, it mutarotates, and it exists in three forms, alpha, beta, and open.

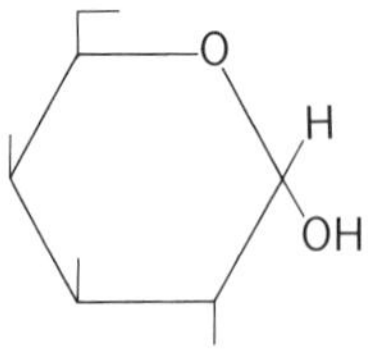

α-GALACTOSE

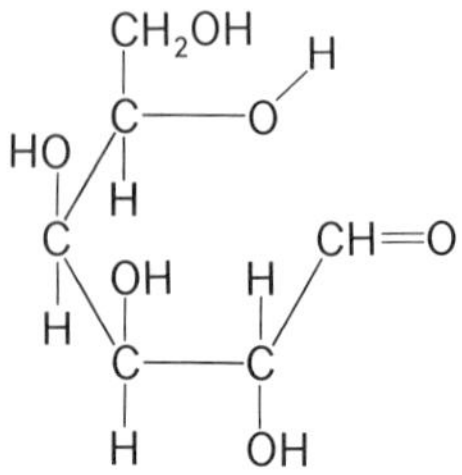

GALACTOSE, OPEN FORM

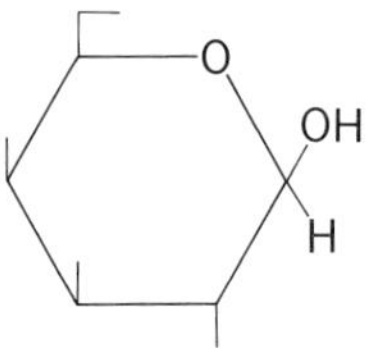

β-GALACTOSE

14.16 Ketohexoses. Fructose

Fructose, the most important ketohexose, is found together with glucose and sucrose in honey and fruit juices. It can exist in more than one form, and as a building unit in sucrose it exists in a cyclic, five-membered hemiketal form.

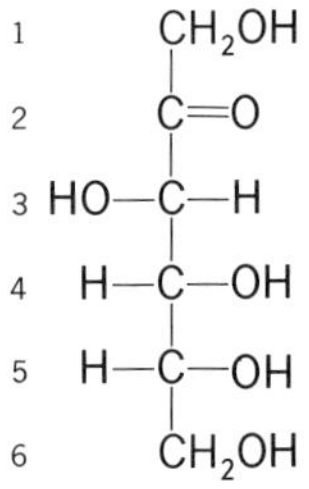

D(−)-FRUCTOSE

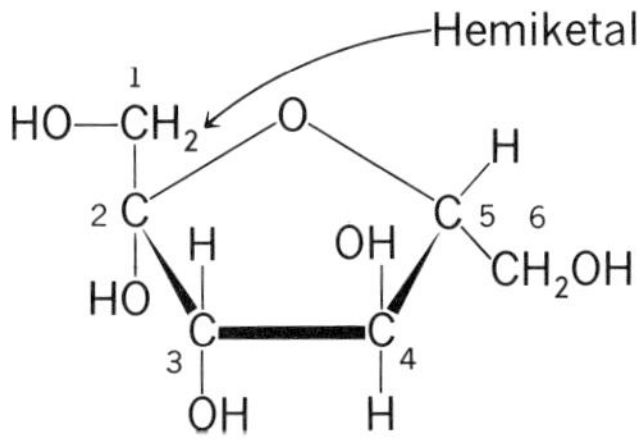

Hemiketal carbon, C-2

ONE OF THE CYCLIC FORMS OF FRUCTOSE

Because fructose is strongly levorotatory, we sometimes call it levulose; the specific rotation is $[\alpha]_D$ −92.4°.

14.17 Amino Sugars

Amino derivatives of glucose and galactose occur widely in nature as building units for some important polymeric substances. Glucosamine is a structural unit in chitin, the principal component of the shells of lobsters, crabs, and certain insects and in heparin, a powerful blood anticoagulant. Galactosamine is a structural unit of chondroitin, a polymeric substance that is important to the structure of cartilage, skin, tendons, adult bone, the cornea, and heart valves.

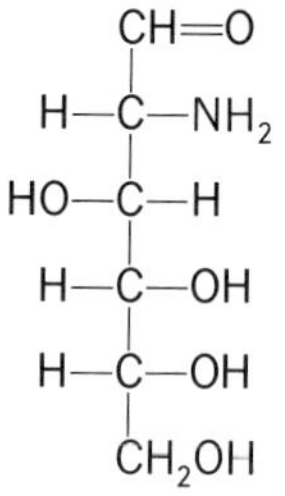

OPEN FORM 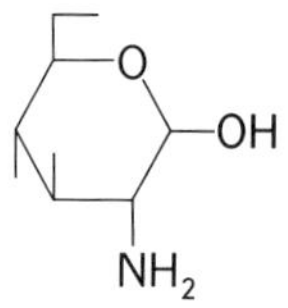 CYCLIC FORM

D-GLUCOSAMINE

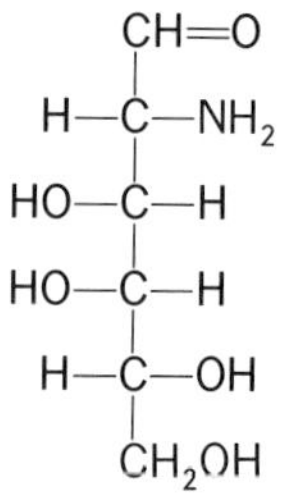

OPEN FORM 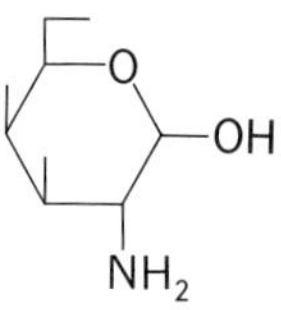 CYCLIC FORM

D-GALACTOSAMINE

Disaccharides

14.18 General Features

Nutritionally, the three important disaccharides are maltose, lactose, and sucrose. A fourth will be mentioned, cellobiose, which is obtained by partial hydrolysis of cellulose. All are in the D-family. How these are related to the monosaccharides just discussed may be seen in the following word equations.

$$\text{Maltose} + H_2O \longrightarrow \text{glucose} + \text{glucose}$$
$$\text{Lactose} + H_2O \longrightarrow \text{glucose} + \text{galactose}$$
$$\text{Sucrose} + H_2O \longrightarrow \text{glucose} + \text{fructose}$$
$$\text{Cellobiose} + H_2O \longrightarrow \text{glucose} + \text{glucose}$$

All these disaccharides are glycosides and are formed by one monosaccharide acting as a hemiacetal and the other as an alcohol to make an acetal.

14.19 Maltose

Although present in germinating grain, maltose or malt sugar does not occur widely in the free state in nature. It can be prepared by partial hydrolysis of starch. As with glucose, two forms of maltose are known and each mutarotates. Maltose is a reducing sugar and hydrolyzes to two molecules of glucose. The enzyme *maltase* catalyzes this reaction, and it is known to act only on α-glucosides, not on β-glucosides.

Because the bridging oxygen leaves the first ring in an alpha orientation, the acetal "bridge" is an alpha bridge. Therefore, maltose is an α-glucoside.

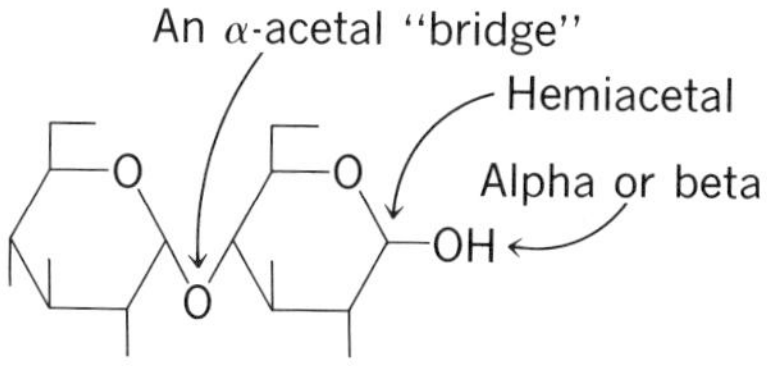

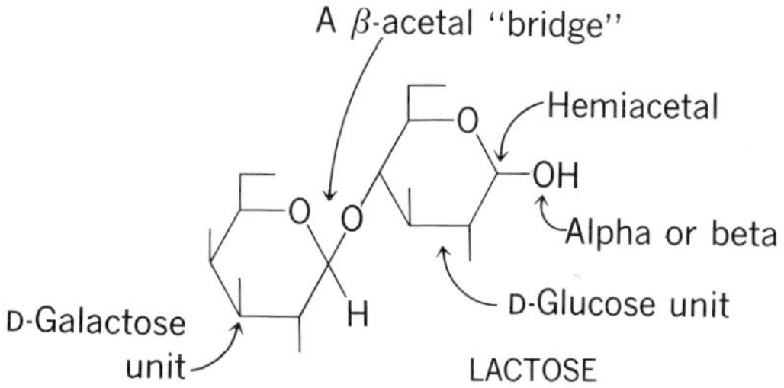

14.20 Lactose

Lactose or milk sugar occurs in the milk of mammals, 4 to 6% in cow's milk, 5 to 8% in human milk. It is obtained commercially as a by-product in the manufacture of cheese. Like maltose, it exists in alpha and beta forms, it is a reducing sugar, and it mutarotates. Its hydrolysis yields galactose and glucose.

14.21 Sucrose

The juice of sugar cane, which contains about 14% sucrose, is obtained by crushing and pressing the canes. It is then freed of proteinlike substances by the precipitating action of lime. Evaporation of the clear liquid leaves a semisolid mass from which raw sugar is isolated by centrifugation. (The liquid that is removed is *blackstrap molasses.*) Raw sugar (95% sucrose) is processed to remove odoriferous and colored contaminants. The resulting white sucrose, our table sugar, is probably the largest volume *pure* organic chemical produced. Much of our supply of sucrose now comes from sugar beets, and so-called beet sugar and cane sugar are chemically identical. Structurally, a molecule of sucrose is derived from one glucose unit and one fructose unit:

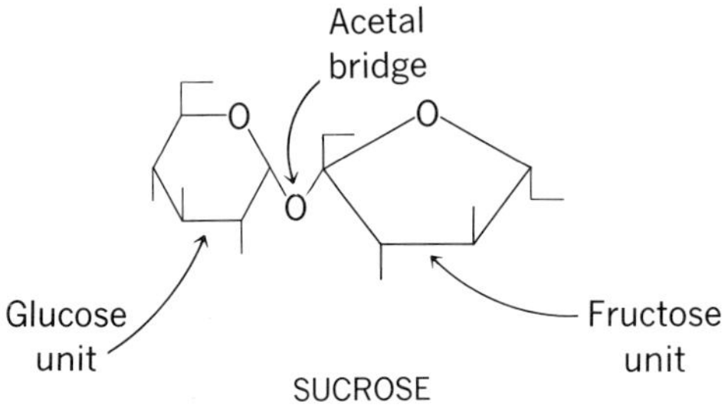

Acetal bridge

Glucose unit

Fructose unit

SUCROSE

An acetal oxygen bridge links the two units. No hemiacetal group is present in the sucrose molecule. Sucrose is not a reducing sugar.

The 50 : 50 mixturc of glucose and fructose that forms when sucrose is hydrolyzed is often called *invert sugar.* Honey, for example, consists of this sugar for the most part. The term "invert" comes from the change or inversion in sign of optical rotation that occurs when sucrose, $[\alpha]_D + 66.5°$, is converted into the 50:50 mixture, $[\alpha]_D - 19.9°$.

14.22 Cellobiose

Cellobiose ("cello-," cellulose; "bi-," two; "-ose," sugar) is the disaccharide unit in cellulose (e.g., cotton fiber). Like maltose, it delivers two glucose molecules when it is hydrolyzed. It differs structurally from maltose only in the orientation of the acetal oxygen bridge.

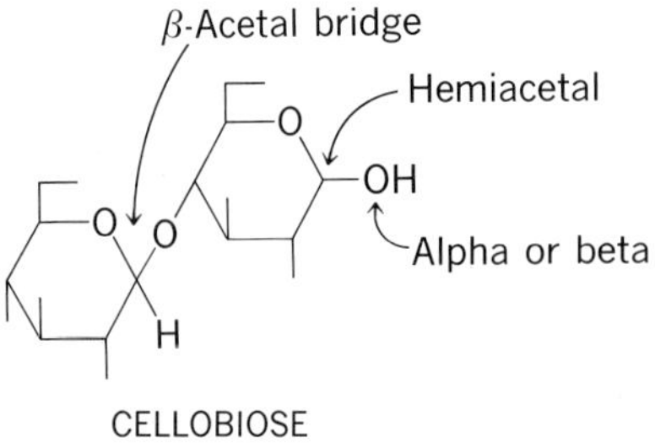

β-Acetal bridge

Hemiacetal

OH

Alpha or beta

CELLOBIOSE

Polysaccharides

14.23 General Features

Much of the glucose that is produced in a plant by photosynthesis (Section 14.38) is used to make its cell walls and its rigid fibers. Much is also stored for the food needs of the plant. But instead of being stored as glucose molecules, which are too soluble in water, it is converted to a much less soluble form, starch. This polymer of glucose is particularly abundant in plant seeds.

Starch is used for food. During digestion we break it down eventually to glucose, and what we do not need right away for energy we store. We do not normally excrete excess glucose but convert it to a starchlike polymer, glycogen, or to fat. In this section we concentrate on the three types of glucose polymers: starch, glycogen, and cellulose.

14.24 Starch

The skeletal outline of what is believed to constitute the basic structure of starch is shown in Figure 14.10. It is actually a mixture of polyglucose molecules, some linear and some branched. One type, *amylose,* consists of long, unbranched polymers

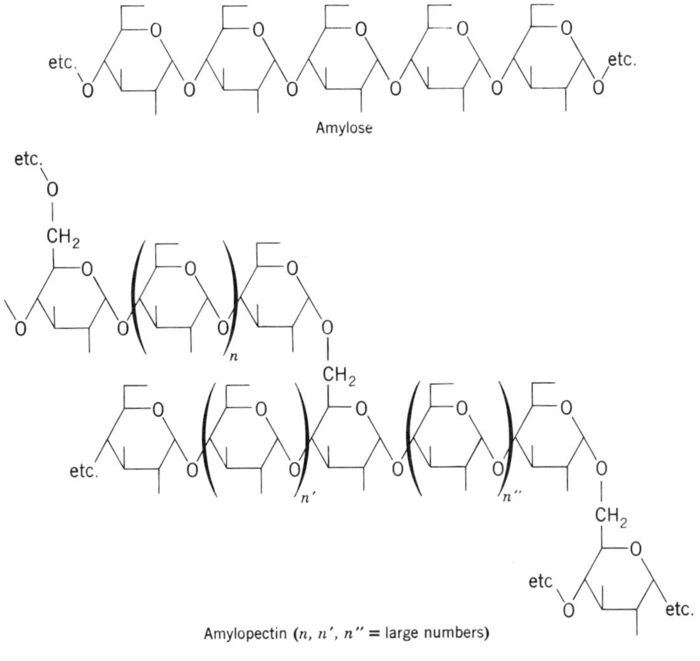

Amylose

Amylopectin (n, n', n'' = large numbers)

Figure 14.10
Basic structural features of the types of glucose polymers found in starches. Formula weights of various kinds of starch range from 50,000 to several million. A formula weight of 1 million means that about 6000 glucose units are present in one molecule.

of α-glucose. The other, *amylopectin,* is the branched polymer of α-glucose. Natural starches are about 10 to 20% amylose and 80 to 90% amylopectin.

The acetal oxygen bridges linking glucose units together in starch are easily hydrolyzed, especially in the presence of acids or certain enzymes. We have learned to expect this behavior of acetals and, when we digest starch, digestive juices do nothing more than hydrolyze it.

The partial breakdown products of amylopectin are still large molecules called the *dextrins.* They are used to prepare mucilages, pastes, and fabric sizes.

Starch is not a reducing carbohydrate. The potential aldehyde groups would be at the ends of chains only in percentages too low for detection by Benedict's reagent. Starch does, however, give an intense, brilliant blue-black color with iodine.[1] This **iodine test** for starch can detect extremely minute traces of starch in solution. The chemistry of the test is not definitely known, but iodine molecules are believed to become trapped within the vast network of starch molecules. Should this network disintegrate, as it does during the hydrolysis of starch, the test will fail.

14.25 Glycogen

Liver and muscle tissue are the main sites of glycogen storage in the body. Under the control of enzymes, some of which are in turn controlled by hormones, glucose molecules can be mobilized from these glycogen reserves to supply chemical energy for the body.

Glycogen differs from starch by the apparent absence of any molecules of the unbranched, amylose type. It is branched very much like amylopectin, perhaps even more so. Molecular weights of various glycogen preparations have been reported over a range of 300,000 to 100,000,000, corresponding roughly to 1800 to 60,000 glucose units. During the digestion and absorption of a meal containing carbohydrates, the body builds up its glycogen deposits. Between meals, during fasts, the deposits are made to deliver glucose.

14.26 Cellulose

Starch and glycogen are polymers of the alpha form of glucose. Cellulose is a polymer of the beta form. Its molecules are unbranched and resemble amylose. A portion of the structure of cellulose is shown in Figure 14.11.

The structural difference between amylose and cellulose is one of the tremendous trifles in nature. In humans amylose is digestible but cellulose is not. Yet the only difference between them structurally is the orientation of the oxygen bridges. Starch has alpha bridges; cellulose, beta.

14.27 Cellulose Ethers

When aqueous sodium hydroxide acts on cellulose and then an alkyl halide is added, one or more of the hydroxyl groups (per glucose unit) are converted into

[1]The starch-iodine reagent is iodine, I_2, dissolved in an aqueous solution of sodium iodide, NaI. Iodine, by itself, is only slightly soluble in water. When iodide ion is present, however, it combines with iodine molecules to form a triiodide ion, I_3^-, which liberates iodine easily on demand.

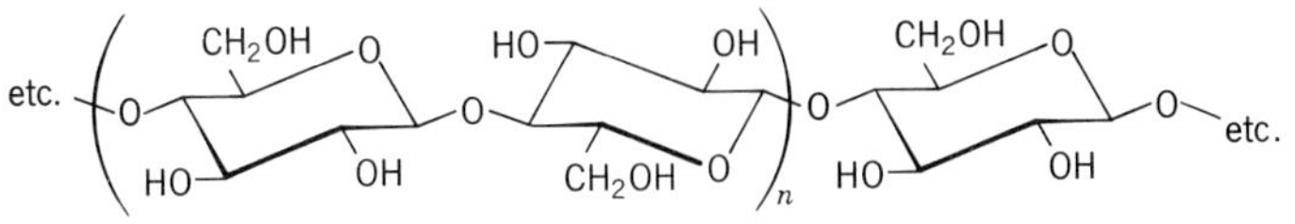

Figure 14.11
Cellulose, a linear polymer of β-glucose. Using the symbols that best correspond to the geometry of the glucose units we see better how these long molecules are like flattened tubes, almost ribbonlike. (Note that every other glucose unit is "flipped over" from our normal way of writing such units.) In cotton, the cellulose molecules have from 2000 to 9000 glucose units (n = 1000 to 4500, above). Because these long molecules are so flat they can nestle close to each other. Because of the many —OH groups they can attract each other strongly by thousands of hydrogen bonds between the strands. The cotton fiber thus gets its strength.

an ether. Methyl ethers (Methocel), ethyl ethers (Ethocel), and benzyl ethers are the commonest. A cellulose ether averaging less than one ether group per glucose unit is soluble in alkali and serves as sizes and finishing agents for textiles, as paper coatings, and as emulsifying and thickening agents for creams, lotions, shampoos, and toothpaste. If the cellulose ether averages more ether groups per glucose unit, it is used in paints, lacquers, varnishes, enamels, films, molded plastics, and packaging sheets.

14.28 Cellulose Nitrates

The action of a mixture of nitric acid and sulfuric acid on cellulose converts it into cellulose nitrate ("nitrocellulose"). By controlling the conditions it is possible to obtain a product with an average of one, or two, or three nitrate ester groups per glucose unit in cellulose.

The trinitrate is called *gun cotton,* and its use in blasting powder and smokeless artillery shells indicates its highly explosive nature.

14.29 Cellulose Acetate

Under carefully controlled conditions cellulose can be acetylated to an average of two acetate units per glucose unit. This cellulose diacetate is widely used in making the backing for photographic film.

The diacetate is also used to make "acetate rayon," a material that can be spun into fibers and used in fabrics.

Cellulose esters of propionic and butyric acids, or combinations of these also have wide commercial application in molded items.

14.30 Regenerated Cellulose. Viscose Rayon

"Rayon" refers to any fiber made by modifying native cellulose. Woody materials contain cellulose that cannot easily be used as fiber. By the action of carbon disulfide and sodium hydroxide cellulose is changed into a more soluble form. (The solution is quite viscous; hence the name, "viscose rayon.") After aging, acid is added and the cellulose is regenerated as fine filaments that are caught up on a rotating drum, stretched and twisted to make a fiber as strong as silk. The equation for this is:

$$\underset{\substack{\text{RAW CELLULOSE}\\ [\text{R}\cdot \text{ is } C_6H_9O_4\\ \text{in } (C_6H_{12}O_5)_n,\\ \text{cellulose}]}}{R{-}O{-}H} + S{=}C{=}S + NaOH \longrightarrow \underset{\substack{\text{CELLULOSE}\\ \text{XANTHATE}\\ \text{(A viscous solution)}}}{R{-}O{-}\overset{\overset{\displaystyle S}{\|}}{C}{-}S^{-}Na^{+}} \xrightarrow{H^+}$$

$$\underset{\substack{\text{REGENERATED}\\ \text{CELLULOSE}\\ \text{(Viscose rayon)}}}{R{-}O{-}H} + CS_2 + Na^+$$

Metabolism of Carbohydrates

14.31 Carbohydrates in Nutrition

For adults virtually all the carbohydrates in the diet are starches, sucrose, and small amounts of lactose (for milk drinkers) and maltose (e.g., in syrups). When these are fully digested, the products are glucose and small amounts of galactose and fructose. The catabolism[2] of all three "merge" two or three steps into their pathways of chemical use. The sugar we use, for all practical purposes, is therefore all glucose. We use glucose molecules almost entirely for energy. What we do not need right away we change into fat that we store, saving chemical energy in this form for later needs, if any.

14.32 Energy for Living from Glucose

When we use glucose for energy the final products from its molecules are simply carbon dioxide and water. These are the same products we get if we merely burn glucose in oxygen. Hence, we can use data from the combustion of glucose for an idea of the maximum amount of energy available from it. It's 673 kcal/mole (or, in SI units, 2.82×10^6 joules/mole). Of course, we do not run on the same principle as a steam engine. We cannot convert heat directly to work. Hence, we "burn" glucose in a very indirect way, one small step at a time, to keep down the release of heat and retain the energy of glucose in a form we can convert to whatever work we want—mechanical, chemical, electrical, or sound.

14.33 Adenosine Triphosphate, ATP

The principal form of chemical energy for immediate bodily use is another chemical called adenosine triphosphate, or ATP. Its structure is shown in Figure 14.12. It has been common in biochemistry to call the bonds between phosphate units

[2]Catabolism includes reactions that break molecules down. Anabolism includes the reactions that make larger and larger molecules. Metabolism is the sum of catabolism and anabolism—that is, anything that happens chemically in an organism.

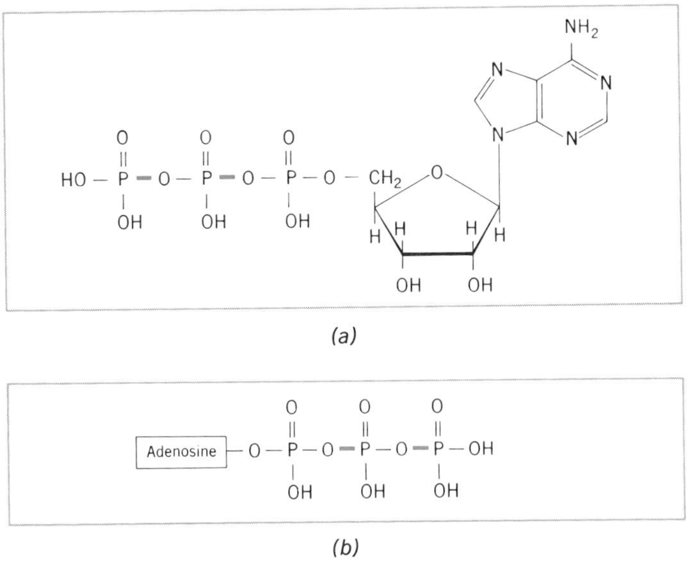

$$HO-\overset{\overset{\displaystyle O}{\|}}{\underset{\underset{\displaystyle OH}{|}}{P}}-O-\overset{\overset{\displaystyle O}{\|}}{\underset{\underset{\displaystyle OH}{|}}{P}}-O-\overset{\overset{\displaystyle O}{\|}}{\underset{\underset{\displaystyle OH}{|}}{P}}-O-CH_2-\text{(ribose ring: O, H, H, H, H, OH, OH)}-\text{adenine }(NH_2,\ N,\ N,\ N,\ N)$$

(a)

$$\boxed{\text{Adenosine}}-O-\overset{\overset{\displaystyle O}{\|}}{\underset{\underset{\displaystyle OH}{|}}{P}}-O-\overset{\overset{\displaystyle O}{\|}}{\underset{\underset{\displaystyle OH}{|}}{P}}-O-\overset{\overset{\displaystyle O}{\|}}{\underset{\underset{\displaystyle OH}{|}}{P}}-OH$$

(b)

Figure 14.12
Adenosine triphosphate, ATP. (*a*) Full structure (*b*) Abbreviated structure.

(shown in boldface in Figure 14.12) "high-energy phosphate bonds." The energy is not confined solely in these bonds; they happen to be the ones that break when ATP reacts to drive some energy-consuming activity of a cell.

One such activity is the flexing of a muscle. We may write this event as an equation, one that grossly oversimplifies the complexities:

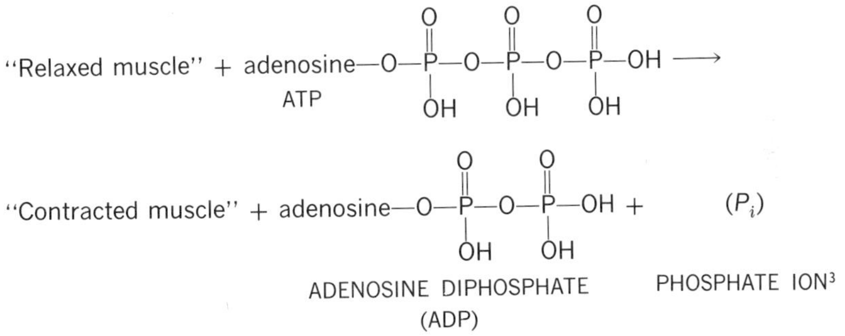

$$\text{“Relaxed muscle”} + \underset{\text{ATP}}{\text{adenosine}}-O-\overset{\overset{\displaystyle O}{\|}}{\underset{\underset{\displaystyle OH}{|}}{P}}-O-\overset{\overset{\displaystyle O}{\|}}{\underset{\underset{\displaystyle OH}{|}}{P}}-O-\overset{\overset{\displaystyle O}{\|}}{\underset{\underset{\displaystyle OH}{|}}{P}}-OH \longrightarrow$$

$$\text{“Contracted muscle”} + \text{adenosine}-O-\overset{\overset{\displaystyle O}{\|}}{\underset{\underset{\displaystyle OH}{|}}{P}}-O-\overset{\overset{\displaystyle O}{\|}}{\underset{\underset{\displaystyle OH}{|}}{P}}-OH + (P_i)$$

ADENOSINE DIPHOSPHATE (ADP) PHOSPHATE ION[3]

ATP, thus, reacts with muscle proteins in their relaxed state. Parts of the protein molecules suddenly coil up causing the contraction. Muscular work, therefore, consumes ATP. It is changed to ADP (adenosine diphosphate) and inorganic phosphate ion (P_i). The chemical job of the cell is now simply stated—it must put ADP and P_i back together again. Otherwise the cell cannot do more work.

14.34 Respiratory Chain

The principal machinery the cell uses to regenerate ATP is a series of reactions called the respiratory chain. This series removes hydrogen (or the pieces of the

[3]The state of ionization of the released phosphate unit, P_i, depends on the pH of the medium. At cell pH, P_i is a mixture largely of HPO_4^{2-} and $H_2PO_4^-$.

molecule of hydrogen, H:⁻ and H⁺) from intermediates of the breakdown of either glucose or fatty acids, sends these pieces along a series of enzymes (Section 15.21), and delivers them to oxygen. In energy terms the whole series is "downhill," and as it goes it drives ADP and P_i up an energy hill to make ATP. What the respiratory chain is and how it works are topics beyond the scope of this book. It does need chemicals that can give to it pieces of the element hydrogen, however, and we shall see how these chemicals are made next.

14.35 The Citric Acid Cycle

The metabolites, MH_2, needed to "fuel" the respiratory chain are mostly made by a series of reactions called the citric acid cycle (Krebs cycle, tricarboxylic acid cycle). See Figure 14.13. Acetyl units are needed as the fuel for this series, and the chief sources are the breakdown of glucose and fatty acids. These acetyl units are bound to an enzyme; therefore we write them as units of acetyl coenzyme A ($CH_3CO—SCoA$).

14.36 Glycolysis

Anaerobic Catabolism of Glucose.

It often happens that a tissue is called upon to do something that uses its ATP when the oxygen supply is temporarily short. When this happens the cell can make ATP by another means, one that needs no oxygen and is therefore called the anaerobic sequence of glucose catabolism. Its other name is glycolysis from "glycose loosening." From glucose the over all result is the production of lactic acid and some ATP.

$$\underset{\text{GLUCOSE}}{C_6H_{12}O_6} + 2ADP + P_i \xrightarrow[\text{steps}]{\text{Several}} \underset{\text{LACTIC ACID}}{2CH_3CH(OH)CO_2H} + 2ATP$$

Once the cell gets a renewed supply of oxygen, the lactic acid is removed, some of it broken to acetyl units for entering the citric acid cycle.

$$\underset{\text{LACTIC ACID}}{CH_3CH(OH)CO_2H} \longrightarrow [H:^- + H^+] + \underset{\text{PYRUVIC ACID}}{CH_3COCO_2H}$$

[H:⁻ + H⁺] → TO RESPIRATORY CHAIN AND ATP

$$CH_3COCO_2H \xrightarrow{H_2O} CO_2 + [H:^- + H^+] + CH_3CO_2H$$

[H:⁻ + H⁺] → TO RESPIRATORY CHAIN AND ATP

CH_3CO_2H → TO CITRIC ACID CYCLE

14.37 Fermentation

Fermentation is a process whereby glucose is broken down to carbon dioxide and ethyl alcohol. It is catalyzed by the enzymes in certain yeasts. The reactions of

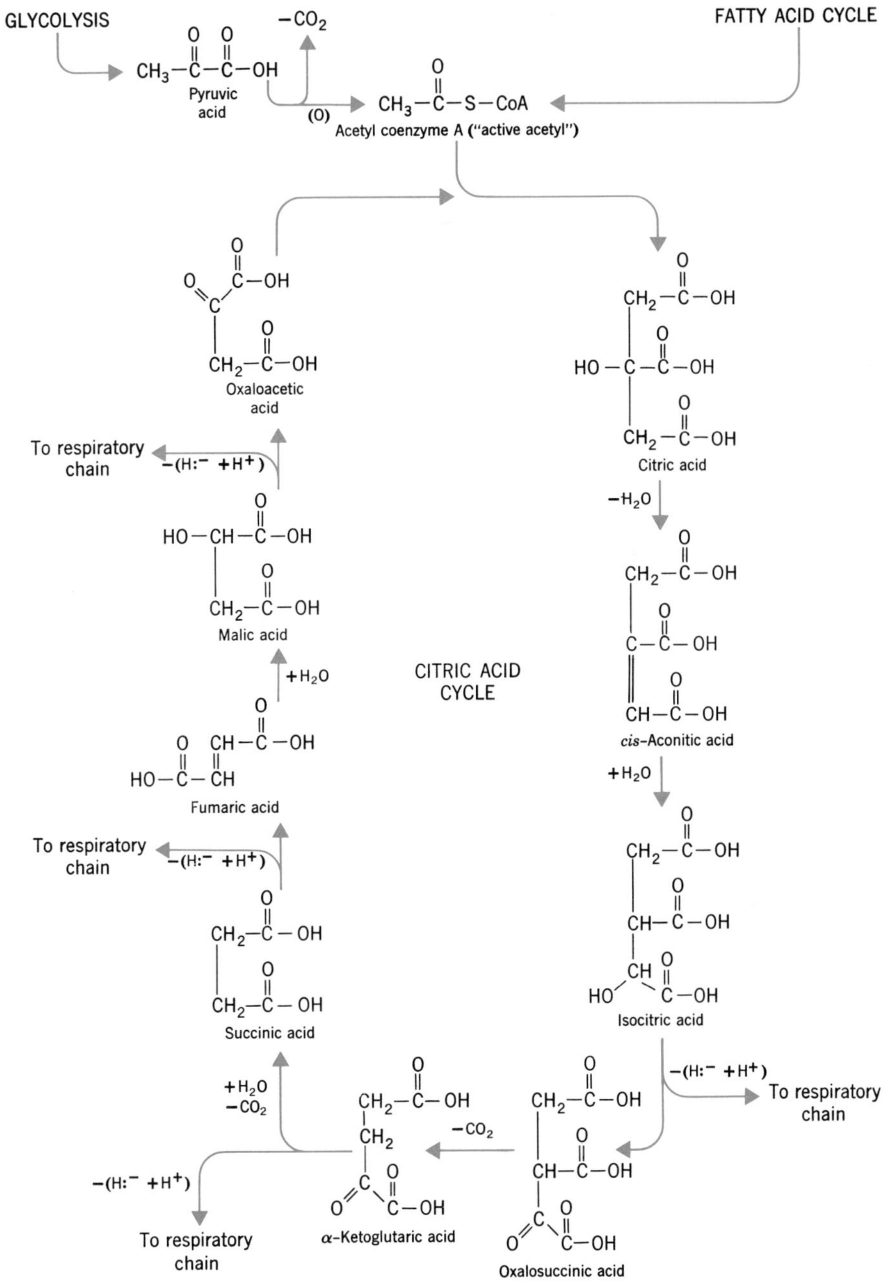

GLYCOLYSIS

$CH_3-C(=O)-C(=O)-OH$

Pyruvic acid

$-CO_2$

(O)

FATTY ACID CYCLE

$CH_3-C(=O)-S-CoA$

Acetyl coenzyme A ("active acetyl")

Oxaloacetic acid

To respiratory chain

−(H:⁻ +H⁺)

Malic acid

$+H_2O$

CITRIC ACID CYCLE

Fumaric acid

To respiratory chain

−(H:⁻ +H⁺)

Succinic acid

$+H_2O$
$-CO_2$

−(H:⁻ +H⁺)

To respiratory chain

α-Ketoglutaric acid

$-CO_2$

Oxalosuccinic acid

−(H:⁻ +H⁺)

To respiratory chain

Isocitric acid

$+H_2O$

cis-Aconitic acid

$-H_2O$

Citric acid

Figure 14.13

The Citric Acid Cycle. Two-carbon acetyl groups are fed in at the top to make citric acid. By a series of dehydrations, hydrations, dehydrogenations and decarboxylations, citric acid is trimmed back to oxaloacetic acid. Pieces of the element hydrogen are transferred to the respiratory chain to reduce oxygen and to drive phosphorylation of ADP to ATP.

fermentation are identical with those of glycolysis except for the latter stages. The enzymes in yeast include those that can catalyze the decarboxylation of pyruvic acid to acetaldehyde, which is then reduced to ethyl alcohol.

$$CH_3-\overset{O}{\overset{\|}{C}}-\overset{O}{\overset{\|}{C}}-O-H \xrightarrow{\text{In certain yeasts}} CH_3-\overset{O}{\overset{\|}{C}}-H + O=C=O\uparrow$$

PYRUVIC ACID → ACETALDEHYDE + CARBON DIOXIDE

Reduction of ACETALDEHYDE → CH_3CH_2OH ETHYL ALCOHOL

14.38 Photosynthesis

The end products of glucose catabolism are carbon dioxide and water, and we've reached something of a molecular "sea level" with these substances. There is no way we can wring more energy out of them. For living organisms in the biosphere to keep going, to live, to reproduce, and thus to keep it all alive century after century we have to have something standing behind our supply of glucose or fatty acids in order constantly to replenish those sources of chemical energy for living. It is here that our sun enters the system of the living.

Green-leafed plants contain chlorophyll, a substance whose highly unsaturated molecules can absorb sunlight and hold its energy in such a form that enzymes unique to a plant can work. They use the solar energy to drive the conversion of carbon dioxide and water up a long energy hill to glucose units and other high energy units. This conversion is called photosynthesis, and the simplest statement we can make for it is the following "equation."

$$nCO_2 + nH_2O \xrightarrow[\text{Plant enzymes}]{\text{Sunlight; chlorophyll}} \underset{\text{ESSENTIAL UNIT IN GLUCOSE}}{(CH_2O)_n} + nO_2$$

Between 60 and 70 billion tons of carbon dioxide are "fixed" annually on our planet by photosynthesis. Over half of this is done by, of all things, microscopic marine organisms such as phytoplankton in the oceans. The earth's forests make about 15 billion tons of new wood each year. The energy for all of this requires about 0.04% of the solar energy that reaches our planet's outer atmosphere.

Brief Summary

1. Carbohydrates are polyhydroxy aldehydes or ketones or substances that can be hydrolyzed to these.

—Monosaccharides are carbohydrates that cannot be hydrolyzed. (Important examples: glucose, galactose, fructose.)

—Disaccharides are those that can be hydrolyzed to two monosaccharides. (Important examples: maltose, lactose, and sucrose.)

—Polysaccharides are carbohydrates that can be hydrolyzed to hundreds and thousands of monosaccharide units. (Important examples: starch, glycogen, cellulose.)

2. Glucose chemistry is central to all carbohydrate chemistry because glucose units figure structurally in most carbohydrate molecules.

3. The important reactions of glucose and all aldoses are:

(a) oxidation—Tollens' reagent or Benedict's reagent. All aldoses and ketoses as well as lactose, cellobiose and maltose are reducing sugars.

(b) osazone formation

(c) formation of glycosides (e.g., glucosides from glucose; galactosides from galactose.)

4. Naturally occurring aldoses and ketoses (e.g., fructose) are in the D-family of optical isomers. This means that the chiral carbon farthest from the carbonyl group is configurationally the same as in D-(+)-glyceraldehyde, the "parent" or reference compound for all members of the D-family.

5. The plane-projection diagram is the most common way the various optical isomers of the monosaccharides are written when attention must be drawn to their internal, relative arrangement of groups.

6. The aldoses as well as lactose and maltose mutarotate. This means that freshly prepared solutions of these undergo slow changes in their optical activity. The mechanism of this change is ring opening and closing involving cyclic, hemiacetal structures.

7. Compounds α-glucose and β-glucose respectively form α-glucosides and β-glucosides with alcohols. These have properties of acetals.

8. If the alcohol used to make a glucoside is another monosaccharide molecule, we obtain disaccharides.

$$\alpha\text{-Glucose} + \underset{(C_4\text{—OH used})}{\text{glucose}} \longrightarrow \text{maltose} + H_2O$$

$$\beta\text{-Glucose} + \underset{(C_4\text{—OH used})}{\text{glucose}} \longrightarrow \text{cellobiose} + H_2O$$

$$\beta\text{-Galactose} + \underset{(C_4\text{—OH used})}{\text{glucose}} \longrightarrow \text{lactose} + H_2O$$

$$\alpha\text{-Glucose} + \beta\text{-fructose} \longrightarrow \text{sucrose} + H_2O$$

9. The digestion of maltose, lactose, or sucrose is simply their hydrolysis back to monosaccharide units.

10. Starch and glycogen are polymers of α-glucose. Because of other oxygen bridges, they are highly branched polymers.

11. Cellulose is a polymer of β-glucose. If the remaining —OH groups on the glucose units (three per unit are left) are modified we have cellulose ethers, cellulose acetates, or cellulose nitrates, all commercially important.

12. The nutritionally important carbohydrates are sucrose, maltose, lactose, and starch. After these are digested we have only glucose, fructose, and galactose entering the circulatory system. After just a few steps into metabolism it is as if we dealt exclusively with glucose.

13. Glucose catabolism involves two main sets of reactions. One is anaerobic (glycolysis); the other is aerobic (citric acid cycle).

14. The citric acid cycle runs primarily when the system needs to use the respiratory chain to make ATP.

15. The anaerobic sequence is a handy short-term source of ATP when oxygen supplies are low.

16. The respiratory chain channels the pieces of the element hydrogen ($H{:}^-$ and H^+) from metabolites of the citric acid cycle (and other sources) to an enzyme that catalyzes the reduction of oxygen to water.

17. The energy made available by this reduction drives the synthesis of ATP from ADP and inorganic phosphate ion (P_i).

18. Photosynthesis is nature's scheme for using solar energy to make energy-rich molecules (e.g., glucose) from energy-poor substances (e.g., CO_2 and H_2O).

Problems and Exercises

1. Write the structure (with plane projection) for the products if galactose is allowed to react with excess phenylhydrazine.
2. When it was discovered that gulose and idose were both aldoses and gave identical phenylosazones, scientists knew exactly how gulose and idose were alike and how they were different. Explain by structures and equations. (Cf. Exercise 14.1.)
3. Fructose also forms a phenylosazone. It is also identical with the osazone from glucose. How must glucose and fructose, therefore, be alike?
4. Write the structure of a monosaccharide that will give the same osazone as threose.
5. Write structures for α-maltose, β-maltose, and the open form of maltose. (*Hint:* only one ring is open.)
6. Write equations for the reaction of maltose with (a) water (acid or enzyme catalyzed), and (b) excess phenylhydrazine.
7. Is lactose a galactoside or a glucoside? Explain.
8. Write a series of structures that would be present at equilibrium after lactose has undergone mutatrotation.
9. Repeat problem 8 for cellobiose.
10. Trehalose is a disaccharide found in young mushrooms and yeast and is the chief carbohydrate in the blood (hemolymph) of certain insects. On the basis of its structural features, answer the following questions about trehalose. If a reaction is predicted, write it.

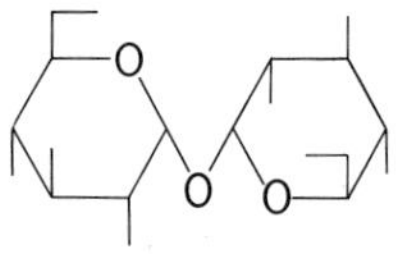

TREHALOSE

(a) Is trehalose a reducing sugar? Why?
(b) Can it mutarotate? Why?
(c) Does it form an osazone? Why?
(d) Can it be hydrolyzed?

11. Write enough of the structure of cellulose trinitrate to illustrate all its essential features.
12. Why would cellulose triacetate be more soluble in nonaqueous solvents than cellulose?
13. In some rare instances galactose appears in the urine. How might an analyst erroneously report the presence of glucose?

chapter fifteen

Amino Acids and Proteins

15.1 Occurrence

Proteins are found in all cells and in virtually all parts of cells. Proteins constitute about half the body's dry weight. Proteins hold a living organism together and run it. As skin they give it a shell. As muscles and tendons they provide levers. As one substance inside bones, like steel in reinforced concrete, they constitute a reinforcing network. Some proteins—in buffers, antibodies, and hemoglobin—serve as policemen and long-distance haulers. Others form the communications network or nerves. Certain proteins, such as enzymes, hormones, and gene regulators direct and control all forms of body repair and construction. No other class of compounds is involved in such a variety of functions, all essential to life.

Proteins have one characteristic in common: they are all polymers. The monomer units are classified as α-amino acids. Of all the proteins ever analyzed, only about 20 different α-amino acids occur commonly. A few others occur here and there in unusual instances.

Amino Acids

15.2 Common Structural Features

The system common to all the α-amino acids is:

$$\underset{\textbf{1}}{NH_2-\overset{\alpha}{\underset{|\atop G}{C}}H-\overset{O\atop \|}{C}-OH} \rightleftharpoons \underset{\textbf{2}}{\overset{+}{N}H_3-\underset{|\atop G}{C}H-\overset{O\atop \|}{C}-O^-}$$

The *G-* stands for an organic group or system that is different in each of the various amino acids.[1] A list of 24 amino acids in Table 15.1 indicates the nature of these side chains.

15.3 Optical Activity among the Amino Acids and Proteins

In all the naturally occurring amino acids except the simplest, glycine, the α-carbon is chiral, and the substances are optically active. Serine is used as the reference for assigning amino acids to configurational families. Naturally occurring serine is in the L-family. Its relation to L-glyceraldehyde shows in the plane

$$\begin{array}{c} CO_2H \\ NH_2 - \!\!\!\!+\!\!\!\! - H \\ CH_2OH \end{array} \qquad \begin{array}{c} CHO \\ HO - \!\!\!\!+\!\!\!\! - H \\ CH_2OH \end{array}$$

L-SERINE (Naturally occurring enantiomer) L-GLYCERALDEHYDE

projection formulas. All of the other naturally occurring amino acids are found to be of the L-configuration, too. Except in extremely rare instances nature does not use the D-forms. The proteins in our bodies are made only from L-amino acids. If we were fed the D-enantiomers, we could not use them. Our enzymes, which themselves are made from the L-forms of amino acids, work with L-forms of amino acids only.

The physiological properties of proteins are related not only to the configurations of their amino acid units but also to the kinds of side chains that are present. We shall therefore examine briefly some generalizations about these side chains and then see how the amino acids are incorporated into proteins. (We say little here about reactions at the amino and carboxyl groups because they are generally those of amines and acids studied earlier.)

15.4 Amino Acids with Nonpolar Side Chains

The first group of amino acids in Table 15.1 are those with essentially nonpolar side chains. They are said to be *hydrophobic* groups ("hydro-," water; "-phobic," hating).[2] When a huge protein molecule folds into its distinctive shape (cf. tertiary structure, Section 15.16), these hydrophobic groups tend to be folded next to each other rather than next to highly polar groups.

15.5 Amino Acids with Hydroxyl-Containing Side Chains

The second set of amino acids in Table 15.1 consists of those whose side chains carry alcohol or phenol groups. In cellular environments they are neither basic nor acidic, but they are polar and *hydrophilic* (water loving). They can be

[1]As we shall learn in Section 15.9, structure **2** is the better representation. In these first sections, however, structure **1** will be easier to use, and we shall do so.

[2]What looks like a 2° amino group (and therefore a polar group) in tryptophan is unable to coordinate with a proton. It is not basic. The nitrogen's unshared pair is delocalized into the aromatic system. Hence tryptophan is put in this group because, for all practical purposes, its side chain is nonbasic and only slightly polar.

Table 15.1

Amino Acids $HO{-}\overset{O}{\overset{\|}{C}}{-}\underset{NH_2}{\underset{|}{CH}}{-}G$

	G (Side Chain)	Name	Symbols Used in Representing Amino Acids in Protein Structures	pI
Side chain is nonpolar	$-H$	Glycine	Gly[a]	5.97
	$-CH_3$	Alanine	Ala	6.00
	$-CH(CH_3)_2$	Valine	Val	5.96
	$-CH_2CH(CH_3)_2$	Leucine	Leu	5.98
	$-\underset{CH_3}{\underset{|}{C}}HCH_2CH_3$	Isoleucine	Ile	6.02
	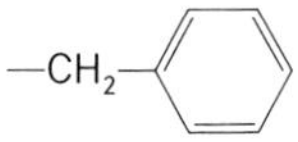 $-CH_2-C_6H_5$ (phenyl ring)	Phenylalanine	Phe	5.48
	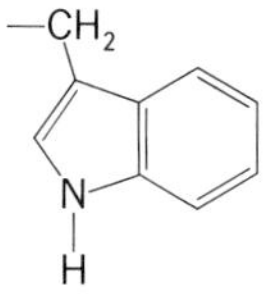 $-CH_2$ (indole ring, N—H)	Tryptophan	Trp	5.89
	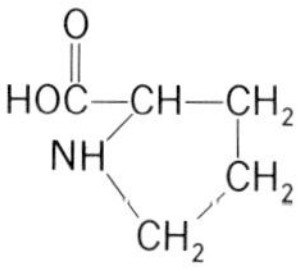 $HO\overset{O}{\overset{\|}{C}}-CH-CH_2$, NH, CH_2, CH_2 (ring) (complete structure)	Proline	Pro	6.30
Side chain has a hydroxyl group	$-CH_2OH$	Serine	Ser	5.68
	$-\underset{CH_3}{\underset{|}{C}}HOH$	Threonine	Thr	5.64
	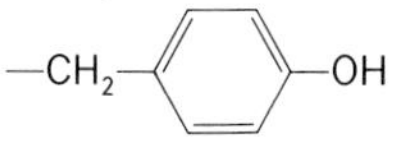 $-CH_2-C_6H_4-OH$	Tyrosine	Tyr	5.66
	$-CH_2-C_6H_2I_2-OH$	Diiodotyrosine	—	—
	$-CH_2-C_6H_2I_2-O-C_6H_2I_2-OH$	Thyroxine	—	—
	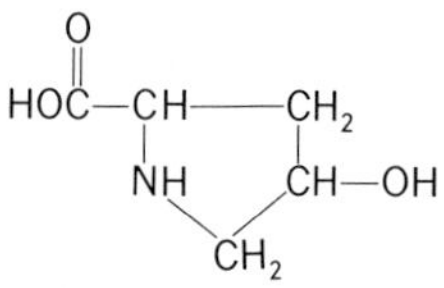 $HO\overset{O}{\overset{\|}{C}}-CH-CH_2$, NH, $CH-OH$, CH_2 (ring) (complete structure)	Hydroxyproline	Hyp	5.74

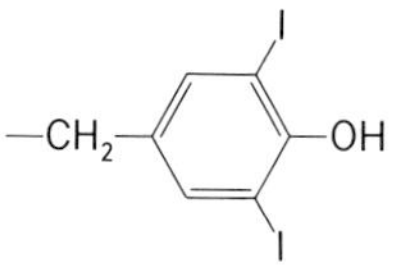

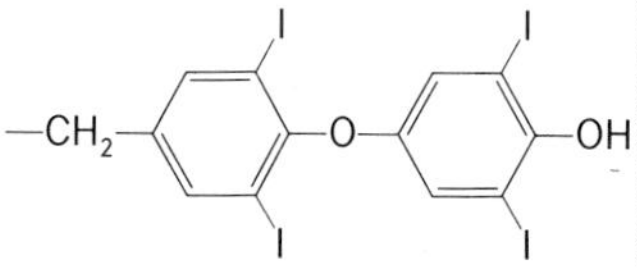

Table 15.1 (*Continued*)

	G (Side Chain)	Name	Symbols Used in Rapresenting Amino Acids in Protein Structures	pI
Side chain has a carboxyl group (or the corresponding amide)	$-CH_2CO_2H$	Asparatic acid	Asp	2.77
	$-CH_2CH_2CO_2H$	Glutamic acid	Glu	3.22
	$-CH_2CONH_2$	Asparagine	Asn	5.41
	$-CH_2CH_2CONH_2$	Glutamine	Gln	5.65
Side chain has a basic amino group	$-CH_2CH_2CH_2CH_2NH_2$	Lysine	Lys	9.74
	$-CH_2CH_2CH_2NH-C(=NH)-NH_2$	Arginine	Arg	10.76
	$-CH_2-$C (ring: CH—N, CH, N—H)	Histidine	His	7.59
Side chain contains sulfur	$-CH_2S-H$	Cysteine	Cys	5.07
	$-CH_2-S$ / $-CH_2-S$ (S—S bonded)	Cystine	Cys or Cys[b]	5.03
	$-CH_2CH_2SCH_3$	Methionine	Met	5.74

[a]These three-letter symbols are recommended by the Joint Commission on Biochemical Nomenclature of IUPAC IUB.
[b]For half the cystine unit.

either hydrogen-bond donors or acceptors. Hence they can help bind neighboring groups into ordered structures as a long protein chain folds into its final shape.

15.6 Amino Acids with Acidic Side Chains

Molecules of aspartic acid and glutamic acid have an extra carboxyl group each. When these amino acids are included in a protein, their side chains make the protein more acidic than otherwise. They are proton donors.

In proteins these amino acids frequently occur as asparagine and glutamine, that is, in forms where the sidechain carboxyl groups are the corresponding amides. Although with still very polar side chains, in these forms they no longer contribute acidic groups to proteins.

15.7 Amino Acids with Basic Side Chains

A lysine molecule has an extra 1° amino group that makes its side chain basic, hydrophilic, and a hydrogen-bond donor or acceptor. The side chain in an arginine molecule has the guanidine group, $-NH-C(=NH)-NH_2$. One of the most powerful proton-accepting groups found in organisms, it exists almost exclusively in its protonated form, $—NH—C(=NH_2^+)—NH_2$.

15.8 Amino Acids with Sulfur-Containing Side Chains

The side chain in a molecule of cysteine has an —SH group, called the mercaptan group or the sulfhydryl group. A whole family of organic compounds,

CH_3SH METHYL MERCAPTAN

$CH_3CH_2CH_2CH_2SH$ *n*-BUTYL MERCAPTAN

C_6H_5—SH THIOPHENOL

the mercaptans, is based on it. (The name comes from the fact that molecules with the mercaptan group react with ionic mercury forming a precipitate. The mercaptan group, in other words, is a "mercury-capturing" group.) For our needs only one reaction of the mercaptan group merits attention—its oxidation to a disulfide. In general:

$$\underset{\text{MERCAPTAN}}{2\ RSH} \xrightarrow[\text{agents}]{\text{Mild oxidizing}} \underset{\text{DISULFIDE}}{R—S—S—R} + H_2O$$

The one reaction of disulfides important to us now is their easy reduction. In general:

$$R—S—S—R \xrightarrow[\text{agents}]{\text{Mild reducing}} 2\ R—SH$$

Because its mercapto group makes it especially reactive toward mild oxidizing agents, cysteine is a particularly important amino acid. Cysteine and cystine, of course, are interconvertible, a property of far-reaching importance in some proteins.

$$\underset{\text{TWO MOLECULES OF CYSTEINE}}{HOOCCH(NH_2)CH_2S—H + HOOCCH(NH_2)CH_2S—H} \underset{\text{Reduction}}{\overset{\text{Oxidation}}{\rightleftarrows}} \underset{\text{ONE MOLECULE OF CYSTINE}}{HOOCCH(NH_2)CH_2S—SCH_2CH(NH_2)COOH} + (H:^- + H^+)$$

The disulfide linkage in cystine is common to several proteins, as we shall see. It is especially prevalent in proteins having a protective function, such as those forming hair, fingernails, and shells.

15.9 Dipolar Ionic Character of Amino Acids

Most carboxylic acids have acid dissociation constants K_a on the order of 10^{-5}, but for glycine K_a is 1.6×10^{-10}. This value is low enough to be in the range of phenols. Similarly, most primary aliphatic amines have basic dissociation constants K_b on the order of 10^{-5}, but for glycine K_b is 2.5×10^{-12}, making it so weak a proton acceptor that is is virtually neutral in this respect.

Amino acids have other properties at variance with the simple structure, **1.** Even the simplest, those with lowest formula weights, are nonvolatile crystalline solids without true melting points. When heated, they decompose and char before melting can occur. Furthermore, the amino acids tend to be insoluble in nonpolar solvents and soluble in water. These facts do not correlate well with structure **1.** The *dipolar ionic structure,* **2,** is more consistent.

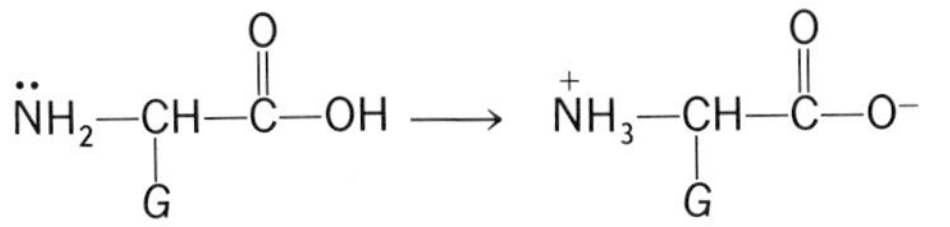

$$\ddot{N}H_2{-}\underset{\underset{G}{|}}{CH}{-}\overset{\overset{O}{\|}}{C}{-}OH \longrightarrow \overset{+}{N}H_3{-}\underset{\underset{G}{|}}{CH}{-}\overset{\overset{O}{\|}}{C}{-}O^-$$

1 **2**

DIPOLAR IONIC FORM
OF AN AMINO ACID

Evidently the proton-accepting end in **1,** $—NH_2$, has taken a proton from the proton-donating end, $—CO_2H$. In a crystal made of dipolar ionic particles, forces of attraction between them will be strong, as in any ionic crystal (e.g., a salt). Before heating supplies enough energy to melt such crystals, covalent bonds are disrupted within the dipolar ions, and decomposition occurs. As for solubility, ionic substances, of course, are known for their insolubility in nonpolar solvents.

As for their low K_a and K_b values, the dipolar ionic structure offers a ready explanation. In an amino acid we are not dealing with a free carboxyl group, —COOH, as the proton donor, but rather with a substituted ammonium ion, $—NH_3^+$. The proton is more strongly bound within this group than it would be on the —COOH group. Similarly, the basic group, the proton-accepting group, in an amino acid is not the free $—NH_2$ group. Instead it is the carboxylate group, COO^-. The latter is a weaker proton acceptor. The net effect is that amino acids are both very weak acids and very weak bases.

15.10 Isoelectric Points of Amino Acids

When an amino acid is dissolved in water, proton transfers occur, as illustrated with glycine:

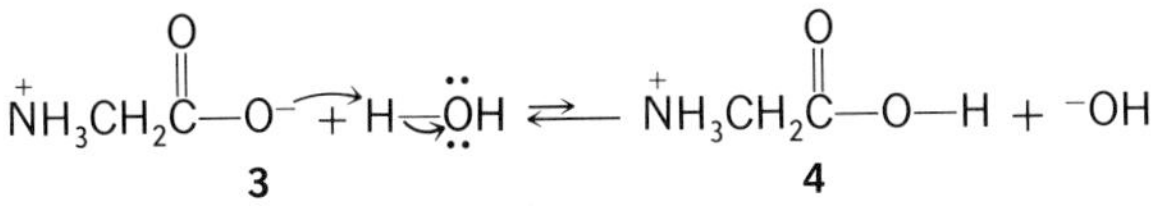

$$\text{I} \quad \overset{+}{N}H_3CH_2\overset{\overset{O}{\|}}{C}{-}O^- + H{-}\ddot{\underset{..}{O}}H \rightleftharpoons \overset{+}{N}H_3CH_2\overset{\overset{O}{\|}}{C}{-}O{-}H + {}^-OH$$

3 **4**

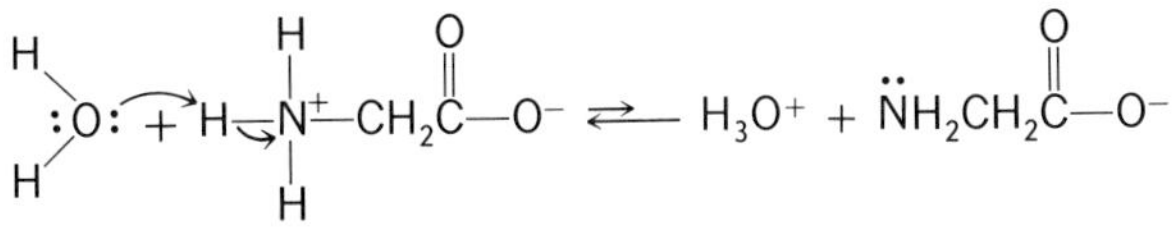

$$\text{II} \quad H_2\ddot{O}: + H{-}\overset{+}{N}H_2{-}CH_2\overset{\overset{O}{\|}}{C}{-}O^- \rightleftharpoons H_3O^+ + \ddot{N}H_2CH_2\overset{\overset{O}{\|}}{C}{-}O^-$$

3 **5**

(The OH^- ions made by I are mostly neutralized by the H_3O^+ ions made by II.) In its dipolar ionic form, **3,** glycine is a slightly better proton donor (equilibrium II) than proton acceptor (equilibrium I). All three forms, **3, 4,** and **5,** coexist at equilibrium, although **4** and **5** are present only in trace amounts. Form **5** is in slight excess of **4.** If a trace of mineral acid were added, equilibrium II would shift to the left, and I to the right. The concentration of 5 could be made equal to that of **4.** At this new pH glycine would exist almost exclusively as 3. If an electric current were passed through the solution, there would be essentially no migration of glycine units in either direction, toward either electrode. Whatever units do exist as 4 and 5 equilibrate rapidly via I and II, and each glycine unit that isn't 3 spends an *equal* amount of time as 4 and 5. Before 4 is attracted very far by the negative electrode, it changes to 5 by the proton transfers of equations I and II. Similarly 5 changes to 4. Since the rates of these changes are now *identical,* there is no *net* migration. *The pH at which an amino acid exhibits no net migration in an electric field is called the isoelectric point.* (Note carefully that the isoelectric point is a pH-value, a number.) See Table 15.1 for values of *p*I.

Primary Structure of Proteins

15.11 The Peptide Bond

If the α-amino group of one amino acid is acted upon by the carboxyl group of another amino acid, an amide link forms between the two. In general:

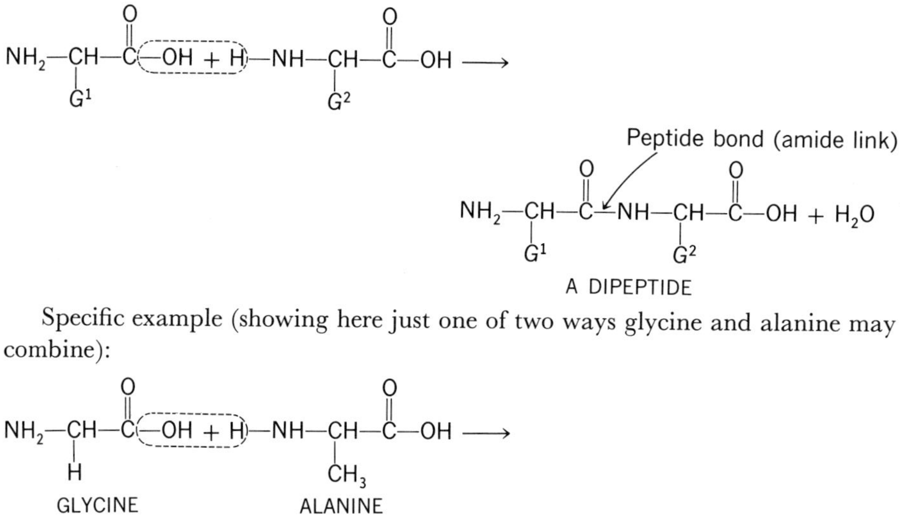

$$\mathrm{NH_2{-}\underset{\underset{G^1}{|}}{CH}{-}\overset{\overset{O}{\|}}{C}{-}OH + H{-}NH{-}\underset{\underset{G^2}{|}}{CH}{-}\overset{\overset{O}{\|}}{C}{-}OH} \longrightarrow$$

Peptide bond (amide link)

$$\mathrm{NH_2{-}\underset{\underset{G^1}{|}}{CH}{-}\overset{\overset{O}{\|}}{C}{-}NH{-}\underset{\underset{G^2}{|}}{CH}{-}\overset{\overset{O}{\|}}{C}{-}OH + H_2O}$$

A DIPEPTIDE

Specific example (showing here just one of two ways glycine and alanine may combine):

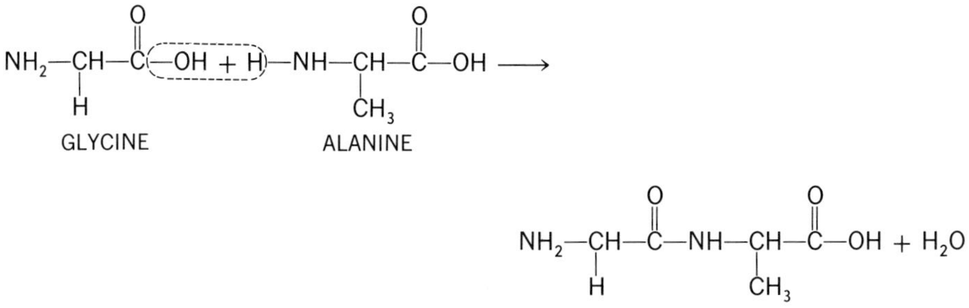

$$\mathrm{NH_2{-}\underset{\underset{H}{|}}{CH}{-}\overset{\overset{O}{\|}}{C}{-}OH + H{-}NH{-}\underset{\underset{CH_3}{|}}{CH}{-}\overset{\overset{O}{\|}}{C}{-}OH} \longrightarrow$$

GLYCINE ALANINE

$$\mathrm{NH_2{-}\underset{\underset{H}{|}}{CH}{-}\overset{\overset{O}{\|}}{C}{-}NH{-}\underset{\underset{CH_3}{|}}{CH}{-}\overset{\overset{O}{\|}}{C}{-}OH + H_2O}$$

GLYCYLALANINE

Gly · Ala[3]

[3]By convention the symbol Gly · Ala means that the first designated amino acid unit, Gly for glycine, has a free amino group and that the last designated unit, Ala for alanine, has the free carboxyl group. We say that Gly is the *N*-terminal amino acid and Ala is the *C*-terminal amino acid. In accordance with this convention amino acids are often written with the $-NH_2$ to the left and the $-CO_2H$ to the right. In naming a dipeptide the "-ine" ending of the name of the amino acid unit that comes first (i.e., left side in this order) is changed to "-yl." This principle will carry over to polypeptides as we shall soon see.

In protein chemistry the amide link is called the *peptide bond.* The product in the example, glycylalanine, is a *dipeptide,* a molecule that can be hydrolyzed to two molecules of amino acids.

The other sequence of the two amino acids could just as well have been used. In general:

$$NH_2{-}\underset{G^2}{CH}{-}\overset{O}{\overset{\|}{C}}{-}OH + H{-}NH{-}\underset{G^1}{CH}{-}\overset{O}{\overset{\|}{C}}{-}OH \longrightarrow NH_2{-}\underset{G^2}{CH}{-}\overset{O}{\overset{\|}{C}}{-}NH{-}\underset{G^1}{CH}{-}\overset{O}{\overset{\|}{C}}{-}OH + H_2O$$

Specific example:

$$\underset{\text{ALANINE}}{NH_2{-}\underset{CH_3}{CH}{-}\overset{O}{\overset{\|}{C}}{-}OH} + \underset{\text{GLYCINE}}{H{-}NH{-}\underset{H}{CH}{-}\overset{O}{\overset{\|}{C}}{-}OH} \longrightarrow \underset{\substack{\text{ALANYLGLYCINE} \\ \text{Ala} \cdot \text{Gly}}}{NH_2{-}\underset{CH_3}{CH}{-}\overset{O}{\overset{\|}{C}}{-}NH{-}\underset{H}{CH}{-}\overset{O}{\overset{\|}{C}}{-}OH} + H_2O$$

Just as the two letters N and O can be arranged to give two different words, NO and ON, so two different amino acids can be joined to give two isomeric dipeptides that have identical "backbones" and differ only in the sequences of side chains.

Dipeptides, of course, are still amino acids although they are no longer α-amino acids. A third α-amino acid can react at one end or the other. In general:

$$NH_2{-}\underset{G^1}{CH}{-}\overset{O}{\overset{\|}{C}}{-}NH{-}\underset{G^2}{CH}{-}\overset{O}{\overset{\|}{C}}{-}OH + H{-}NH{-}\underset{G^3}{CH}{-}\overset{O}{\overset{\|}{C}}{-}OH \longrightarrow$$

$$\underset{\text{A TRIPEPTIDE}}{NH_2{-}\underset{G^1}{CH}{-}\overset{O}{\overset{\|}{C}}{-}NH{-}\underset{G^2}{CH}{-}\overset{O}{\overset{\|}{C}}{-}NH{-}\underset{G^3}{CH}{-}\overset{O}{\overset{\|}{C}}{-}OH} + H_2O$$

Specific example:

$$\underset{\substack{\text{GLYCYLALANINE} \\ \text{Gly} \cdot \text{Ala}}}{NH_2{-}\underset{H}{CH}{-}\overset{O}{\overset{\|}{C}}{-}NH{-}\underset{CH_3}{CH}{-}\overset{O}{\overset{\|}{C}}{-}OH} + \underset{\substack{\text{PHENYLALANINE} \\ \text{Phe}}}{H{-}NH{-}\underset{CH_2C_6H_5}{CH}{-}\overset{O}{\overset{\|}{C}}{-}OH} \longrightarrow$$

$$\underset{\substack{\text{GLYCYLALANYLPHENYLALANINE} \\ \text{Gly} \cdot \text{Ala} \cdot \text{Phe}}}{NH_2{-}\underset{H}{CH}{-}\overset{O}{\overset{\|}{C}}{-}NH{-}\underset{CH_3}{CH}{-}\overset{O}{\overset{\|}{C}}{-}NH{-}\underset{CH_2C_6H_5}{CH}{-}\overset{O}{\overset{\|}{C}}{-}OH} + H_2O$$

The tripeptide shown in the last example is only one of six possible isomers involving these three different amino acids. The set of all possible sequences for a tripeptide made from glycine, alanine, and phenylalanine is as follows.

$$NH_2CH(H)C(=O){-}NHCH(CH_3)C(=O){-}NHCH(CH_2C_6H_5)COOH \qquad \text{Gly} \cdot \text{Ala} \cdot \text{Phe}$$

$$NH_2CH(H)C(=O){-}NHCH(CH_2C_6H_5)C(=O){-}NHCH(CH_3)COOH \qquad \text{Gly} \cdot \text{Phe} \cdot \text{Ala}$$

$$NH_2CH(CH_3)C(=O){-}NHCH(H)C(=O){-}NHCH(CH_2C_6H_5)COOH \qquad \text{Ala} \cdot \text{Gly} \cdot \text{Phe}$$

$$NH_2CH(CH_3)C(=O){-}NHCH(CH_2C_6H_5)C(=O){-}NHCH(H)COOH \qquad \text{Ala} \cdot \text{Phe} \cdot \text{Gly}$$

$$NH_2CH(CH_2C_6H_5)C(=O){-}NHCH(H)C(=O){-}NHCH(CH_3)COOH \qquad \text{Phe} \cdot \text{Gly} \cdot \text{Ala}$$

$$NH_2CH(CH_2C_6H_5)C(=O){-}NHCH(CH_3)C(=O){-}NHCH(H)COOH \qquad \text{Phe} \cdot \text{Ala} \cdot \text{Gly}$$

Each of these structures represents a different compound with its own unique set of physical properties. Their chemical properties will be quite similar, however, for the same functional groups are present in all.

Each tripeptide is of course an amino acid, although none is an α-amino acid. Each can combine with still another amino acid at one end or the other. With a repetition of this process we can envision how a protein molecule is structured. All protein molecules have this repeating sequence in the backbone:

$$\underbrace{H{-}N{-}C{-}C(=O)}_{N\text{-TERMINAL UNIT}}{-}\left(N{-}C{-}C(=O)\right)_n{-}\underbrace{N{-}C{-}C(=O){-}OH}_{C\text{-TERMINAL UNIT}}$$

(*n* may vary from dozens to thousands)

This repeating sequence, together with the sequence of side chains, constitutes the *primary structure* of all proteins. The peptide bond is thus the chief covalent linkage holding amino acid units together in proteins. The only other covalent bond at this level of protein structure is the disulfide linkage made possible by cystine units.

15.12 Determining the Structure of a Protein

By long, laborious processes the structures of a few proteins have been determined. Both chemical methods and X ray diffraction techniques have been used. We shall say nothing about the second technique and give only the general idea for the chemical method.

Using letters to represent amino acid units, let us suppose that a protein has the "molecular formula" A,B,C,D,E. That is, analysis shows that a molecule of this protein is a pentapeptide made up of amino acids A, B, C, D, and E, but the sequence is to be determined. (There are 125 possible isomers.) To have proceeded this far, the protein must have been hydrolyzed and the mixture of amino acids (plus some di- and tripeptides) separated.

A mixture such as this can be separated and its components identified by several variations of chromatography. For example, the amino acids could be converted to volatile esters and analyzed by gas-liquid chromatography. Ion exchange chromatography or thin layer chromatography (Section 14.5) are often used. Paper chromatography, a variation of thin layer chromatography, is described in Figure 15.1. When the components of the mixture have separated and are at various spots on the paper, something has to be done to find them. (Seldom are the components themselves colored.) To do this, the paper is dried and sprayed with a "visualizing" chemical. If that is picked correctly, it reacts with the mixture's components, where they now are located, to produce visible spots. Using R_f values (Figure 15.1) and color tests, we can identify the components of an unknown mixture of amino acids and di- and tripeptides.

To return to our pentapeptide problem, suppose that hydrolysis were only partial, and that the mixture contains individual amino acids together with dipeptides and tripeptides resulting from random hydrolyses of peptide bonds. A huge number of di- and tripeptides are known and can be readily identified. By means of paper chromatography we next identify fragments AD, CB, DC, BE, and DCB, in addition to individual amino acids. Our question now is what sequence in the original pentapeptide would give these fragments. If the fragments are lined up as follows, the pattern quickly appears:

```
AD
 DC
 DCB
   BE
  CB
ADCBE = the sequence in the original pentapeptide
```

Only this sequence could give the fragments observed. In principle this is a simple operation. In practice teams of scientists have needed years of effort to determine the structures of low-formula-weight proteins (fewer than 300 amino acid units). In the late 1960s methods for the sequencing of proteins by automated operations were developed.

Among the proteins whose structures are completely known are the pituitary hormones, oxytocin and vasopressin, Figure 15.2; the adrenocorticotropic hormone (ACTH); the hormone of the pancreas, insulin, Figure 15.3; the enzyme, ribonuclease; and the blood's oxygen carrier, hemoglobin.

Proteins that perform like functions in different species (e.g., insulins from

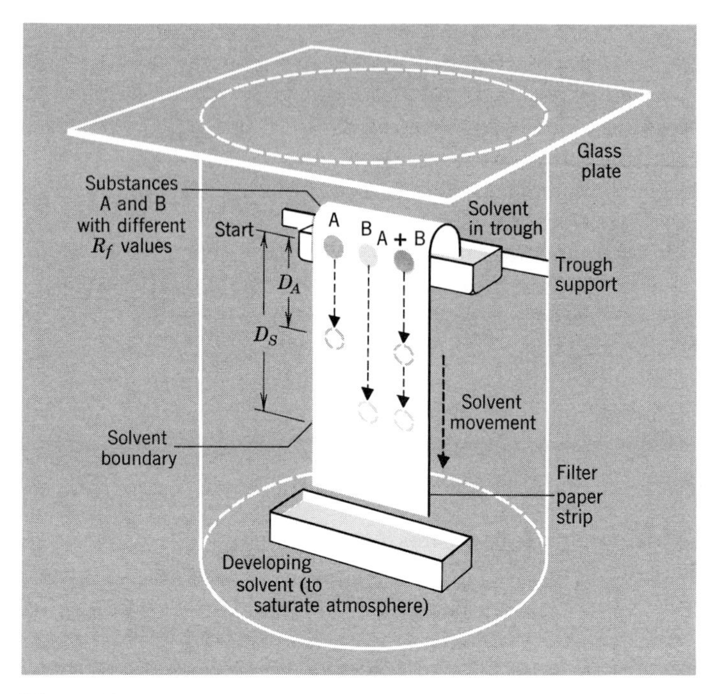

Figure 15.1
Paper chromatography. A drop of the mixture (A + B) to be separated is spotted on the paper at the starting line. The paper consists largely of cellulose molecules which have water molecules adhering strongly to them by means of hydrogen bonds. (The paper feels dry, however.) The components of the mixture, A + B, are attracted to this stationary phase of water and cellulose, with (as illustrated) molecules of A absorbing more strongly than those of B. The paper is then positioned to permit a solvent (e.g., a butanol-water mixture) to creep by capillary action past the starting line. As it does, molecules of A and B distribute themselves between the moving phase and the stationary phase. As shown, molecules of B are more soluble in the moving phase, and gradually, as this phase advances, component B pulls away from component A. The R_f value for component A is the ratio of distance traveled by component A(D_A) to distance traveled by the moving solvent (D_S). When determined under very carefully controlled conditions, R_f values are important physical constants which identify components of unknown mixtures. (Adapted, with permission, from L. C. Conn and P. K. Stumpf, *Outlines of Biochemistry,* third edition, John Wiley and Sons, New York, 1972, page 502.)

various animals) have similar but not identical amino acid sequences. Yet, what seem to be small changes in other sequences result in considerable differences in function. Oxytocin and vasopressin illustrate this. Oxytocin is a hormone that stimulates contraction of peripheral blood vessels and a rise in blood pressure. Vasopressin, another hormone, helps regulate the composition of the blood. Where oxytocin uses leucine (Leu), vasopressin has arginine (Arg); where oxytocin has isoleucine (Ile), vasopressin has phenylalanine (Phe). *Protein function* is highly sensitive to *protein structure* in ways we are just beginning to understand. One of the most dramatic and well-authenticated examples of this sensitivity is the hemoglobin change of sickle-cell anemia.

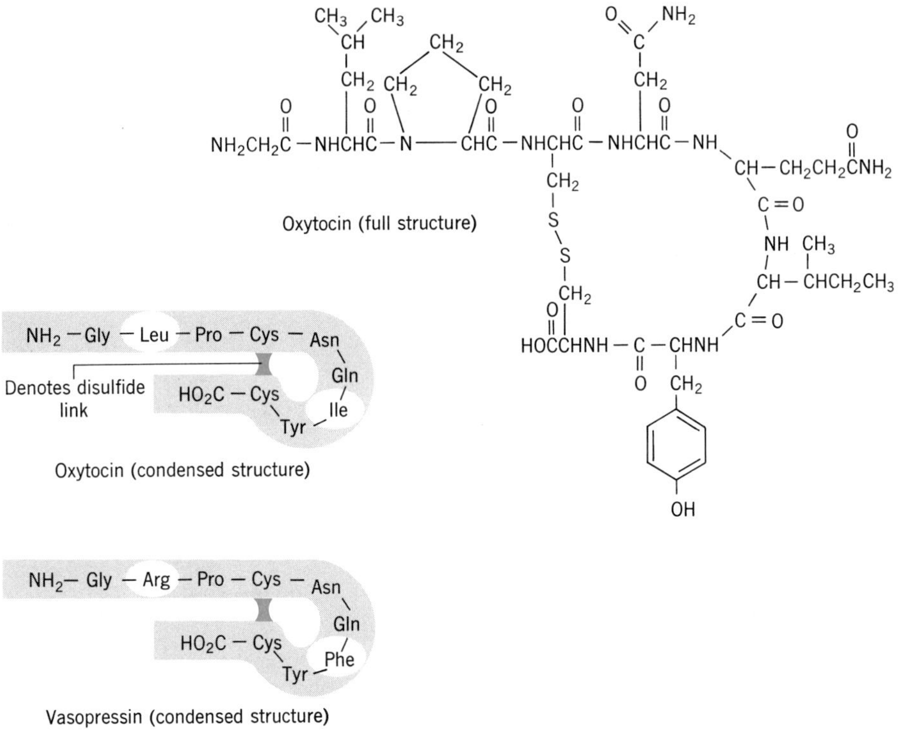

Oxytocin (full structure)

NH_2 — Gly — Leu — Pro — Cys — Asn — Gln — Ile — Tyr — Cys — CO_2H

Denotes disulfide link

Oxytocin (condensed structure)

NH_2 — Gly — Arg — Pro — Cys — Asn — Gln — Phe — Tyr — Cys — CO_2H

Vasopressin (condensed structure)

Figure 15.2
Oxytocin and vasopressin, two low-formula-weight proteins. Sites of differences between the two are indicated in the two condensed structures.

15.13 Sickle-Cell Anemia and Altered Hemoglobin

The hemoglobin molecule, as it occurs in adult human beings, has a molecular weight of about 65,000. It consists of four long polypeptide chains, each intricately folded together with a large, flat, nonprotein molecule called heme (see Figure 15.4) in which a ferrous ion is held.

The polypeptide chains of normal hemoglobin contain about 300 amino acid units. In the hemoglobin of those suffering from sickle-cell anemia, *only one of these is changed.* A glutamic acid unit sixth in from the *N*-terminus of the β-chain is replaced by a valine unit. Glutamic acid has a carboxyl group, —COOH, on its side chain; the side chain of valine is the isopropyl group. The —COOH can ionize, but isopropyl cannot. Normal and sickle-cell hemoglobins are therefore capable of having different electric charges which affects their relative solubilities as well as other properties.

When the oxygen supply is high, as in arterial blood, both normal hemoglobin, HHb, and sickle-cell hemoglobin, HbS, have about the same solubility in the bloodstream. When the oxygen supply is lower, as in venous blood, the solubility of HbS is less than that of HHb. Consequently, when circulating blood has delivered its oxygen, the altered hemoglobin tends to precipitate in red cells that contain it. This precipitate distorts the shapes of the red cells, the blood becomes more difficult to pump, and a greater strain is placed on the heart. The red cells

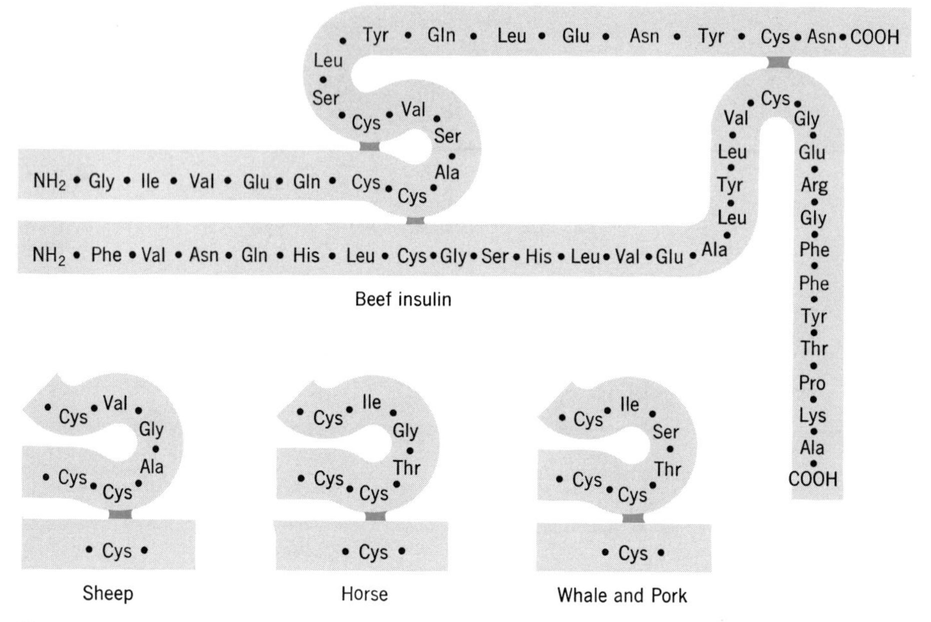

Figure 15.3
The primary structure of insulin. F. Sanger, British biochemist, and a team of coworkers at Cambridge worked from 1945 to 1955 on the determination of this structure. Sanger received the 1958 Nobel prize in chemistry for this work and the development of the techniques. Insulin from different species have the slight differences noted. Fortunately, these are not serious for diabetics on insulin therapy. (Usually, when an alien protein gets into circulation, the body makes antibodies to combine with it. The rejection of skin grafts from one person to another and the extraordinary problems of transplanting organs are related to this formation of antibodies.) Sheep insulin, however, can be given to human diabetics. Some antibodies form, but the insulin activity is reduced only slightly.

Figure 15.4
The heme molecule. The iron in heme holds its Fe^{2+} state in *oxygenated* hemoglobin, that is, hemoglobin which has picked up oxygen molecules. In *oxidized* hemoglobin, however, iron is in the Fe^{3+} state, and the substance is brownish in color. When meat is cooked, its color changes from red to brown largely because Fe^{2+} oxidizes to Fe^{3+}.

may also clump together enough to plug a capillary here and there. Sometimes the red cells split open.

A child with the severe form of the disease has received the genetic trait from both parents. A child with the mild form has only one afflicted parent and shows no symptoms except when in an environment with a low concentration of oxygen (e.g., the depressurized cabin of an airplane). The disease is especially widespread in central and western Africa, and children with the severe form often die before they are two years old. The molecular difference responsible for the anemia, a matter of life or death for the victims, is startingly small. One amino acid in 300 is substituted for another.

Secondary Structure of Proteins

15.14 Order versus Chaos in Proteins

Highly crystalline materials such as inorganic salt crystals, in which there is a considerable degree of order, give distinctive X ray diffraction pictures. Many proteins give X ray diffraction patterns indicating the presence of considerable crystal-like order.

Crystalline features arise because molecules of most proteins take up orderly arrangements, and these kinds of arrangements are called the *secondary structures* of proteins. The α-helix is the most important example and the only one we shall study.

15.15 α-Helix

In many proteins studied thus far, the chains are coiled into a spiral called an α-helix. Pauling and Corey, in 1951, were the first to publish extensive evidence for the helix (Figure 15.5). The opportunity for carbonyl oxygens to form hydrogen bonds to amide hydrogens provides the forces that cause the coiling. Thus hydrogen bonds of the type

$$>C{=}\overset{\delta-}{O}\cdots\overset{\delta+}{H}{-}\overset{|}{N}{-}$$

are bonds of crucial importance in the molecular basis of life.

15.16 Tertiary Structure. Folded Helices

Helix formation is not always the final shaping, for helices may in turn fold, twist, or assume some other final configuration of minimum energy within the particular environment. Once an organism has put together the particular amino acid sequence that makes a protein, folding and coiling apparently follow automatically without any further need for enzymes and other chemicals.

15.17 Denaturation of Proteins

A wide variety of reagents and conditions that do not break peptide bonds will destroy the biological nature and activity of the protein. When this happens,

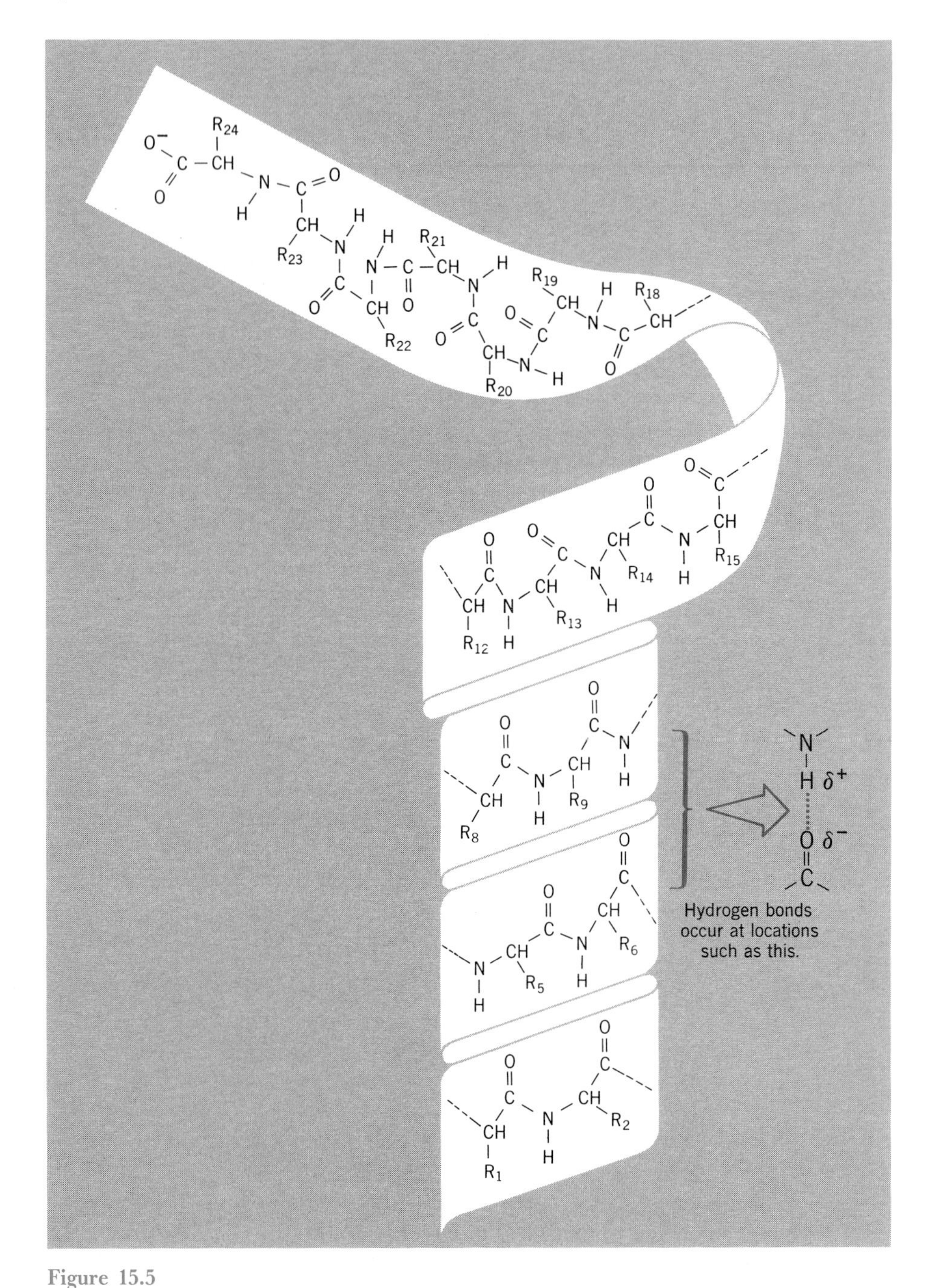

Figure 15.5
This representation of a section of a protein chain as a ribbon shows the right-handed coiling of an α-helix. Hydrogen bonds exist between carbonyl oxygens and hydrogens on the amide nitrogens. Although each represents a weak force of attraction between the turns of the coil, there are many of them, and the total force stabilizing the helix is more than enough to counteract the natural tendency for the chain to adopt a randomly flexing form. It is understandable that the repeating series of hydrogen bonds up and down the helix is called the "zipper" of the molecule. (From G. H. Haggis, D. Michie, A. R. Muir, K. B. Roberts, and P. M. B. Walker, *Introduction to Molecular Biology,* John Wiley and Sons, New York, 1964, page 51.)

the protein is said to have been *denatured*. After denaturation the protein usually coagulates. Several of the more common chemicals and conditions that denature proteins are listed in Table 15.2.

At the molecular level denaturation is a disorganization of the shape of a protein. It can occur as an unfolding or uncoiling of a structure or as the separation of the protein into subunits that then unfold or uncoil. (Figure 15.6).

Mercuric, silver, and lead salts are dangerous primarily because these metallic ions wreak havoc among important proteins of the body, particularly enzymes. If they reach the brain they can cause permanent mental impairment.

Denaturing agents differ widely in their action, depending largely on the protein. Some proteins (e.g., skin, hide, hair) strongly resist most denaturing actions, as they must.

15.18 How Proteins Are Classified

Having studied the general principles of protein structure and behavior, let us turn our attention to the several types of proteins. They may be classified in a variety of ways, of which we outline two.

Table 15.2
Chemicals and Conditions That Cause Denaturation

Denaturing Agent	How the Agent May Operate
Heat	Disrupts hydrogen bonds by making molecules vibrate too violently. Produces coagulation as in the frying of an egg
Solutions of urea ($NH_2-\overset{O}{\overset{\|\|}{C}}-NH_2$)	Disrupt hydrogen bonds. Since it is amidelike, urea can form hydrogen bonds of its own
Ultraviolet radiation	Appears to operate the same way that heat operates (e.g., sunburning)
Organic solvents (e.g., ethyl alcohol, acetone, isopropyl alcohol)	May interfere with hydrogen bonds in protein, since alcohol molecules are themselves capable of hydrogen bonding. Quickly denatures the proteins of bacteria, thus killing them (e.g., disinfectant action of ethyl alcohol, 70% solution)
Strong acids or bases	Can disrupt hydrogen bonds. Prolonged action of aqueous acids or bases leads to actual hydrolysis of proteins
Detergents	May affect hydrogen bonds
Salts of heavy metals (e.g., salts of the ions Hg^{2+}, Ag^{+}, Pb^{2+})	Ions combine with SH groups. These ions usually precipitate proteins
Violent whipping or shaking	May form surface films of denatured proteins from protein solutions (e.g., beating egg white into meringue)

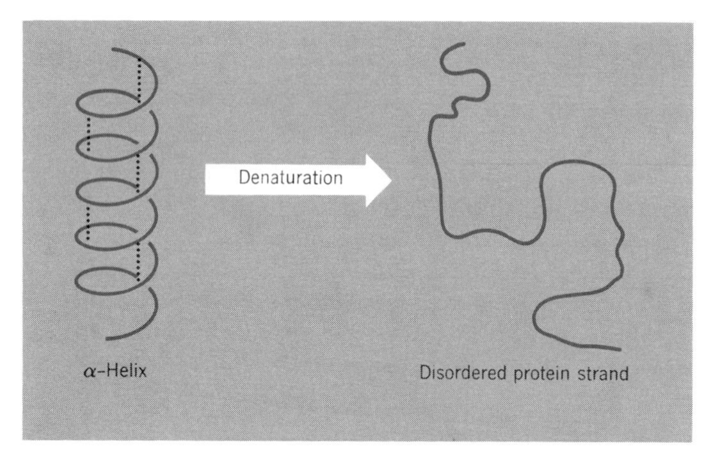

Figure 15.6
Denaturation of a protein is fundamentally a disorganization of its molecular configurations without the rupture of any peptide bonds.

I. Gross Structure and Solubility

A. Fibrous proteins. As the name implies, these proteins consist of fibers. These proteins can be stretched, and they contract when the tension is released. They perform important structural, supporting, and protecting functions and are insoluble in aqueous media.

1. Collagens. These are the proteins of connective tissue, making up about one-third the body's protein. When acted upon by boiling water, they are converted into more soluble *gelatins*. In contrast to collagen, gelatin is readily digestible. Hence the collagen-to-gelatin conversion that takes place when meat is cooked is an important preliminary to digestion. At the molecular level the change is thought to be simply an unfolding or uncoiling of collagen molecules to expose the peptide bonds to the hydrolytic action of water and digestive enzymes.
2. Elastins. Elastic tissues such as tendons and arteries are elastins. They are similar to collagens, but they cannot be changed into gelatin.
3. Keratins. These proteins make up such substances as wool, hair, hoofs, nails, and porcupine quills. They are exceptionally rich in cystine, which has a disulfide link.
4. Myosins. Muscle tissue is rich in myosin, a protein directly involved in the extension and contraction of muscle.
5. Fibrin. Fibrin is the protein that forms from fibrinogen, its soluble precursor, when a blood vessel breaks. The long fibrin molecules tangle together to form a clot.

B. Globular proteins. Members of this broad class are soluble in aqueous media, some in pure water, others in solutions of certain electrolytes. In contrast to the fibrous proteins, the globular proteins are easily denatured. Examples include the following.

1. Albumins. Egg albumin is the most familiar member of this class. Albumins are soluble in pure water (forming colloidal dispersions), and they are easily coagulated by heat. In the bloodstream albumins contribute to osmotic pressure relations and to the pool of buffers.

2. *Globulins.* These are soluble in solutions of electrolytes, and they are also coagulated by heat. The γ-globulins in blood are very important elements in the body's defensive mechanisms against infectious diseases.

II. Function

Another way of classifying proteins, including those already discussed, is by describing the biological functions they serve.

A. *Structural proteins.* Fibrous proteins are examples.

B. *Enzymes.* These are the catalysts of a living organism without which it could not live.

C. *Hormones.* Many, but not all, hormones are proteins.

D. *Toxins.* These proteins produced by bacteria in the living organism act as poisons to that organism.

E. *Antibodies.* The body makes these proteins to destroy foreign proteins that invade it during an attack by an infectious agent.

F. *Oxygen-transporting protein.* Hemoglobin is the name of this important protein.

Laboratory Synthesis of Polypeptides

15.19 The Need for Protecting Groups

"A historic feat in biochemistry," according to the 1955 Nobel chemistry prize citation, occurred when Vincent DuVigneaud and coworkers synthesized the first two polypeptide hormones, oxytocin and vasopressin (Figure 15.2). These are nonapeptides. Their molecules are so small and simple that we might be excused for wondering why their syntheses could then have been regarded with such awe. The actual problems are many and difficult, however.

To get the right length and the correct sequence of side chains, the amino acid units have to be put on one at a time. Even if the yield of putting on each unit were high—say 90%— the *net* result of getting 90% in each of eight successive steps (for making a nonapeptide) is less than 1%. Making matters worse, each of the steps is actually a series of steps; we cannot just mix, say, glycine and alanine, heat the mixture, and expect to get one dipeptide. Amino acids are difunctional; they can react at either end with any neighboring amino acids in the mixture. The dipeptides we would get would be Gly·Gly, Gly·Ala, Ala·Gly, and Ala·Ala. Besides these we would produce many tri-, tetra- and higher polypeptides. Moreover, several amino acids have functional groups on their side chains that can enter into reactions of their own. We have to devise a means of protecting these, and we have to protect either the α-amino group or the carboxyl group at one stage or another.

While it is well beyond the intended scope of this book to go into all the details of available protecting groups, the basic strategy is this: change the functional group you want to protect to a different one that will survive all the peptide-bond forming reactions elsewhere on the molecule. When you are finished with those reactions, remove the protecting group, using a reaction that will not break up the polypeptide. For example, an amino group might be protected this way:

$$\underset{\substack{t\text{-BUTYL CHLOROCARBONATE}\\\text{(An acid chloride-ester}\\\text{of carbonic acid)}}}{CH_3-\overset{\overset{CH_3}{|}}{\underset{\underset{CH_3}{|}}{C}}-O-\overset{\overset{O}{\|}}{C}-Cl} + \underset{\text{GLYCINE}}{\overset{\text{Amine}}{NH_2}-CH_2CO_2H} \longrightarrow \underset{\text{PROTECTING GROUP}}{CH_3-\overset{\overset{CH_3}{|}}{\underset{\underset{CH_3}{|}}{C}}-O\overset{\text{Ester bond}}{-}\overset{\overset{O}{\|}}{C}\overset{\text{Amide bond}}{-}NH-CH_2CO_2H}$$

The carboxyl end, $-CO_2H$, of the altered amino acid can now be operated on to join it to the amino group of another amino acid. (We shall omit the steps.) This particular protecting group—there are many other types available—has the virtues of (1) being very easily hydrolyzed by dilute acid (much more easily hydrolyzed than ordinary peptide bonds); and (2) the products of this hydrolysis—isobutylene and carbon dioxide—are both gases and easily and cleanly removed.

$$\underbrace{CH_3-\overset{\overset{CH_3}{|}}{\underset{\underset{CH_3}{|}}{C}}-O-\overset{\overset{O}{\|}}{C}}_{\text{PROTECTING GROUP}}-NH-\underset{\underset{G^n}{|}}{CH}-\overset{\overset{O}{\|}}{C}\underset{\text{PEPTIDE}}{\sim\sim\sim\sim} \xrightarrow{H^+} \underset{\text{ISOBUTYLENE}}{CH_2=C\begin{matrix}CH_3\\CH_3\end{matrix}} + CO_2 + NH_2-\underset{\underset{G^n}{|}}{CH}-\overset{\overset{O}{\|}}{C}\sim$$

15.20 The Merrifield Solid-Phase Technique

R. B. Merrifield and coworkers in the late 1960s prepared the hormone bradykinin, a nonapeptide, in a 27-hour period with an overall yield of 85%. This amazing result came after years of perfecting an automated procedure for handling the innumerable repeating steps. At the heart of the Merrifield technique is the use of a synthetic resin, in the form of small insoluble beads, as the supplier of the protecting group for the $-CO_2H$ end of the first amino acid of the chain. The carboxyl group is anchored by ester links to molecules in the resin beads. The amino acid's molecular "tails" waggle in the solution where new amino acids are successively attached. The polypeptide "grows" outward from the bead all the while, up to the last step, anchored to it covalently. The use of the small beads means that the growing polypeptide is always in the solid state. After each step, the beads with their attached molecules are simply filtered—quantitatively—and thoroughly washed. All unused reagents are removed—quantitatively. (The method uses the protecting group for $-NH_2$ units described in Section 15.19.) The sequence is as follows:

$$\underbrace{(CH_3)_3COC}_{\text{PROTECTING GROUP}}\!\overset{O}{\overset{\|}{}}NH-\underset{G^2}{CH}-\overset{O}{\overset{\|}{C}}-OH$$

$$\xrightarrow{-H_2O}$$

$$NH_2-\underset{G^1}{CH}-\overset{O}{\overset{\|}{C}}-O-\bigcirc \quad \text{Resin bead}$$

$$\downarrow$$

$$(CH_3)_3C-O-\overset{O}{\overset{\|}{C}}-NH-\underset{G^2}{CH}-\overset{O}{\overset{\|}{C}}-NH\underset{G^1}{CH}\overset{O}{\overset{\|}{C}}-O-\bigcirc$$

$$\xrightarrow{H^+} CH_2{=}C(CH_3)_2 + CO_2$$

$$NH_2-\underset{G^2}{CH}-\overset{O}{\overset{\|}{C}}-NH\underset{G^1}{CH}\overset{O}{\overset{\|}{C}}-O-\bigcirc$$

DIPEPTIDE STAGE

1. $(CH_3)C-O-\overset{O}{\overset{\|}{C}}-NH-\underset{G^3}{CH}-\overset{O}{\overset{\|}{C}}-OH$

2. $\xrightarrow{H^{\oplus}} CH_2{=}C(CH_3)_2 + CO_2$

$$NH_2-\underset{G^3}{CH}-\overset{O}{\overset{\|}{C}}-NH\underset{G^2}{CH}\overset{O}{\overset{\|}{C}}-NH\underset{G^1}{CH}\overset{O}{\overset{\|}{C}}-O-\bigcirc$$

At the end the chain is removed from the resin by a reaction that does not hydrolyze peptide bonds.

The Merrifield method has been used to make insulin with a total of 51 amino acid units (Figure 15.3), and an enzyme, ribonuclease, with 124 amino acid units. The importance of the Merrifield procedure is severalfold. Structures of small polypeptides can be verified by synthesis. Polypeptide hormones can be made with innumerable variations in structure to find out how the hormone works—what exactly in its structure is critical to its function and why? The same can be done for some enzymes. Some day it may be easier and cheaper to make insulin or an insulin substitute this way than by letting sheep do it for us.

15.21 Proteins and Enzymes

Enzymes, the large family of catalysts in living systems, are substances whose molecules are mostly or entirely proteins. The names of enzymes nearly always end in "-ase"—for example, dehydrogenase, esterase, lipase, transaminase. The rest of the name indicates the kind of reaction the enzyme catalyzes or the kind of functional group it acts on.

A complete enzyme, sometimes called a holoenzyme, usually involves a joining of a nonprotein group, called a *cofactor,* to the wholly protein (or polypeptide) unit, called the *apoenzyme*.

Cofactor + Apoenzyme ⟶ Enzyme

The cofactor may be an organic compound, called a *coenzyme,* or it may be a metal ion, one of the several "trace minerals" in the diet.

Many important coenzymes are phosphate esters of certain vitamins. The vitamins thiamine, riboflavin, nicotinamide (niacinamide), and pyridoxal are

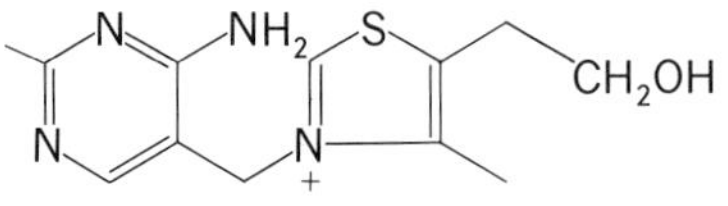

THIAMINE (B_1)

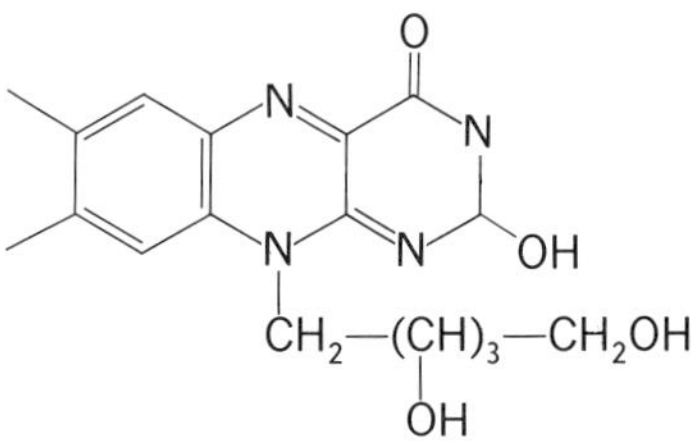

RIBOFLAVIN (B_2)

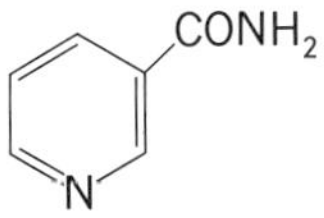

NICOTINAMIDE (NIACINAMIDE)

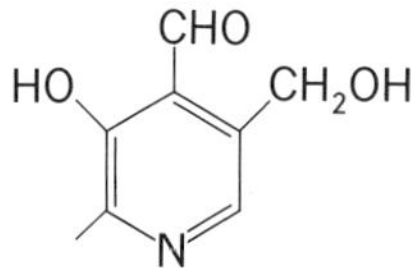

PYRIDOXAL (B_6)

known examples involved in coenzymes. A **vitamin** is any substance meeting the following criteria. It is an organic compound we cannot make at all or can make, but too slowly to help. Its absence causes a specific disease—a vitamin deficiency disease. Its presence is essential to normal growth and health. And it is found in small concentrations in various foods, or something similar to it—a provitamin—is present that we can change to the vitamin.

Coenzymes often supply the *catalytic site* to the enzyme, the place where actual catalysis occurs. The coiled and folded apoenzyme unit furnishes binding sites, places where molecules of the compound being acted on—the *substrate*—nestle in and briefly cling to the enzyme while the catalytic site acts.

Nowhere is the importance of secondary and tertiary structural features of proteins more dramatically evident than in enzyme chemistry. The shape of the apoenzyme must be complementary to that of the substrate. If anything throws this off, the enzyme cannot act, and the reaction does not go rapidly enough. Enzyme molecules, therefore, can discriminate among reactant molecules, and this feature of enzyme action accounts for the extraordinary specificity of enzymes. Urease, for example, catalyzes the hydrolysis of urea and only urea, not any other amide no matter how seemingly closely related structurally. That is why poisons that bind themselves to enzymes are among our most dangerous. They include nerve gases, toxins in food poisoning (botulism), "arsenic" (arsenate ion), cyanide, and some of our most potent pesticides. At its binding sites the shape of the enzyme must be complementary to that of the substrate in the way a key's grooves and channels must be complementary to that of the tumbler lock. In fact, the theory of enzyme action is called the "lock and key" theory. When poisons bind to an enzyme, the fitness of the enzyme is ruined.

Proteins and Nutrition

15.22 Digestion of Proteins

The end products of protein digestion are amino acids. These are rapidly absorbed through the walls of the small intestine, and they enter general circulation.

Although the body has the capacity to use some amino acids to make others, a few cannot be synthesized this way. We cannot form these amino acids from the carbon skeletons and the amino groups of other intermediates, and for this reason they are called **essential amino acids** (Table 15.3). Nonessential amino acids, also listed in Table 15.3, can be made in the body. (Nonessential, in this context, means *temporarily* dispensable. Obviously all the amino acids listed are used by the body, and protein synthesis could not proceed without them.) Dietary proteins that contain all the essential amino acids are called *adequate proteins*. Gelatin without tryptophan and zein (protein in corn) without lysine are inadequate proteins.

Modern research in nutrition has proven that there is literally nothing of a chemical nature besides the genes themselves that is more important to the later health of an individual than the diet it receives in both the prenatal and postnatal periods. Besides the absence of birth-defect causing drugs, the presence of enough protein well-balanced in all essential amino acids is absolutely essential for both physical and mental growth and resistance to infectious diseases. Those who will be mothers and those who will care for mothers are well advised to take these facts with utmost seriousness. The disease called kwashiorkor is but one part of the evidence.

Table 15.3

Amino Acids, Classified as Nutritionally Essential or Nonessential in Maintaining Nitrogen Equilibrium in an Adult Man

Essential	Nonessential
Isoleucine	Alanine
Leucine	Arginine
Lysine	Aspartic acid and asparagine
Methionine	Cystine
Phenylalanine	Glutamic acid and glutamine
Threonine	Glycine
Tryptophan	Histidine
Valine	Hydroxyproline
	Proline
	Serine
	Tyrosine

Data from W. C. Rose, *Federation Proceedings*, Vol. 8, 1949, page 546.

15.23 Kwashiorkor

In parts of Latin America, Asia, and Africa the death rate among children is several times that in developed, industrialized societies. Children by the thousands are doomed to short lives with bloated bellies, patchy skin, and discolored hair. As long as they are nourished at their mother's breast, they enjoy health. When the second child comes and displaces the first, the symptoms appear in the first child. The disease is called kwashiorkor, a name taken from two words of an African dialect meaning "first" and "second"—the disease that the first child contracts when the second one is born. The diet of the firstborn, instead of milk, is now starchy and contains inadequate protein. Hardly recognized until the 1940s, the ailment is now known to be a protein deficiency disease. Both undernutrition and malnutrition are responsible. The initial symptoms are a loss of appetite and diarrhea—both of which lead the mother to reduce the amount of food she gives the child, thus hastening the onset of additional complications. Weakened, the child is even more susceptible to the diseases that are a constant hazard in the tropics.

Brief Summary

1. Proteins are polymers of α-amino acids of the L-configurational family (except the nonchiral glycine).

2. The primary structural feature of a protein is the repeating system:

$$\left(\text{-N-}\overset{|}{\underset{|}{\text{C}}}\text{-}\overset{\text{O}}{\overset{\|}{\text{C}}}\text{-} \right)_n$$

where individual units are furnished by about 20 amino acids. This feature is common to all proteins. What makes proteins different is the length of the basic chain and the particular sequence in which the sidechains of the α-carbons appear.

3. In addition to the peptide bond proteins occasionally have another covalent bond—the disulfide linkage contributed by cystine units.

4. Between or within proteins strands innumerable hydrogen bonds $\rangle C{=}O\cdots H{-}\overset{|}{N}{-}$ help to hold the strands in particular conformations.

5. Hydrogen bonds are important noncovalent bonds acting to stabilize protein molecules in particular shapes.

6. The α-helix is one of the important secondary structures of proteins.

7. Proteins with secondary structures often coil or twist further to give distinctive final shapes called tertiary structures.

8. Many properties of proteins are related to their side-chain units, and these may either be hydrophobic or hydrophilic. The hydrophobic groups are the nonpolar or hydrocarbon groups that are not attracted by water molecules. Hydrophilic groups are those with amino, carboxyl, hydroxyl or amide systems, or other polar units.

9. Amino acids exist largely in dipolar ionic forms that are weakly acidic as well as weakly basic.

10. The isoelectric point of an amino acid is the pH of the solution in which the amino acid units are in the isoelectric condition. They will not migrate in either direction if electrodes are placed in the solution.

11. In determining the structure of a protein X-ray analysis may be used; or amino acids may be removed one at a time (sequencing methods); or the protein may be hydrolyzed to a mixture of di- and tripeptides (and some higher peptides) that are then "mapped" in a manner that accounts for their presence.

12. In synthesizing a protein in a laboratory protecting groups must be put on during certain procedures and later removed.

13. A few amino acids—the "essential" ones—must be present in the diet since the body cannot make them from other molecules in the diet.

14. "Nonessential" amino acids are called that not because we do not need them but because our bodies can make them, if need be, from other molecules.

Problems and Exercises

1. Describe in structural terms and symbols what all proteins have in common.
2. What are the structural factors that are the basis of differences among proteins?
3. Write the structures of the isomeric dipeptides that could be hydrolyzed to a mixture of serine and methionine.
4. Draw the structures of the following pentapeptides: (a) Ala•Gly•Phe•Val•Leu, (b) Glu•Lys•Tyr•Thr•Ser. Make the best judgment about the placement of protons and the location of charged sites. (c) Which of the two compounds would tend to be more soluble in a nonaqueous medium? Why? (d) Which would tend to be more soluble in water? Why? (e) If, in the pentapeptide of part (b), the serine residue (Ser) were replaced by arginine, would the new compound be more or less basic?
5. The side chain in arginine is derived from guanidine, $\ddot{N}H_2—C(=\ddot{N}H)—\ddot{N}H_2$. It is a very strong base, having a K_b value of about 1, in comparison with aliphatic amines having K_b values on the order of 1×10^{-5}. This strength suggests that the protonated form of guanidine, $\ddot{N}H_2—C(=\overset{+}{N}H_2)—\ddot{N}H_2$, is especially stable compared to the nonprotonated form. As an exercise in resonance theory, offer an explanation for this unusual stability. (Try to write contributing structures for this ion. Are they equivalent? What does resonance theory say about this?)
6. The following peptides were subjected to partial hydrolyses giving the fragments shown. Deduce the amino acid sequences in the peptides. (Commas in the "formulas" indicate the sequence is unknown.)

   (a) Glu, Gly, His, Phe, Tyr gave Phe•Glu + Gly•Tyr + Glu•His + Tyr•Phe.
   (b) Asp, Ile, Met, Pro, Tyr gave Tyr•Asp + Met•Ile + Asp•Pro + Pro•Met.
7. Describe what would happen, chemically, to the following pentapeptide under these conditions: (a) digestion, (b) action of a reducing agent.

$$\overset{+}{N}H_3-\underset{\displaystyle CH_3}{CH}-\overset{O}{\overset{\|}{C}}-NH-\underset{\displaystyle CH_2C_6H_5}{CH}-\overset{O}{\overset{\|}{C}}-NH-\underset{\displaystyle CHOHCH_3}{CH}-\overset{O}{\overset{\|}{C}}-NH-\underset{\displaystyle CH(CH_3)CH_3}{CH}-\overset{O}{\overset{\|}{C}}-NH-\underset{\displaystyle CH_2-S-S-CH_2-CH(NH_3^+)-\overset{O}{\overset{\|}{C}}-O^-}{CH}-\overset{O}{\overset{\|}{C}}-O^-$$

8. Discuss the role of the hydrogen bond in protein structure.
9. Explain why there is no net migration of amino acid molecules in an electric field when they are in a medium at their isoelectric point.
10. In terms of events at the molecular level, what happens during denaturation?
11. What is the relation between collagen and gelatin?
12. Describe functions for the following: (a) elastins, (b) keratins, (c) fibrin, (d) collagen.
13. To help the food situation of countries where corn is a major part of the diet, plant scientists have been interested in developing a corn hybrid that is relatively high in lysine. (Older varieties are low in lysine.) Explain this interest in lysine.

chapter sixteen

Heterocyclic Systems in Nucleic Acids and Drugs

16.1 Occurrence

Cyclic compounds are either homocyclic or heterocyclic. In the molecules of the homocyclic compounds all ring atoms are carbon. In heterocyclic molecules at least one ring atom is something other than carbon, and the most common hetero atoms are oxygen, nitrogen, and sulfur.

These two families of cyclic systems, of course, are extremely broad. Either on the rings or on side chains we may find carbonyl, alcohol, amino groups—the list is as long as the list of functional groups—that have most and sometimes all of their regular properties. We studied those in earlier chapters. In this chapter we shall look at what is unique about some principal types of heterocyclic ring systems. We touch here only on highlights, and we do so because a huge number of natural substances have heterocyclic rings. We have seen some already. The epoxides (Section 6.37) are heterocyclic. All the sugars in their closed forms are heterocyclic. A few amino acids are, too. Heme (Figure 15.4) contains the pyrrole system. Besides these are a great number of drugs, dyes, polymers, vitamins, and the awesome chemicals of heredity, DNA and RNA, the nucleic acids.

16.2 Five-Membered Heterocyclic Compounds—One-Hetero Atom

The three most common compounds are pyrrole, **1**, furan, **2**, and thiophene, **3**, three with unsaturated rings. (The standard numberings of the ring atoms are shown.)

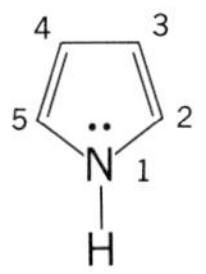

4 3 5 2 N 1 H
1
PYRROLE
(bp 131°)

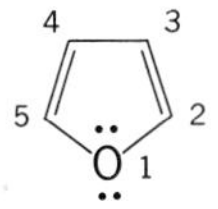

4 3 5 2 O 1
2
FURAN
(bp 32°)

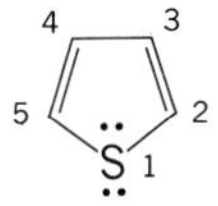

4 3 5 2 S 1
3
THIOPHENE
(bp 84°)

We shall sidetrack only briefly to note that the saturated analogs of these are all known compounds: pyrrolidine, **4**, tetrahydrofuran, **5**, and tetrahydrothiophene, **6**. The first (**4**) is simply a 2° amine, and it has the typical reactions of any aliphatic 2° amine. The second (**5**) is a typical aliphatic ether. The third (**6**) is typically aliphatic, too. Here, then, is the first major point to learn about heterocyclic

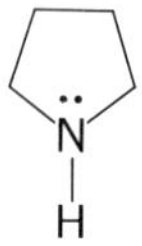

4
PYRROLIDINE
(bp 89°)

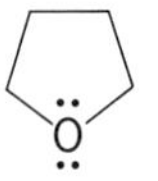

5
TETRAHYDROFURAN
(bp 66°)

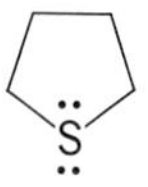

6
TETRAHYDROTHIOPHENE
(bp 121°)

compounds. Those whose rings are five membered or higher and are saturated do not have properties that are significantly different from their aliphatic, open-chain counterparts. Hence, we really need to say little if anything more about them.

The unsaturated, heterocyclic ring systems are the especially interesting types, and we shall consider them almost exclusively in this chapter.

16.3 Some Typical Syntheses

Each parent heterocycle can be made in a variety of special ways. By "parent" heterocycle we mean those whose rings have no substituents. Thiophene, for example, can be made simply by heating butane and sulfur at 650°! Pyrrole can be isolated from coal tar. It also forms by the action of heat on di-ammonium salts of glycaric acids, dicarboxylic acids made from aldohexoses:

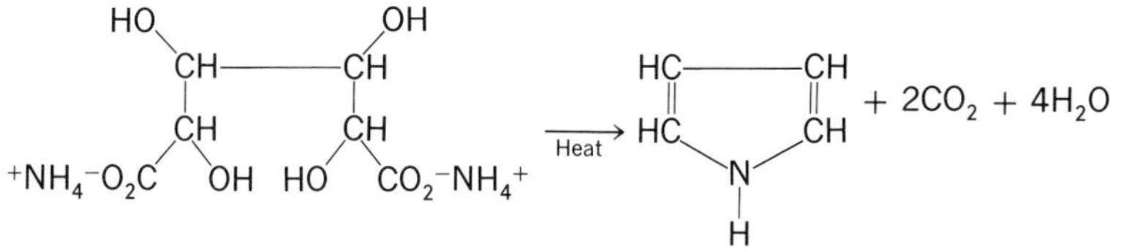

$^{+}NH_4{}^{-}O_2C$–CH(OH)–CH(OH)–CH(OH)–CH(OH)–$CO_2{}^{-}NH_4{}^{+}$ $\xrightarrow{\text{Heat}}$ pyrrole + $2CO_2 + 4H_2O$

A GLYCARIC ACID
(As its ammonium salt)

PYRROLE

The chief commercial route to furan is from the aldehyde, furfural, **7**, which is made from a mixture of pentoses that can be obtained from very inexpensive agricultural wastes: bran, corn cobs, straw and oat hulls. In these materials there are polymeric pentoses called pentosans, and the overall synthesis involves these reactions:

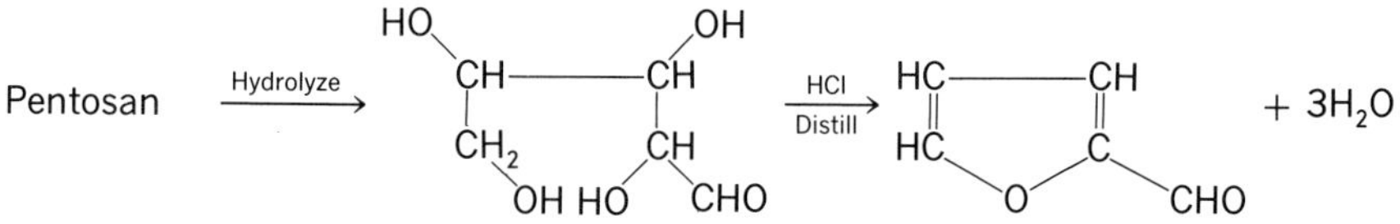

Pentosan $\xrightarrow{\text{Hydrolyze}}$ A PENTOSE $\xrightarrow[\text{Distill}]{\text{HCl}}$ furfural + $3H_2O$

(A poly-pentose)

A PENTOSE

7
FURFURAL

A very general route to a variety of these three heterocycles is from 1,4-dialdehydes. If 1,4-diketones are used then substituted compounds are made:

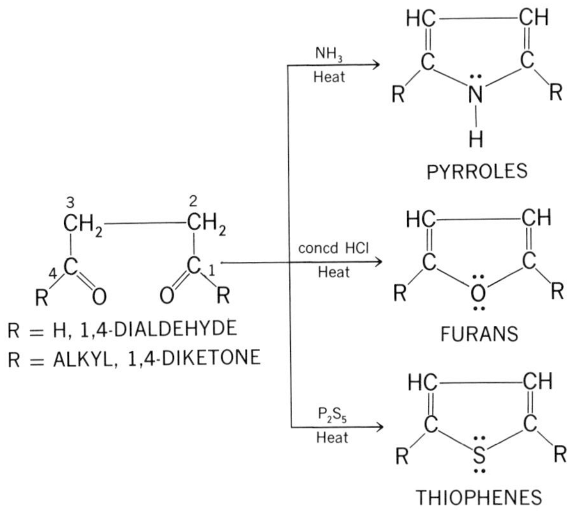

R = H, 1,4-DIALDEHYDE
R = ALKYL, 1,4-DIKETONE

PYRROLES

FURANS

THIOPHENES

16.4 Some Reactions

Thiophene, pyrrole, and furan stand at the borderline between being dienelike and benzenelike (i.e., aromatic). As Figure 16.1 shows, each ring system has an aromatic sextet of electrons, six electrons that are delocalized in a flat ring. The rings, of course, lack the symmetry of benzene. In each ring, one pair of electrons of the sextet comes from a hetero atom. Heteroatoms are electronegative. They attract electrons. Because oxygen is much more electronegative than nitrogen or sulfur, a carbon-oxygen bond is more polarized than either a carbon-nitrogen or a carbon-sulfur bond. Hence, the ring in furan is more unsymmetrical in terms of the distribution of electron density than the ring in thiophene. Of furan, pyrrole and thiophene molecules, those of thiophene are least polarized. They are more benzenelike.

In accordance with this, thiophene gives most of the reactions typical of benzene—substitution rather than addition reactions. The following illustrate this:

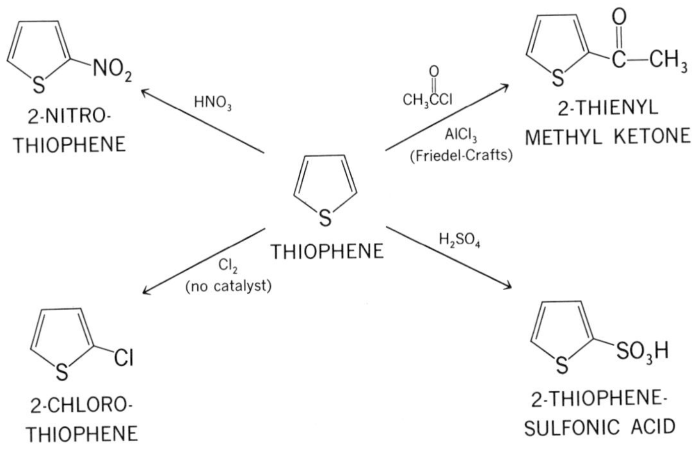

2-NITRO-THIOPHENE

2-THIENYL METHYL KETONE

THIOPHENE

2-CHLORO-THIOPHENE

2-THIOPHENE-SULFONIC ACID

While the kinds of reactions of thiophene are the same as with benzene, their ease is different. The thiophene ring is much more reactive than is the benzene ring

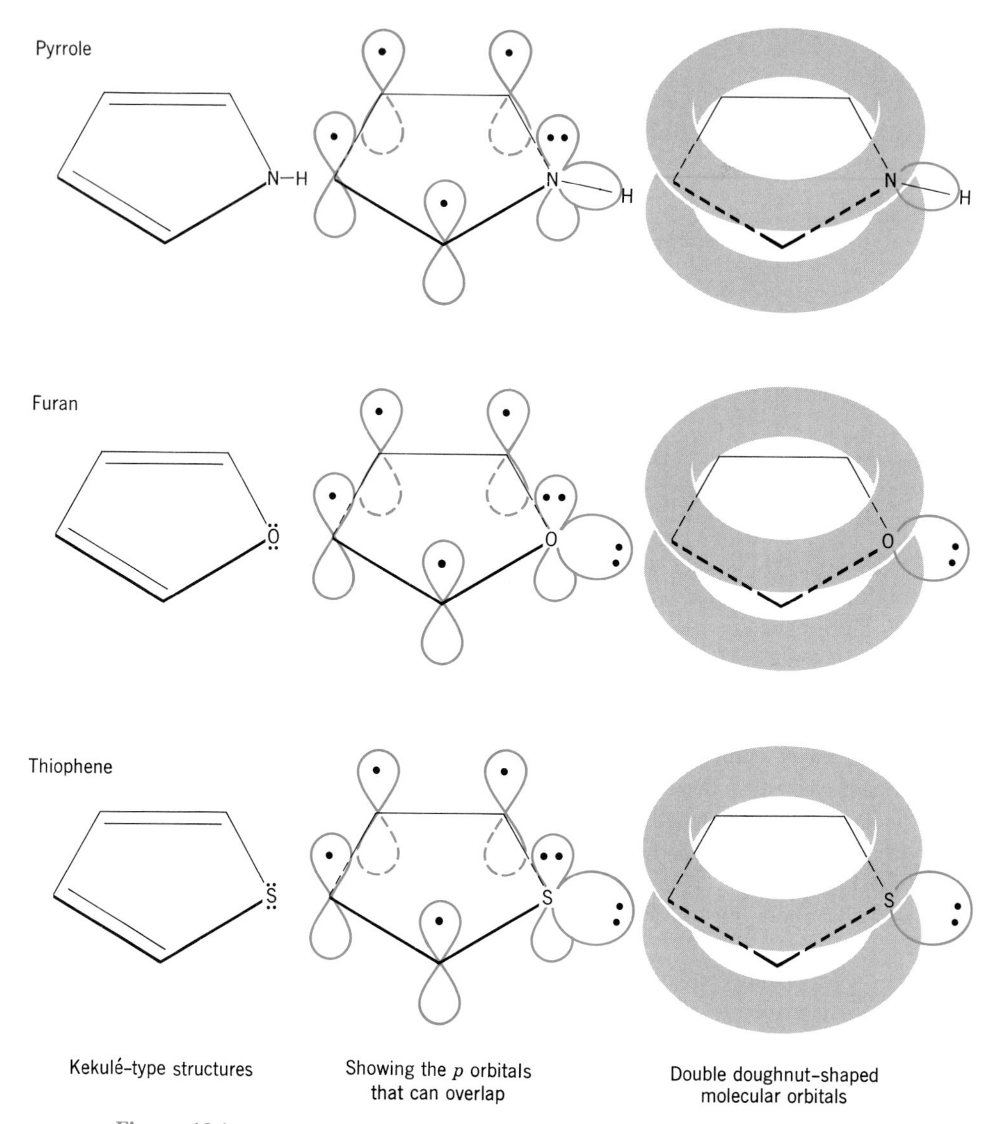

Figure 16.1
Each carbon of each ring provides a *p* orbital. There is one electron in each of these for a total of four. The remaining two to make up the "aromatic sextet" are provided by a *p* orbital holding two electrons furnished by the heteroatom. Each heteroatom still has a pair of unused electrons, except for nitrogen.

in these substitution reactions. Quite mild conditions work. To cite just one example, thiophene is sulfonated at room temperature by sulfuric acid (and it is best to use an inert solvent); benzene is efficiently sulfonated only at elevated temperatures. The ease of a reaction is partly related to what else is on the ring. The same groups that deactivate the benzene ring (Section 4.16) deactivate the rings of thiophene, pyrrole, and furan. Activators for benzene are activators for these heterocycles. We shall not pursue this further, nor shall we go into the reasons why thiophene, for example, substitutes in the 2- rather than the 3-position. (The reasons are not difficult; your instructor may discuss them if time allows.)

Pyrrole, in contrast to thiophene, is very sensitive to the strong acids that are used for many electrophilic substitution reactions. Polymeric red resins form when pyrrole is mixed with mineral acids. The pyrrole ring does undergo those aromatic substitutions that can be accomplished by weak electrophilic reagents in neutral or basic conditions. To cite just one example, pyrrole rings couple with benzene-diazonium salts (Section 8.15):

$$\text{PYRROLE} + C_6H_5N_2^+Cl^- \longrightarrow \text{2-PHENYLAZOPYRROLE } (N{=}N{-}C_6H_5) + HCl$$

The pyrrole ring, therefore, resembles the highly reactive benzene ring in a phenol, only it is even more reactive. (And, like phenol, pyrrole darkens on standing.)

There is an interesting consequence of the fact that the unshared pair of electrons on nitrogen in a molecule of pyrrole is delocalized. It is the very low basicity of pyrrole. For an amine to be basic (Section 8.8) its pair of electrons on nitrogen must be available to a proton. This is not the case with pyrrole. Compared with aliphatic amines, pyrrole is a neutral compound. Molecules of its reduced form, pyrrolidine, however, have an unshared pair of electrons localized on nitrogen. Hence, it is basic, like any 2° aliphatic amine. (The K_b values are diethylamine, 9.6×10^{-4}; pyrrole, 2.5×10^{-14}; and pyrrolidine, 1×10^{-3}.)

Furan is the least benzenelike compound of the three simple heterocycles so far mentioned. Acids cause it to change to a resin. It can be brominated at low temperature to 2-bromofuran:

$$\text{FURAN} + Br_2 \xrightarrow{0^\circ} \text{2-BROMOFURAN} + HBr$$

As this reaction illustrates, by being very careful about the temperature of the reaction furan will give substitution reactions typical of aromatic systems. The difference is that the ring is so reactive that the reactions are hard to control and side reactions also become serious.

Our purpose here is not to survey all the reactions of these substances but simply to demonstrate some general trends. These three parent heterocycles do have aromatic character. They all give aromatic substitution reactions, mostly at their 2-positions. All, however, are much more reactive than benzene.

16.5 Other Unsaturated Heterocycles with Five-Membered Rings

There are hundreds of ring systems with heteroatoms. Many have two or more heteroatoms in their rings. Many have extra rings fused to the parent heterocycle. Just a few examples where the heterocyclic ring is still five membered are the following.

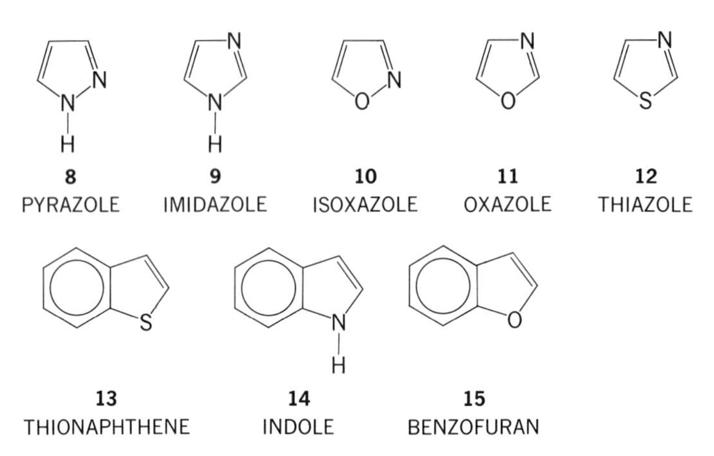

The imidazole system is found in histidine, an amino acid (Table 15.1). The indole ring is a part of tryptophan, and it occurs widely among some of nature's most physiologically active compounds. LSD is an example (Section 16.18). The reduced form of thiazole is part of penicillin. A reduced form of benzofuran is part of the molecules of morphine, codeine, and heroin (Section 16.16).

16.6 Unsaturated Heterocycles with Six-Membered Rings

The most common families whose parent compounds are also well known are those with heterocyclic nitrogen atoms. Of those with one ring-nitrogen, pyridine, **16,** and its close relatives, quinoline, **17,** and isoquinoline, **18,** are the most important.

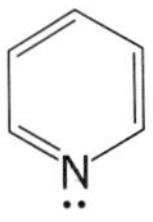

N N N

16 PYRIDINE (bp 116°)

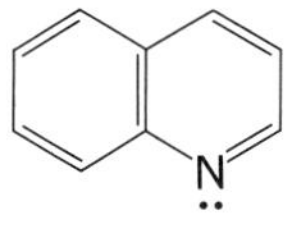

17 QUINOLINE (bp 238°)

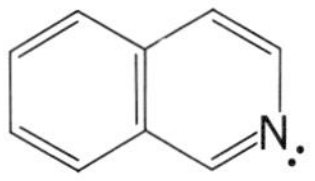

18 ISOQUINOLINE (bp 240°)

All three ring systems are found in various alkaloids. Molecules of nicotine include the pyridine ring; those of quinine, the quinoline system; and those of papaverine, the isoquinoline ring. (Alkaloids are further described in Section 16.14.)

16.7 Chemical Properties of Pyridine, Quinoline and Isoquinoline

These three are aromatic compounds. However, strikingly unlike the five-membered heterocycles, they are vastly more resistant to ring substitution than even benzene itself. We can make bromobenzene by heating benzene and bromine together in the presence of an iron salt. To brominate pyridine, we must get bromine and pyridine into the vapor state at a temperature of 300°. 3-Nitropyridine is made only in low yield when pyridine is heated with sulfuric acid and fuming nitric acid in the presence of iron salts at 300°! In quinoline and isoquinoline it is the benzene rings that are attacked, not the nitrogen-containing pyridine rings.

The source of these properties is the ring nitrogen. Note in **16** that the nitrogen is bound to just two other atoms, not to three as in amines, or even as in pyrrole. Two electrons of nitrogen are delocalized, but two are not. The latter pair, therefore, is available to bind an acid—a proton, for example, or an iron ion. That is why these three heterocycles are basic. That is why nitration is so difficult. The medium is strongly acidic; the ring nitrogen cannot help but accept a proton; and this makes the ring electron poor—positively charged. It no longer is attractive to an electrophile, which is also electron poor. Hence, an electrophilic substitution is difficult.

One important lesson here is that while unsaturated heterocyclic compounds are different in many ways from other cyclic systems, their actual properties finally do make sense in terms of the same principles we use for other systems.

16.8 Other Nitrogen Heterocycles

Of the huge number we could mention, we shall consider only two families, the pyrimidines and the purines. Both are important in the chemicals of heredity, DNA and RNA (Section 16.11). Pyrimidine, **19,** is a pyridinelike compound, only even more stable toward electrophilic substitution. Purine, **20,** has two rings, one pyrimidinelike, the other imidazolelike (cf., **9**).

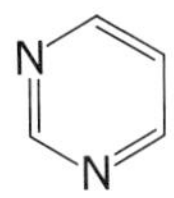

19
PYRIMIDINE

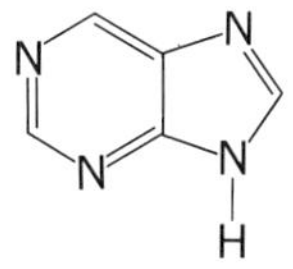

20
PURINE

Barbituric acid, **21,** the parent compound for the barbiturate drugs (Section 16.9), has the pyrimidine ring system, at least in its "enol" form, **21**(a). In this form it is actually like a phenol, only it is a stronger acid than phenol. It is, in fact, slightly stronger than acetic acid. Its "keto" form, **21**(b), is shown because some barbiturates are at least partly in that form, and because their structures are often shown that way.

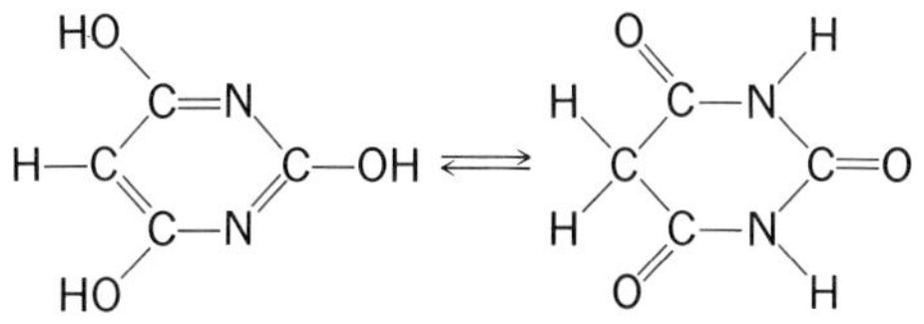

(a)
"ENOL" FORM

(b)
"KETO" FORM

21
BARBITURIC ACID

16.9 Barbiturates

These are the most widely prescribed and the most frequently abused of a large family of drugs that act as sedatives. They are made by a very simple reaction, the condensation of urea (or thiourea) with a disubstituted malonic ester by the action of sodium ethoxide:

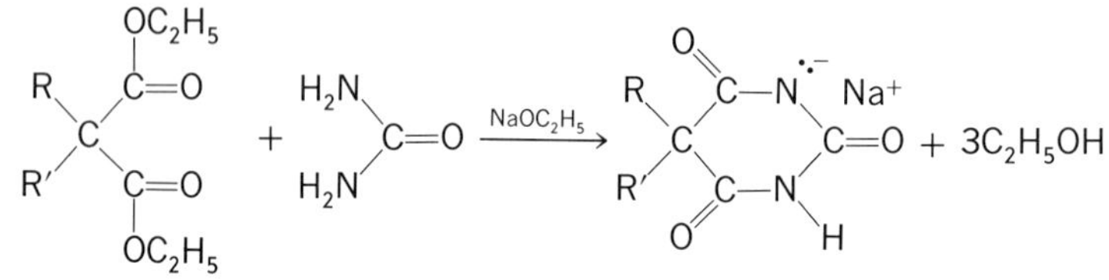

$$\underset{\text{DISUBSTITUTED MALONIC ESTER}}{RR'C(C(=O)OC_2H_5)_2} + \underset{\text{UREA}}{(H_2N)_2C{=}O} \xrightarrow{NaOC_2H_5} \underset{\substack{\textbf{22}\\ \text{A SODIUM BARBITURATE}}}{RR'C(C(=O))_2N^{-}Na^{+}C(=O)NH} + 3C_2H_5OH$$

Because one ring carbon has two alkyl groups, the ring cannot become fully benzenelike (aromatic), as in **21**(a). These disubstituted barbituric acids are still more acidic than phenol, but their sodium salts are normally written in the manner of structure **22**. Both the sodium salts and the free acids are used medicinally, but the salts are much more soluble in water and therefore easier to administer. Some examples are shown in Table 16.1.

In properly supervised doses, the barbiturates act mildly on the central nervous system to cause both breathing and the heart beat to slow. The system's response is sleep. At higher doses, the results are like those of alcohol drunkenness. (In street lingo they are called "goof balls" or "downers" or simply "barbs.") At still higher doses breathing and heart action may be so slowed that death results. That is what happens in suicides caused by an overdose of "sleeping pills." The combination of alcohol and barbiturates can be especially lethal—it's a danger greater than the sum of the individual effects. Many accidental deaths have happened because people who have had too much to drink decide to take barbiturate sleeping pills.

Table 16.1
Barbiturates

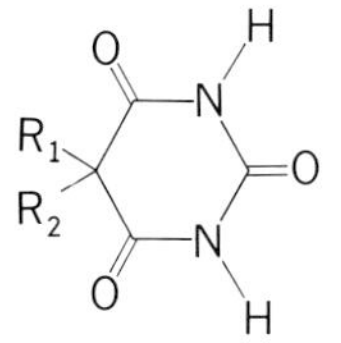

BARBITURIC ACIDS (ring: R_1R_2C, two C=O, two N–H, C=O) | SODIUM BARBITURATES (ring: R_1R_2C, two C=O, N^-Na^+, N–H, C=O)

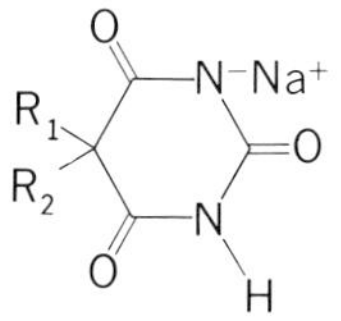

R_1 =	R_2 =	Name of Acid	Name of Salt
H—	H—	Barbituric acid	Sodium barbiturate
CH_3CH_2—	C_6H_5	Phenobarbital (Luminal®)	Phenobarbital sodium
CH_3CH_2—	$CH_3CH_2CH_2CH(CH_3)$—	Pentobarbital (Nembutal®)	Pentobarbital sodium
$CH_2{=}CHCH_2$—	$CH_3CH_2CH_2CH(CH_3)$—	Secobarbital (Seconal®)	Secobarbital sodium

Marilyn Monroe was an example. This phenomenon—two drugs being in combination more dangerous than one would predict from their individual properties—is fairly common. We say that one drug *potentiates* the other—makes it more potent. *Potentiation* is also observed among pesticides and poisons.

Nucleic Acids and the Chemicals of Heredity

16.10 The Main Types of Nucleic Acids

The nucleic acids are one of the great families of biochemical polymers. Their monomers are called nucleotides of which there are five principal ones. The nucleotides can be hydrolyzed, and three kinds of products are produced: a pentose sugar (either ribose or deoxyribose), phosphoric acid, and a small set of pyrimidine and purine bases. Figure 16.2 outlines these relations.

Nucleic acids built on ribose are called ribonucleic acids or RNA. Those based on deoxyribose are called deoxyribonucleic acids or DNA. Molecules of either kind are high-formula-weight polymers having long chains. Sections of DNA molecules are genes, the individual units of heredity.

16.11 The Pyrimidines and Purines in Nucleic Acids

There are principally five: two purines—adenine (or A) and guanine (or G)—and three pyrimidines—thymine (T), uracil (U), and cytosine (C). One-letter symbols are used here in the same way we used three-letter symbols for amino acids (Table 15.1). A, T, G, and C are obtained from DNA; A, U, G, and C from RNA. (U takes the place of T in going from DNA to RNA.) Except for adenine each one has at least one oxygen function on a ring, and this introduces a complication. We shall use uracil to illustrate. Uracil can be written as a fully aromatic, dihydroxypyrimidine, **23**(a) (where in the Kekule structures used we can see two enol systems) or as a partly keto form, **23**(b) or the full keto form, **23**(c).

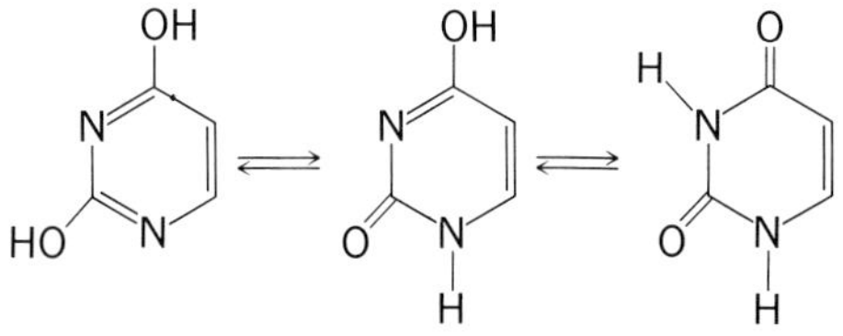

(a) (b) (c)

23

URACIL

The forms shown in Figure 16.2 are those believed to be present in nucleic acids. These forms permit two things to happen. First, each has at least one hydrogen attached to a ring nitrogen. In nucleic acids, these hydrogens have been removed, replaced by attachments between their ring nitrogens and the nucleic acid "backbones." Second, these forms leave the molecules ways both for accepting and donating hydrogen bonds between units of the purines and pyrimidines. Perhaps nowhere in all nature does the existence of hydrogen bonds seem to be so dramatically important as in the nucleic acids.

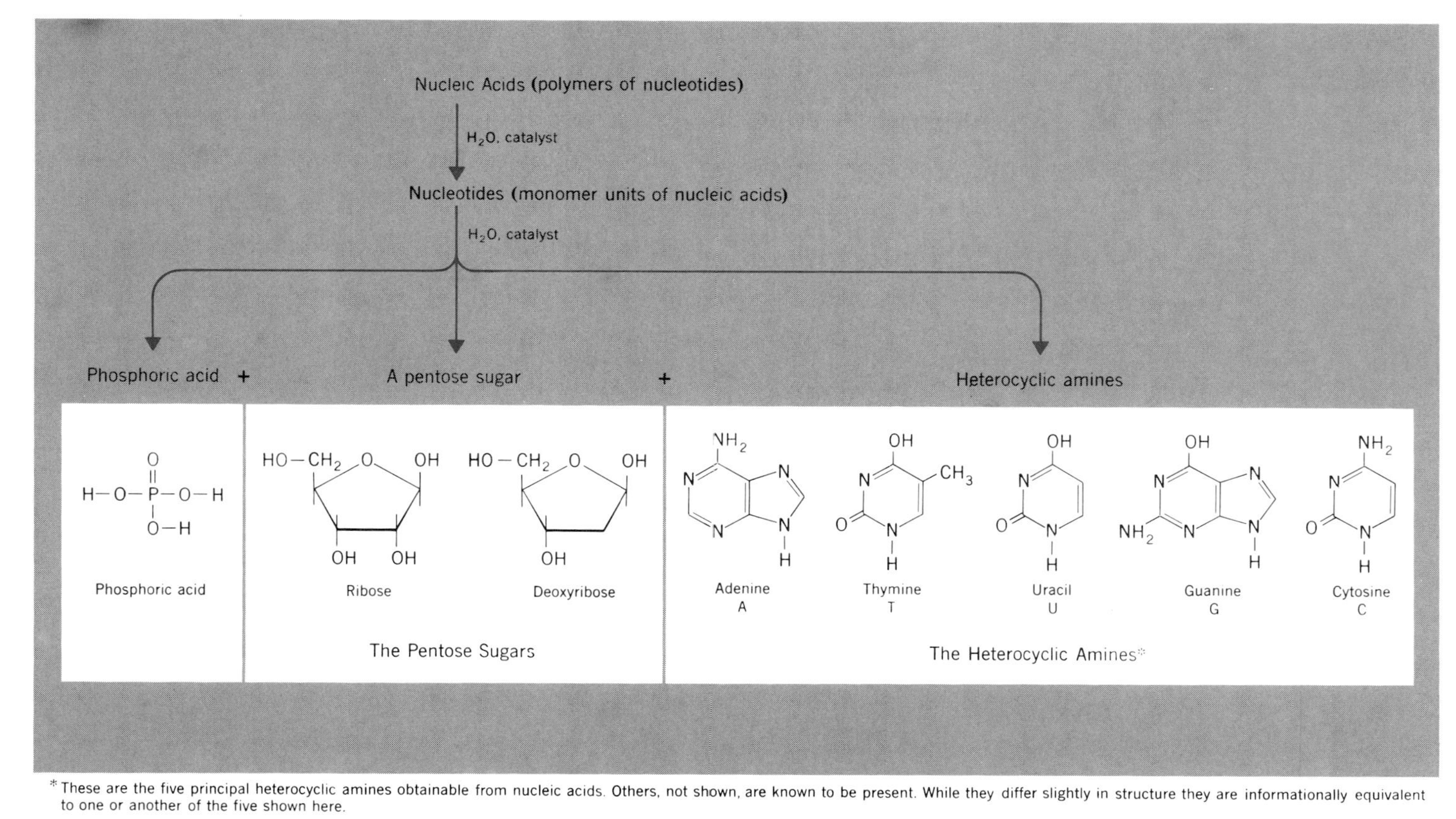

* These are the five principal heterocyclic amines obtainable from nucleic acids. Others, not shown, are known to be present. While they differ slightly in structure they are informationally equivalent to one or another of the five shown here.

Figure 16.2
Hydrolysis products of nucleic acids.

16.12 The DNA Double Helix. Crick-Watson Theory

The structures of four nucleotides are shown in Figure 16.3. Each is an ester of phosphoric acid and deoxyribose, the latter modified by the attachment of a purine or a pyrimidine base. On paper (as distinguished from how the cell does it), the formation of a DNA strand—the polymerization of nucleotides—is simply the splitting out of water to make new phosphate-deoxyribose ester links. This is also shown in Figure 16.4. A simpler view of the main features of the DNA polymer is shown in Figure 16.4.

When DNA is hydrolyzed and the heterocyclic bases analyzed, there is found a 1:1 ratio of adenine to thymine (A to T) and a 1:1 ration of cytosine to guanine (C to G). This fact, first noted by E. Chargaff, and many other facts—especially X ray studies by Maurice Wilkins—led Francis Crick and James Watson to propose a structure for DNA that could be used to explain how it stored and transmitted biological information. The justly famous Crick-Watson theory opened floodgates of research and established what is now known as the field of molecular biology.

The functional groups and molecular geometries (both bond angles and bond lengths) of adenine and thymine are such that the two fit to each other and hold by hydrogen bonds. The same can be said about the pair cytosine and guanidine. Figure 16.5 illustrates how this works.

Crick and Watson proposed that DNA in cells normally exists as a double helix. Two nucleic acid strands, running in opposite directions, align themselves side by side. The bases on one strand are complementary to the bases directly opposite them on the other strand. By "complementary" we mean that they can pair with each other by hydrogen bonds and thereby the two strands cling together. They then twist into a right-handed helix, as illustrated in Figure 16.6.

According to the Crick-Watson theory, the genetic code is the sequence of heterocyclic bases along the DNA chain. A set of three side-chain amines in a row is one code word. The code has been deciphered, and each code word is translated by the cell into a particular amino acid. The series of code words on DNA specifies an analogous series of amino acids in a polypeptide that the cell makes, a polypeptide that is or becomes part of an enzyme. Each gene, therefore, can specify one polypeptide. The sequence, in broad outline, is:

$$\text{DNA} \xrightarrow[\text{(message is transferred)}]{\text{Transcription}} \text{RNA} \xrightarrow{\text{Translation}} \text{Protein} \xrightarrow{\text{Catalysis}} \text{Other compounds and reactions in cells}$$

The side-chain amine sequence in DNA determines a complementary sequence of sidechains in RNA. This controls an amino acid sequence in polypeptides.

When a cell is about to divide all its genes must be duplicated. Each new daughter cell must get exact copies of the original genes. The process whereby genes are duplicated is called replication. In broad outline, it involves an uncoiling of a double helix. Along each DNA strand a complementary strand is synthesized, as sketched in Figure 16.7.

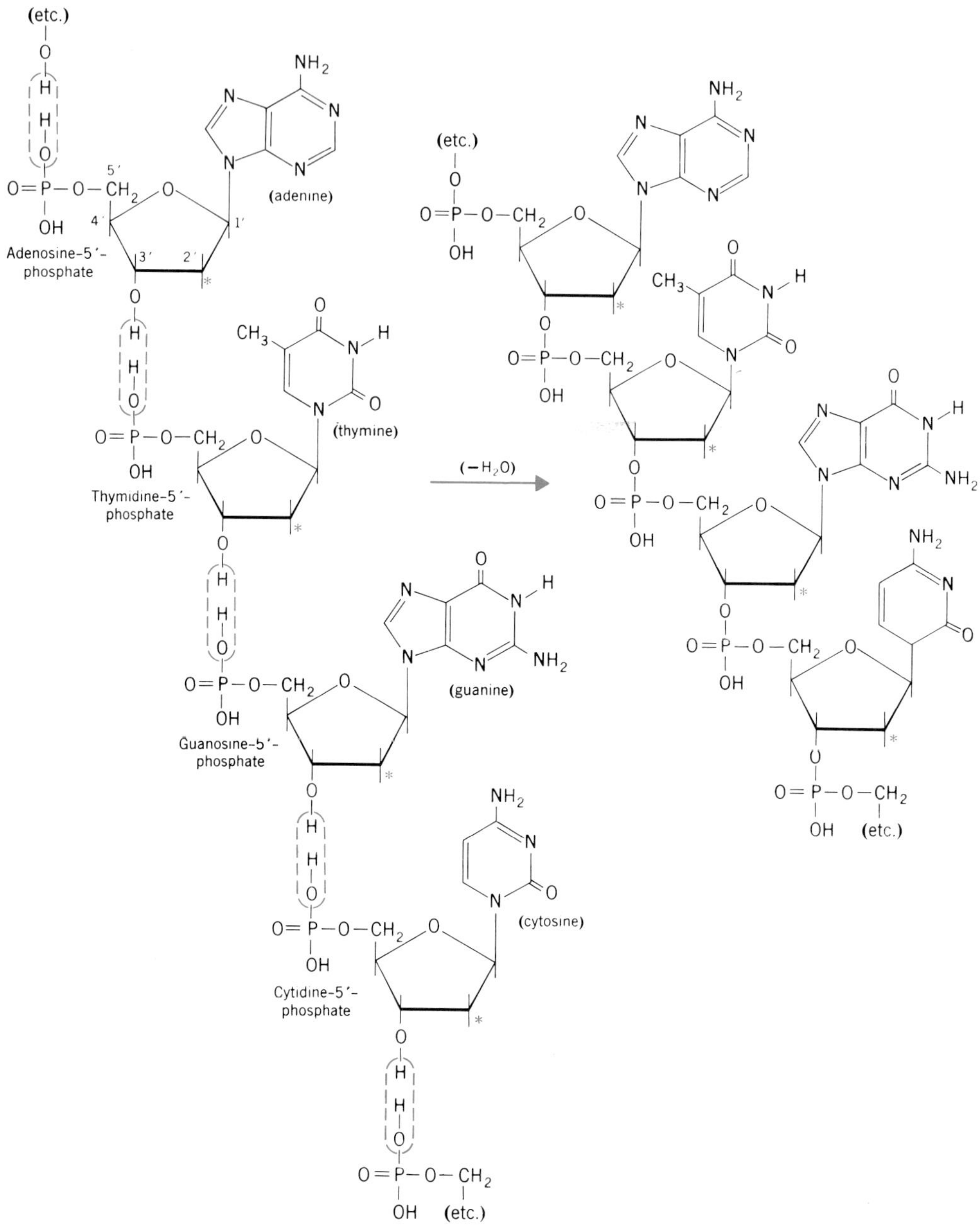

Figure 16.3
The relation of nucleotides to nucleic acids. Shown here are the chief nucleotides for a segment of a DNA chain. The splitting out of water between them joins the chain. (The asterisks mark sites where there would be other —OH groups if we wanted to visualize the formaton of a segment of an RNA chain. The other variation would be the substitution of uracil for thymine.)

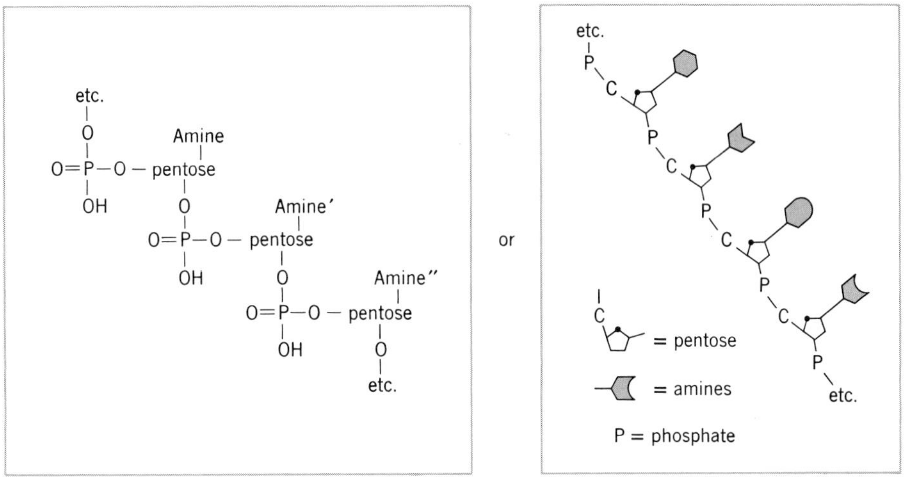

etc.
O
O=P—O — pentose
OH
Amine
O
O=P—O — pentose
OH
Amine′
O
O=P—O — pentose
OH
Amine″
O
etc.

or

etc.
P
C
P
C
P
C
P
C
P
etc.
C
= pentose
= amines
P = phosphate

Figure 16.4
Two "condensed" ways of depicting the main structural features of nucleic acids.

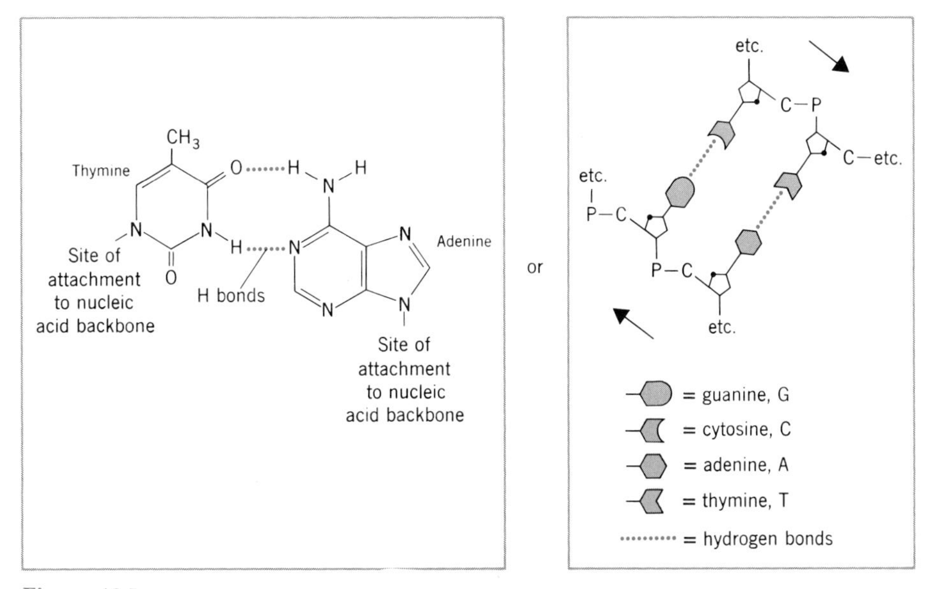

Figure 16.5
Thymine "pairs" with adenine; T with A. Guanine pairs with cytosine; G and C. Adenine can also pair with uracil; A with U. Pairing involves hydrogen bonds, as illustrated here.

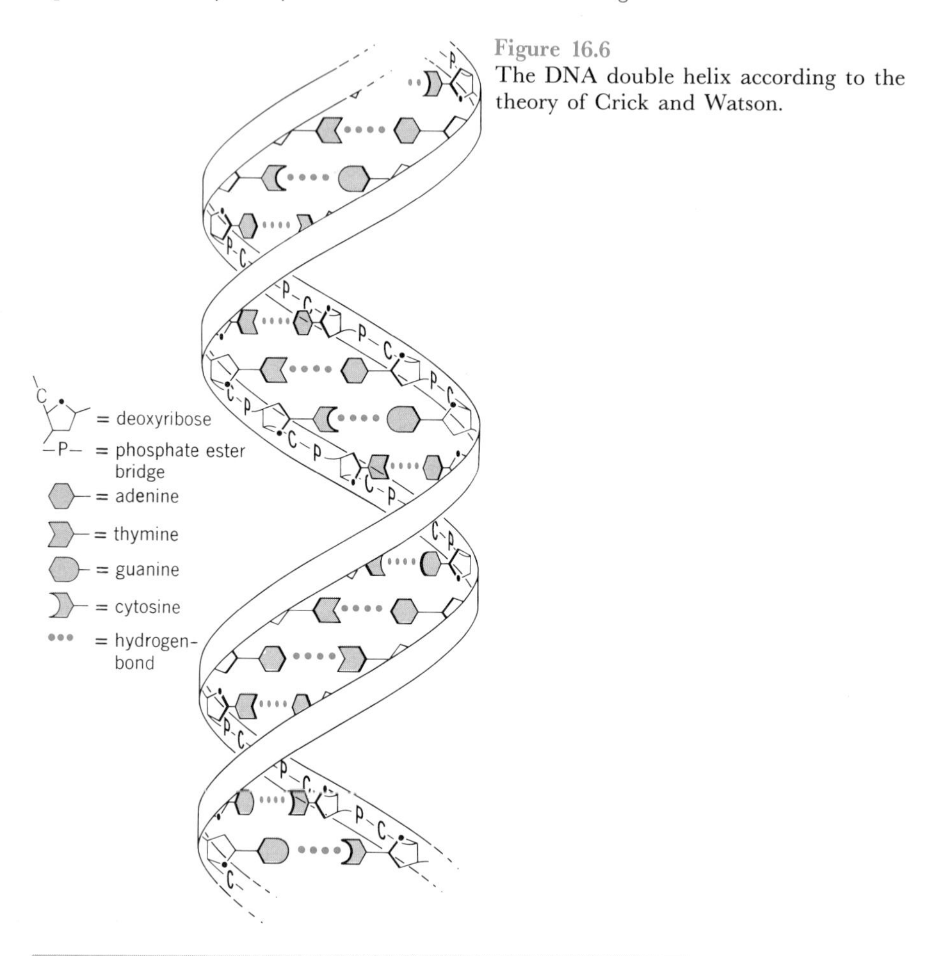

Figure 16.6
The DNA double helix according to the theory of Crick and Watson.

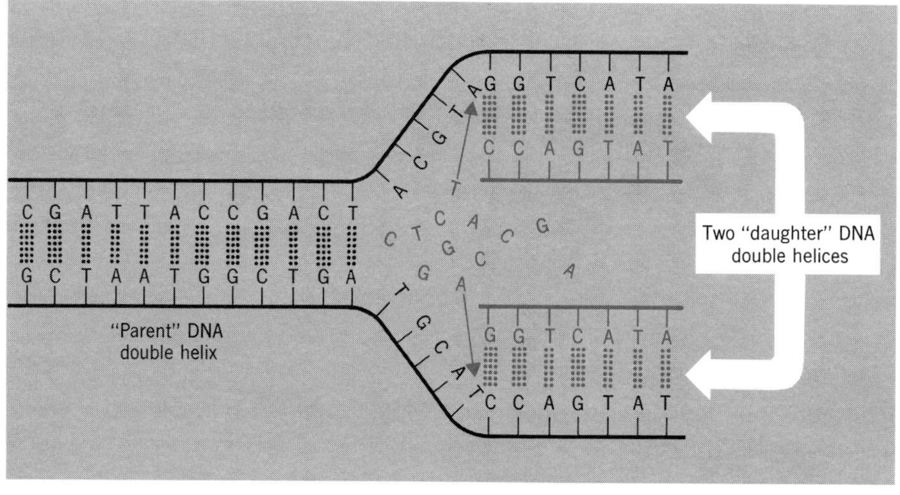

Figure 16.7
Replication of DNA. In this very highly schematic diagram we see the basic plan. The monomer units, shown here simply by letters, join to each other in an order determined by the complementary order on the untwining DNA double helix.

16.13 Errors in Replication

There is no chemical in the body whose safety should be more carefully guarded than DNA. Many poisons act by damaging this substance. X rays, gamma rays, cosmic rays, and other atomic radiations—all of which are ionizing radiations—also damage DNA. If the damage is relatively slight, the DNA is altered in such a way that when it replicates, copying errors occur. The new DNA for the daughter cells is different. A mutation has occurred. It might or might not lead to serious trouble because the system does have ways to repair DNA. But it could lead to a tumor, to a cancer, or to a birth defect if it occurs in a fetus. One of the fears, as yet not fully founded, of some of the alkaloid drugs is that they may be able to damage DNA.

16.14 Alkaloids

The alkaloids are physiologically active compounds, obtained from plants, whose molecules are amines or derivatives of amines. Most are heterocyclic, and their rings vary both in size and degrees of unsaturation. In most the heteroatom is nitrogen and is the part of the molecule responsible for the basic properties. ("Alkaloid" means "alkali-like.") They exert their physiological action in one or more ways on the nervous system. (Long-term effects may involve other systems in the body.) They include drugs of inestimable value in relieving human suffering. Morphine (for severe pain) and quinine (antimalaria) are but two examples. They also include drugs whose abuse is the cause of enormous human misery. Cocaine is an example. Heroin is another example, although it actually is a synthetic derivative of a natural alkaloid.

Our goals in this unit are to survey some of the principal alkaloids and, more importantly, to see examples of an important fact in drug chemistry and drug application. That fact is that very small variations in molecular structure can give huge changes in drug action. The main strategy for rational drug research (as opposed to the accidental discovery of new drugs) is to use the occasional discovery of a new kind of drug to suggest the new lines of investigation. The discovered drug may (and usually does) have unfortunate side effects. The research team then asks for substances whose molecules differ from the "parent" in small ways. These are tested by other members of the team. At least two animals have to be used in the testing, and various ways to administer the drug must be tried. Finally human volunteers are sought. Millions of dollars of research go into the development of one new drug, and during the effort scores of compounds may be tried, tested, and abandoned. The team effort brings together not only pharmacologists, but also synthetic organic chemists, toxicologists, zoologists, microbiologists, physicians, medical statisticians, computor programmers, and the like, to say nothing of investment teams.

16.15 β-Phenylethylamines

These are borderline members of the alkaloid family; not all are from plants, and the basic amine function is not part of a ring. Nonetheless, they are usually treated in the alkaloid group. Moreover, they illustrate how drug activity can vary dramatically with structure. Important examples are in Table 16.2.

Table 16.2
Some β-Phenylethylamine Drugs

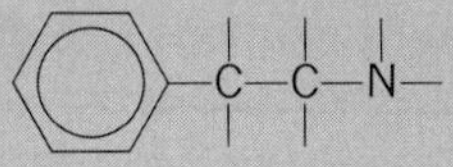

The parent molecular framework.[a]	
Two hormones HO, HO—C$_6$H$_3$—CH(OH)—CH$_2$—NH—CH$_3$ (−)-Epinephrine (adrenaline) HO, HO—C$_6$H$_3$—CH(OH)—CH$_2$—NH$_2$ (−)-Norepinephrine (arterenol)	**Two bronchial aids** C$_6$H$_5$—CH(OH)—CH(CH$_3$)—NH—CH$_3$ (−)-Ephedrine HO—C$_6$H$_4$—CH(OH)—CH$_2$—NH—CH$_3$ (−)-Neo-Synephrine
Two "pep pills"—amphetamines C$_6$H$_5$—CH$_2$—CH(CH$_3$)—NH$_2$ (±)-Benzedrine (+)-Dexedrine ("speed") C$_6$H$_5$—CH$_2$—CH(CH$_3$)—NH—CH$_3$ (+)-Methedrin ("crystal," "meth")	**An antibiotic** NO$_2$—C$_6$H$_4$—CH(OH)—CH(CH$_2$OH)—NH—C(=O)—CHCl$_2$ Chloramphenicol (chloromycetin) **An hallucinogen** (CH$_3$O)$_3$C$_6$H$_2$—CH$_2$—CH$_2$—NH$_2$ Mescaline

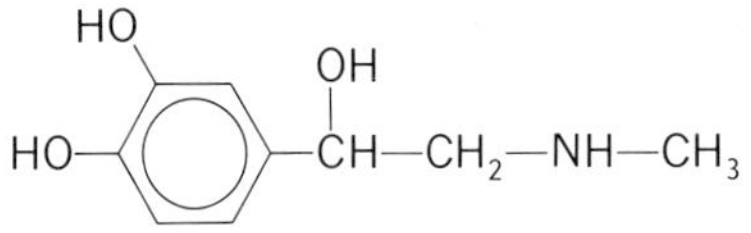

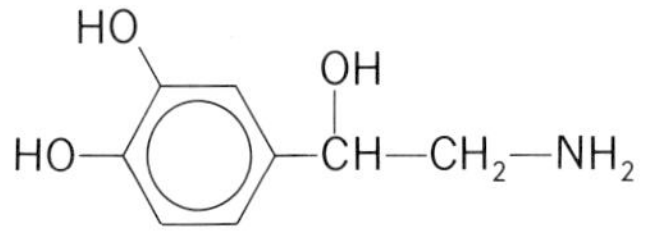

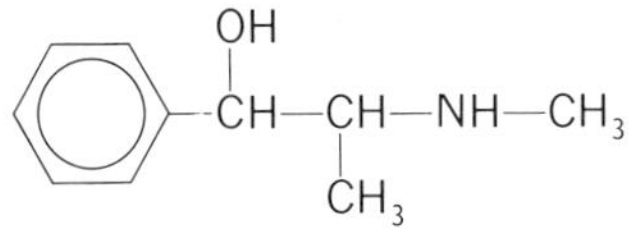

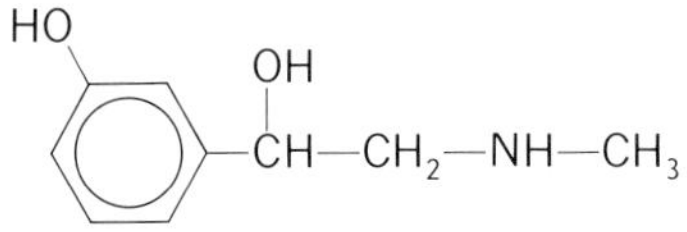

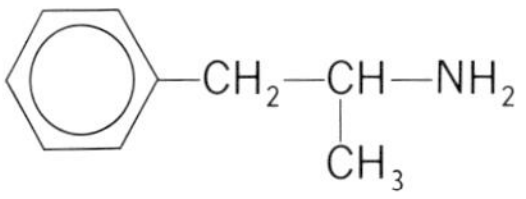

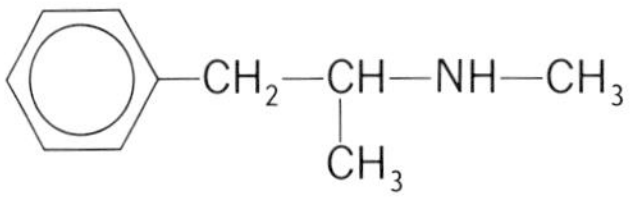

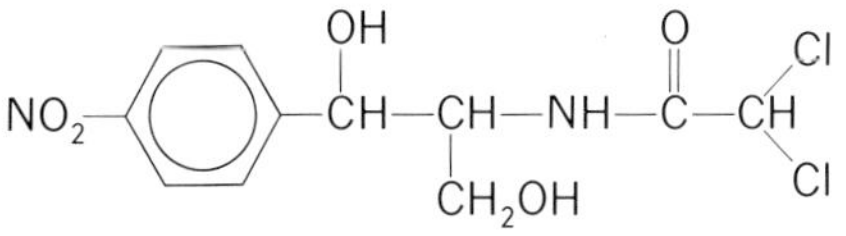

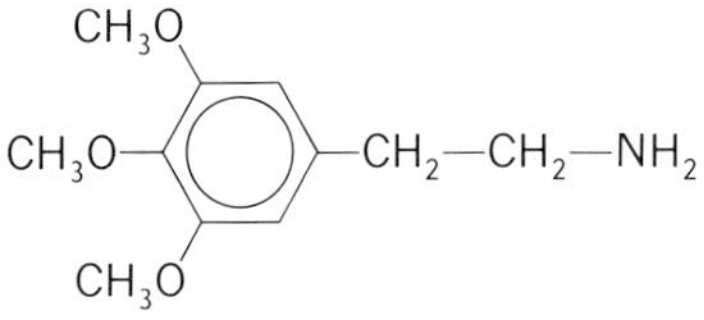

[a] The drugs shown here are usually obtained as hydrogen sulfate or hydrogen chloride salts.

Epinephrine and norepinephrine are two hormones produced in the adrenal gland. Both are levorotatory; they are roughly 20 times more active than their dextrorotatory enantiomers. Thus, even a variation in "handedness" greatly affects activity. They act on the sympathetic nerve fibers to strengthen the heartbeat, raise the blood pressure, and dilate the pupils of the eyes.

The natural actions of these two hormones, which are released in times of stress, fright, or awareness of danger, are to increase your alertness and make you feel less fatigued for awhile. In short, they help get you ready to handle emergencies or other kinds of stress, such as depression. Some drugs mimic these effects. Unfortunately, especially when abused, they also bring on sleeplessness, increased irritability followed, ironically, by mental depression and fatigue. The **amphetamines** are notorious examples (Figure 16.8). Medically, they are classified (and prescribed) as stimulants and antidepressants. Sometimes they are prescribed in weight-control programs or to treat certain mild mental disorders. While they are legally available only by prescription, they are available illegally in enormous amounts. Billions of amphetamine pills are made annually. Known generally as "pep pills" or "uppers," individual compounds are called "bennies" (Benzedrine), "dexies" (Dexedrine), or "speed" (methedrine). (Cf. Table 16.2.) The dangers of abuse include suicide, hostility and belligerence, paranoia and hallucinations, and hepatitis. Some scientists consider them vastly more dangerous than heroin.

Also included in Table 16.2 are substances long used to treat bronchial asthma. Ephedrine, obtained from a Chinese herb, is one. Neo-Synephrine® used in nose drops, is synthetic.

Mescaline ("peyote," "mescal") is isolated from the mescal button, a growth on the top of the peyote cactus. Indians of northern Mexico and the southwestern United States have long used it in religious ceremonies. Medically, it is one of the family of **hallucinogens,** substances that cause illusions of time and place, make unreal experiences or things seem real, or distort the qualities of objects.

Chloromycetin® (chloramphenicol), also in Table 16.2, is unusual in that it contains both a nitro group and chlorine, neither of which had ever been observed before in naturally occuring organic compounds. It is an antibiotic effective against Rocky Mountain fever and typhus.

16.16 Opium Alkaloids

When unripe seed capsules of the opium poppy (papaver somniferum) are scored with a knife, a milky sap oozes out and dries. Roughly 10% is morphine, **24** (Figure 16.9). Medically, morphine is used as a painkiller, and for severe pain there is essentially nothing as effective. Overuse unfortunately leads to addiction. Codeine, **25,** is another (but trace) alkaloid in poppy sap. It is used as a cough suppressant in cold medicines.

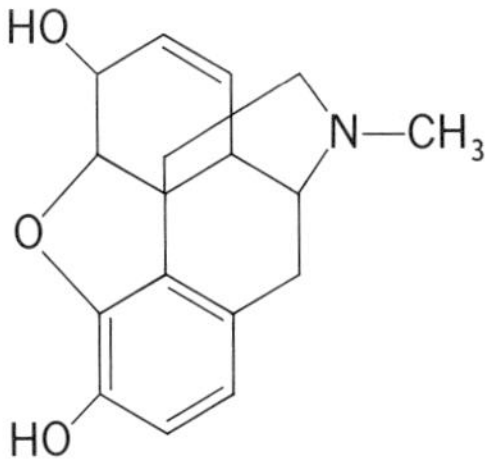

24 MORPHINE

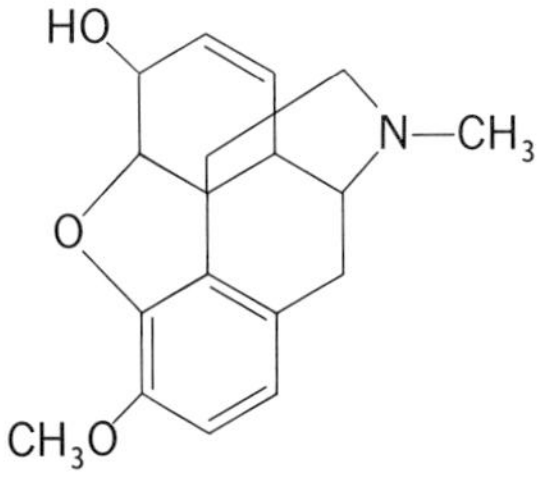

25 CODEINE

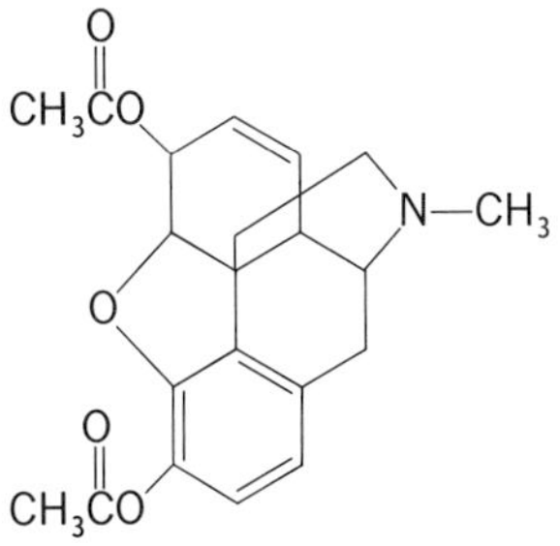

26 HEROIN

Heroin, **26,** is made from morphine by acetylating the two —OH groups. Known by a variety of slang names—horse, H, smack, junk, and dope, to name just a few—it has no legal use in the United States. While it produces a brief period of euphoria, it also causes lethargy, confusion, nausea, constricted pupils, and respiratory depression. Those who become addicts face great difficulties in stopping.

Stimulants

Called by such names as "speed," "dexies," "pep pills," "ups" "A's," "bennies," "drivers," "crossroads," "footballs," etc.

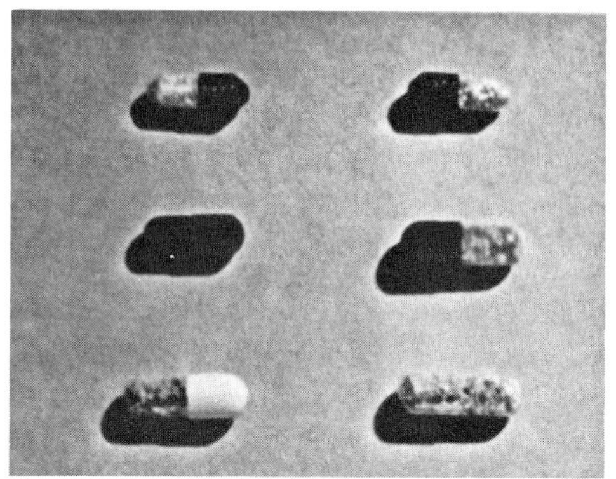

AMPHETAMINE CAPSULES

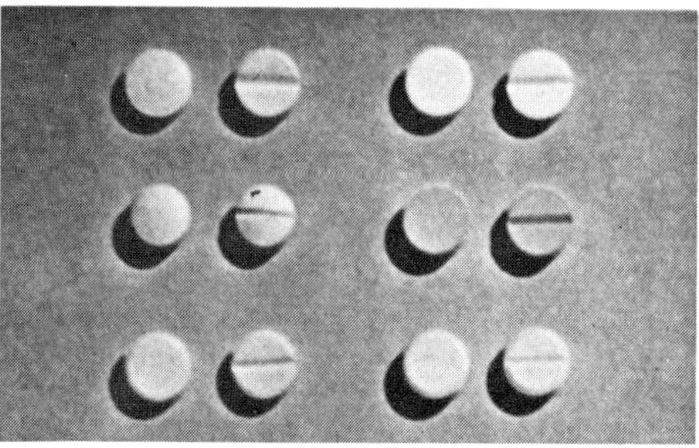

AMPHETAMINE TABLETS

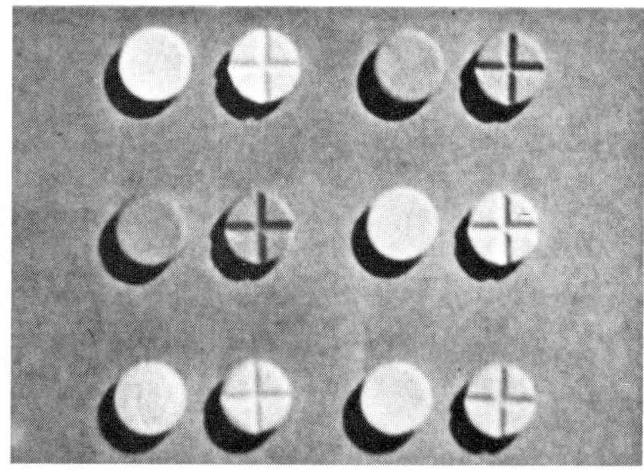

AMPHETAMINE TABLETS

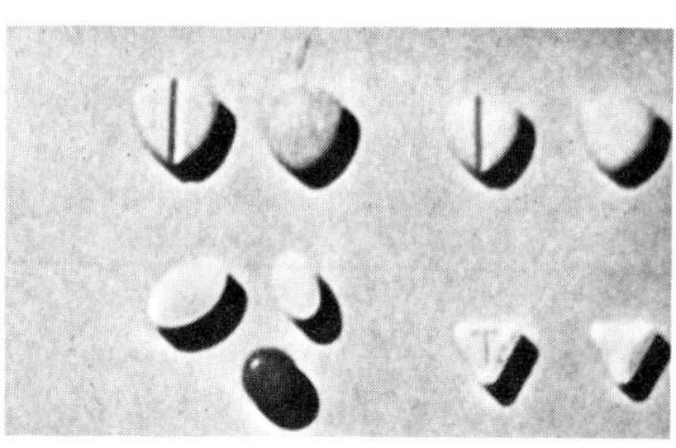

AMPHETAMINE TABLETS

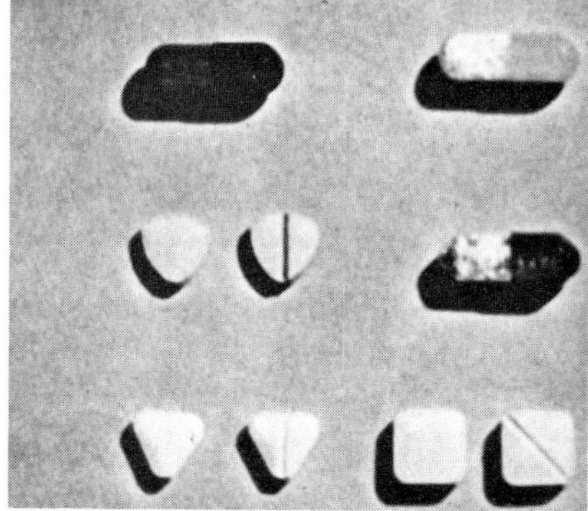

AMPHETAMINE–BARBITURATE COMBINATION

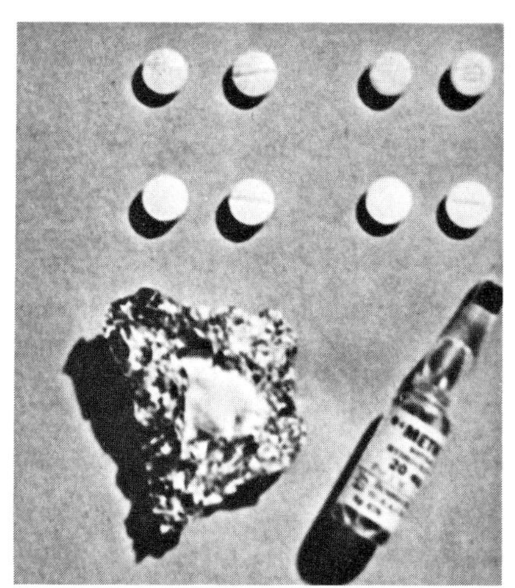

DOSAGE FORMS OF METHAMPHETAMINE

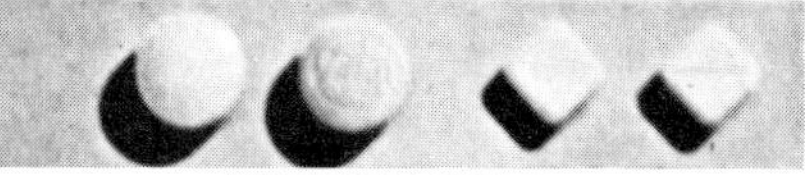

PHENMETRAZINE TABLETS

Figure 16.8
Amphetamines are among the most widely abused of the drugs that are legally available. (Courtesy National Audiovisual Center.)

Figure 16.9
Opium is obtained as a milky exudate from the unripened capsule of the opium poppy. Morphine is the chief, active component. Seen in the background is a block of morphine base. (Courtesy National Audiovisual Center.)

The controversial methadone (**27**) treatment program substitutes a legal drug for an illegal one, a drug, so it is claimed, that blocks out the heroin addict's hunger for heroin and a drug that can be taken orally.

$C_2H_5C(=O)$—$C(C_6H_5)_2$—$CH_2CH(CH_3)N(CH_3)_2$

27
METHADONE

C_6H_5—C(=O)—O, C(=O)—OCH_3, N—CH_3

28
COCAINE

16.17 Coca Alkaloids

The leaves of the coca bush found in the Andes Mountains of Peru and Bolivia contain cocaine ("snow," "coke," "c"). It has apparently been a native practice for centuries to chew these leaves for their local anesthetic action as well as to promote physical endurance. Overconfidence in one's abilities often results, and large doses cause hallucinations, long periods of depression, and sometimes convulsion and death. In some addicts, antisocial acts become violent and murderous. The image of the drug addict, held by some, as being a dangerous maniac probably originated when cocaine became one of the first drugs to be widely abused in nineteenth century Europe. Abusers of cocaine sometimes went on maniacal binges of killing.

16.18 Ergot Alkaloids

Various amides of lysergic acid, alkaloids with an indole system, are obtained by extracting the ergot fungus that grows on diseased rye grain. Because they possess some medicinal properties, these derivatives of lysergic acid have been extensively studied by drug companies. A. Stoll and A. Hofmann of Switzerland prepared a large number of derivatives of lysergic acid to test as therapeutic agents. Among them was the diethylamide, LSD, **29** (from the German, "*l*yserg*s*äure*d*iäthylamid"). In 1943 Hofmann accidentally discovered that LSD is a powerful hallucinogen. A speck of a few micrograms in size can produce the trancelike state reported by users. Just a trace more, still less than a grain of salt, can cause severe psychotic reactions that lead the user to self-destructive acts. How it works, biochemically, is not known.

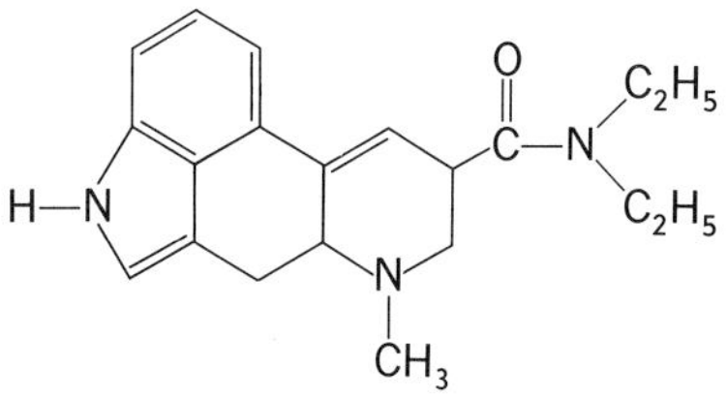

29
LYSERGIC ACID DIETHYLAMIDE
LSD

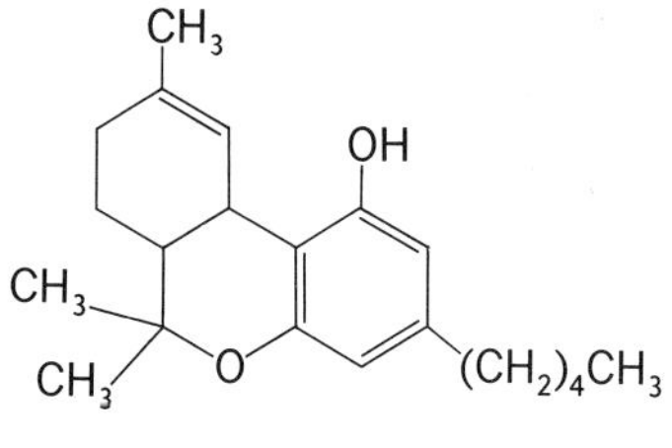

30
TETRAHYDROCANNABINOL
THC

16.19 Marijuana

While the active principle in marijuana is a heterocyclic compound, it is not an alkaloid. Since it is a mild hallucinogen, we include a brief mention of it here. Marijuana is made by chopping the dried leaves and flowers of the Indian hemp plant, *cannabis sativa.* The active principle is THC or tetrahydrocannabinol, **30.**

Marijuana or its related preparations—hashish, dagga, bhang, charas, etc.—is the most widely used illicit drug in the world. Widespread disagreement exists over the wisdom of making it legal. Some facts are well established. THC or its metabolites persist in circulation far longer than those of alcohol. Marijuana is not addictive; it is not a narcotic (except that it is defined that way in federal law); it does not lead to an increase in sexual activity. Physiologically, its biochemical workings do not create a craving for heroin. It creates fantasies of increased creativity but not actual increases. Whether it is no worse than alcohol (which is bad enough when abused) has not been established to the point of widespread acceptance among physical and medical scientists. Low doses of either alcohol or marijuana have sedative-depressant action on the central nervous system, including muscular weakness, poor coordination, and loss of balance. Overdoses of marijuana have been reported to cause depression, paranoid reactions, catatonic excitement, and various hallucinations. In habitual users of marijuana, according to one study at Columbia University, the system's ability to maintain its white cell count in blood is significantly reduced. (White cells are needed to ward off infectious diseases.) Moreover, some metabolites of marijuana accumulated in the gonads, but the long-term effect on genes has not been discovered.

Selected References

Any textbook for a two-term course in organic chemistry will include a chapter on heterocyclic compounds. The following are references to special-interest topics, and represent a tiny fraction of the enormous volume of literature available.

1. *Organic Chemistry of Life*. Readings from *Scientific American*, W. H. Freeman and Company, San Francisco. Articles of interest—all reprints from the parent magazine—include the following: "Alkaloids," "Barbiturates," "The Hallucinogenic Drugs," "The Structure of the Hereditary Material," and "The Genetic Code: II."
2. J. D. Watson. *Molecular Biology of the Gene,* 2nd edition, 1970. W. A. Benjamin, Inc., New York.

Brief Summary

1. Heterocyclic systems tend to display the properties of aliphatic compounds when they are saturated and the properties of aromatic compounds when they are unsaturated.

2. The three chief five-membered heterocycles are pyrrole, furan, and thiophene.

3. These three are all aromatic, but they are all much more reactive than benzene.

4. Pyrrole is not basic because the "unshared pair" of electrons on its nitrogen actually is shared into the aromatic sextet.

5. Pyridine, quinoline, and isoquinoline are three important nitrogen heterocycles having six-membered rings and one nitrogen. All are basic. The heterorings are very resistant to any kind of reaction, including electrophilic substitution.

6. The pyrimidine system occurs in barbiturates and in the purine and pyrimidine bases of nucleic acids.

7. The barbiturates are sedatives that are widely abused. They potentiate the toxicity of alcohol (and vice versa).

8. The nucleic acids are polymers whose "backbones" consist of repeating units of phosphate-pentose. Projecting from the pentose units are purine or pyrimidine systems.

9. In DNA the pentose is deoxyribose, and the heterocyclic amines are adenine, thymine, cytosine, and guanine.

10. In RNA the pentose is ribose, and the heterocyclic amines are adenine, uracil, cytosine, and guanine.

11. DNA is the chemical of genes.

12. The genetic code is the sequence of sidechain amines on DNA. Three amines in a row make up one code word. Each code word specifies a particular amino acid.

13. DNA directs the synthesis of RNA, meaning that the sequence of sidechain amines on DNA determines that the sequence of sidechain amines on the RNA will be exactly complementary.

14. RNA directs the synthesis of polypeptide, which becomes part of an enzyme. Thus, we have the one-gene, one-enzyme relation.

15. The normal or "resting" state of DNA is as a double helix, as proposed by Crick and Watson.

16. When DNA replicates—duplicates itself exactly—the sequences of side-chain amines in the double helix are duplicated as two new double helices form. If not duplicated exactly, a mutation—or worse—may result.

17. The alkaloids are basic, nearly always heterocyclic compounds that are isolated from plants and that have unusual physiological activity.

18. Important families of alkaloids include those with the β-phenylethylamine system, the opium alkaloids, the coca alkaloids, and the ergot alkaloids.

Questions

1. Define each term.
   (a) heterocyclic (b) homocyclic
   (c) barbiturate (d) potentiation
   (e) nucleic acid (f) "downer"
   (g) "upper" (h) alkaloid
   (i) replication (j) hallucinogen
2. What is the name of the principal, active substance in each of the following?
   (a) coca alkaloids
   (b) ergot alkaloids
   (c) opium alkaloids
   (d) marijuana
3. What are two structural differences between DNA and RNA?
4. What is meant by "base pairing" in nucleic acid chemistry?
5. If this sequence appeared on a DNA strand: A G T C G G A what sequence would appear in each case:
   (a) the DNA strand opposite it in a double helix?
   (b) the RNA strand put together under the direction of the one shown? (Remember that thymine does not occur in RNA. Uracil takes its place.)
6. If the sequence G G C is the genetic code "word" for a glycine unit, write the structure of the short peptide that would be made under the direction of this gene segment: G G C G G C G G C
7. Illicit manufacturers of heroin use large quantities of acetic anhydride. Why?
8. Write the structure(s) of the organic product(s) that would form in each case. If no reaction occurs, write "none."

(a)  thiophene + H_2SO_4 $\longrightarrow$ (b) 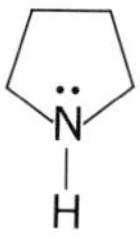 pyrrolidine (N–H) + HCl_{aq} $\longrightarrow$ (c) 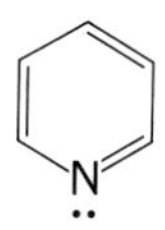 pyridine + HNO_3 $\longrightarrow$

(d) 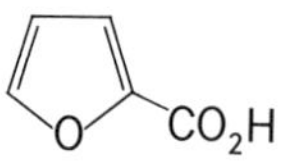 2-furoic acid (CO_2H) + $NaOH_{aq}$ $\longrightarrow$ (e) $CH_3\overset{O}{\overset{\|}{C}}CH_2CH_2\overset{O}{\overset{\|}{C}}CH_2CH_3 + P_2S_5 \longrightarrow$

chapter seventeen

Spectroscopy and Identifying Organic Compounds

Suppose you are an organic chemist in charge of a team working for an industrial or government laboratory. You are handed a vial containing an unknown substance and told that it has to be identified. You will have to be prepared to testify under oath in court as to your identification of the unknown. How do you organize the team for the job? What do you do to identify the compound and be able to testify that it is such and such a substance beyond the slightest shadow of a doubt? We shall look at these questions in this chapter.

17.1 The Strategy of Identification—the Rational Approach

The opposite of "rational approach" is "dumb luck." Naturally, if you are lucky, if you make a shrewd guess that happens to be right and you can prove it is right, accept it with all good cheer. An identification is an identification. But it seldom works that way. Then what do you do?

There are several preliminary questions to put to the unknown substance.

"Is the substance organic or not?" See if it will burn! And if it does, see if carbon dioxide is among the products. (Let some of the gases bubble into limewater, $Ca(OH)_2$. Carbon dioxide will react to produce a milky precipitate of calcium carbonate, $CaCO_3$.) From here on we shall assume that the substance is organic.

"Is the substance actually a mixture of substances, or is it a single substance (which may still have trace impurities)?" This is a harder question to answer experimentally, but at least on paper the strategy is straightforward. See if you can find a physical operation that separates the substance into two or more other substances. You might try thin-layer chromatography, TLC. (See Section 14.5.) If the first attempt gives only one spot, you have to be careful. Closely related compounds might move so closely together on the thin layer that they look like one spot. You have to try other adsorbants and other solvents. You might try gas-liquid chromatography (GLC), Section 2.10, using the same precautions if you get only one peak. There are other techniques one could use, but we shall now assume that if the substance is a mixture, you have found that out and have been able to separate it into its pure components. From here on we assume that we have a substance that is pure and we have data to prove it. (The data might be a sharp melting point; one spot on a thin-layer in TLC; one peak on a GLC chart.)

The next step depends on whether you can be sure—from the history of the sample—that the substance has been prepared before (by others). Is it a known compound, one reported in the literature of chemistry together with some of its physical constants? The history of the sample might be such that you are quite certain it is one of a dozen or more vitamins. Or it might be a steroid, a flavoring agent, a dye, a plastic, an antibiotic, an illicit drug or some other drug, etc. There are well-established ways to handle these—often by some form of chromatography or spectroscopy. However, let's assume you are not at all sure about the nature of the substance, and that, in fact, you suspect the substance is new—never reported before in the literature. Then what?

You need a molecular formula. There are two strategies for getting this. In the one, the traditional approach, one still very important, a small, accurately weighed sample of the compound is burned in oxygen. The carbon dioxide and water are separately collected and weighed. These weights can, by simple calculations, be translated into the percent carbon and the percent hydrogen in the substance. If other elements are present, their percents must also be determined. After that, the formula weight has to be measured. From the analytical data and the formula weight you can calculate a molecular formula. Since we are interested in basic strategy at this point, we shall not describe in detail how these operations are done.

The other approach to getting a formula weight is mass spectroscopy. It will not work for all kinds of substances, however, and we shall see why later. In any case, part of the strategy of identifying an unknown compound is to determine its molecular formula. We shall now assume that you have successfully done this. You know that the compound is, for example, $C_{10}H_{14}O_2$, and you have a few physical constants for it—melting point or boiling point, solubilities in various solvents, for example.

Your next questions are "How are the atoms in $C_{10}H_{14}O_2$ put together?" "Which of the many possible isomers of this formula is the substance?" Your next step, assuming that your laboratory has the equipment, is to use one or more tools of molecular spectroscopy. These are used to find out which functional groups are present. If the instruments are not available, you would use chemical tests to seek out these groups. Quite often, a chemist uses both. They supplement each other, and the chemical tests have the advantage of being quickly and inexpensively carried out. By way of a review, Table 17.1 lists the chemical tests we have presented throughout this text. We shall concentrate in this chapter on the methods of spectroscopy.

17.2 Mass Spectroscopy

The basic features of one kind of instrument for this technique are shown in Figure 17.1. In this method, molecules of the substance in a high vacuum are bombarded

Table 17.1
Summary of Chemical Tests for Functional Groups Included in This Text

	Name of Reagent or Test	Used in Distinguishing	Section
Solvents	Water	Monofunctional, low-formula-weight compounds, or polyhydroxy or poly amino compounds, or salts from others	1.5; 2.9; 6.13; 6.34; 7.3; 8.6; 9.5; 10.3;
	$NaOH_{aq}$ (but not H_2O)	Water-insoluble acids	10.7; 10.24
	HCl_{aq} (but not H_2O)	Water-insoluble amines	8.9
	concd H_2SO_4	Alkenes from alkanes[a]	3.18
Other reagents	Br_2 in CCl_4	Alkenes or alkynes from alkanes	3.15
	$KMnO_{4aq}$	Easily oxidized compounds from others Alkenes or alkynes from alkanes Alcohols	 3.24 6.29–6.30; 9.12; 10.2
	Lucas test ($ZnCl_2$ + HCl)	1° from 2° from 3° alcohols	6.24
	Nitrous acid	1° from 2° from 3° amines	8.12
	2,4-Dinitrophenylhydrazine	Aldehydes and ketones from other carbonyl compounds	9.20–9.21
	Tollens' test	Aldehydes from ketones	9.13
	Benedict's test (Fehling's)	Certain aldehydes or -hydroxyketones from others	9.14
	Iodoform test	—CH(OH)CH$_3$ or —C(=O)CH$_3$ from other systems	9.15
	Iodine test	Starch	14.24

[a]Concd sulfuric acid will dissolve (or decompose) any compound with oxygen, nitrogen, or sulfur plus those with alkene or alkyne groups or easily sulfonated aromatic rings.

by high-energy electrons. Many things can happen. A molecule may have *one* of its electrons knocked off. Let M be our symbol for the parent molecule. This event changes M to M^+, called the *molecular ion*. Oppositely charged disks in the instrument will attract these positive ions out of the collision chamber. The ions accelerate, travel faster and faster. Those "on target" for the center slits of the disks go on through. The molecular ions, however, are usually so unstable that many do not long survive. Supposing that we can represent M^+ simply as $[NOPQRS]^+$. (Between

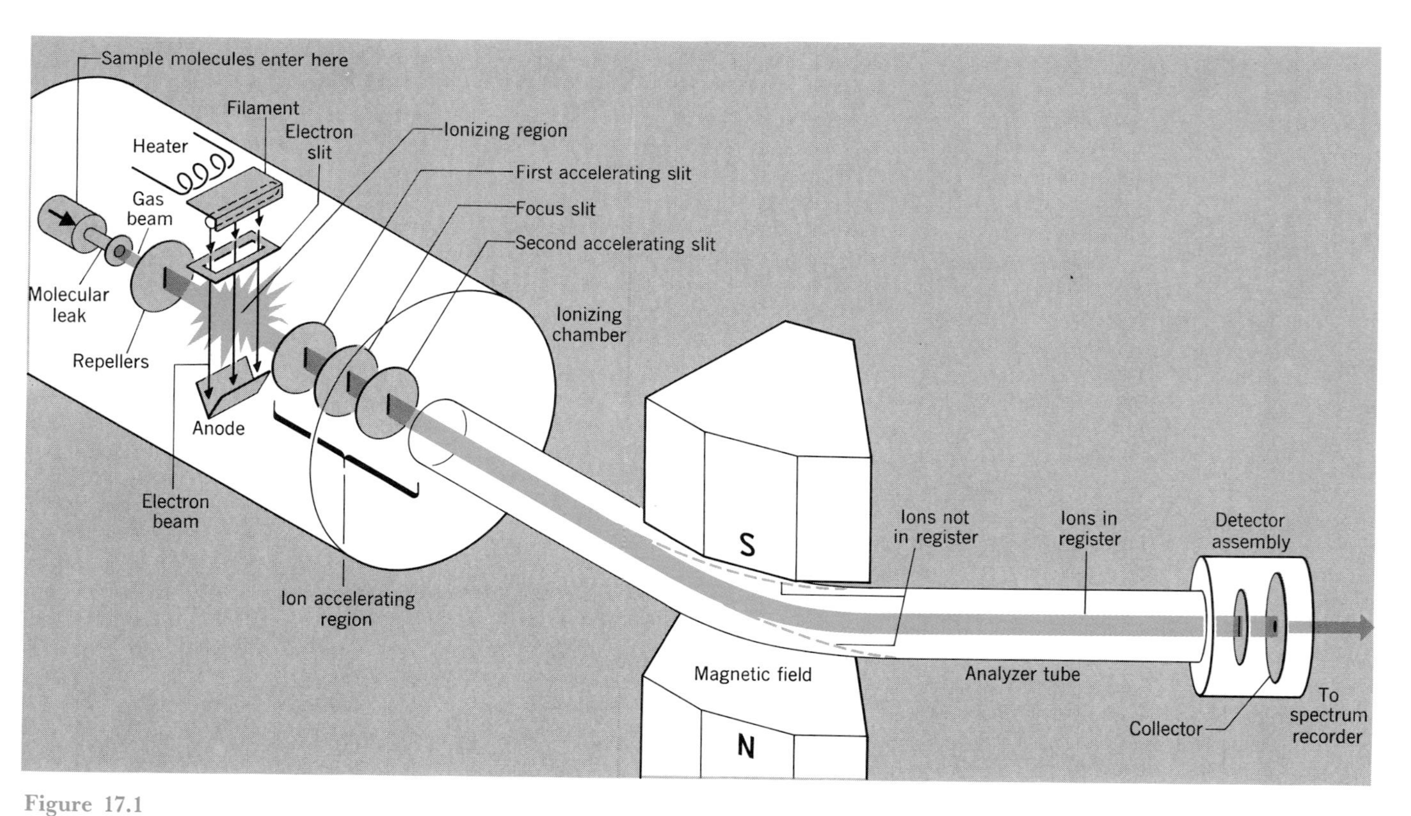

Figure 17.1
Mass spectrometer. Schematic diagram of CEG model 21-103. The magnetic field that brings ions of varying mass-to-charge ratios into register is perpendicular to the page.

each letter we imagine there is a covalent bond.) It can fragment in any one of the following ways:

$$[NOPQRS]^+ \longrightarrow [NOPQR\cdot]^+ + S\cdot \text{ (a free radical; uncharged)}$$
$$\text{or, } [NOPQRS]^+ \longrightarrow [NOPQ\cdot]^+ + RS\cdot$$
$$\text{or, } [NOPQRS]^+ \longrightarrow [NOP\cdot]^+ + QRS\cdot$$

and so on.

The breaking up of M^+ might have gone by another series, which we illustrate by showing one possibility:

$$[NOPQRS]^+ \longrightarrow NOPQR\cdot + S\cdot^+$$

A huge number of breakup modes occur, and the many molecular ions made in the collision chamber produce a complex mixture of other, smaller positive ions and free radicals. All of the positive ions, including survivors of M^+, will continue to be accelerated by the attracting force, and they enter a curved tube. The crucial step is now about to occur.

The instrument includes a powerful magnet whose strength can be varied. This magnetic force makes the path of the traveling ions bend around the curve. How much change in path occurs depends on both the charge and the mass of the ion; on the ratio: mass/charge, or m/e. At one particular magnetic strength, and only one, the path of the molecular ions having the ratio m_{M^+}/e will be bent to match exactly the curvature of the tube. These ions are said to be "in register" (Figure 17.1). Only those ions will reach a detecting device at the other end of the tube. How many get there, of course, can be measured and recorded (in relative, not absolute values). The force of the magnet is now varied slightly. (The instrument is so sophisticated that all of this is done automatically. The operator turns it on, adjusts it, injects the sample, and waits for all the data to be recorded on a chart—the mass spectrogram.) At the new magnetic strength, a different ion with a different mass-to-charge ratio will be in register and in focus on the detector.

The instrument, in summary, makes positive ions out of the molecules of the substance. It sorts these ions according to their mass-to-charge ratios. The chance of having ions of more than one unit charge happens to be very small. For all practical purposes, therefore, the instrument sorts and records the ions simply according to their formula weights. (If, in m/e, e = one unit, then the ratio is equivalent, numerically, to m, mass or formula weight.) If the molecular ions survive the trip in enough numbers to be "seen" by the detector, we get an exact formula weight of the substance. The best instruments are so sensitive they can, to give an example, distinguish between formula weights of 310.360 ($C_{22}H_{46}$) and 310.335 ($C_{20}H_{42}N_2$). (These numbers are calculated by using the best values for the masses of the most common *isotopes* C-12, H-1, and N-14.)

What of the other ions? Figure 17.2 is a typical mass spectrogram presented in graph form. The molecular ion is marked. Note the two small peaks at higher formula weights, marked $[M^+ + 1]$ and $[M^+ + 2]$. How can they be there if the marked peak is, in fact, the molecular ion? How can you have a peak for an ion heavier than the molecule? The answer is that some molecules will have a carbon-13 atom in place of a carbon-12. Some will have hydrogen-2 (deuterium) instead of hydrogen-1. Because of the natural presence of trace amounts of isotopes of carbon, hydrogen, oxygen and other elements, there must always be small peaks at

m/e	Percent of Base Peak		Isotope Abundances *m/e*	Isotope Abundances Percent of M^+
38	4.4		92 M^+	100
39	16		93 $M^+ + 1$	7.37
45	3.9		94 $M^+ + 2$	0.29
50	6.3			
51	9.1			
62	4.1			
63	8.6			
65	11		(Recalculated to base on M^+)	
91	100	(Base)		
92	68	M^+		
93	5.3	$M^+ + 1$		
94	0.21	$M^+ + 2$		

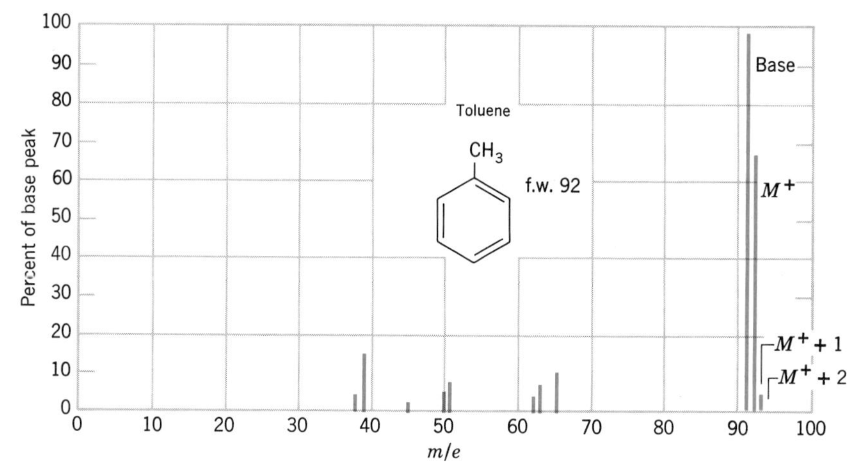

Figure 17.2
Mass spectrogram, graphical, and tabular forms. The most intense peak in any spectrum is called the base peak. All values of *m/e* are rated according to their relative intensities compared to this base peak. Isotope peaks above M^+ are recalculated on the basis of the intensity of M^+ in order to use the isotope abundance tables (cf. Table 17.2) in determining possible molecular formulas. (From R. M. Silverstein, G. C. Bassler and T. C. Morrill. *Spectrometric Identification of Organic Compounds,* third edition, 1974. John Wiley & Sons, Inc., New York. Used by permission.)

$[M^+ + 1]$ and $[M^+ + 2]$. With sulfur and certain halogens, where masses of important isotopes are two units above the common isotopes, there will be significant peaks at $[M^+ + 2]$. Sulfur-34 is an important isotope of sulfur, which is mostly sulfur-32. Chlorine-37 is present with chlorine-35. Specialists in mass spectroscopy, knowing the natural abundances of the isotopes, have calculated what *ought* to be the relative intensities of $[M^+ + 1]$ and $[M^+ + 2]$ if the formula of M^+ is, for example, C_7H_8. Using computors, they have done this for all the combinations of carbon, hydrogen, oxygen and nitrogen that add up to a formula weight of, say, 92. Table 17.2 lists just a few examples. Thus the specialist has data from the mass spectrogram—the value of the formula weight for M^+ and the intensities of $[M^+ + 1]$ and $[M^+ + 2]$ relative to it, and he has the computed tables. With

Table 17.2
Relative Intensities of [M^+ + 1] and [M^+ + 2] Peaks Caused by Natural Isotopes of C, N, O and H in the Molecular Ion, M^+.

M^+	Some Formulas of Mass 92	Percent of Intensity of M^+	
		[M^+ + 1]	[M^+ + 2]
92	CH_2NO_4	1.65	0.81
	$CH_4N_2O_3$	2.03	0.61
	$CH_6N_3O_2$	2.40	0.42
	CH_8N_4O	2.77	0.23
	$C_2H_4O_4$	2.38	0.82
	$C_2H_8N_2O_2$	3.13	0.44
	$C_3H_8O_3$	3.49	0.64
	$C_5H_4N_2$	6.23	0.16
	C_6H_4O	6.59	0.38
	C_7H_8	7.69	0.26
	N_2O_4	9.10	0.80

Data from J. H. Beynon, *Mass Spectrometry and Its Applications to Organic Chemistry*, 1960, Elsevier, Amsterdam.

these he can usually write a precise molecular formula for the unknown. It is important, of course, that these ions do survive and register at the detector. Not all compounds will give M^+, [M^+ + 1], and [M^+ + 2], that survive the treatment. Only fragment ions survive. Now the formula weight and molecular formula cannot be determined (except in special cases). Still, an analysis of fragment ions yields a wealth of valuable information.

The ions likeliest to survive will be the most stable. Here is where the specialist draws on all that is known about the relative stabilites of carbonium ions. A molecule whose mass spectrum has a very intense peak at 91 (e.g., Figure 17.2) very likely has a benzyl group. The benzyl cation, $C_6H_5CH_2^+$, has a formula weight of 91; and this cation is among the most stable carbonium ions that we know. The specialist has learned to expect relatively low intensities for such ions as CH_3^+ and $CH_3CH_2^+$. Primary carbonium ions are among the least stable. These statements only hint at the way the specialist goes about interpreting a mass spectrogram; but they support this generalization about the technique.

> Mass spectroscopy is a way of determining a number of the molecular pieces that a substance must have. Often, the technique gives an exact formula weight—an exact molecular formula—and (by putting all these together) it gives the complete structure of the molecule.

17.3 Infrared Spectroscopy

This technique is used principally to discover the presence or absence of a large number of functional groups. Looking at an infrared spectrum the specialist can tell almost at a glance if the molecules of the compound have an —OH group, a carbonyl group, a benzene ring, a 1° or 2° amine, and a double bond, to name some. If there is a carbonyl group he can tell, usually, if it is caused by the presence of an acid chloride, an acid anhydride, an ester, a carboxylic acid, an amide, or an aldehyde or ketone. If the molecule also has an alkene group, he can tell (usually) if it is conjugated with the carbonyl group or not. If there is a benzene ring he can tell if it is monosubstituted or not. If it is disubstituted, there is a good chance that he can tell if the groups are ortho, meta, or para to each other. This is apparently a tool and technique of tremendous power, and some research labs have infrared spectrophotometers as readily available as their melting point apparatus.

Infrared radiation spans a section of the electromagnetic spectrum (Figure 17.3) at wavelengths just above visible light. It includes what is sometimes called "heat rays." The eye cannot detect this radiation, but special photocells can. When infrared radiation is sent through a sample of an organic compound, different functional groups will absorb different wavelengths. When that happens, the instrument records an absorption band. The location of such bands, their relative intensities, and sometimes their shapes are the raw data that the specialist interprets in terms of functional groups. Figure 17.4 shows the infrared spectra of a few simple compounds together with the interpretations.

The basis for the absorption of infrared light by a functional group is very simple. Atoms that are joined by a covalent bond behave as if they were connected by tiny springs. The groups vibrate constantly. An important kind of vibration is stretching. The two groups stretch out and back. See Figure 17.5. The important point is that each functional group vibrates at just one precise natural frequency. It can vibrate at a higher amplitude at that frequency, however, not just any higher amplitude but a very definite higher one if it receives and absorbs just the right amount of energy. As it happens, the right amount of energy for most functional groups lies between 4000 cm^{-1} and 660 cm^{-1}, exactly where depends on the group itself.

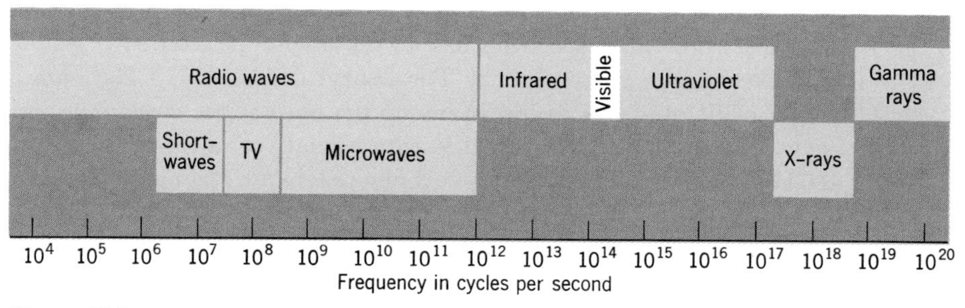

Figure 17.3
The electromagnetic spectrum.

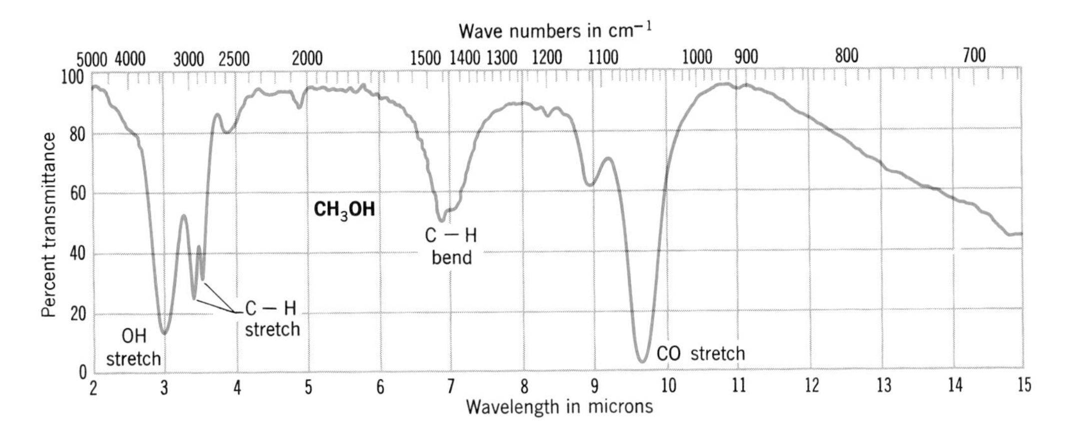

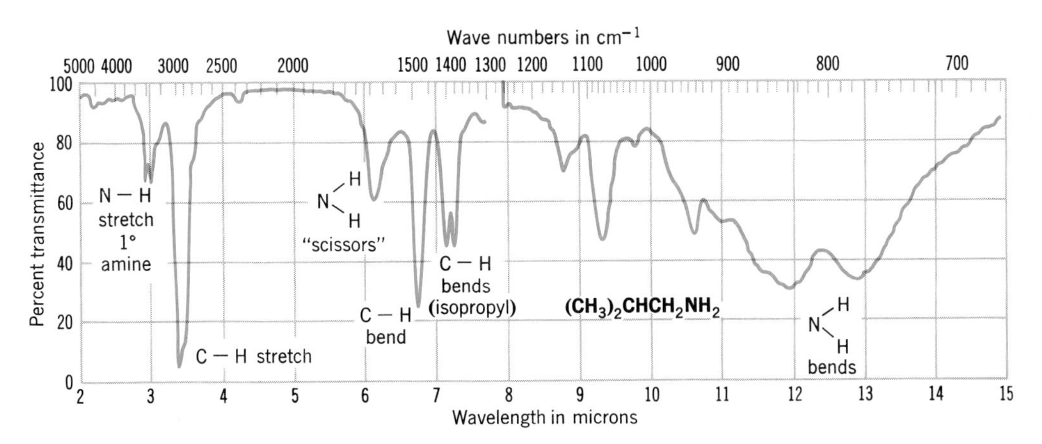

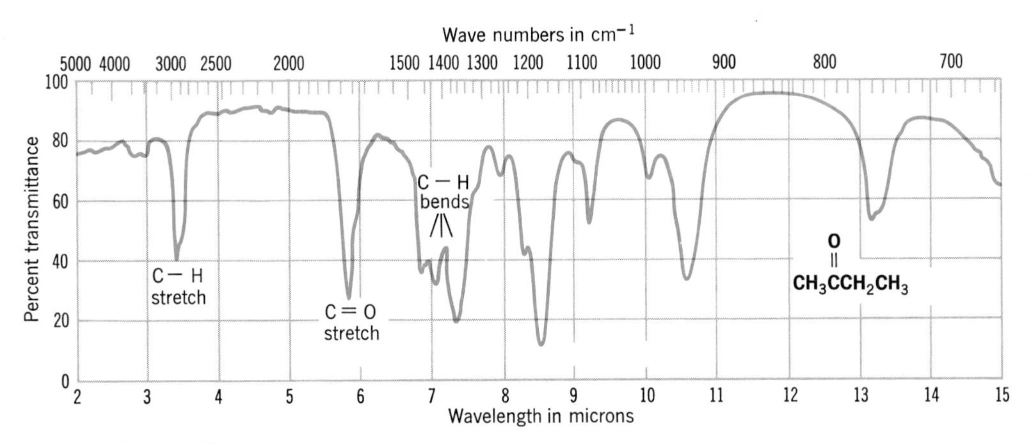

Figure 17.4
Three infrared spectra: methanol (top), isobutylamine (middle) and butanone (bottom). (Courtesy Sadtler Research Laboratories, Inc., Philadelphia, Pa.)

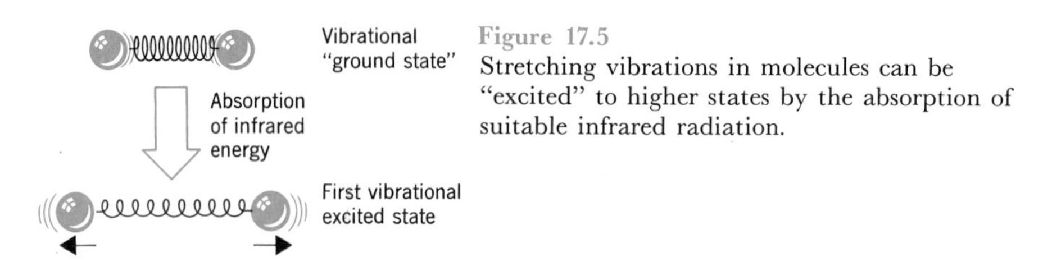

Figure 17.5
Stretching vibrations in molecules can be "excited" to higher states by the absorption of suitable infrared radiation.

17.4 The Units Used In Infrared Analysis

The basic relation between wave length, frequency and the velocity of light is given by:

$$\frac{\text{Centimeters}}{\text{Cycle}} \times \frac{\text{Cycles}}{\text{Second}} = \frac{\text{Centimeters}}{\text{Second}} = 3 \times 10^{10}\,\frac{\text{cm}}{\text{sec}}$$

or:

$$\text{Wavelength} \times \text{Frequency} = \text{Velocity}$$

Energy is associated with each particular frequency of light. The amount of energy is directly proportional to that frequency. The higher the frequency the higher the associated energy. This is true for the entire spectrum (Figure 17.3). In equation form, we have:

$$\begin{array}{ccc} \text{energy} \ldots\ldots & \text{is proportional to} \ldots\ldots & \text{frequency} \\ E & \propto & v \end{array}$$

The proportionality constant is one of the universal constants of nature, called Planck's constant, h, named for Max Planck, a pioneer of modern physics. The equation, then, is:

$$E = hv$$

The location of an infrared absorption band in a spectrum may be given either as a particular wave length or as a frequency. Those who prefer to use wave lengths usually state them in units of microns, μ, one-millionth of a meter. Those who prefer frequency units use the unit of "reciprocal centimeter" defined by:

$$\text{Number of reciprocal centimeters} = \text{cm}^{-1} = \frac{10{,}000}{\mu} = \frac{10{,}000}{\text{wave length in microns}}$$

Because it is frequency and not wavelength that is *directly* proportional to the associated energy, units of frequency, cm^{-1}, are preferred. If we accept this, then an important rule of thumb is easily stated: "the stronger the bond the higher its frequency in the infrared."

Beside stretching frequencies, there are bending frequencies. Groups not only can stretch out and in from each other but also bend side to side with respect to each other. It is easier to bend than to stretch, and bending frequencies involve lower energies, and they lie at lower frequencies.

17.5 Group Frequencies

Table 17.3 contains some of the most useful frequencies of absorption in infrared analysis. The very fact that we can construct such a table implies an important point about infrared spectroscopy. Over a huge number of substances having a given group, the frequency of that particular group's vibration nearly always occurs within a rather narrow range. Any molecule with many C—H bonds of an alkane, for example, will always have two or more strong absorptions between 2850 and 2950 cm^{-1}. The O—H stretching frequency of all alcohols will absorb strongly between 3500 and 3650 cm^{-1}.

The infrared spectrum of a compound, if carefully taken, is so uniquely that compound's that we compare it to an individual's fingerprints. An infrared spectrum is "the fingerprint of a compound."

Table 17.3
Some Important Group Frequencies in the Infrared

	Group	Frequency Range (cm^{-1})	Notes
Stretching frequencies	O—H (alcohols, phenols)	3200–3550	Very strong; broad
	N—H In 1° amines	3500 3400	Two weak bands
	In 2° amines	3300–3350	Weak, single band
	C—H In alkenes and aromatics	3000–3100	Relatively weak, sharp
	In alkanes	2840–3000	Relatively strong (because there usually are many C—H bonds in a molecule); May be up to 4 bands
	C=O (aliphatic)		
	Acid chlorides	1800–1815	All carbonyl bands are strong and quite sharp
	Esters	1735	
	Aldehydes	1725	
	Carboxylic acids (dimers)	1710	
	Ketones	1715	
	Amides	1650	
	C=C In $-\overset{|}{C}=CH_2$	1640–1650	Weak; sharp
	In $-\overset{|}{C}=\overset{|}{C}-$	1660–1670	Weak; sharp
	C—O Esters, alcohols, ethers	1000–1300	Intense; variable location; often broad
Bending frequencies	C—H Aliphatic	1350–1400	Moderate
	NH_2	1580–1650	Moderate
	N—H and NH_2	670–900	Moderate; broad

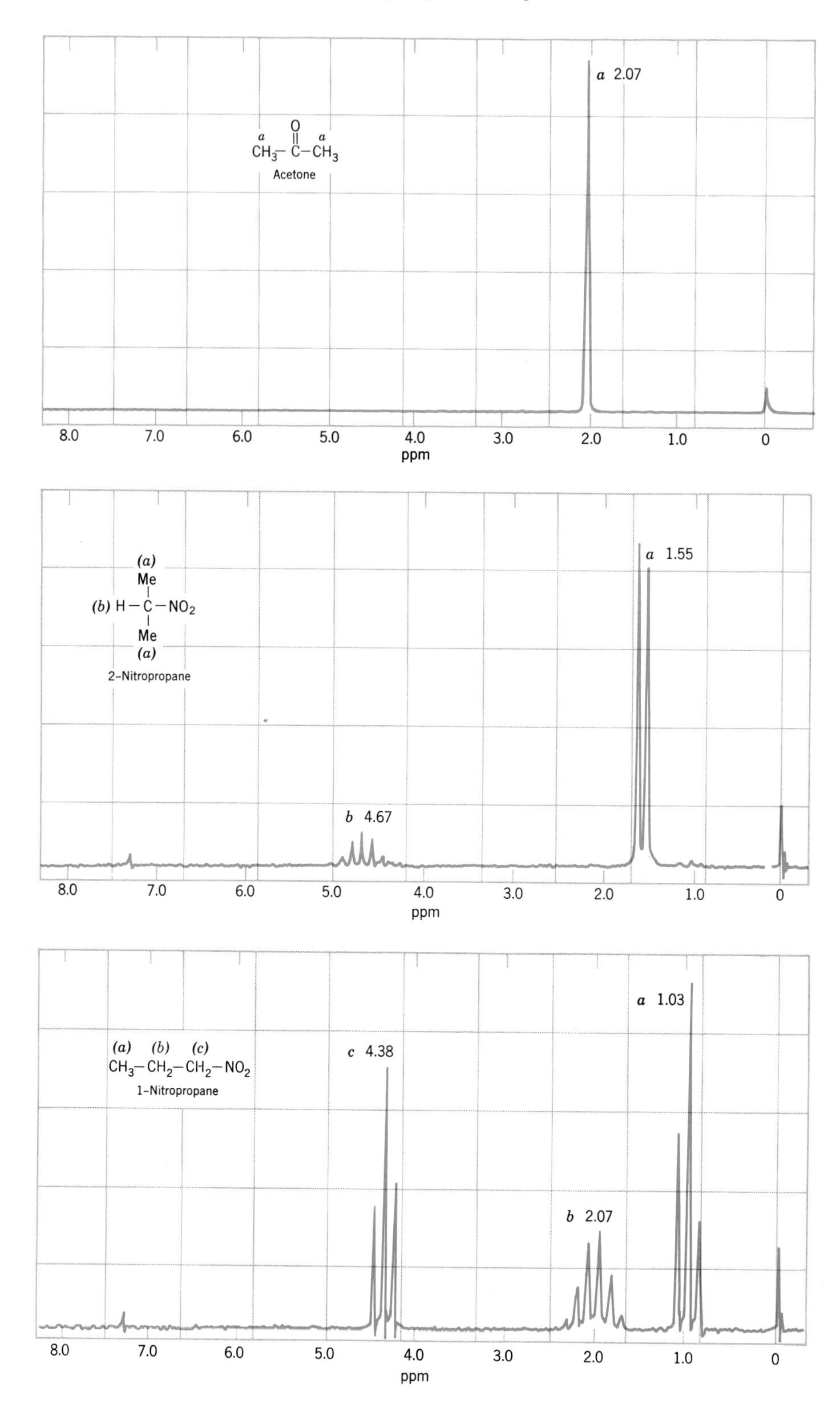
a 2.07
a a
CH3–C(=O)–CH3
Acetone
8.0
7.0
6.0
5.0
4.0
ppm
3.0
2.0
1.0
0
(a)
Me
(b) H–C–NO2
Me
(a)
2-Nitropropane
a 1.55
b 4.67
8.0
7.0
6.0
5.0
4.0
ppm
3.0
2.0
1.0
0
(a) (b) (c)
CH3–CH2–CH2–NO2
1-Nitropropane
a 1.03
c 4.38
b 2.07
8.0
7.0
6.0
5.0
4.0
ppm
3.0
2.0
1.0
0

17.6 Nuclear Magnetic Resonance Spectroscopy. NMR

In its most common application, this technique is used to get information about the hydrogens that are present in a molecule. Several other elements can be studied, but we shall stay with hydrogen. We need no fancy instruments to tell us that an organic compound has hydrogen, but are the hydrogens on —CH_3 groups or —CH_2— positions, alkene systems, benzene rings, or oxygen, nitrogen, or sulfur, etc.? Are the hydrogens alpha or beta to a carbonyl group, a halogen, etc.? NMR spectroscopy, therefore, is a powerful supplement to infrared analysis. The latter tells us much about functional groups. NMR tells about the nonfunctional groups—the alkanelike portions, rings, and the like. The NMR will tell us the following:

1. How many sets of different kinds of hydrogens there are in the molecule. (There is one set in acetone, $CH_3\overset{\overset{\displaystyle O}{\|}}{C}CH_3$, two sets in 2-nitropropane, $(CH_3)_2CHNO_2$, and three sets in 1-nitropropane, $CH_3CH_2CH_2NO_2$.
2. The relative number of hydrogens in each set. (In 1-nitropropane the proportions are 3:2:2. In 2-nitropropane the proportion is 6:1.)
3. The number of hydrogens in the sets that are neighbors to the one in question. (From the appearance of the absorption for the —CH_3 hydrogens in 2-nitropropane, the specialist can, with total confidence, tell that each CH_3 group is attached to a carbon holding one hydrogen. Similarly, from the appearance of the absorptions for the hydrogen in the $-\overset{|}{C}H-$ group, in 2-nitropropane the specialist knows that attached to its carbon are groups holding six hydrogens.

Figure 17.6 shows the NMR spectra of these simple compounds together with their interpretation.

17.7 Nuclear "Magnets"

This technique is possible because the nuclei of hydrogens (but not carbons or oxygens) behave like tiny magnets that can point (in a relative sense, of course) either south or north—that is, either in one direction or the opposite. Half—exactly half—of all the nuclear magnets in the hydrogens in a sample of an organic compound point one way and half the other way—except when the sample is put between the poles of a very powerful magnet. Some of the tiny nuclear magnets "line up" with the magnet's field, something like the needle of a compass lining up with the earth's magnetic field. Only some magnets can, on balance, get lined up. The molecules are tumbling and twisting violently, as they always do. The point, however, is that the powerful magnet creates two magnetic states for the hydrogens. One state is at a lower energy (the "lined up" state), and the other state is at a higher energy. In other words, the magnet creates two magnetic energy

Figure 17.6
Three NMR spectra: acetone (top), 2-nitropropane (middle), and 1-nitropropane (bottom). The sharp peak at 0 ppm is the reference peak given by tetramethylsilane, $(CH_3)_4Si$. (Spectra courtesy of Varian Associates, Palo Alto, California.)

levels for the tiny proton magnets. Now all we have to do, in principal, is send in suitable energy to get some lined-up magnets flipped from the lower to the higher energy state. If the energy we send in makes this happen—if it is of just the right frequency—then, of course, it is absorbed. Then, we have an absorption band in a spectrum. Energy of very low frequency—in the microwave region—is used. When the instrument is tuned so that this happens, we say that a condition of resonance exists. That is why the technique is called nuclear magnetic resonance spectroscopy. Figure 17.7 shows the essential parts of the instrument.

17.8 Chemical Shift

Hydrogens in different chemical environments within molecules (CH_3— versus —CH_2—, versus —CH_2—NO_2, etc.) respond slightly differently to the magnetic field. The absorptions for one kind of hydrogen are shifted relative to absorptions for others. The amount of the shift is called the *chemical shift* for that particular kind of hydrogen. But "shift relative to what?" In current practice, the sample is treated with a small amount of an inert chemical called tetramethylsilane, $(CH_3)_4Si$, which is the answer to that question. Chemical shifts are measured relative to the band caused by the 12 equivalent hydrogens in this compound. This band always (well, nearly always) is the first band on the right side of the

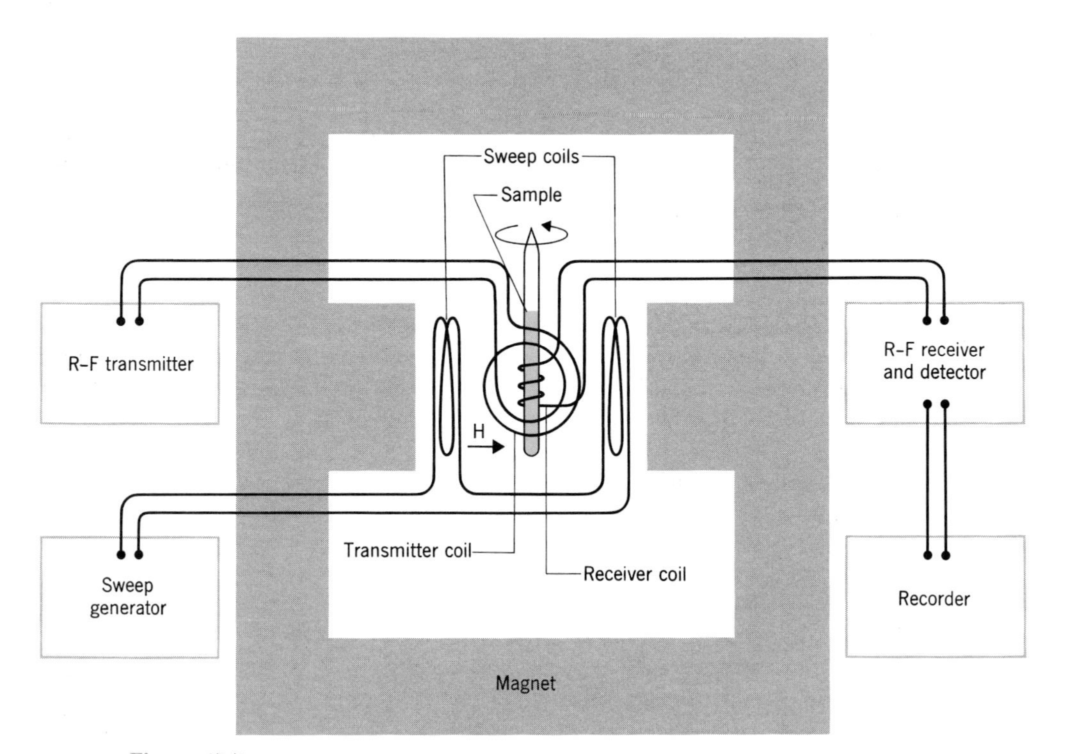

Figure 17.7
NMR spectrometer. The sample is in a vial surrounded by the coils of the receiver and of the radiofrequency transmitter. The assembly is positioned between the poles of a powerful electromagnet whose magnetic strength can be adjusted by the sweep generator. (Courtesy of Varian Associates, Palo Alto, California.)

Table 17.4
Chemical Shifts (δ-values) of Hydrogens

	Functional Group X =	CH_3—X	—CH_2—X	—CH—X (with one bond up)
α-Hydrogens on saturated carbons	H	0.23	0.9	1.25
	CH_3 or —CH_2—	0.9	1.25	1.5
	Cl	3.1	3.4	4.0
	Br	2.7	3.3	4.1
	I	2.2	3.2	4.2
	OH	3.5	3.6	3.6
	O—R	3.3	3.4	
	O—Ar	3.7	3.9	
	O—C(=O)—R	3.6	4.1	5.0
	—C(=O)—H	2.2	2.3	2.4
	—C(=O)—R	2.1	2.4	2.5
	—C(=O)—OH	2.1	2.3	2.6
	—C(=O)—O—R	2.1	2.3	2.6
Hydrogens at unsaturated centers	—C(—)=CH_2	4.6–5.0		
	—C(—)=C(—)—H	5.0–5.7		
	Aromatic	6.5–8.3		
	—C(=O)—H	9.5–9.8		
Hydrogens on oxygen and nitrogen	O—H (aliphatic alcohols)	0.5–5.0 (Varies with concentration)		
	NH_2 (1° aliphatic amine)	0.6–1.6		
	NH (2° aliphatic amine)	0.3–0.5		

spectrum. See Figure 17.6. Table 17.4 gives selected values of some common chemical shifts. They are expressed as δ-values ("delta values"); and the units are parts per million, ppm. (The "parts" referred to are units of magnetic field strength.) Usually, but not always, hydrogens in the most electron-rich environments have the smallest δ-values. As electron density is drained away from hydrogens by electronegative groups, the δ-values get larger. There are important exceptions to this, but they are beyond the scope of our study.

17.9 Coupling

As seen in the NMR spectra of Figure 17.6 a particular set of hydrogens gives rise not just to a single peak but usually to a cluster of two or more peaklets. (Chemical shifts, therefore, are measured to the *middle* of a cluster.) The fact that there are clusters of peaklets means that the energy levels available to the nuclear magnets in a given set of hydrogens (e.g., —CH_3) are not just one upper and one lower level. If it meant that, there would be only one peak. A cluster of three peaklets must mean three pairs of energy levels, all three pairs belonging to or being possibilities for the set of hydrogens in question. What causes these different levels? The answer is the hydrogens, the tiny nuclear magnets, "next door" on the adjacent carbon(s), if there are any. If there are not, then we will not see peaklets; only one peak. That is why we always see only one peak for tetramethylsilane, our reference, $(CH_3)_4Si$, or for neopentane, $(CH_3)_4C$. The carbon "next door" to each CH_3— carries no hydrogen. But if there is one hydrogen "next door" to CH_3, as in the case of 2-nitropropane (Figure 17.6), then the absorption for the hydrogens of the methyl groups is split into two peaklets.

The rule of thumb is: "there are always two peaklets for a set of hydrogens if that set has one and only one hydrogen 'next door.'"

Now examine the NMR spectrum for 1-nitropropane, Figure 17.6. The —CH_3 now has two hydrogens "next door." The absorptions for the —CH_3 hydrogens now occur as a group of three peaklets, a triplet. "If there are two neighbors, then three peaklets." The hydrogens in the —CH_2—NO_2 system also have two neighbor hydrogens and show a triplet.

> In general, the number of peaklets for a particular set of hydrogens equals the number of their immediate next door neighbors plus 1.

The middle —CH_2— has five neighboring hydrogens, and it absorbs as a sextet (5 + 1). The lone hydrogen of the —$\overset{|}{C}H$— in 2-nitropropane has six neighbors. Therefore the NMR absorption for that lone hydrogen should be a set of seven peaklets (6 + 1). The trouble is that this lone hydrogen causes a weak absorption. If a weak absorption is split up into seven peaklets, the two on opposite sides are often too weak to see. (Special experiments, however, bring them out.)

This phenomenon where neighboring nuclear magnets interact with each other and cause observable changes in each other's energy levels is called "coupling." This coupling gives rise to the splitting of peaks. The CH_3— and the —$\overset{|}{C}H$— hydrogens in 2-nitropropane couple with each other and produce a splitting of absorption bands into clusters of peaklets.

There is a great deal more to a full analysis of this phenomenon, but we must leave that to more advanced treatments. What we have left to do is explain how the NMR spectrum tells us what the relative numbers of hydrogens in the various sets are. That is probably the easiest of all. The relative numbers of hydrogens causing the clusters of peaklets are directly proportional to the areas on the spectrum under the clusters. (A crude way to measure this would be carefully to cut out each cluster with a scissors and weigh the different papers! The weights are proportional to the areas, if the paper has uniform density. A better way is to count the squares of the graph paper enclosed by each cluster. Still better is an electronic measuring device attached to the machine which computes those areas automatically as the spectrum is obtained.) The ratio of areas in the NMR spectrum of 2-nitropropane is 6:1. In 1-niropropane the ratio is 3:2:2. We cannot tell if this relative ratio of 3:2:2 is caused by an actual ratio of 3:2:2 or 6:4:4 or 9:6:6. All we get is the relative ratio. We need more information to decide, information that the molecular formula, of course, supplies.

In summary, the NMR will tell us:

1. The number of different sets of hydrogens —by the number of clusters of peaklets.
2. The relative number of hydrogens in the sets —by the relative areas occupied by each cluster
3. The number of neighboring hydrogens that a set has —by the number of peaklets in the set: the number of peaklets = the number of neighbors + 1.

We have assumed all along that we are dealing with substances whose molecules have sets of hydrogens that differ rather widely in chemical shifts. This has been true in our examples. When that is not true, the analysis immediately becomes much more complicated. The number of peaklets begins to change; the shapes of the clusters change. These complications happen often, and, again, we have to leave this to more advanced treatments. Our purpose, remember, has been to present what the technique can do in the hands of a specialist. The problem stated at the start of this chapter had to do with strategy for identifying an unknown substance. Today, no research laboratory in chemistry would be without an NMR instrument. Very often an unknown can be identified beyond the shadow of a doubt from three spectra: the mass spectrum, the infrared spectrum, and the NMR spectrum. If the mass spectrum gives the molecular formula, then the infrared discloses the major functional groups, and the NMR reveals the patterns of carbon-hydrogen sets and their locations relative to the functional groups.

17.10 Ultraviolet-Visible Spectroscopy

This method is useful in identifying compounds but, more importantly, in analyzing them quantitatively. Colorimetry is one example of the use of a visible spectrum of a compound to measure its concentration. Since analytical aspects are taught in other courses we shall leave such topics to them and their texts. Our treatment here will be extremely brief.

The basis for ultraviolet-visible spectroscopy is the fact that electrons in molecules, as in atoms, occupy certain energy levels. There are also unoccupied energy levels higher up the energy scale. If the right frequency of light is sent in, an electron will be promoted to a higher level, and the light is absorbed. To excite electrons we need energy more powerful than the infrared. We need it in the visible and ultraviolet regions (Figure 17.3). Infrared excites a molecule vibrationally. Ultraviolet or visible light can excite a molecule electronically.

The range of wavelengths for this method is from 180 mμ to 800 mμ; for the ultraviolet: from 180 mμ to 400 mμ; and for the visible: from 400 mμ to 800 mμ (mμ = millimicron).

An unsaturated molecule will usually absorb in the ultraviolet range. If it has several double bonds in conjugation, it will absorb in the visible and be a colored compound. Hence this technique can be of value in identifying a compound if the specialist wants particular information about the nature of the double bonds. Are they isolated or conjugated? If conjugated, how many in a row? The location of absorption bands and their intensities can be correlated in a way that tells or at least suggests to the specialist answers to these questions. He can distinguish, for example, between these two unsaturated ketones:

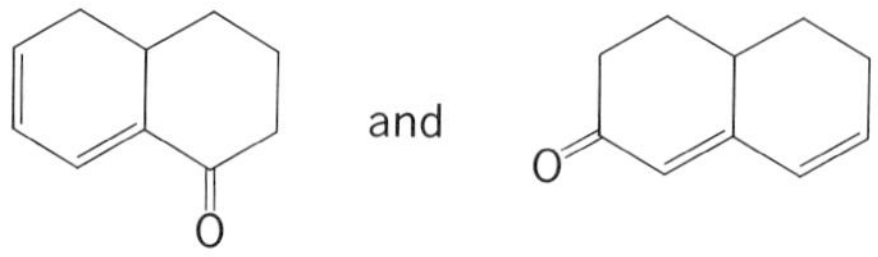

and

O O

In summary, the ultraviolet-visible spectrum of a compound provides some information about the nature of the unsaturation in its molecules. This kind of spectroscopy is also excellent for routine quantitative analysis of known compounds.

Selected References

Most texts in organic chemistry for two-term courses give more extensive treatment of spectroscopy and provide many more actual spectra that have been interpreted. The following books deal with the four types of spectra and include many worked examples and problems.

1. R. M. Silverstein and G. C. Bassler. *Spectrometric Identification of Organic Compounds,* 3rd ed. 1974. John Wiley & Sons, Inc., New York.
2. C. J. Creswell, O. Runquist, and M. M. Campbell. *Spectral Analysis of Organic Compounds,* 2nd edition, 1970. Burgess Publishing Company, Minneapolis, Minnesota. A programmed learning text.
3. D. J. Pasto and C. R. Johnson. *Organic Structure Determination,* 1969. Prentice-Hall, Inc., Englewood Cliffs, N. J. Besides surveying the various kinds and uses of spectra, this text includes a study of all of the important techniques for purifying compounds, proving their purity, and applying a large number of chemical tests as an aid in identifying an organic "unknown."

Brief Summary

To identify an unknown substance you must do the following in approximately this order.

1. Find out if it is organic or not.

2. Find out if it is a mixture or not. If it is, separate the mixture into individual pure substances and analyze each as follows.

3. Find out what elements are present. If it is organic, carbon is present and hydrogen almost certainly so. Besides that there may be oxygen, nitrogen, sulfur, one of the halogens, maybe phosphorus. (We left to other books details of how such qualitative analyses are made.)

4. Determine the molecular formula of the substance. This could involve:

(a) Quantitative analysis for each element and a measurement of molecular weight.

(b) Use of the mass spectrum.

5. Find out what functional groups are present. Such information may be obtained by:

(a) Doing one or more chemical tests, including studying the solubility of the substance in various solvents, and seeing if it changes the color of litmus paper.

(b) Interpreting its infared spectrum.

(c) Interpreting its mass spectrum.

(d) Interpreting its NMR spectrum.

(e) Using the ultraviolet-visible spectrum if there is evidence of unsaturation and the other spectra do not appear to provide enough information about it.

Questions

1. Suppose you have narrowed the choice down to benzene and cyclohexane in identifying an unknown. Then you obtain its infrared spectrum—No. 1, shown below. What is the unknown compound? How do you know?

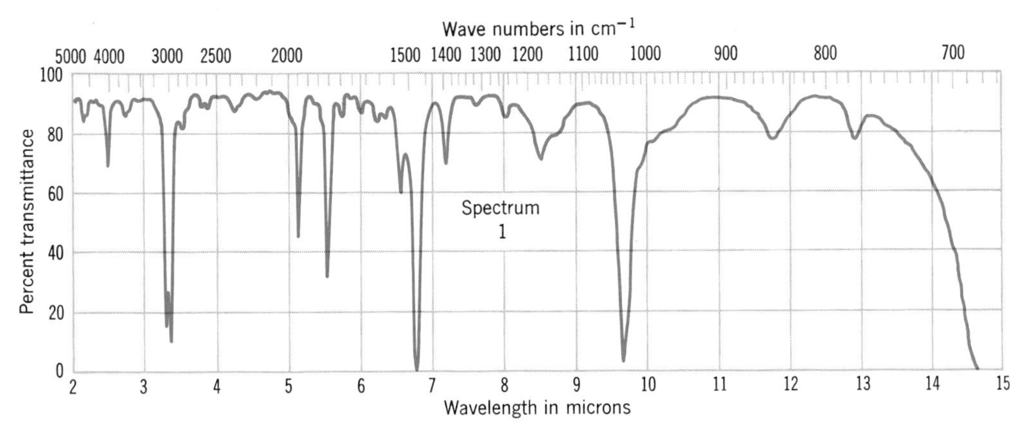

(Spectrum courtesy of Sadtler Research Laboratories, Philadelphia, Pa.)

2. The choice now is between ethyl acetate and ethyl propionate. Spectrum 2 is the NMR spectrum of the compound. Which is it? How do you know?

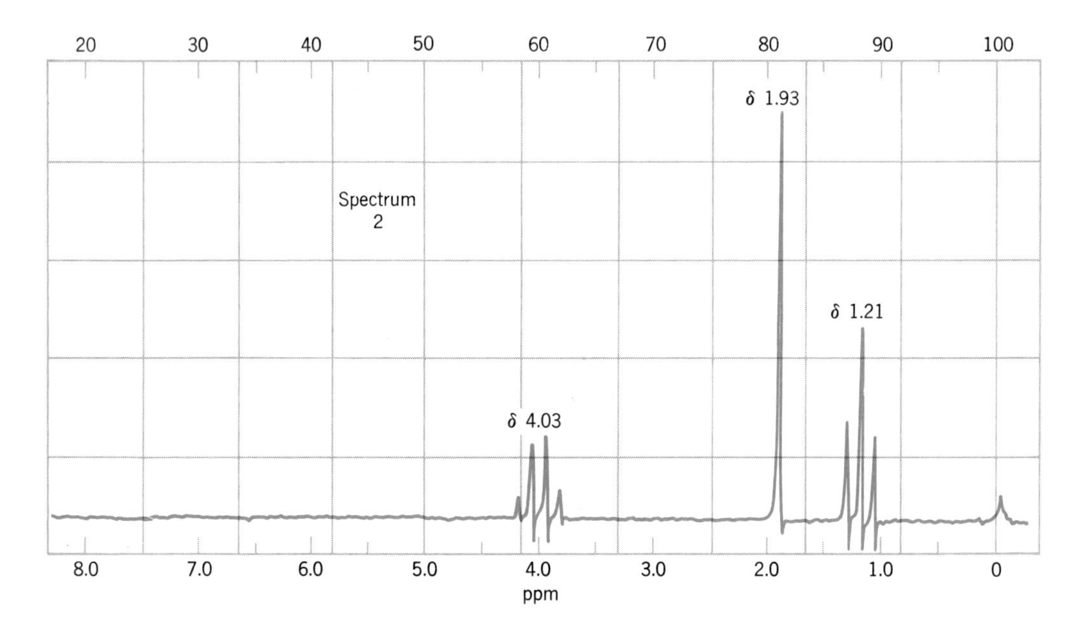

(Spectrum courtesy of Varian Associates, Palo Alto, Calif.)

3. The choice is between ethyl acetate and *n*-butyl alcohol. Spectrum 3 is the infrared spectrum of the compound. Which is it? How do you know?

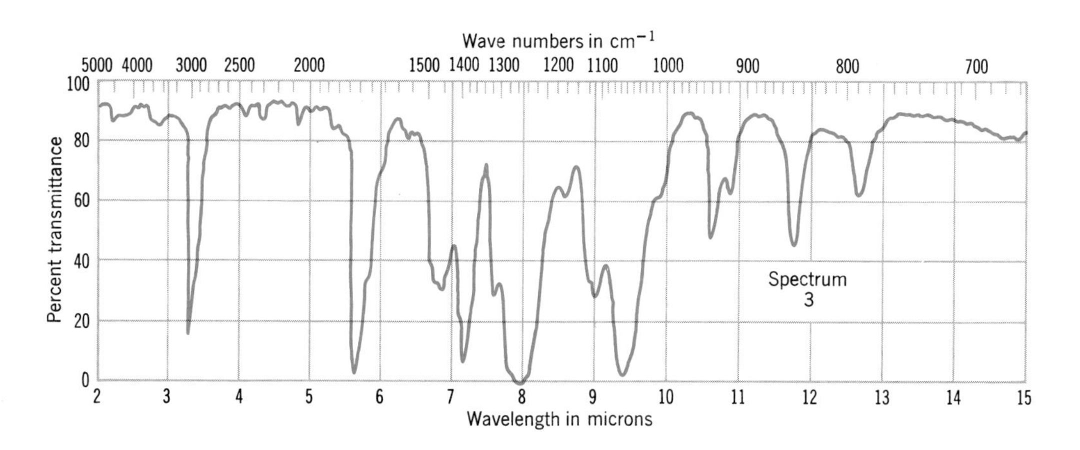

(Spectrum courtesy of Sadtler Research Laboratories, Philadelphia, Pa.)

4. Spectrum 4 is the NMR spectrum of either 2,2-dichloropropane or 1,3-dichloropropane. Which is it? Why?

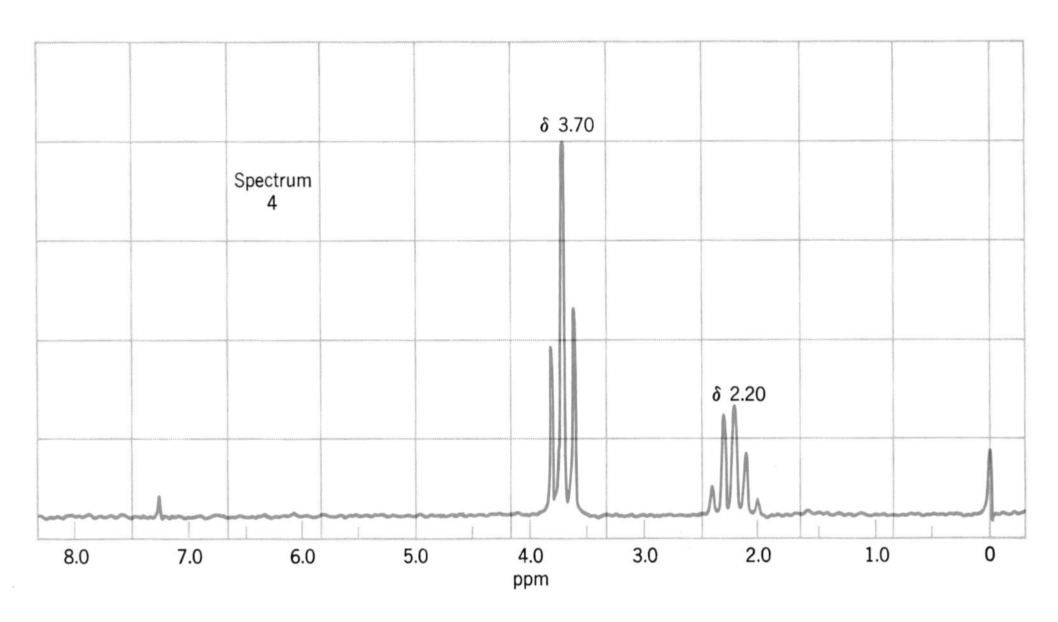

(Spectrum courtesy of Varian Associates, Palo Alto, Calif.)

Appendix

The R/S System of Configurational Families

In 1956 R. S. Cahn, C. K. Ingold, and V. Prelog proposed a system for describing the absolute configuration at a chiral carbon that overcomes the major difficulty of the D/L system, its dependence on a particular reference compound. In time, the R/S system may be used exclusively. Because the D/L system is used in some current technical publications and papers (and in all of them prior to 1956), it, too, must be learned. It is especially common in carbohydrate and amino acid chemistry. The rules developed for the R/S system have also made possible an excellent means of handling *cis/trans* isomers: the E/Z system. We shall study the R/S system first.

The absolute configuration at a chiral carbon in the R/S system is described with reference to a set of rules for arranging the four different groups at that carbon in a *priority sequence*. Lowest priority is given the lowest number—1. Highest priority has the highest number—4. (We'll get to the rules for assigning priority numbers soon.) After you make priority assignments, you then hold up the molecular model, or draw it on paper, or imagine that the atom or group with lowest priority—1—is behind the chiral carbon. Imagine it being behind a steering wheel and along the steering column with you seated in the driver's seat. This places the remaining three groups around the rim of the steering wheel. See Figure A.

Now you have to imagine tracing a circle by moving your finger from the group of priority 4, then to 3 and then to 2. Remember, first 4, then 3, then 2. If your finger moves clockwise, we say the groups are in a *rectus* or R-configuration. ("Rectus" means "to the right" in Latin.) But if your hand has to move counter-clockwise to go from 4 to 3 to 2, then the configuration is said to be *sinister,* or S-configuration. ("Sinister" means "to the left" in Latin.)

The Priority Rules

1. Consider the atoms directly attached to the chiral center. The lower the atomic number, the lower the priority. Hence, a hydrogen atom, if present, will always have priority number 1. ("Low priority—low number.")
2. If two or more first atoms are the same, look at the atoms attached directly to them; "the lower their atomic numbers, the lower the priority."

   For example, the whole ethyl group at a chiral carbon will have a higher priority than a methyl group.

$—CH_2—H \qquad —CH_2—CH_3$

In each case the first atom is carbon. The "second atom"—actually, of course,

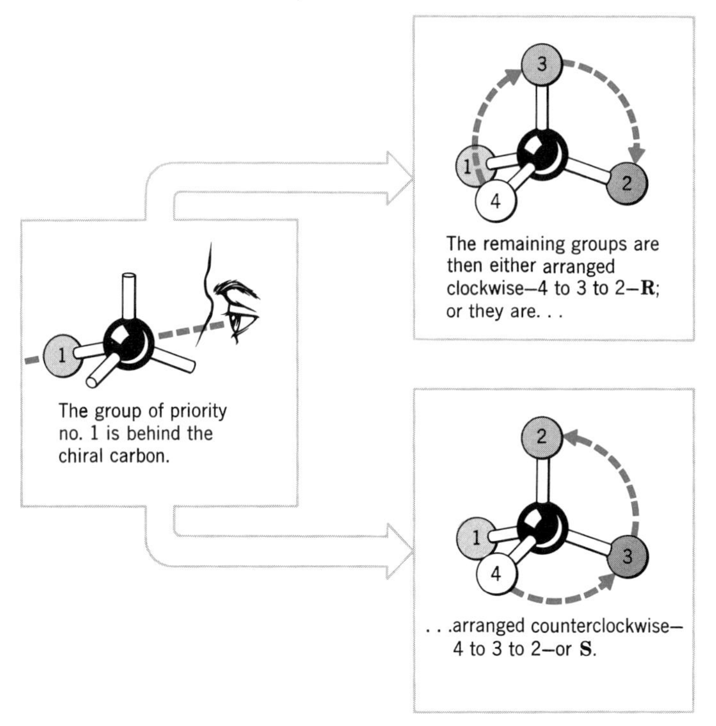

Figure A
The R/S System of configurational families.

a set of three second-atoms—are —H, —H, —H for methyl and —H, —H, and $—CH_3$ for ethyl. Its set includes an atom of higher atomic number than any in the CH_3 set. Hence, the ethyl group has a higher priority than the methyl group. Thus in 2-chlorobutane, $CH_3\underset{\displaystyle Cl}{\underset{|}{C}}HCH_2CH_3$, the groups have these priority numbers: 1, H; 2, CH_3; 3, CH_2CH_3 and 4, Cl. If you cannot make a decision at the second atom, go to the third, etc. Thus, in 3-methylhexane, $CH_3CH_2CH_2—\underset{\displaystyle CH_3}{\underset{|}{C}}H—CH_2CH_3$, the groups at the chiral carbon have these priority numbers: 1, H; 2, CH_3; 3, CH_2CH_3 and 4, $CH_3CH_2CH_2$.

3. If you come to a second (or higher) atom with a double bond or a triple bond, rewrite the bond at that point to "double" or "triple" the atom at the other end. For example, imagine that $—CH{=}CH_2$ is, instead, $—CH\begin{matrix} CH_2—\overset{|}{\underset{|}{C}}— \\ \underset{|}{\overset{|}{C}}— \end{matrix}$. The ketone group, $\rangle C{=}O$, is imagined to be $—\overset{|}{\underset{|}{C}}—O—\overset{|}{\underset{|}{C}}—$ (with the C bonded to $O—C\langle$). Thus in glyceraldehyde, $CH_3—\underset{\displaystyle OH}{\underset{|}{C}}H—\overset{\displaystyle O}{\overset{\|}{C}}\ H$,

the priority numbers for the groups at the chiral carbon are: 1, H; 2, CH_3—; 3, —$\overset{\overset{\displaystyle O}{\|}}{C}$H; and 4, —OH.

In a molecule having two or more chiral centers, each center is handled individually and separately. Each chiral center is either in the R or the S family.

These are the principle rules. They will not handle every problem, but those that the rules do not cover are few. (The complete set of rules is in the original article: *Experientia,* Vol. 12, 1956, page 86). Table A is a list of several groups arranged in their correct priority sequence.

If you have trouble visualizing perspective drawings, and you are shown one with the atom or group of lowest priority number not in the rearward location where you would like it to be, there is a very simple solution. If you interchange *any* two groups at a chiral carbon you change its configuration from R to S or from S to R. Each time you change any two groups you make this simple change from one family to the other. For example:

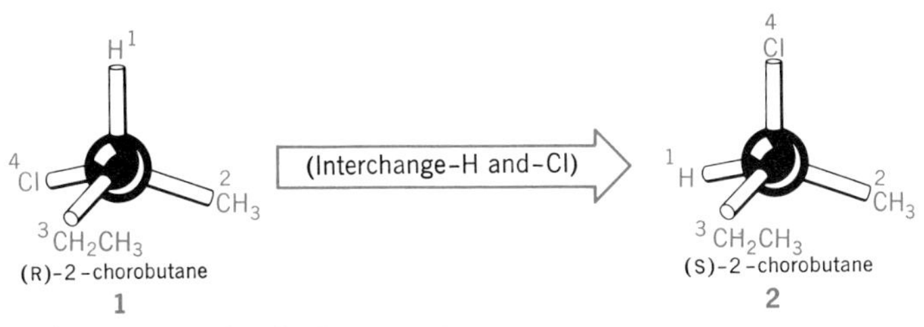

(R)-2-chorobutane
1

(S)-2-chorobutane
2

In **2** the group of priority number 1 is behind the chiral carbon and it is easy to see that the configuration of **2** is S. Because we developed **2** from **1** by exchange of just two groups, **1** must be R.

The following are examples of structures that have been correctly assigned to configurational families.

CHO; H, OH; CH_2OH ≡ CHO; $HOCH_2$, OH; H

D-(+)-Glyceraldehyde
(plane projection)

R-(+)-Glyceraldehyde
(perspective)

CO_2H; H_2N, H; CH_2OH ≡ CO_2H; $HOCH_2$, H; H_2N

L-(−)-Serine
(plane-projection)

S-(−)-Serine
(perspective)

The E/Z System for Geometric Isomers

The priority rules for R/S configurational families are also used for describing *cis/trans* relations. The advantage is the complete lack of ambiguity. The procedure is first to consider separately the two sets of groups, one at one end of the double bond, the other at the opposite end. Each set has two groups.

The isomer whose higher priority groups are on the same side of the double bond (formerly, in a *cis* relation) is called the Z-isomer. The isomer in which the higher priority groups are on the opposite side of the double bond (formerly, the *trans* relation) is called the E-isomer. ("Z" is from the German, *zusammen*, or "together;" "E" is from the German, *entgegen*, or "opposite.") These examples illustrate how the system works. (The higher priority groups of each set are marked by an

CH_3 F
C=C
Cl Br

(Z)-1-FLUORO-1-BROMO-2-CHLOROPROPENE

CH_3 Br*
C=C
*Cl F

(E)-1-FLUORO-1-BROMO-2-CHLOROPROPENE

*CH_3
CH_3 $CHCH_2CH_3$
C=C
*CH_3CH_2 CH_2CHCH_3
CH_3

(E)-3,5-DIMETHYL-4-ISOBUTYL-3-HEPTENE

asterisk.) In none of these examples could you use either *cis* or *trans* in the name and have it mean anything without adding a statement that specified which groups you personally have picked to be *cis* or *trans* to each other.

Table A. Sequence Rule Priorities—The R/S System

Priority No.	Group	Priority No.	Group	Priority No.	Group	Priority No.	Group
1	Hydrogen	11	Benzyl	21	Methylamino	31	Phenoxy
2	Methyl	12	Isopropyl	22	Ethylamino	32	Acetoxy
3	Ethyl	13	Vinyl	23	Phenylamino	33	Benzoyloxy
4	*n*-Propyl	14	*sec*-Butyl	24	Acetylamino	34	Fluoro
5	*n*-Butyl	15	*t*-Butyl	25	Dimethylamino	35	Sulfhydryl (HS—)
6	*n*-Pentyl	16	Phenyl	26	Nitro	36	Sulfo (HO_3S—)
7	Isopentyl	17	*p*-Tolyl	27	Hydroxy	37	Chloro
8	Isobutyl	18	Acetyl	28	Methoxy	38	Bromo
9	Allyl	19	Benzoyl	29	Ethoxy	39	Iodo
10	Neopentyl	20	Amino	30	Benzyloxy		

(From *J. Org. Chem.* Vol. 35, 1970; page 2866. This is a partial list of the groups listed in this paper.)

Answers to Selected Problems

The following are the answers to most of the problems found within the chapters.

Chapter 1

1.1

```
        H  H                                  H  H          H  H
        |  |                                  |  |          |  |
(a) H—C—C—H      (b) H—O—O—H      (c) H—C—N—H    (d) H—N—N—H
        |  |                                  |
        H  H                                  H
```

1.2

```
                      O                               F   F
                      ‖                               |   |
(a) H—C≡N      (b) H—C—H      (c) H—O—N=O      (d) F—C=C—F
```

1.3 The bonds are polar, $\overset{\delta-}{O}=\overset{\delta+}{C}=\overset{\delta-}{O}$, but their polarities are in opposite directions. They cancel each other out, and the molecule as a whole is nonpolar.

Chapter 2

2.1

```
                                              CH3  CH3  CH3
                                              |    |    |
(a) Br—CH2CHCH2CH2CH3              (b) CH3C————C————C————CHCH2CH2CH3
          |                                   |    |    |    |
          NO2                                 CH3  CH3  CH3  CH(CH3)2

         I          CH3CHCH2CH3                            CH3
         |             |                                   |
(c) CH3C—CH—CH—CH—CHCH2CH2CH3           (d) Br—CH—CH—CH3
         |  |   |     |                            |
         I  CH3 CH(CH3)2 C(CH3)3                   Cl
```

2.2

(a) 3-methylhexane
(b) 2,3-dimethyl-4-*t*-butylheptane
(c) 2,4-dimethyl-5-*sec*-butylnonane
(d) 1-bromo-3-chloro-2-iodopropane (This follows the alphabetical order; other orders and the other sequence of numbers should be accepted.)
(e) 4-*t*-butyl-5-isopropyloctane or 4-isopropyl-5-*t*-butyloctane
(f) 1,3-dinitro-2,2-dimethylpropane
(g) 1-chloro-2,3-dimethylpentane
(h) 2,6-dimethyl-3-ethylheptane
(i) 2-methylpentane
(j) 5-methyl-4-ethylnonane

2.3 Use a large molar excess of methane over chlorine.

2.4 Use a large molar excess of chlorine over methane.

2.5

(a) Two: 1-chlorobutane (*n*-butyl chloride)
2-chlorobutane (*sec*-butyl chloride)

(b) Two: 1-chloro-2-methylpropane (isobutyl chloride)
2-chloro-2-methylpropane (*t*-butyl chloride)

2.6 (X = I, Br, or Cl)

(a) $2\ CH_3X$ (b) $CH_3X + CH_3CH_2X$ (c) $(CH_3)_2CHX + CH_3X$

(d) $(CH_3)_2CHX + CH_3CH_2X$ or $CH_3\underset{X}{\underset{|}{C}}HCH_2CH_3 + CH_3X$

Chapter 3

3.1 (At this stage, only common names are expected.)
(a) *trans*-β-butylene (b) α-butylene
(c) isobutylene (d) *cis*-β-butylene

3.2 (a) not possible (b) possible
(c) not possible (d) possible
(e) not possible

3.3 (a) $CH_3CH{=}CHCH(CH_3)_2$ (b) 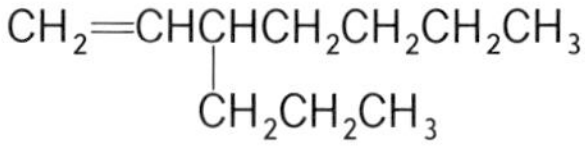 $CH_2{=}CHCH(CH_2CH_2CH_3)CH_2CH_2CH_2CH_3$

(c) 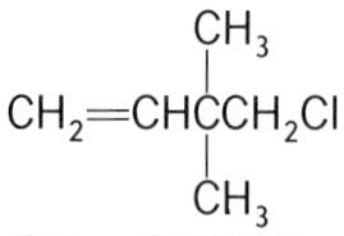 $CH_2{=}CHC(CH_3)_2CH_2Cl$ (d) $(CH_3)_2C{=}C(CH_3)_2$

(e) $CH_2{=}CHCH_2I$ (f) $CH_2{=}CHBr$

3.4 (a) 2-methylpropene (isobutylene)
(b) 3,6-dimethyl-4-isobutyl-3-heptene
(c) 1-chloro-1-propene
(d) 3-bromo-1-propene (allyl bromide)
(e) 4-methyl-1-hexene
(f) 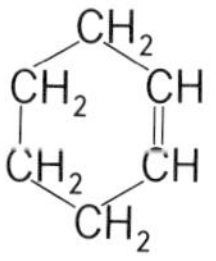  cyclohexene ring: CH_2, CH_2, $CH{=}CH$, CH_2, CH_2; cyclopentene
(g) 3,4-dimethylcyclohexene

3.5 (a) 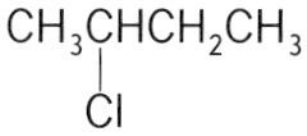 $CH_3CH(Cl)CH_2CH_3$ (b) $(CH_3)_3CBr$

(c) 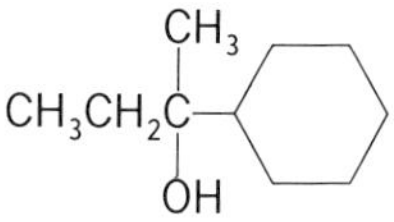 $CH_3CH_2C(CH_3)(OH){-}C_6H_{11}$ (cyclohexyl) (d) 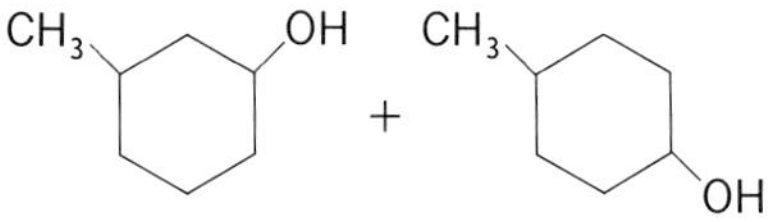 3-methylcyclohexanol + 4-methylcyclohexanol (CH₃, OH on cyclohexane rings)

3.6 (a) $CH_3CH{=}CHCH_3$ or $CH_2{=}CHCH_2CH_3$
(b) cannot be made this way
(c) 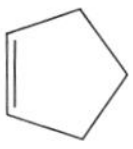 cyclopentene
(d) 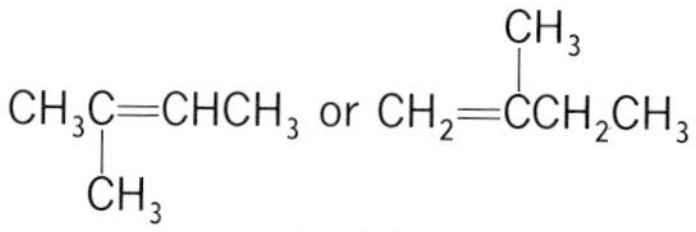 $CH_3C(CH_3){=}CHCH_3$ or $CH_2{=}C(CH_3)CH_2CH_3$
(e) cannot be made this way

3.7 (a) 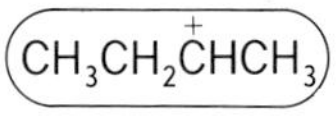 $\boxed{CH_3CH_2\overset{+}{C}HCH_3}$ $CH_3CH_2CH_2CH_2^+$ $CH_3CH_2CH(Cl)CH_3$

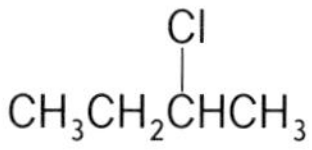

(b) $\boxed{(CH_3)_3C^+}$ $(CH_3)_2CHCH_2^+$ $(CH_3)_3CCl$

(c) 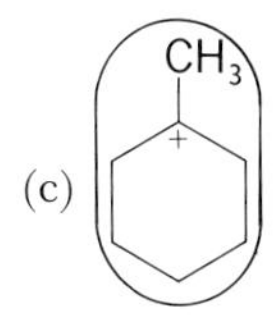

(d) $CH_3CH_2\overset{+}{C}HCH_3$ (only possible carbonium ion) $CH_3CH_2CH(Cl)CH_3$

(e) $CH_3\overset{+}{C}HCH_2CH_2CH_3$ $CH_3CH_2\overset{+}{C}HCH_2CH_3$ (both form)

$CH_3CH(Cl)CH_2CH_2CH_3 + CH_3CH_2CH(Cl)CH_2CH_3$

(f) $CH_3CH(Cl)CH_2CH_2CH_3$ (2-pentanol, from $CH_3\overset{+}{C}HCH_2CH_2CH_3$)

and

$CH_3CH_2CH(Cl)CH_2CH_3$ (3-pentanol, from $CH_3CH_2\overset{+}{C}HCH_2CH_3$)

From 2-pentene the two possible carbonium ions are both secondary and they are therefore of comparable stabilities. Hence, both may be expected to form. From propylene the two possible carbonium ions are in different classes; the secondary carbonium ion, being more stable than the primary, will be the one that forms.

3.8 (a) $CH_3CO_2H + CO_2$ (b) $CH_3\overset{O}{\overset{\|}{C}}CH_3$

(c) $CH_3\overset{O}{\overset{\|}{C}}(CH_2)_4\overset{O}{\overset{\|}{C}}CH_3$ (d) $CH_3CH{=}CHCH_3$ (Cis or trans 2-butene)

3.9 (a) $CH_2{=}CH_2$ (b) $CH_3CH{=}CH_2$ (c) 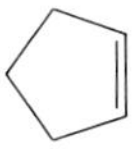 cyclopentene

Chapter 4

(a) *o*- and *p*-nitrotoluene (b) *m*-chloronitrobenzene
(c) *o*- and *p*-phenolsulfonic acid (d) *o*- and *p*-bromoacetanilide
(e) *o*- and *p*-dichlorobenzene

1. a mixture of *o*- and *p*-bromonitrobenzene
2. mostly *m*-bromonitrobenzene

3. (a) 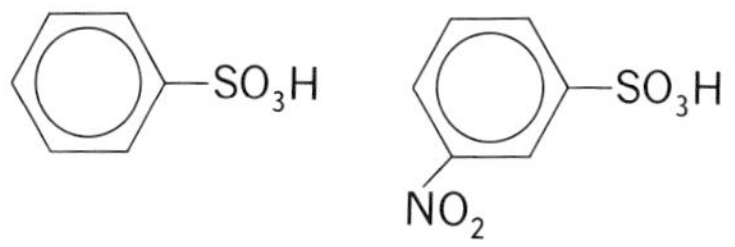 C_6H_5–SO_3H → *m*-$NO_2C_6H_4$–SO_3H

(b) C_6H_5–Br → CH_3–C_6H_4–Br (*para*) + *o*-$CH_3C_6H_4$–Br

(c) C_6H_5–CH_3 → NO_2–C_6H_4–CH_3 (*para*) + *o*-$NO_2C_6H_4$–CH_3

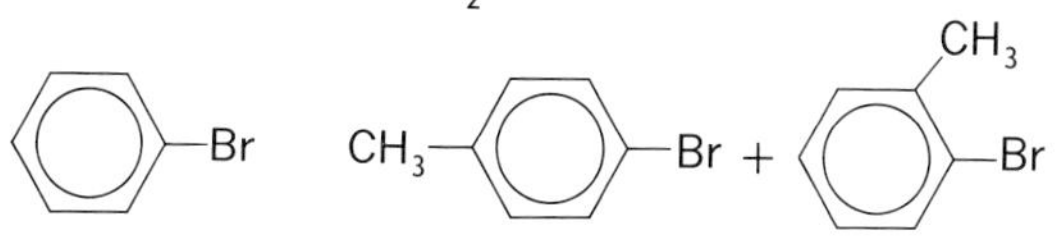

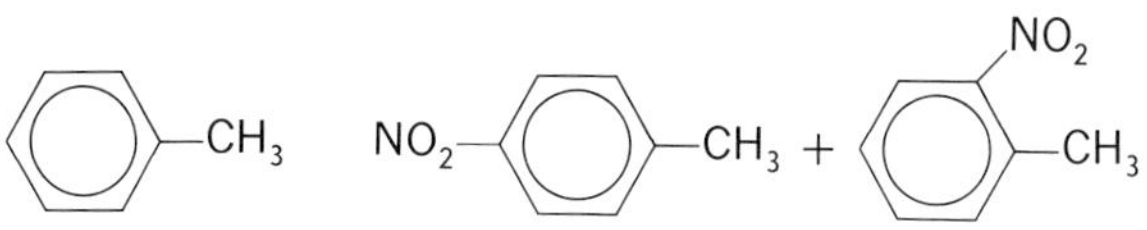

4.3 (Only details for the second steps are shown here.) If the attack is *ortho:*

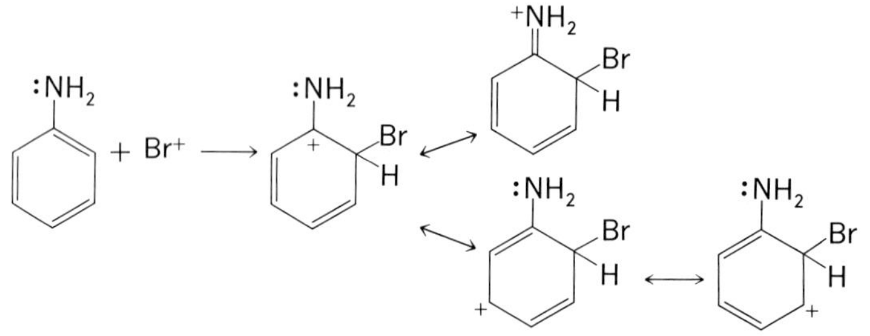

If the attack is *para:*

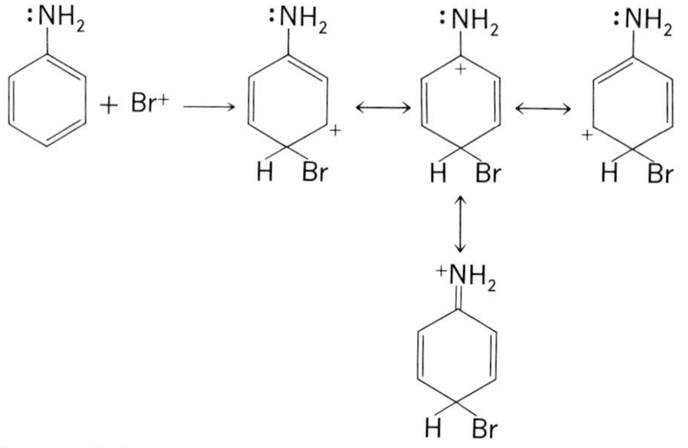

If the attack is *meta:*

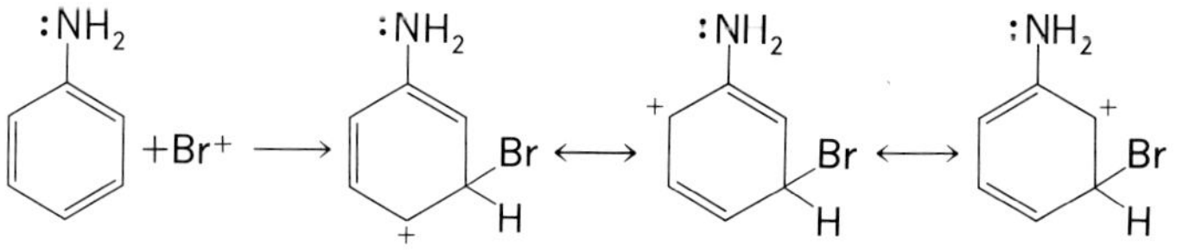

Chapter 6

6.1 (a) CH_3X and $CH_3CH{=}O$

(b) $C_6H_5CH{=}O$ and CH_3X

or

C_6H_5Br and $CH_3CH{=}O$

(Aryl bromides are generally preferred to aryl chlorides, but you were not expected to remember that. If you wrote C_6H_5Cl or C_6H_5X, count it as correct.)

(c) CH_3X and $CH_2{=}O$

(d) CH_3CH_2X and $CH_3C(=O)CH(CH_3)_2$

or

CH_3X and $CH_3CH_2C(=O)CH(CH_3)_2$

or

$(CH_3)_2CHX$ and $CH_3CH_2C(=O)CH_3$

(e) $C_6H_5CH_2X$ and $CH_2{=}O$

or

C_6H_5Br and $CH_2{-}CH_2$ (bridged by O; ethylene oxide)

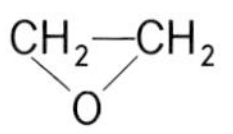

Chapter 8

8.1

(a) $(CH_3CH_2)_3N$

(b) $CH_3CH_2N(CH_3)_2$

(c) 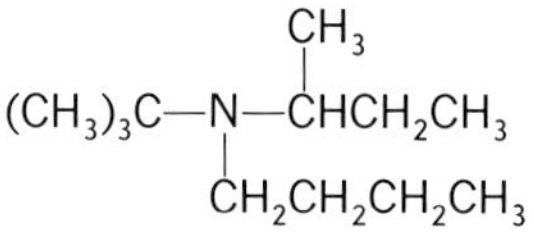 $(CH_3)_3C—N(CH_2CH_2CH_2CH_3)—CH(CH_3)CH_2CH_3$

(d)  $NO_2—C_6H_4—NH_2$ (para)

(e) 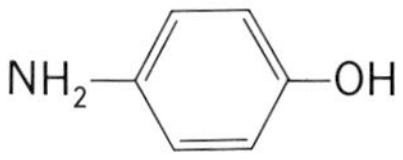 $NH_2—C_6H_4—OH$ (para)

(f) $(CH_3CH_2)_4N^+Cl^-$

(g) 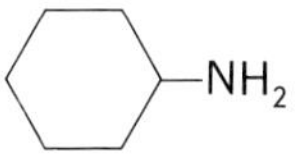 $C_6H_{11}—NH_2$ (cyclohexyl)

(h) $(C_6H_5)_2NH$

(i) 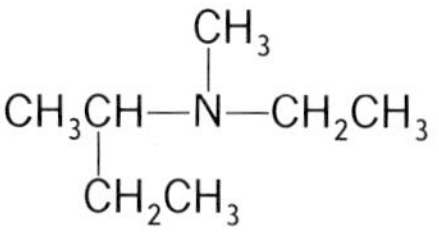 $CH_3CH(CH_2CH_3)—N(CH_3)—CH_2CH_3$

(j) $C_6H_5CH_2NHCH_3$

8.2

(a) aliphatic, 3° (b) aliphatic, 3°
(c) aliphatic, 3° (d) aromatic, 1°
(e) aromatic, 1° (f) aliphatic, 4° salt
(g) aliphatic, 1° (h) aromatic, 2°
(i) aliphatic, 3° (j) aliphatic, 2°

8.3

(a) di-*n*-butylethylamine (b) 1,2-diaminoethane
(c) *N*, *N*-diethylaniline (d) cyclopentylamine
(e) tetramethylammonium bromide

8.4

(a) $(CH_3)_2NNO$

(b) 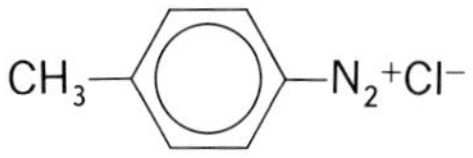 $CH_3—C_6H_4—N_2^+Cl^-$ (para)

(c) $C_6H_5—CH_2OH$ (also some $C_6H_5—CH_2Cl$)

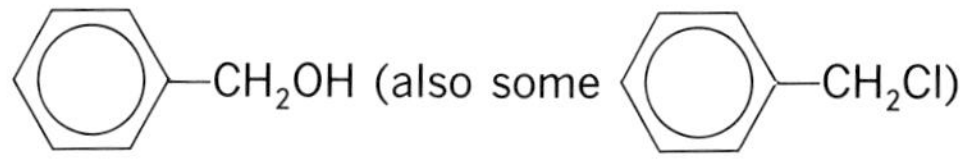

(d) 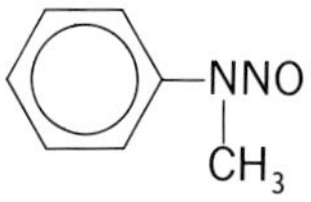 $C_6H_5—N(CH_3)NO$

(e) 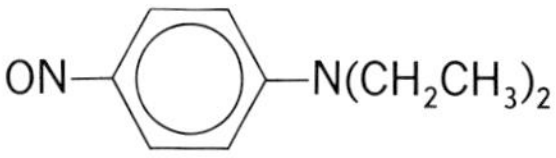 $ON—C_6H_4—N(CH_2CH_3)_2$ (para)

Chapter 9

9.1

(a)	acid	(b)	aldehyde	(c)	acid chloride
(d)	amide	(e)	ester	(f)	ketone
(g)	alcohol	(h)	amide	(i)	aldehyde
(j)	ketone	(k)	acid	(l)	ester
(m)	alcohol, ketone	(n)	alcohol, ketone	(o)	acid
(p)	aldehyde, ketone	(q)	ester	(r)	ether, ketone
(s)	ester, aldehyde	(t)	amide	(u)	acid
(v)	amide	(w)	diamide	(x)	anhydride
(y)	acid chloride	(z)	phenol, aldehyde	(aa)	alkene, ester
(bb)	ester, alkene	(cc)	diester	(dd)	amide, acid

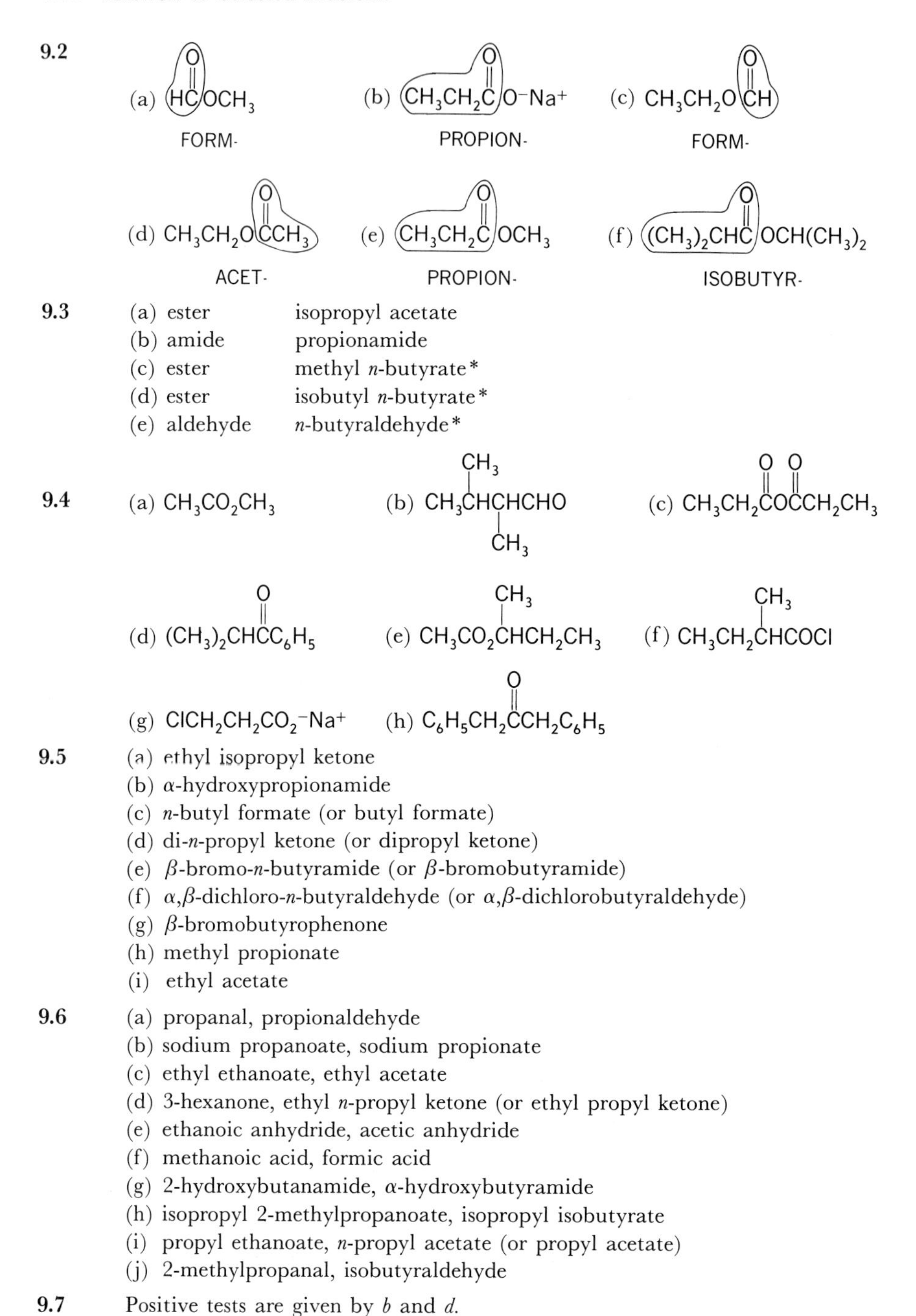

9.2

(a) $HCOOCH_3$ (C=O) — FORM-

(b) $CH_3CH_2C(=O)O^-Na^+$ — PROPION-

(c) $CH_3CH_2OCH(=O)$ — FORM-

(d) $CH_3CH_2OC(=O)CH_3$ — ACET-

(e) $CH_3CH_2C(=O)OCH_3$ — PROPION-

(f) $(CH_3)_2CHC(=O)OCH(CH_3)_2$ — ISOBUTYR-

9.3

(a)	ester	isopropyl acetate
(b)	amide	propionamide
(c)	ester	methyl *n*-butyrate*
(d)	ester	isobutyl *n*-butyrate*
(e)	aldehyde	*n*-butyraldehyde*

9.4

(a) $CH_3CO_2CH_3$

(b) $CH_3CH(CH_3)CH(CH_3)CHO$

(c) $CH_3CH_2C(=O)OC(=O)CH_2CH_3$

(d) $(CH_3)_2CHC(=O)C_6H_5$

(e) $CH_3CO_2CH(CH_3)CH_2CH_3$

(f) $CH_3CH_2CH(CH_3)COCl$

(g) $ClCH_2CH_2CO_2^-Na^+$

(h) $C_6H_5CH_2C(=O)CH_2C_6H_5$

9.5

(a) ethyl isopropyl ketone
(b) α-hydroxypropionamide
(c) *n*-butyl formate (or butyl formate)
(d) di-*n*-propyl ketone (or dipropyl ketone)
(e) β-bromo-*n*-butyramide (or β-bromobutyramide)
(f) α,β-dichloro-*n*-butyraldehyde (or α,β-dichlorobutyraldehyde)
(g) β-bromobutyrophenone
(h) methyl propionate
(i) ethyl acetate

9.6

(a) propanal, propionaldehyde
(b) sodium propanoate, sodium propionate
(c) ethyl ethanoate, ethyl acetate
(d) 3-hexanone, ethyl *n*-propyl ketone (or ethyl propyl ketone)
(e) ethanoic anhydride, acetic anhydride
(f) methanoic acid, formic acid
(g) 2-hydroxybutanamide, α-hydroxybutyramide
(h) isopropyl 2-methylpropanoate, isopropyl isobutyrate
(i) propyl ethanoate, *n*-propyl acetate (or propyl acetate)
(j) 2-methylpropanal, isobutyraldehyde

9.7 Positive tests are given by *b* and *d*.

*These same names without the "*n*-" are also acceptable. You will see both usages. In other words, "butyrate" means "*n*-butyrate"; if you mean to write the isomer you must write "isobutyrate."

9.8 (a) $CH_3CH(OH)(OCH_2CH_3)$ (b) $CH_3CH_2CH(OH)(OCH_3)$ (c) $(CH_3)_2CHCH(OH)(OCH_3)$

9.9 (a) $CH_3CH(OCH_2CH_3)_2$ (b) $CH_3CH_2CH(OCH_3)_2$ (c) $(CH_3)_2CH(OCH_3)_2$

9.10 (a) $CH_2O + 2CH_3OH$ (b) $CH_3CHO + 2CH_3CH_2OH$
(c) $CH_3CHO + 2CH_3OH$ (d) not an acetal

9.11 (a) $CH_3CH_2CH(OH)CH(CH_3)CHO$ (b) $CH_3CH_2CH_2CH(OH)CH(CH_2CH_3)CHO$ (c) $C_6H_5CH_2CH(OH)CH(C_6H_5)CHO$

9.12 (a) $C_6H_5CH{=}C(CH_3){-}CHO$ (b) $C_6H_5CH{=}C(CH_2CH_3){-}CHO$ (c) $C_6H_5CH{=}C(CH_3){-}C(=O){-}C_6H_5$

Chapter 10

10.1 (a) $CH_3CH_2OH \xrightarrow[\text{or } Cr_2O_7^{2-}]{MnO_4^-} CH_3CO_2H$

(b) $CH_3CH_2OH + HBr \longrightarrow CH_3CH_2Br \xrightarrow[\text{2. } CO_2 \text{ 3. } H^+]{\text{1. Mg, ether}} CH_3CH_2CO_2H$

or $CH_3CH_2Br \xrightarrow{NaCN} CH_3CH_2CN \xrightarrow{\text{Hydrolyze}} CH_3CH_2CO_2H$

(c) $(CH_3)_3COH + HBr \longrightarrow (CH_3)_3CBr \xrightarrow[\text{2. } CO_2 \text{ 3. } H^+]{\text{1. Mg, ether}} (CH_3)_3CCO_2H$

(d) $C_6H_5NH_2 \xrightarrow[\text{HCl,cold}]{NaNO_2,} C_6H_5N_2^+Cl^- \xrightarrow[Cu_2(CN)_2]{} C_6H_5CN \xrightarrow{\text{Hydrolyze}} C_6H_5CO_2H$

(e) $(CH_3)_2CHCl \xrightarrow[\text{2. } CO_2 \text{ 3. } H^+]{\text{1. Mg, ether}} (CH_3)_2CHCO_2H$

10.2 (a) $CH_3CO_2CH_2CH_3$, ethyl acetate
(b) $C_6H_5CO_2CH(CH_3)_2$, isopropyl benzoate
(c) $(CH_3)_2CHCO_2CH_2CH_2CH_3$, propyl isobutyrate (or *n*-propyl isobutyrate)
(d) $CH_3CH_2CO_2CH_3$, methyl propionate

10.3 (a) $CH_3CONHCH_3$, N-methylacetamide
(b) $CH_3CH_2CH_2CON(CH_2CH_3)_2$, N,N-diethylbutyramide
(c) $C_6H_5CONHC_6H_5$, N-phenylbenzamide (more often called benzanilide)
(d) no amide can form

10.4 (a) CH_3CO_2H
(b) $C_6H_5CO_2CH_3$
(c) $CH_3CO_2CH_2CH_3 + CH_3CO_2H$
(d) $CH_3CH_2CONH_2$
(e) $C_6H_5CH_2CON(CH_3)_2 + (CH_3)_2NH_2^+Cl^-$
(f) no reaction

10.5

$$R{-}\overset{\overset{\displaystyle :\ddot{O}}{\|}}{C}{-}Cl \;(\uparrow :\ddot{O}H_2) \longrightarrow R{-}\overset{\overset{\displaystyle :\ddot{O}:^-}{|}}{\underset{\underset{\displaystyle H\;\;\ddot{O}^+\;\;H}{|}}{C}}{-}Cl \longrightarrow R{-}\overset{\overset{\displaystyle :O:}{\|}}{C}{-}\overset{+}{O}(H){-}H + Cl^- \xrightarrow{H_2O} R{-}CO_2H + H_3O^+$$

$$R{-}\overset{\overset{\displaystyle :\ddot{O}}{\|}}{C}{-}O{-}\overset{\overset{\displaystyle :O:}{\|}}{C}{-}R \;(\uparrow :\ddot{O}(R')H) \longrightarrow R{-}\overset{\overset{\displaystyle :\ddot{O}:^-}{|}}{\underset{\underset{\displaystyle R'\;\;\ddot{O}^+\;\;H}{|}}{C}}{-}O{-}\overset{\overset{\displaystyle O}{\|}}{C}{-}R \longrightarrow R{-}\overset{\overset{\displaystyle O}{\|}}{C}{-}O{-}R' + H{-}O{-}\overset{\overset{\displaystyle O}{\|}}{C}{-}R$$

$$R{-}\overset{\overset{\displaystyle :\ddot{O}}{\|}}{C}{-}Cl \;(\uparrow H_3N:) \longrightarrow R{-}\overset{\overset{\displaystyle :\ddot{O}:^-}{|}}{\underset{\underset{\displaystyle {}^+NH_2{-}H \leftarrow :NH_3}{|}}{C}}{-}Cl \longrightarrow R{-}\overset{\overset{\displaystyle O}{\|}}{C}{-}NH_2 + NH_4^+ + Cl^-$$

10.6

(a) $CH_3CH_2CO_2H + CH_3OH$
(b) $CH_3CO_2^-Na^+ + CH_3CH_2OH$
(c) $C_6H_5CONH_2 + (CH_3)_2CHOH$
(d) $CH_2{=}\overset{\overset{\displaystyle CH_3}{|}}{C}CO_2CH_2CH_2CH_3 + CH_3OH$
(e) $CH_3(CH_2)_{14}CO_2CH_3 + CH_3(CH_2)_{12}CO_2CH_3 + CH_3(CH_2)_{16}CO_2CH_3 + HOCH_2\underset{\underset{\displaystyle OH}{|}}{C}HCH_2OH$
(f) $(CH_3)_2CHCH_2CO_2H + CH_3CH_2OH$
(g) $(CH_3)_2CHCO_2H + CH_3CH_2CH_2OH$
(h) $2CH_3CO_2^-Na^+ + HOCH_2CH_2OH$

10.7

(a) $CH_3CO_2H + NH_3$
(b) $CH_3CH_2CO_2H + CH_3NH_2$
(c) $CH_3CO_2H + C_6H_5NH_2$
(d) $CH_3CH_2CO_2H + (CH_3)_2NH$
(e) $NH_2CH_2CO_2H + NH_2\underset{\underset{\displaystyle CH_3}{|}}{C}HCO_2H$

Chapter 12

12.1 Alkyl bromides are used; chlorides or iodides might also be used.
(a) methyl bromide (c) isopropyl bromide
(b) ethyl bromide (d) benzyl bromide

12.2 (a) $CH_3CO_2C_2H_5$ (b) $C_6H_5CH_2CO_2C_2H_5$

12.3 (X = Cl, Br or I)
(a) CH_3X (b) $(CH_3)_2CHX$
(c) $C_6H_5CH_2X$ (d) cyclopentyl$-CH_2X$

Chapter 13

13.1 (a) $CH_3CH(OH)CH_3$ (None) (b) $CH_3\overset{*}{C}H(NH_2)COOH$ (c) $HOOCCH_2CH_2\overset{*}{C}H(NH_2)COOH$

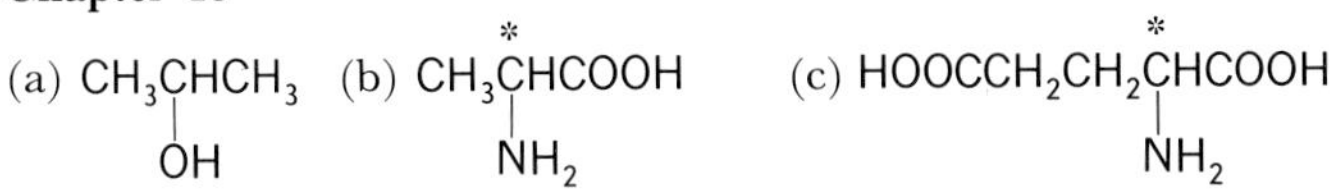

(d) CH_2OCH_3–$\overset{*}{C}H(OH)$–CH_2OH (e) $CH_3CH(CH_3)\overset{*}{C}H(OH)CH_3$

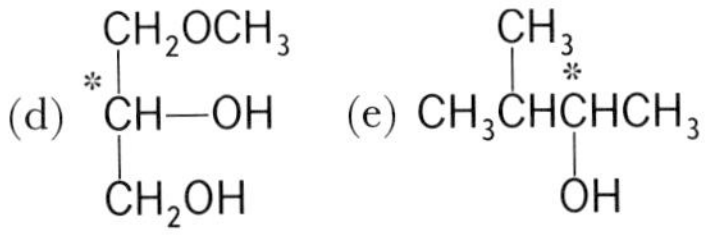

Chapter 14

14.1 (a)

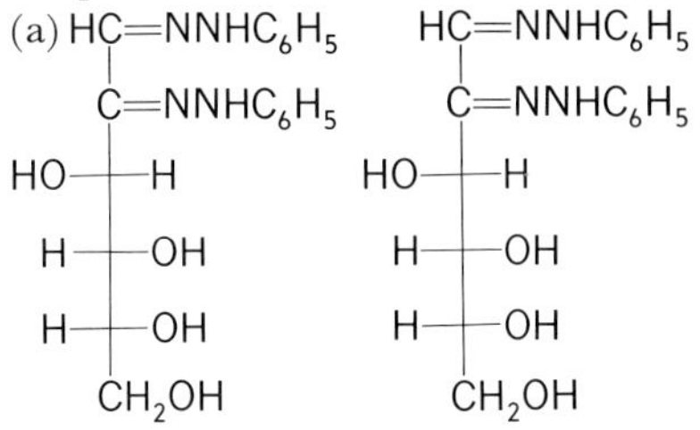

D-Glucosazone		D-Mannosazone	
	$HC{=}NNHC_6H_5$		$HC{=}NNHC_6H_5$
	$C{=}NNHC_6H_5$		$C{=}NNHC_6H_5$
HO	H	HO	H
H	OH	H	OH
H	OH	H	OH
	CH_2OH		CH_2OH

(b) They are identical
(c) D-allosazone and D-altrosazone are identical.
(d) D-Talose
(e) L-Idose

Index

Entries in italics are in tables.
In indexing names of organic compounds
preference has usually been given to common names.